Ontology Theory, Management and Design: Advanced Tools and Models

Faiez Gargouri
Higher Institute of Informatics and Multimedia of Sfax, Tunisia

Wassim Jaziri
Higher Institute of Informatics and Multimedia of Sfax, Tunisia

Information Science REFERENCE

INFORMATION SCIENCE REFERENCE
Hershey · New York

Director of Editorial Content:	Kristin Klinger
Director of Book Publications:	Julia Mosemann
Acquisitions Editor:	Lindsay Johnston
Development Editor:	Joel Gamon
Publishing Assistant:	Sean Woznicki
Typesetter:	Myla Harty
Production Editor:	Jamie Snavely
Cover Design:	Lisa Tosheff
Printed at:	Yurchak Printing Inc.

Published in the United States of America by
Information Science Reference (an imprint of IGI Global)
701 E. Chocolate Avenue
Hershey PA 17033
Tel: 717-533-8845
Fax: 717-533-8661
E-mail: cust@igi-global.com
Web site: http://www.igi-global.com/reference

Library of Congress Cataloging-in-Publication Data

Ontology theory, management, and design : advanced tools and models / Faiez Gargouri and Wassim Jaziri, editors.
p. cm.
Includes bibliographical references and index.
Summary: "The focus of this book is on information and communication sciences, computer science, and artificial intelligence and provides readers with access to the latest knowledge related to design, modeling and implementation of ontologies"--Provided by publisher. ISBN 978-1-61520-859-3 (hardcover) -- ISBN 978-1-61520-860-9 (ebook) 1. Ontologies (Information retrieval) 2. Knowledge representation (Information theory) 3. Artificial intelligence. I. Gargouri, Faiez, 1965- II. Jaziri, Wassim, 1975-
TK5105.88815.O588 2010
004--dc22

2009053383

British Cataloguing in Publication Data
A Cataloguing in Publication record for this book is available from the British Library.

Editorial Advisory Board

Catherine Roussey, *LIRIS, France*
Florence Sedes, *Toulouse, France*
Sahbi Sidhom, *Nancy, France*
Jacques Teller, *Liège, Belgium*
Christelle Vangenot, *Lausanne, Switzerland*

List of Reviewers

Rim Faiz, *IHEC, Tunisia*
Frédéric Fürst, *Amiens, France*
M. Mohsen Gammoudi, *Tunisia*
Fabien Gandon, *INRIA, France*
Chirine Ghedira, *Lyon1, France*
Lamia Hadrich, *FSEGS, Tunisia*
Mohamed Jemni, *CCK, Tunisia*
Gilles Kassel, *Picardie, France*
Thérèse Libourel, *LIRMM, France*
Michel Mainguenaud, *Rouen, France*
Mimoun Malki, *Sidi Belabes, Algeria*
Brahim Medjahed, *Dearborn, USA*
Thierry Paquet, *LITIS, France*
Guy Pierra, *ENSMA, France*
Yann Pollet, *CNAM, France*
Chantal Reynaud, *Paris, France*
Christophe Roche, *Annecy, France*
Catherine Roussey, *LIRIS, France*
Florence Sedes, *Toulouse, France*
Sahbi Sidhom, *Nancy, France*
Jacques Teller, *Liège, Belgium*
Christelle Vangenot, *Lausanne, Switzerland*
Rafik Bouaziz, *FSEGS, Tunisia*

Table of Contents

Preface xv

Section 1
Introduction and Overview: Theory, Concepts and Foundations

Chapter 1
Ontologies in Computer Science: These New "Software Components" of Our Information Systems 1
Fabien L. Gandon, INRIA, France

Chapter 2
Ontology Theory, Management and Design: An Overview and Future Directions 27
Wassim Jaziri, MIRACL Laboratory, Tunisia
Faiez Gargouri, MIRACL Laboratory, Tunisia

Section 2
Theoretical Models and Aspects: Formal Frameworks

Chapter 3
Exceptions in Ontologies: A Theoretical Model for Deducing Properties from Topological Axioms 78
Christophe Jouis, Université Paris III, France
Julien Bourdaillet, Université de Montréal, Canada
Bassel Habib, LIP6, France
Jean-Gabriel Ganascia, LIP6, France

Chapter 4
An Algebra of Ontology Properties for Service Discovery and Composition in Semantic Web 98
Yann Pollet, CEDRIC Laboratory, France

Chapter 5
Approaches for Semantic Association Mining and Hidden Entities Extraction in Knowledge Base 119
Thabet Slimani, ISG of Tunis, Tunisia
Boutheina Ben Yaghlane, IHEC of Carthage, Tunisia
Khaled Mellouli, IHEC of Carthage, Tunisia

Chapter 6
Reusing the Inter-Organizational Knowledge to Support Organizational Knowledge Management Process: An Ontology-Based Knowledge Network 142
Nelson K. Y. Leung, RMIT International University Vietnam, Vietnam
Sim Kim Lau, University of Wollongong, Australia
Joshua Fan, University of Wollongong, Australia

Chapter 7
Building and Use of a LOM Ontology 162
Ghebghoub Ouafia, University of Jijel, Algeria
Abel Marie-Hélène, Compiègne University of Technology, France
Moulin Claude, Compiègne University of Technology, France
Leblanc Adeline, Compiègne University of Technology, France

Section 3
Ontology Management: Construction, Evolution and Alignment

Chapter 8
Ontology Evolution: State of the Art and Future Directions 179
Rim Djedidi, Supélec – Campus de Gif, France
Marie-Aude Aufaure, MAS Laboratory, France

Chapter 9
Large Scale Matching Issues and Advances 208
Sana Sellami, LIRIS, France
Aicha-Nabila Benharkat, LIRIS, France
Youssef Amghar, LIRIS, France

Chapter 10
From Temporal Databases to Ontology Versioning: An Approach for Ontology Evolution 225
Najla Sassi, MIRACL Laboratory, Tunisia
Zouhaier Brahmia, MIRACL Laboratory, Tunisia
Wassim Jaziri, MIRACL Laboratory, Tunisia
Rafik Bouaziz, MIRACL Laboratory, Tunisia

Section 4
Ontology Applications and Experiences

Chapter 11
Ontology Learning from Thesauri: An Experience in the Urban Domain 247
Javier Nogueras-Iso, Universidad de Zaragoza, Spain
Javier Lacasta, Universidad de Zaragoza, Spain
Jacques Teller, Université de Liège, Belgium
Gilles Falquet, Université de Genève, Switzerland
Jacques Guyot, Université de Genève, Switzerland

Chapter 12
Applications of Ontologies and Text Mining in the Biomedical Domain 261
A. Jimeno-Yepes, European Bioinformatic Institute, UK
R. Berlanga-Llavori, Universitat Jaume I, Spain
D. Rebholz-Schuchmann, European Bioinformatic Institute, UK

Chapter 13
Ontology Based Multimedia Indexing 284
Mihaela Brut, Institut de Recherche en Informatique de Toulouse, France
Florence Sedes, Institut de Recherche en Informatique de Toulouse, France

Chapter 14
Semantic Enrichment of Web Service Architecture 303
Aicha Boubekeur, University of Tiaret, Algeria
Mimoun Malki, University of Sidi Bel-Abbes, Algeria
Abdellah Chouarfia, University of Oran, Algeria
Mostefa Belarbi, University of Tiaret, Algeria

Compilation of References 322

About the Contributors 354

Index 361

Detailed Table of Contents

Preface xv

Section 1
Introduction and Overview: Theory, Concepts and Foundations

Chapter 1
Ontologies in Computer Science: These New "Software Components" of Our Information Systems 1
Fabien L. Gandon, INRIA, France

Ironically the field of computer ontologies suffered a lot from ambiguity. The word ontology can be used and has been used with very different meanings attached to it. We will introduce in this chapter two faces of ontologies in computer science: (1) the ontology object: focusing on the nature and the characteristics of the ontology object, its core notions and its lifecycle. (2) ontology engineering: the branch of knowledge modeling that develops ontology-oriented computable models of some domain of knowledge, focusing here on the design rationale and the assisting tools.

Chapter 2
Ontology Theory, Management and Design: An Overview and Future Directions 27
Wassim Jaziri, MIRACL Laboratory, Tunisia
Faiez Gargouri, MIRACL Laboratory, Tunisia

Ontologies now play an important role in providing a commonly agreed understanding of a domain and in developing knowledge-based systems. They intend to capture the intrinsic conceptual and semantic structure of a specific domain. Many methodologies, tools and languages are already available to help anthologies' designers and users. However, a number of questions remain open: what ontology development methodology provides the best guidance to model a given problem, what steps to be performed in order to develop an ontology? which techniques are appropriate for each step? how ontology' lifecycle steps are upheld by the software tools? how to maintain an ontology and to evolve it in a consistent way? how to adapt an ontology to a given context? To provide answers to these questions, we review in this chapter the main methodologies, tools and languages for building, updating and representing ontologies that have been reported in literature.

Section 2
Theoretical Models and Aspects: Formal Frameworks

Chapter 3
Exceptions in Ontologies: A Theoretical Model for Deducing Properties from Topological Axioms 78
Christophe Jouis, Université Paris III, France
Julien Bourdaillet, Université de Montréal, Canada
Bassel Habib, LIP6, France
Jean-Gabriel Ganascia, LIP6, France

This chapter is a contribution to the study of formal ontologies. It addresses the problem of atypical entities in ontologies. The authors propose a new model of knowledge representation by combining ontologies and topology. In order to represent atypical entities in ontologies, the four topological operators of interior, exterior, border and closure are introduced. These operators allow to specify whether an entity, belonging to a class, is typical or not. The authors define a system of topological inclusion and membership relations into the ontology formalism, by adapting the four topological operators with the help of their mathematical properties. These properties are used as a set of axioms which allows to define the topological inclusion and membership relations. Further, the authors define combinations of the operators of interior, exterior, border and closure that allow the construction of an algebra. They model is implemented in AnsProlog, a recent logic programming language that allows negative predicates in inference rules.

Chapter 4
An Algebra of Ontology Properties for Service Discovery and Composition in Semantic Web.......... 98
Yann Pollet, CEDRIC Laboratory, France

The authors address in this chapter the problem of the automated discovery and composition of Web Services. Now, Service-oriented computing is emerging as a new and promising paradigm. However, selection and composition of Services to achieve an expected goal remain purely manual and time consuming tasks. Basing our approach on domain concept definitions thanks to an ontology, the authors develop here an algebraic approach that enables to express formal definitions of Web Service semantics as well as user information needs. Both are captured by the means of algebraic expressions of ontology properties. They present an algorithm that generates efficient orchestration plans, with characteristics of optimality regarding Quality of Service. The approach has been validated by a prototype and an evaluation in the case of an Health Information System.

Chapter 5
Approaches for Semantic Association Mining and Hidden Entities Extraction in Knowledge Base 119
Thabet Slimani, ISG of Tunis, Tunisia
Boutheina Ben Yaghlane, IHEC of Carthage, Tunisia
Khaled Mellouli, IHEC of Carthage, Tunisia

Due to the rapidly increasing use of information and communications technology, Semantic Web technology is being increasingly applied in a large spectrum of applications in which domain knowledge is represented by means of an ontology in order to support reasoning performed by a machine. A semantic association (SA) is a set of relationships between two entities in knowledge base represented as graph paths consisting of a sequence of links. Because the number of relationships between entities in a knowledge base might be much greater than the number of entities, it is recommended to develop tools and invent methods to discover new unexpected links and relevant semantic associations in the large store of the preliminary extracted semantic association. Semantic association mining is a rapidly growing field of research, which studies these issues in order to create efficient methods and tools to help us filter the overwhelming flow of information and extract the knowledge that reflect the user need. The authors present, in this work, an approach which allows the extraction of association rules (SWARM: Semantic Web Association Rule Mining) from a structured semantic association store. Then, we present a new method which allows the discovery of relevant semantic associations between a preliminary extracted SA and predefined features, specified by user, with the use of Hyperclique Pattern (HP) approach. In addi-tion, the authors present an approach which allows the extraction of hidden entities in knowledge base. The experimental results applied to synthetic and real world data show the benefit of the proposed methods and demonstrate their promising effectiveness.

Chapter 6
Reusing the Inter-Organizational Knowledge to Support Organizational Knowledge Management Process: An Ontology-Based Knowledge Network 142
Nelson K. Y. Leung, RMIT International University Vietnam, Vietnam
Sim Kim Lau, University of Wollongong, Australia
Joshua Fan, University of Wollongong, Australia

Various types of Knowledge Management approaches have been developed that only focus on managing organizational knowledge. These approaches are inadequate because employees often need to access knowledge from external knowledge sources in order to complete their works. Therefore, a new inter-organizational Knowledge Management practice is required to enhance knowledge sharing across organizational boundaries in their business networks. In this chapter, an ontology-based Inter-organizational knowledge Network that incorporates ontology mediation is developed so that heterogeneity of knowledge semantic in the ontologies could be reconciled. The reconciled inter-organizational knowledge could be reused to support organizational Knowledge Management process semi- or automatically. The authors also investigate the application of ontology mediation that provides mechanisms of reconciling inter-organizational knowledge in the network.

Chapter 7
Building and Use of a LOM Ontology 162
Ghebghoub Ouafia, University of Jijel, Algeria
Abel Marie-Hélène, Compiègne University of Technology, France
Moulin Claude, Compiègne University of Technology, France
Leblanc Adeline, Compiègne University of Technology, France

The increasing number of available resources that may be used during e-learning can raise problems of access, management and sharing. An e-learning application therefore shares the same problem of relevance to the Web concerning the access to learning resources. Semantic web technologies provide promising solutions to such problems. One main feature of this new web generation is the shared understanding based on ontologies. This chapter presents an approach to index learning resources semantically using a LOM ontology. This ontology was developed to clarify the concepts, and to describe the existing relations between elements of the LOM standard. The author present also our tool based on this ontology which allows to describe learning objects and helps retrieving them.

Section 3
Ontology Management: Construction, Evolution and Alignment

Chapter 8
Ontology Evolution: State of the Art and Future Directions 179
Rim Djedidi, Supélec – Campus de Gif, France
Marie-Aude Aufaure, MAS Laboratory, France

Ontologies evolve continuously throughout their lifecycle to respond to different change requirements. Several problems emanate from ontology evolution: capturing change requirements, change representation, change impact analysis and resolution, change validation, change traceability, change propagation to dependant artifacts, versioning, etc. The purpose of this chapter is to gather research and current developments to manage ontology evolution. The authors highlight ontology evolution issues and present a state-of-the-art of ontology evolution approach by describing issues raised and the ontology model considered (ontology representation language), and also the ontology engineering tools supporting ontology evolution and maintenance. Furthermore, they sum up the state-of-the-art review by a comparative study based on general characteristics, evolution functionalities supported, and specificities of the existing ontology evolution approaches. At the end of the chapter, the authors discuss future and emerging trends.

Chapter 9
Large Scale Matching Issues and Advances 208
Sana Sellami, LIRIS, France
Aicha-Nabila Benharkat, LIRIS, France
Youssef Amghar, LIRIS, France

Nowadays, the Information technology domains (semantic web, E-business, digital libraries, life science, etc) abound with a large variety of data (e.g. DB schemas, XML schemas, ontologies) and bring up a hard problem: the semantic heterogeneity. Matching techniques are called to overcome this challenge and attempts to align these data. In this chapter, the authors are interested in studying large scale matching approaches. They survey the techniques of large scale matching, when a large number of schemas/ontologies and attributes are involved. They attempt to cover a variety of techniques for schema matching called Pair-wise and Holistic, as well as a set of useful optimization techniques. They compare the different existing schema/ontology matching tools. One can acknowledge that this domain is on top of effervescence and large scale matching needs many more advances. Then the authors provide conclusions concerning important open issues and potential synergies of the technologies presented.

Chapter 10

From Temporal Databases to Ontology Versioning: An Approach for Ontology Evolution 225

Najla Sassi, MIRACL Laboratory, Tunisia
Zouhaier Brahmia, MIRACL Laboratory, Tunisia
Wassim Jaziri, MIRACL Laboratory, Tunisia
Rafik Bouaziz, MIRACL Laboratory, Tunisia

The problem of versioning is present in several application areas, such as temporal databases, real-time computing and ontologies. This problem is generally defined as managing changes in a timely manner without loss of existing data. However, ontology versioning is more complicated than versioning in database because of the usage and content of ontology which incorporates semantic aspects. Consequently, ontology data models are much richer than those of database schemas. In this chapter, the authors are interested in developing an ontology versioning system to express, apply and implement changes on the ontology.

Section 4
Ontology Applications and Experiences

Chapter 11

Ontology Learning from Thesauri: An Experience in the Urban Domain.. 247

Javier Nogueras-Iso, Universidad de Zaragoza, Spain
Javier Lacasta, Universidad de Zaragoza, Spain
Jacques Teller, Université de Liège, Belgium
Gilles Falquet, Université de Genève, Switzerland
Jacques Guyot, Université de Genève, Switzerland

Ontology learning is the term used to encompass methods and techniques employed for the (semi-) automatic processing of knowledge resources that facilitate the acquisition of knowledge during ontology construction. This chapter focuses on ontology learning techniques using thesauri as input sources. Thesauri are one of the most promising sources for the creation of domain ontologies thanks to the richness of term definitions, the existence of a priori relationships between terms, and the consensus provided by their extensive use in the library context. Apart from reviewing the state of the art, this chapter shows how ontology learning techniques can be applied in the urban domain for the development of domain ontologies.

Chapter 12

Applications of Ontologies and Text Mining in the Biomedical Domain 261

A. Jimeno-Yepes, European Bioinformatic Institute, UK
R. Berlanga-Llavori, Universitat Jaume I, Spain
D. Rebholz-Schuchmann, European Bioinformatic Institute, UK

Ontologies represent domain knowledge that improves user interaction and interoperability between applications. In addition, ontologies deliver precious input to text mining techniques in the biomedical domain, which might improve the performance in different text mining tasks. This chapter will explore on the mutual benefits for ontologies and text mining techniques. Ontology development is a time consuming task. Most efforts are spent in the acquisition of terms that represent concepts in real life. This process can use the existing scientific literature and the World Wide Web. The identification of concept labels, i.e. terms, from these sources using text mining solutions improves ontology development since the literature resources make reference to existing terms and concepts. Furthermore, automatic text processing techniques profit from ontological resources in different tasks, for example in the disambiguation of terms and the enrichment of terminological resources for the text mining solution. One of the most important text mining tasks that exploits ontological resources consists of the mapping of concepts to terms in textual sources (e.g. named entity recognition, semantic indexing) and the expansion of queries in information retrieval.

Chapter 13
Ontology Based Multimedia Indexing 284
Mihaela Brut, Institut de Recherche en Informatique de Toulouse, France
Florence Sedes, Institut de Recherche en Informatique de Toulouse, France

The chapter goal is to provide responses to the following question: how the ontologies could be used in order to index and manage the multimedia collections? Alongside with reviewing the main standard formats, voca-bularies and ontology categories developed especially for multimedia content description, the chapter emphasis the existing techniques for acquiring ontology-based indexing. Since a fully automatic such technique is not possible yet, the chapter also proposes a solution for indexing a multimedia collection by combining technologies from both semantic Web and multimedia indexation domains. The solution considers the man-agement of multimedia metadata based on two correlated dictionaries: a metadata dictionary centralizes the multimedia metadata obtained through an automatic indexation process, while the visual concepts dictionary identifies the list of visual objects contained in multimedia documents and considered in the ontology-based annotation process. This approach facilitates as well the multimedia retrieval process.

Chapter 14
Semantic Enrichment of Web Service Architecture 303
Aicha Boubekeur, University of Tiaret, Algeria
Mimoun Malki, University of Sidi Bel-Abbes, Algeria
Abdellah Chouarfia, University of Oran, Algeria
Mostefa Belarbi, University of Tiaret, Algeria

The SOA: Service Oriented Architecture is a paradigm which allows the unification in the approaches of integration of the information systems. This data integration of shared semantic description of handled by the services. This integration of the data is more flexible considering the limited number of the concepts used by the services. Therefore, architecture is suggested in order to reduce domain ontologies development and integration complexity. It allows also finding and automatic invocation of the services. Ontologies are integrated without doing major changes in operating mode of web services like HTTP, SOAP. This chapter presents an architecture which is a step towards its automation through the semantic web services without redefining the information system completely.

Compilation of References 322

About the Contributors 354

Index 361

Preface

The aim of this book *Ontology Theory, Management and Design: Advanced Tools and Models which* is to gather the latest advances in various topics of ontologies and their applications.

Ontologies, as formal representations of knowledge, are currently widely in use in computer science research. So far they have mainly been used as support for modeling and managing large applications in several domains such as knowledge engineering, semantic web, information retrieval, database design, e-Business, data warehousing, data mining, etc.

The focus of this book is on Information and Communication Sciences, Computer Science, and Artificial Intelligence. The audience for this book is extensive and will include variety of audiences. In fact, this book will be of a great value to academic and professional organisations and will be instrumental in providing researchers, scientists, academics, postgraduate students, practitioners and professionals with access to the latest knowledge related to design, modeling and implementation of ontologies.

This book is organized in self-contained chapters to provide greatest reading flexibility. We have received 44 chapters from researchers of various disciplines (Artificial Intelligence, Knowledge Engineering, Information Systems) and from different countries. All submitted chapters have been reviewed on a double-blind review basis, by at least three reviewers. After an evaluation process by the PC members, 14 chapters have been selected. Acceptance was based on relevance, technical soundness, originality, and clarity of presentation. Chapters are divided into four parts:

- **Section 1:** Introduction and Overview: Theory, Concepts and Foundations
- **Section 2:** Theoretical Models and Aspects/ Formal Frameworks
- **Section 3:** Ontology Management: Construction, Evolution and Alignment
- **Section 4:** Ontology Applications and Experiences

This book is organised as follows.

The purpose of chapter 1: *Ontologies in Computer Science: These New "Software Components" of our Information Systems*, is an introduction to the branch of knowledge modeling that develops ontology-oriented computable models of some domain of knowledge.

Chapter 2: *Ontology Theory, Management and Design: An Overview and Future Directions*, reviews the main methodologies, tools and languages for building, updating and representing ontologies that have been reported in the literature.

Chapter 3: *Exceptions in Ontologies: A Theoretical Model for Deducing Properties from Topological Axioms*, is a contribution to the study of formal ontologies. It addresses the problem of atypical entities in ontologies and proposes a new model of knowledge representation by combining ontologies and topology.

Chapter 4: *An Algebra of Ontology Properties for Service Discovery and Composition in Semantic Web*, addresses the problem of the automated discovery and composition of Web Services. The authors present an algorithm that generates efficient orchestration plans, with characteristics of optimality regarding quality of service.

In Chapter 5: *Approaches for Semantic Association Mining and Hidden Entities Extraction in Knowledge Base*, an approach for the extraction of association rules from a structured semantic association store is proposed.

Chapter 6: *Reusing the Inter-Organizational Knowledge to Support Organizational Knowledge Management Process: An Ontology-Based Knowledge Network*, describes three main meditation methods used to reconcile mismatches between heterogeneous ontologies.

Chapter 7: *Building and Use of a LOM Ontology*, presents an approach to index learning resources semantically using a LOM ontology. A tool is also developed to allow describing and retrieving learning objects.

The purpose of Chapter 8: *Ontology Evolution: State of the Art and Future Directions*, is to gather research and current developments to manage ontology evolution. The authors highlight ontology evolution issues and present a state-of-the-art of ontology evolution approaches and tools.

In Chapter 9: *Large Scale Matching Issues and Advances*, the authors survey the techniques of large scale matching and compare existing schema matching tools.

Chapter 10: *From Temporal Databases to Ontology Versioning: An Approach for Ontology Evolution*, focuses in developing an ontology versioning system to express, apply and implement changes on the ontology. The proposed ontology versioning approach, based on three steps, aims to assist users in expressing evolution requirements, observing their consequences on the ontology and comparing ontology versions.

Chapter 11: *Ontology Learning from Thesauri: An Experience in the Urban Domain*, focuses on ontology learning techniques using thesauri as input sources. Apart from reviewing the state of the art, this chapter shows how ontology learning techniques can be applied in the urban domain for the development of domain ontologies.

In Chapter 12: *Applications of Ontologies and Text Mining in the Biomedical Domain*, the authors present the possible interactions between ontologies and text mining. The contents of this chapter are specially aimed to the state-of-the-art and the new opportunities that are arising from the combination of text mining and ontology-based technology.

The goal of Chapter 13: *Ontology Based Multimedia Indexing*, is to provide responses to the following question: how the ontologies could be used in order to index and manage the multimedia collections? This chapter also proposes a solution for indexing a multimedia collection by combining technologies from both semantic Web and Multimedia indexation domains.

Chapter 14: *Semantic Enrichment of Web Service Architecture*, presents a flexible architecture dedicated to services semantic integration, based on Web Services Modeling Ontology approach. The developed architecture is compared with some research works.

Faiez Gargouri
Wassim Jaziri
Editors

Section 1

Introduction and Overview:

Theory, Concepts and Foundations

Chapter 1
Ontologies in Computer Science:
These New "Software Components" of Our Information Systems

Fabien L. Gandon
INRIA, France

ABSTRACT

Ironically the field of computer ontologies suffered a lot from ambiguity. The word ontology can be used and has been used with very different meanings attached to it. The authors will introduce in this chapter two faces of ontologies in computer science: (1) the ontology object: focusing on the nature and the characteristics of the ontology object, its core notions and its lifecycle. (2) ontology engineering: the branch of knowledge modeling that develops ontology-oriented computable models of some domain of knowledge, focusing here on the design rationale and the assisting tools.

INTRODUCTION

"When I use a word," Humpty Dumpty said in rather a scornful tone, "it means just what I choose it to mean - neither more nor less." – Lewis Carroll, *Through the Looking-Glass*, Chapter VI

Knowledge engineering is a broad research domain where the core issues include the acquisition and the modeling of knowledge. Modeling knowledge, consists in representing it in order to store it, to communicate it or to externally manipulate it. Automating the manipulation of knowledge leads to the design of knowledge-based systems *i.e.*, systems which behavior relies on the symbolic manipulation of formal models of knowledge pieces in order to perform meaningful operations that simulate intelligent capabilities. As such, knowledge engineering is a branch of artificial intelligence.

Knowledge representation raises the problem of the form *i.e.*, the choice of a representation formalism that allows us to capture the semantics at play in the targeted pieces of knowledge. One approach that emerged in the late 80s is based on the concept of ontologies. An ontology, as we shall see, is that part of the knowledge model that captures the semantics of primitives used to make formal assertions in a domain of application. Computer science ontologies

DOI: 10.4018/978-1-61520-859-3.ch001

are children of Artificial Intelligence that recently came to maturity and powerful conceptual tools of Knowledge Modeling. Ontologies provide a coherent base to build on, and a shared reference to align with, in the form of a consensual conceptual vocabulary on which one can build descriptions and communication acts.

In this introduction chapter, we shall focus on the branch of knowledge modeling that develops ontology-oriented computable models of some domain of knowledge.

The word *ontology* can be used and has been used with very different meanings attached to it. Ironically, and as we will see, the ontology field suffered a lot from ambiguity. We will introduce here two faces of ontologies in computer science:

- The *ontology object*: focusing on the nature and the characteristics of the ontology object, its core notions and its lifecycle.
- The *ontology engineering*: focusing on the design rationale and the assisting tools.

FROM ONTOLOGY TO ONTOLOGIES

The term "ontology" was constructed from the Greek Ontos ("what is", "what exists") and Logos ("the discourse", "the study"). It first appeared in 1606 in "Ogdoas scholastica" from Jacob Lorhard. In philosophy, ontology is a fundamental branch of metaphysics, concerned with the concept of existence, the basic categories of existing and the most general properties of being. As a branch of philosophy, Ontology is the metaphysical study of the nature and relations of existence.

As computer scientists, at first sight this term and its definition might not seem very useful for our daily work. However when one considers the work a software engineer performs when designing a class hierarchy in an object-oriented programming language or a database schema and especially the questions this engineer has to answer for such a task, then the ontological questions such has "what are the categories of the things existing around us?" appear to be very much closer than expected. Software engineers designing object-oriented models are wondering about: the objects that their applications will handle, the classes that combine the characteristics common to all these objects, the relationships that may exist between these objects, etc. In other words, these engineers are questioning what defines these classes of objects, which characteristics can ensure that a given object belongs to a class, what this membership means in terms of content or possible manipulations. Like Molières' character Mr. Jourdain who is amazed to discover that he has been speaking prose all his life without knowing it, we, computer scientists, could be amazed to see how close our modeling rationale can be to ontological questioning. When we question the existential definition of classes of objects used in the scenarios of the applications we develop, we are sometimes the "Mr. Jourdain" of Ontology.

Computer scientists borrowed the term "ontology" from the philosophers in the early 80s: it can be found for instance in an article by McCarthy (McCarthy, 1980) and in the book of John Sowa (Sowa, 1984) before it became famous with the article by Thomas Gruber (Gruber, 1993). To be more precise, the reason why object-oriented programming presents such a resemblance to the notion of ontologies in computer science is that they have common ancestors: the early systems of symbolic artificial intelligence. The beginnings of this branch of artificial intelligence are intertwined with the beginnings of computer science, because since its beginnings, computer science has perpetuated the dream of the designers of automata to simulate or exceed human intelligence with artificial systems.

A NOTION LOOKING FOR A NAME

The branch of artificial intelligence on which we just focused is said symbolic because it is based on formal representations of knowledge in the form of symbols that the system can store and manipulate (for instance: logical languages and operations, graph-based structures and operations). Unlike in other non-symbolic artificial intelligence approaches, these representations are both understandable by humans (through interpretation) and manipulated by the systems. By applying rules manipulating symbols and defined on these representations and with the right interpretation one can simulate, for example, human reasoning.

The notion and the software artifact that we now name "ontologies" existed in computer science far before the term "ontology" was imported from philosophy. Indeed, back in the 70s, the notion of ontology was already used without being named as such and under various names in different knowledge representation frameworks of symbolic artificial intelligence e.g.: the TBox description logics (Baader *et al.*, 2003), which describes the types of terms that exist in our representation and their characteristics; the support of conceptual graphs (Sowa, 1984), which describes hierarchies of multi-inheritance between types of concepts or types of relationships; the schemas in the Frame based formalisms; the class hierarchies of the object-oriented languages; etc. Even the relation schema of a database is a kind of ontological knowledge.

Yet, we had to wait the 90s for the word "ontology" to be adopted by the entire community and its definition is still causing much ink to flow.

THE RIGHT WORD AT THE RIGHT TIME

The height of ontology in computer science is that, this domain so concerned with the precise definition and naming of notions has been itself searching for a consensual name and definition for a long time. This delay is probably largely due to the very abstract nature of the concept of ontologies. In an attempt to define the notion of ontology in informatics, it is worth recalling that Ontology means the study of properties of what exists. By importing that term in computer science, we have moved from a field (<u>O</u>ntology as a sub-field of Metaphysics) to an object (several <u>o</u>ntologies as software components in computer science). The word ontology was imported because it matched the need for abstracting the notions of ontologies across their different instances in various software and frameworks.

In computer science, an ontology is a software artifact. It is a computer representation of chosen properties of existing things; this representation being usually done in a formalism allowing some rational and automated processing. An ontology is the result of an exhaustive and rigorous formulation of the conceptualization of a domain. This conceptualization is often said to be partial because it is illusory to believe that one could capture the full complexity of a domain in such formalisms. Note also that the degree of formalization of an ontology varies with its intended use. Because of the description of the existing categories they include, ontologies have borrowed their computer name to the philosophical domain, but this multidisciplinary link is also an opportunity to adapt the methods of philosophy in order to propose methodological guidelines to engineer ontologies in computer science.

To make it simple, consider Figure 1. Consider a scene of reality, a certain state of affairs, where la light cube was placed on the right of a dark cube. The description of this scene requires two things: (1) a statement of the facts describing the scene, (2) an unambiguous vocabulary used to make these statements. In knowledge representation, the conceptual descriptions are called facts and the conceptual vocabulary is called an ontology. Thus an ontology provides a conceptual vocabulary to make descriptions.

As we shall see in the next sections, leveraging the work of symbolic artificial intelligence on knowledge-based systems and inference engines,

Figure 1. An ontology provides a conceptual vocabulary to describe a reality

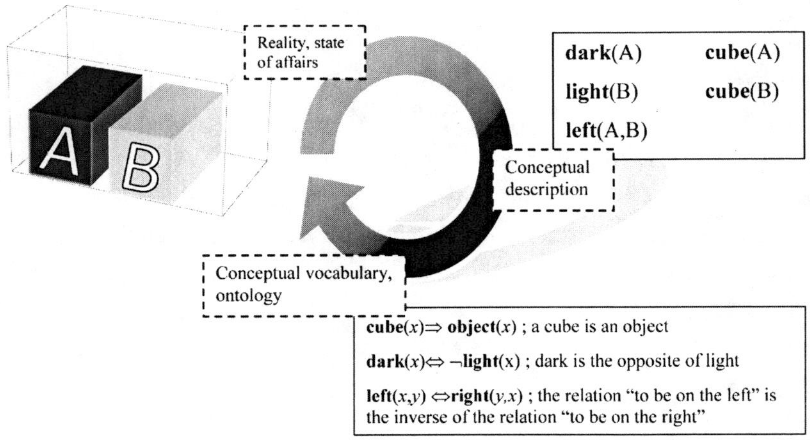

an ontology allows us to introduce mechanisms for reasoning, automatic classification, information retrieval, etc. and when shared among different systems an ontology can also ensure their interoperability.

CONTENT OF ONTHOLOGIES: CHARACTERIZING ONTOLOGICAL KNOWLEDGE

An ontology defines concepts (principles, ideas, categories of objects, potentially abstract concepts) and relationships. It usually includes a hierarchical organization of the relevant concepts and the relevant relationships between these concepts, as well as rules and axioms that constrain these representations.

The set of all the properties of a concept is called the intension of the concept, and the set of all the objects or beings that are instances of this concept is called the extension of the concept. Let us consider an example of a concept that we will intentionally name "Concept C#1". We can associate to C#1:

- An intension: a set of qualitative and functional properties common to all the instances, occurrences, of that concept, and allowing us to define that concept, for instance: "This concept C#1 is a sub-category of motorized transport vehicles, having at least three wheels and designed and arranged for the transport of a small number of people (8 or less) as well as objects of small dimensions";
- An extension: a set of entities that fall into this category, e.g.: (the station-wagon of Tim, the coupé of Bernard, the minivan of Lee, etc.).

To express a concept, we choose a symbolic representation, often linguistic and verbal, sometimes iconic. Peirce distinguished three types of signs (Peirce, 1867): the index, the icon, the symbol. This typology distinguishes among three different ways in which the sign refers to its object: the symbol by a habit or rule for its interpretant; the icon by a quality of its own (e.g. depiction); the index by real connection to its object.

In the case of concept C#1, we can give as examples of linguistic representations used in different contexts: car, auto, automobile, bucket, buggy, bus, clunker, compact, convertible, conveyance, coupe, gas guzzler, hardtop, hatchback, heap, jalopy, jeep, junker, motorcar,

Figure 2. Three types of signs in semiotics

pickup, ride, roadster, sedan, truck, van, wagon, wreck, etc.

We therefore dissociate the concepts and their linguistic manifestations. A term is not a concept, and vice versa, a concept is not a term. A term may be ambiguous, while a concept has only one meaning, a single definition in a given ontology. Applications handling linguistic representations of concepts must then address the problems of synonymy (a concept denoted by several terms) and ambiguity and homonymy (a term denoting a number of concepts).

In the same way as for the concepts, the ontology defines relations that may exist between instances of these concepts. Consider a relation we will intentionally name "Relation R#1". We can associate to R#1:

- An intension, for example: "R#1 is a relation between a person or group who created a document, and the intellectual content, arrangement or form of that document";
- An extension, i.e. all the lists of entities for which this relation holds, for example: ((Orson Welles, The War of the Worlds), (William Shakespeare, Macbeth), etc.);
- Semiotic representations of that relation, in particular linguistic representations, for example: "wrote", "author of", etc.
- A signature: a list specifying the types of the entities that the relation connects, for example: (Person or Group, Document).

STRUCTING THE CONTENT OF AN ONTHOLOGY

In an ontology, the intensions are organized, structured and constrained to represent our conception of the world and its constraints (e.g., a car is always a vehicle). The ontology captures the intensions and the laws that govern them to reflect aspects of our reality. These aspects are chosen for their relevance for the application scenarios considered in the software project using the ontology. Representations of intensions and ontologies use languages more or less formal (graphs, logics, restricted natural language), according to the intended use of the ontology. The formal construction of the intension gives an accurate and unambiguous account of its meaning, which allows its manipulation by software and its use as a knowledge representation primitive to describe and structure, for instance, data, objects, software, users, places, communities, etc.

In ontologies, the intensions are usually organized in a taxonomy or hierarchy of types. We call *subsumption* the act of placing a class below another. Subsumption is also the name of the link between a sub-category and a parent category. The importance of taxonomic organization is justified by the fact that classification and identification (the act of determining whether something belongs to a class) and categorization (the act of identifying the existing categories) are very common inferences that we use all day long. Consider this simple example of a conversation between two people:

- Do you know a good restaurant?
- There's a pizzeria around the corner.
- Thanks

Even in such an ordinary conversation the first speaker generalized his query using the concept of *restaurant*, which represents the most abstract class covering all forms of acceptable answers. The second speaker, probably without even paying attention, used his taxonomy of concepts to infer that a *pizzeria* **is a** *restaurant*, and identify relevant answers. The fact that the taxonomic knowledge is shared between the two speakers is implicit here, since the second person assumes that his answer will be understood without specifying that a pizzeria is a restaurant, and it is indeed the case. The use of shared conceptualizations and inferences they support is at the heart of activities as simple as this exchange of information. Making explicit the ontological knowledge and ensuring their consensual representation are two of the major problems of ontological engineering.

Thus, in an information system, the simple addition of this knowledge can dramatically improve the capabilities of machines. Take, for instance, the very simple case where you are looking for books of Victor Hugo. If your information system works at text level and if you use the keywords "Hugo" and "Book", you may encounter at least two problems:

- The problem of noise: the system might not be able to make the difference between the family name "Hugo" used for an author, the name "Hugo" used for a street name, the first name "Hugo", the brand name "Hugo", etc.;
- The problem of silence: the system might not be able to understand that the word "novel" stands for a subtype of books and that answers in that category are relevant to your query.

By designing an ontology, one can provide the system with some representations of these aspects of our reality e.g. about humans (Man and Woman are sub-types of Human, which is itself a subtype of Living Being), about documents (Novel and Short Story are subtypes of Book, which is itself a subtype of Document) and the relations between the two, with their signatures (for example, there exists a relation Author, which can be established between a document and a Human) (Figure 3).

Using that ontology, one has the vocabulary to describe pieces of reality, for instance to capture the fact that there is a man whose name is "Hugo" and who wrote a novel entitled "Notre Dame de Paris" (Figure 4).

Using the same ontology one can formulate unambiguous queries for instance a query to for search documents written by a man named "Hugo" (Figure 5).

Using the logic of this language, the system can infer that a novel is a book, a book is a document, so a novel is a document, and that the answer "Hugo wrote the novel Notre Dame de Paris" is valid.

Figure 3.

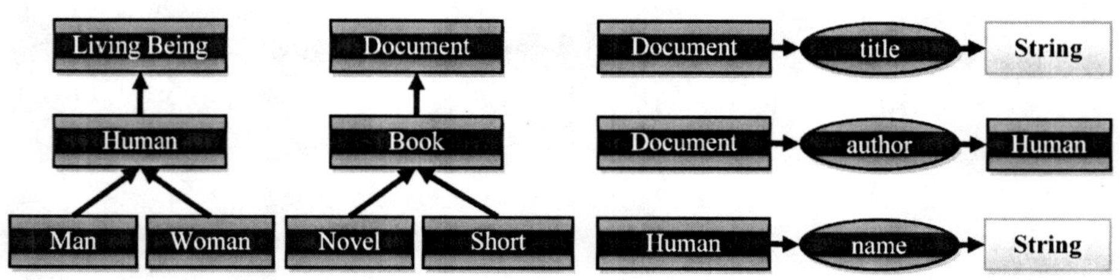

BEYOND THE TAXONOMICAL SKELETON OF ONTHOLOGIES

Considering what was said in the previous sections, it is important not to confuse ontologies and taxonomies. Ontological knowledge goes far beyond the taxonomical knowledge (see Figure 6). As a non limitative list of examples, one can find in an ontology:

- knowledge about composition, e.g. in chemistry (categories of chemical elements), in production (categories of pieces), medicine (anatomical categories), etc.
- complete formal definitions, for example: a person is a manager *if and only if* there exists an group of people managed by this person;
- integrity constraints, e.g. a published book has a single ISBN, parents cannot be younger than their children;
- formulas, e.g. the recommended heart rate for a person during a cardio-vascular exercise is equal to (220 - age) x 0.65;
- algebraic properties, e.g. the relationship "married" is symmetrical, this means that if Thomas is married to Stephanie, then the system must infer that Stephanie is married to Thomas, and vice versa;

Figure 4.

Figure 5.

Figure 6. Example of ontological knowledge in the chemical composition of molecules; a partonomy / meronymy in organic chemistry is ontological knowledge

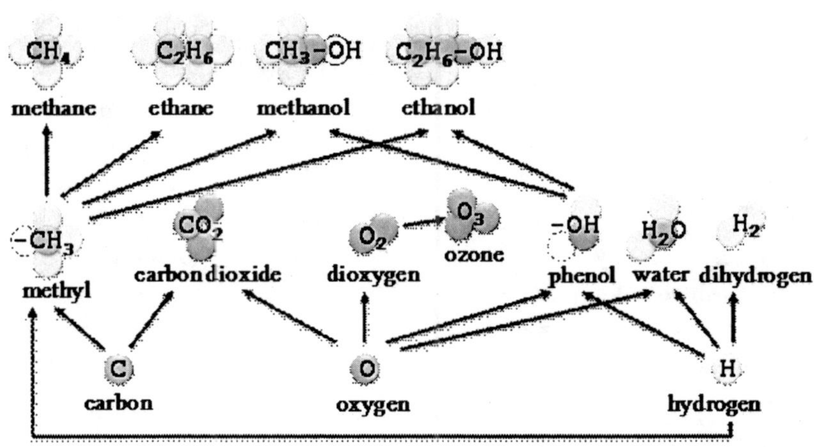

- knowledge by default, for example: by default a car has four wheels;
- inverse relationships, e.g. "to be part of" is the opposite of "includes", i.e. if a door is part of a car then the car includes the door, and vice versa;
- rules specific to a domain, e.g. in biology, for each receptor that activates a molecular function, if that function plays a role in the functioning of the body, then the receptor plays the same role.

The content of an ontology also varies with the type of ontology considered. A domain ontology contains knowledge about a given application field (e.g., aviation). A task ontology contains knowledge about an activity (e.g., diagnosis). A high-level ontology contains very general abstract knowledge, to gather other ontologies (e.g. notions of entity, event, role, etc.).

The content of an ontology also depends on the degree of formalization (natural language, restricted language, simple formalism, complex logics). In particular a very common difference is made between:

- **Lightweight ontologies:** ontologies which typically provide no formal definitions or very little and often focus on the representation of hierarchies of types that do not require very expressive languages (e.g. RDFS);
- **Heavy ontologies:** ontologies that give precise formal definitions for the primitives they define, using representation languages more expressive than lightweight ontologies (eg OWL DL).

There are several reasons why ontological knowledge is separated from other kinds of knowledge. First of all, an ontology allows to factorize knowledge. In a model, ontological knowledge is a knowledge that is always true, whatever the state of the system and the descriptions it contains are. The ontology factorizes knowledge that no longer has to be repeated for each description. For example, we say in an ontology that a car is a vehicle (because it is always true in our application), but we do not set the color of the cars, because it changes from one car to another.

Another advantage is the ability to reuse and share knowledge. Being isolated, ontological knowledge can be reused in different applications, and this reuse (full or partial) can provide the basis for interoperability between different systems.

As separated modules, it is also possible to "compile ontologies" and optimize the inferences they support. Ontological knowledge may be processed to provide efficient structures, to certify their coherence and to optimize the inferences they support. For example, an ontology enables us the calculate transitive closure: if a *coupé* is a *car* and a *car* is a *vehicle*, then a *coupé* is a *vehicle*. The transitive closures of ontologies can be precomputed, indexed and cached efficiently.

Finally and to illustrate the diversity of ontologies, let us arbitrarily list some existing ontologies and their subjects: ASBRU, provides an ontology for guideline-support tasks and problem-solving methods in order to represent and to annotate clinical guidelines in standardized form; the Bibliographic Ontology reuses data types taken from ISO standard; the FIPA Agent Communication Language contains an ontology describing speech acts for artificial agents communication; CHEMICALS Ontology contains knowledge within the domain of chemical elements and crystalline structures; CoreLex is an ontology for lexical semantic database and tagset for nouns; EngMath mathematics and engineering ontologies including ontologies for scalar quantities, vector quantities, and unary scalar functions; Gene Ontology an ontology of molecular functions, biological processes and cellular components; Gentology: genealogy ontology for data interchange between different applications; Open Cyc, an upper ontology containing concepts of common knowledge; PLANET is an ontology for representing plans that is designed to accommodate a diverse range of real-world plans, both manually and automatically created; ProPer, an ontology to manage skills and competencies of people.; SurveyOntology to describe large questionnaires; The Enterprise Ontology is a collection of terms and definitions relevant to business enterprises; The Dublin Core Element Set Ontology for cataloging library items and other electronic resources; the TOVE project develops a set of integrated ontologies for the modeling of both commercial and public enterprises; DOLCE, a Descriptive Ontology for Linguistic and Cognitive Engineering; FOAF (Friend of a friend) is an ontology to describe persons, their activities and their relations to other people and objects; RELATIONSHIPS is an ontology extending FOAF with more details for describing relationships between people; Academic Institution Internal Structure Ontology (AIISO); The Creative Commons ontology lets you describe copyright licenses; etc.

Many other ontologies are available and search engines like Swoogle, Sindice or Watson can help you locate ontologies you might want to reuse.

SUMMARIZING ALL THESE NOTIONS AROUND ONTOLOGIES

At this point, let us compile in table 1 a number of definitions often used in the field of ontologies in computer science.

APPLICATIONS OF THESE EXPLICIT REPRESENTATIONS OF ONTOLOGICAL KNOWLEDGE

Many social entities have to manage and maintain knowledge, be it their *raison d'être* (networks of interest, research teams, schools, etc.) or a side effect of their core operations (manufacturers, administrations, associations, etc.). The agility of these entities to detect, store, recall and activate their knowledge depends on their ability to respond to the outside world (research innovation, response time to market, quality of training, etc.). In this context, knowledge is now part of the capital and resources of organizations, and an efficient information system is a vital asset. In the kingdom of information, knowledge is king.

But information systems are collective applications radically constrained. Differences in experience, different educations, different cultures, different needs, different perspectives, different languages or jargons, different media and formats, different contexts of use, different access rights, etc. set a variety of constraints which may drive an information system to run in virtuous circles or in vicious circles, depending on the importance of these constraints in the daily usages of the system and on how the system meets them.

Ontologies are to semantics, what grounding is to electronics: a common base to build on, and a shared reference to align with. Ontologies are considered as a powerful tool to lift ambiguity: one of the main roles of ontologies is to disambiguate, providing a semantic ground, a consensual conceptual vocabulary, on which one can build descriptions and communication acts. (Bachimont, 2001) explains that on the one hand, ontologies provide notional resources to formulate and make explicit knowledge and on the other hand, they constitute a shared framework that different actors can mobilize.

Ontologies can represent the meaning of different contents exchanged in information systems. The introduction of an ontology in an information system aims at reducing or even eliminating the conceptual and terminological confusion and at aligning our understanding in order to improve communication, sharing, interoperability and the degree of possible reuse. Ontologies in computer science offer a unifying framework and provide primitives *i.e.* basic elements, building blocks for improving communication between people, between people and systems, and between systems. The integration of an ontology into an information

Table 1. A compilation of definitions for a number of core notions in the field of ontologies in computer science

notion	something formed in the mind, a constituent of thought; it is used to structure knowledge and perceptions of the world. || an idea, a principle, which can be semantically valued and communicated.
concept	notion usually expressed by a term (or more generally by a sign) || a concept represents a group of objects or beings sharing characteristics that enable us to recognize them as forming and belonging to this group.
relation	notion of an association or a link between concepts usually expressed by a term or a graphical convention (or more generally by a sign)
extension / intension	distinction between ways in which a notion may be regarded: its extension is the collection of things to which the notion applies; its intension is the set of features those things are presumed to have in common. There exists a duality between intension and extension: to included intensions $I_1 \subset I_2$ correspond included extensions $E_1 \supset E_2$.
concept in intension / intension of a concept	set of attributes, characteristics or properties shared by the object or beings included in or to which the concept applies. e.g. for the concept of a *car* the intension includes the characteristics of *a road vehicle with an engine, usually four wheels and seating for between one and six people.*
concept in extension / extension of a concept	set of objects or beings included in or to which the concept applies. e.g. for the concept of a *car* the extension includes: *the Mazda MX5 with the registration 2561 SH 45, the green car parked at the corner of the road in front of my office*, etc.
relation in intension / intension of a relation	set of attributes, characteristics or properties that characterizes every realization of a relation. e.g. for the relation *parenthood* the intension includes the characteristics of *the raising of children and all the responsibilities and activities that are involved in it.*
signature of a relation	set of concepts that can be linked by a relation, this constraint is a characteristic of the relation that participate to the definition of its intension. e.g. for the relation *parenthood* the signature says it is a *relation between two members of the same species.*
relation in extension / extension of a relation	set of effective realizations of a relation between object or beings. e.g. for the relation *parenthood* the extension includes: *Jina and Toms are the Parents of Jim, Mr Michel Gandon is my Father*, etc.
Ontology	that branch of philosophy which deals with the nature and the organization of reality (Guarino & Giaretta, 1995). || a branch of metaphysics which investigates the nature and essential properties and relations of all beings as such.
Formal Ontology	the systematic, formal, axiomatic development of the logic of all forms and modes of being (Guarino & Giaretta, 1995).
conceptualisation	an intensional semantic structure which encodes the implicit rules constraining the structure of a piece of reality (Guarino & Giaretta, 1995) || the action of building such a structure.
ontology	a logical theory which gives an explicit, partial account of a conceptualization (Guarino & Giaretta, 1995) (based on (Gruber, 1993)); the aim of ontologies is to define which primitives, provided with their associated semantics, are necessary for knowledge representation in a given context. (Bachimont, 2000)
ontological commitment	a partial semantic account of the intended conceptualization of a logical theory (Guarino & Giaretta, 1995) || practically, an agreement to use a vocabulary (*i.e.,* ask queries and make assertions) in a way that is consistent with respect to the theory that specifies the ontology. Software pieces are built so that they commit to ontologies and ontologies are designed so that they enable us to share knowledge with and among these software pieces. (Uschold & Gruninger, 1996)
ontological theory	a set of formulas intended to be always true according to a certain conceptualization (Guarino & Giaretta, 1995).

continued on the following page

Table 1. continued

ontological engineering	the branch of knowledge engineering which exploits the principles of (formal) Ontology to build ontologies (Guarino & Giaretta, 1995). || defining an ontology is a modeling task based on the linguistic expression of knowledge. (Bachimont, 2000)
ontologist	a person who builds ontologies or whose job is connected with ontologies' science or engineering.
state of affairs	the general state of things, the combination of circumstances at a given time. The ontology can provide the conceptual vocabulary to describe a state of affairs. Together this description and the state of affair form a model.
taxonomy	a classification based on similarities.
Mereology	The study of part-whole relationships.
partonomy	a classification based on part-of relation.

system allows us to formally declare a number of knowledge primitives used to characterize the information managed by the system and to rely on these characterizations and the formalization of their meaning to automate processing tasks of the information.

In a search engine, for example, one can aim at improving information retrieval in terms of precision (avoiding ambiguities due to homonymy) or in terms of recall (by incorporating more specific concepts or equivalent using the synonymy, the hyponyms) or by deducing tacit knowledge (e.g., production rules) or by relaxing the constraints that are too strict when no answer is found (using generalization inferences) or by grouping too numerous results according to their similarity in order to present them in a user-friendly way (conceptual grouping or clustering).

ONTOLOGIES ARE LIVING OBJECTS: LIFE-CYCLE OF ONTOLOGIES

Ontologies are living objects, and each stage of their life cycle raises research and development problems. The life cycle of an ontology comprises seven activities: detection of needs, management and planning, design, evolution, dissemination, use and evaluation (Figure 7).

Requirements and Evaluation: the detection of needs, prior to the design, and the evaluation, once an ontology is used, raise the methodological problems of collection (analysis of interviews, questionnaires and surveys, study of ergonomics and usages) and identification (e.g., scenarios-based analysis). In addition, the detection phase requires an initial in-depth state of affairs, because it cannot be based on previous studies or return on experience, as in the case of evaluation.

Design and Evolution: the initial design stage and the evolution stage also share a number of problems:

- Specification of solutions (participatory design, mockups and modeling, prototyping)
- Acquisition of knowledge (text analysis, natural language processing, collaborative platforms, data mining, knowledge extraction tools and methods in general)
- Conceptualization and modeling (ontological design pattern, meta-ontologies, interview with experts)
- Formalization (methods and tools of formal ontology, description logics and tableau algorithms, formal concept analysis,

Figure 7. Main stages in the life-cycle of an ontology

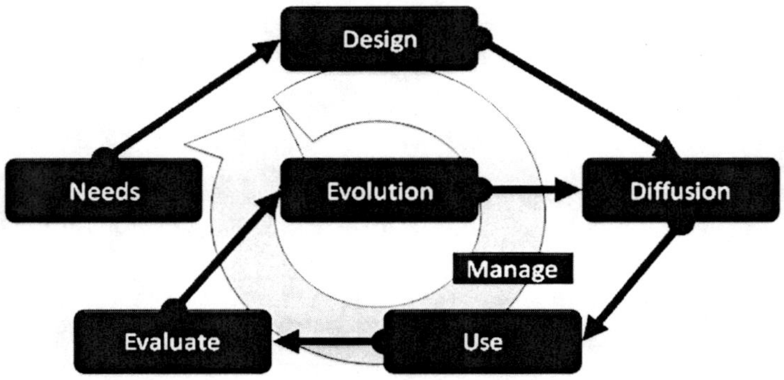

conceptual graphs, semantic web formalisms RDF/S and OWL, RIF)
- Integration of existing resources (ontologies alignment, translation)
- Implementation (engines and stores for conceptual graphs, description logics, object-oriented formalisms, rules, RDFS, OWL, RIF, etc.)

Another problem in design and evolution is to obtain and maintain a consensus on the ontology, on the representation rationale and its underlying conceptualization. Depending on the context, this problem may call upon "groupware" tools and solutions to manage different points of view, different conceptualizations and different terminologies.

Finally the evolution of an ontology raises the problem of maintaining what was built on top of this ontology. Indeed, an ontology is both a living object interesting in itself and a set of primitives to describe facts of the world and algorithms running on these facts. When the ontology changes, its changes impact everything that has been built on top of it. Maintaining the consistency inside an ontology and outside of it, managing history and versions, re-engineering and spreading modifications, are problems to be considered when building ontology-based solutions. Maintenance of the ontology raises problems of technical integration and also of usage integration.

Dissemination: the dissemination phase focuses on the deployment and setup of ontology. The problems of this stage are strongly constrained by the software architectures of the solutions. In a web application context, one can rely on W3C standards. For file sharing, peer-to-peer architectures or other distributed architectures may be used. For the integration of applications, web services architecture can be a solution. In all these architectures (web servers, web services, peer to peer, agents, etc.) the distribution of resources (data, models, applications and users) and their heterogeneity (syntax, semantics, protocols, etc.) raise research problems on interoperability (mediation and alignment) and to scale (large databases, optimization of inference, propagation of queries, data syndication, composition of services, etc.).

Use: the use phase includes all activities based more or less directly on the availability of the ontology, for example, the annotation of resources (multimedia analysis, natural language processing, reverse engineering of database, social tagging, etc.), the resolution of queries (projection algorithm of graphs with constraints, approximate search using semantic distances defined on the ontology to quantify the approximations made), the deduction of knowledge and decision support (rule-based inference engines), assisted navigation and contextualized services (analysis of context, identification and composition of services), the analysis of large volumes of knowledge (clustering, search for frequent patterns, notification, monitoring and intelligence).

All these activities have in common the problem of interaction design: the design of means of interaction with the users and their ergonomics (dynamic interfaces, links between semiotics and semantics, profiles and contexts of use). On this aspect, the ontology brings both new solutions (e.g., inferences exploiting ontologies for the generation of dynamic interface elements) and new problems (e.g., complex data models generates problems for their representation and the interaction with these representations).

Management: The parallel activity of management and planning stresses the importance of a follow-up work and a comprehensive policy to detect, or trigger, prepare and evaluate the iterations of the life-cycle of ontologies and ensure that the solution remains in the virtuous cycle of information systems where contributions bring usefulness and usefulness brings contribution.

ONTOLOGY ENGINEERING

(Mizoguchi *et al.*, 1997) explained the challenge that ontology engineering must face: "Most of the conventional software is built with an implicit conceptualization (…) systems should be built based on a conceptualization represented explicitly". In fact, by making at least some aspects of our conceptualizations explicit to the systems, we can improve their behavior through inferences exploiting this explicit partial conceptualization of our reality. "the ultimate purpose of ontology engineering is: 'To provide a basis of building models of all things, in which information science is interested, in the world'." (Mizoguchi *et al.*, 1997). As the scientific discipline of Ontology is evolving towards an engineering discipline, it develops principled methodologies (Guarino & Welty, 2000).

Scoping an Ontology

One should not start the development of an ontology without knowing its purpose and scope (Fernandez *et al.*, 1997). In order to identify these goals and the limits, one has to clearly state why the ontology is being built, what its intended uses are and who are the stakeholders (Uschold & Gruninger, 1996). Then, one should use the answers to write a requirements specification document. Adapting the characteristics of indexes given in information retrieval (Korfhage, 1997) and the notion of granularity illustrated above, we will propose three characteristics of the scope of an ontology:

- **Exhaustively**: breadth of coverage of the ontology, *i.e.,* the extent to which the set of concepts and relations mobilized by the application scenarios are covered by the ontology. Beware, a shallow ontology (e.g. one concept 'entity' and one relation 'in relation with') can be exhaustive.
- **Specificity**: depth of coverage of the ontology *i.e.,* the extent to which specific concept and relation types are precisely identified. The example given for exhaustivity had a very low specificity; an ontology containing exactly 'german shepherd', 'poodle' and 'golden retriever' may be very specific, but if the scenario concerns all dogs then its exhaustivity is very poor.
- **Granularity**: level of details of the formal definitions of the notions in the ontology *i.e.,* the extent to which concept and relation types are precisely defined with formal primitives. An ontology relying only on subsumption hierarchies has a very low granularity while an ontology in which the notions systematically have a detailed formal definition has a very high granularity.

An interesting technique to capture the application requirements in context, is the one of scenario analysis as presented for example in (Caroll, 1997) and used for software engineering. Scenarios are used as the entrance point in the project (Giboin *et al.*, 2002), they are usually information-rich stories capturing problems and wishes. (Uschold & Gruninger, 1996) uses the notion of motivating scenarios and competency questions placing expressiveness requirements on the envisioned ontology.

Knowledge Acquisition

Data collection or knowledge acquisition is a collection-analysis cycle where the result of a required collection is analyzed and this analysis triggers new collections. Elicitation techniques help elicit knowledge. Several techniques exist for data collection and benefit from two decades of work in the knowledge acquisition community (see, for example, (Dieng, 1990), (Dieng, 1993) and (Dieng *et al.*, 1998) as well as (Sebillotte, 1991), (La France, 1992) and (Aussenac, 1989)). Experts, books, handbooks, figures, tables and even other ontologies are sources of knowledge from which the knowledge can be elicited using in techniques such as: brainstorming/brainwriting, interviews, observations, document analysis, questionaires, data mining.

The data collection is not only a source of raw material: for example interviews of experts might help to build concept classification trees (Fernandez *et al.*, 1997). Knowledge acquisition and modeling is a dialog and a joint construction work with the stakeholders (users, partners, managers, providers, administrators, customers, etc.). They must be involved in the process of ontology engineering, and for these reasons semi-formal/natural language views (scenarios, tables, lists, informal figures) of the ontology must be available at any stage of the ontology lifecycle to enable interaction between and with the stakeholders.

Data collection is a goal-driven process. People in charge of the data collection must always have an idea of what they are looking for and what they want to do with the collected data. It is essential to consider the end product that one desires right from the start (scenarios, models, ontologies…) and from that to derive what information should be identified and extracted during the data collection.

The reuse of ontologies is both seductive (it should save time, efforts and would favor standardization) and difficult (commitments and conceptualizations have to be aligned between the reused ontology and the desired ontology). But as (Guarino, 1997) noticed, "a concept may be 'relevant' for a particular task without being necessarily 'specific' of that task". Therefore, reuse should be possible and pursued. On the other hand, (Bachimont, 2000) observed that it emerges from practice that while it is always possible to adapt an ontology, it is rarely possible to reuse it as it is. This is one of the reasons while small and focused ontologies are more viral than others e.g.: Dublin Core, FoaF, Creative Commons. Documenting an ontology is not only interesting for designers, the document can prove to be a strong asset to encourage appropriation by the users of the system exploiting this ontology, and reuse by the designers of other systems.

Conceptualization and Ontology Design Rationale

Defining ontology is about modeling primitives for the problem solving scoped by the motivating scenarios. A usual way to design these primitives is to start from the linguistic expressions of the knowledge of the targeted domain.

One of the first things to do is to capture and fix the context in which the terminology will be established and shared (Mizoguchi *et al.*, 1997). Fixing the context is the very first step of the semantic normalization: among possible meanings for a unit in context, one fixes the one that should always be associated to it; this amounts to choose a referential context in which, in principle, terms must be interpreted (Bachimont, 2000). Semantic normalization is the choice of a reference context, corresponding to the application scenario that motivated knowledge modeling. This point of view enables the modeler to fix what must be the meaning of a wording, *i.e.,* what is the notion behind.

For (Uschold & Gruninger, 1996), an ontology may take a variety of forms, but necessarily it will include a vocabulary of terms and some specifications of their meaning (*i.e.,* definitions). In fact, this lexicon is the main result of a terminological study of the data collection reports; thus the terminological study is usually the next stage of the semantic commitment after fixing the context. The identification of linguistic expressions in the textual resources, aims to include and define these wordings and terms in lexicons to prepare modeling. The expressions are included and defined if and only if they are relevant to the progress of the application scenarios that are being considered. (Uschold & Gruninger, 1996) propose guidelines to generate the definitions.

As we said before, throughout the ontology design process, it is vital to reach and maintain an agreement with the scenario stakeholders. Being a commitment, the normalization is a joint work between the knowledge engineer and the stakeholders.

The second stage identified in (Bachimont, 2000) is the ontological commitment specifying the formal meaning of the notions. Conceptualizing leads ontology designers to organize the notions and to do so they look for relations between these notions that could be used to structure them. The conceptualization process is a refinement activity, iteratively producing new refined representations of the notions being modeled. Each new intermediary representation, as they are called in (Gómez-Pérez *et al.*, 1996), is a step toward the required degree of formalization.

As we have seen ontologies usually include a taxonomy of concepts. To build the taxonomy of concepts, several approaches (Uschold & Gruninger, 1996) have been opposed in literature (Bottom-Up, Top-Down, Middle-Out) and the choice of an approach and its motivations are closely linked to the domain of intervention, the type of data manipulated and the knowledge acquisition techniques available.

(Bachimont, 2000) proposes to determine the meaning of a unit in the tree using four differential principles. When applied to a given node, these principles make explicit similarities and differences with its neighbors.

OntoSpec (Kassel, 2002) is a method for the semi-informal specification of ontologies. It reuses results from (Guarino & Welty, 2000) and proposes a taxonomy of primitives, and a set of design rules to specify and to structure the ontology.

Guarino and Welty made a significant contribution to the theoretical foundations of the field. In 1992, Guarino started by distinguishing natural concept, role, attributes, slots and qualities. He proposed to use the term 'role' only in Sowa's sense; bearing on Husserl's theory of foundation, he distinguishes between roles and natural concepts, and defines a role as a concept which implies some particular 'pattern of relationships', but does not necessarily act as a conceptual component of something. He defines 'attributes' as concepts having an associate relational interpretation, allowing them to act as conceptual components as well as concepts on their own; He proposes a formal semantics which binds these concepts to their corresponding relations, and a linguistic criterion to distinguish attributes from 'slots', i.e., from those relations which cannot be considered as conceptual components. Moreover, he shows how the choice of considering attributes as concepts enforces 'discipline' in conceptual analysis as well as 'uniformity' in knowledge representation. (Guarino, 1992). In (Guarino & Welty, 2000) is presented an additional and exemplified comprehensive set of definitions aiming at providing the ontologists with methodological elements.

To assists the structuring, semi-automatic methods are of great interest. For instance, Formal Concept Analysis can be used to determine and visualize the correlations of concept and attributes.

Formalization and Operationalization of Ontology

The final formal degree of the ontology depends on its intended use. It is important to recognize that the formalization task does not consist of replacing an informal version by a formal one, but to augment an informal version with the relevant formal aspect needed by the operational system. In the context of a given application, the ontologist will stop his progression on the continuum between informal and formal ontology as soon as he has reached the formal level necessary and sufficient for his system.

(Bachimont, 2000) explains that formalizing knowledge is not sufficient: we must use it in an operational system. A system can only exploit a concept according to the operations or rules that it can associate with it. Therefore, the semantics allowing a system to use a concept is the computer specification of the operations applicable to a concept. Bachimont defines the computational commitment as the act of adding a computational semantics to the concepts of the ontology. Inferences and computational semantics are currently mainly buried in the code of software. Their intension and intention are not captured yet they play a vital role in the choices of conceptualization.

To make an ontology computable, it needs to be implemented in a formal language. If this formal language is standardized and if there exist platforms complying with the standard then at least a minimum set of operation is certified and the computational commitment exists.

There are several families of formalization languages. A formalism framework provides primitives with a fixed semantics and manipulation operators with a known behavior. A symbolic system alone means nothing. A formalism provides a symbolic system (syntax, axioms, inference rules, operators...) and the semantics attached to it (rules of interpretation attaching meaning to symbolic expressions). Thus for ontologies, the stake is to find a formalism providing the adequate modeling primitives to capture the aspects of the ontology for which, according to the motivating scenarios, it was deemed relevant to implement a formalization.

Logic develops logical systems (logical languages with authorized manipulations) that are symbolic systems the interpretation of which provides a simulation of some human inferences. The propositional logic is the base for other logic. From ontology formalization point of view, it is too limited: propositions are indivisible symbols. The propositional logic only considers relations between propositions without considering the structure and the nature of the proposition. One of the consequences is that it cannot represent the difference between individuals and categories and the relations between individuals. These differences being at the heart of ontologies and we need a more expressive language.

The logic of predicate or first order logic (FOL) includes the logic of propositions. The addition of the universal quantificator and of the predicates gives us the ability to differentiate among individuals and categories and to express relations between individuals. For instance, in this logic we can now write: $(\forall x)\ (cat(x) \supset animal(x))$ *i.e.,* for every *x* if *x* is a *cat* then *x* is an *animal*, in other words, the concept *animal* subsumes *cat* (the semantic of subsumption being here the one of set inclusion). This logic in only semi-decidable i.e., there does not exist one algorithm to determine, in a finite time, if one expression is provable or not. The knowledge representation languages usually make some restrictions of the expressiveness to keep the expressiveness they desperately need and cut the rest so that the system remains usable.

Traditional logic programming languages can be used to formalize knowledge models and knowledge bases. However unlike the following knowledge engineering dedicated languages, they are far less structured to be used for that application. In particular, there is no native distinction between ontological and assertional knowledge. Any piece of knowledge is represented by a set of Horn clauses *i.e.,* a statement of the form *head* ← *body*. A clause with no head is a goal; a clause with no body and no variable in its head is a fact. Other clauses are rules; they are used by inference engine together with facts to derive if goals can be successfully achieved or not.

Conceptual graphs (CG) (Sowa, 1984) (Sowa, 2002) come from a merging between existential graphs of Peirce (Roberts, 1973) and semantic networks (Quillian, 1966). This formalism was motivated by needs for natural language processing and needs for friendly presentation of logic to human. A CG is a bipartite oriented graph *i.e.,* there are two types of nodes in the graph (concept nodes and relation nodes) and the arcs are oriented and always link a concept node to a relation node (or vice versa). CGs are existential and conjunctive statements. Relations are n-adic *i.e.,* their arity (valence) is an integer *n* giving the number of concepts they can be linked to. Concepts and relations have a type. The types are primitive or defined. A definition is given by a λ-expression *i.e.,* a graph with formal parameters λ_i that give the definitional pattern. Types of relations also have a fixed valence (giving the number of concepts linked by the relation) and a signature (giving the type of concepts linked by the relation). A conceptual graph is a bipartite graph made of concept nodes and relation nodes. The ontological knowledge upon which conceptual graphs are built is represented by the support, made of two subsumption hierarchies structuring concept types and relation types, a set of individual markers for the concepts, and signatures defining domain and range constraints for the relations. The core reasoning operator of Conceptual Graphs is the computation of subsumption relations between graphs, called specialization/generalization relations. It is based on an operation called 'projection' which is a graph homomorphism such that a graph *G* subsumes a graph *G'* *iff* there exists a homomorphism from *G* to *G'*. The projection takes into account specialization of relations and concepts *i.e.,* nodes of the query graph *G* must be of the same type or subsumers types of nodes in the target graph *G'* they are mapped to.

Object Oriented Formalisms (Ducourneau *et al.*, 1998) propose to represent, capture organize and manipulate knowledge through the notion of virtual objects. In these formalisms there exist two basic entities: object classes and object instances. Classes are categories of objects. A class defines the characteristics shared by all the objects of this category. Classes are structured in an inheritance hierarchy defined by the link "a-kind-of". A class can be instantiated *i.e.,* one can create an object belonging to this class. Instances are final objects instantiated from a class. Instances are linked to their class by a link "is-a". An instance has a unique identifier and attributes. Every attribute has a list of facets giving the value and characteristics of the attribute. The semantics of inheritance is the one of inclusion of sets *i.e.,* the instances of a subclass are also instances of its super class. Subclasses inherit attributes and facets; they can enrich these definitions by adding new attributes, new facets or by refining constraints. Facets can be declarative or procedural to specify the nature (type, domain, cardinality, value) or the behavior of an attribute (default value, daemon *i.e.,* procedures to calculate the value, constraints, filters). A mutation is the operation of trying to change the class of an object. This operation is at the heart of the classification algorithms of object-oriented frameworks that try to automatically classify instances according to their characteristics. Points of view can be defined to build different hierarchies of classes capturing different conceptualizations while enabling an object to inherit from all the aspects defined for its class in the different views. Graphic modeling languages (OMT, UML) have been proposed that look like graph-oriented languages. Object oriented data bases also offer interesting schema definition capabilities and additionally provide efficient data storage and retrieval mechanisms based on object query languages.

Description logics (Ducourneau *et al.*, 1998) (Kayser, 1997) (Baader *et al.*, 2003) are drawing upon predicate logic, semantic networks and frame languages. There again there are two levels: the terminological level where concepts and roles are represented and manipulated and the factual level where assertions and manipulations are made about individuals. Assertions constitute the assertional box (A-Box) while the ontological primitives upon which assertions of the A-Box are built are represented by the terminological box (T-Box). A T-Box is made of a set of concepts being either primitive or defined by a term, a set of roles, and a set of individuals. A concept is a generic entity of an application domain representing a set of individuals. An individual is a particular entity, an instance of a concept. A role is a binary relation between individuals. The description of a role can be primitive or defined. A definition uses the constructors of the language to give the roles attached to a concept and the restrictions of the roles (co-domain). Concepts are organized in a subsumption hierarchy which is computed according to concept definitions (this definition is reduced to a subsumption link in the case of elementary concepts). The fundamental reasoning tasks in DLs are the computation of subsumption relations between concepts to build the taxonomies and the classification *i.e.,* the automatic insertion of a concept in the hierarchy, linking it to its most specific subsumer and the most general concepts it subsumes. There exist different families of description logics depending on the set of constructors they use.

In addition to these classical historic languages, one of the next sections will introduce the formalisms used to provide a framework for the semantic web and, among many other contributions, providing means to formalize, exploit and exchange ontologies.

A comparison of several of several of the knowledge modeling languages is given in (Corcho & Gómez-Pérez, 2000).

Knowledge modeling languages that offer automatic classification (e.g. description logics) may be useful to assist building ontology and allow maintenance updates, but the paradox of formalization is that to obtain this assistant the systems need formalization (formal definitions) that may rely on additional notions that will have to be structured too, and so on.

Most of the knowledge engineering languages provide primitives to distinguish the ontological knowledge from the assertional knowledge, and have dedicated primitives for the description of the taxonomic structure of the ontology. The differences come when additional granularity require, for instance, formal definitions.

The definitions enable us to reason over concepts, for classification and logical equivalence inferences. The definition of concepts is a built-in characteristic of Description Logics, to allow concept classification consisting in the deductive inference of the most specific concepts which subsume a given concept and the most general concepts it subsumes. The definition of concepts is also a feature of the original Conceptual Graphs model defined in (Sowa, 1984). Finally, in Logic Programming, partial definitions of concepts (sufficient conditions) may be represented as rules.

Rules are indeed an alternative and explicitly capture inferences that can be used to factorize knowledge in the ontology and discover implicit knowledge in an assertion. This factorization generally consists in capturing patterns that are the logical consequence of other patterns. For instance: if a person x is a male and the child of a person y then x is the son of y. If a formalism does not provide primitives to express algebraic properties of relations (transitive, symmetric and reflexive or inverse relations) rules may be used for that purpose. For instance, for symmetry: if a person x is married with a person y then y

is married *x*. Logic Programming enables the expression of rules, as Horn clauses. Aside from the concept expressions constructs, Description Logics can also be provided with rules similar to Horn clauses. Finally, the classic Conceptual Graphs formalism is not provided with rules, but an extension is proposed in (Salvat & Mugnier, 1996) to handle graph rules with graphs as head and body.

THERE IS AN ONTOLOGY IN YOUR FUTURE: THE GROWING IMPORTANCE OF ONTOLOGIES IN COMPUTER SCIENCE

The notion of ontology, which was used in computer science even before the word for it was imported, is now far from being an endangered species. On the contrary, the range of applications and areas of interest in ontologies is still growing as I write this.

Formerly reserved to expert systems that simulate human reasoning in specific areas, ontologies are now integrated in a large family of information systems. It is used to: describe and deal with multimedia resources; ground the interoperability of network applications; pilot automatic processing of natural language; build multilingual and intercultural solutions; allow integration of heterogeneous sources of information; describe complex interaction protocols; check the consistency of models; support temporal and spatial reasoning; make logical approximations; and so on.

These usages of ontologies can be found in many application areas: integration of spatial information, human resource management, bio-informatics tools, electronic commerce, e-learning and computer-assisted education, digital libraries, B2B, healthcare, industrial design, news and press, cultural heritage management and museums, etc.

As ontologies emerge as an engineering domain, their expansion is far from being over. Among recent developments, ontologies that used to be primarily applied to data providing metadata on documents, images, videos, etc. are now used to describe software (e.g. web services), their functional characteristics (types of input, types of outputs), and non-functional (cost, quality). The ontological descriptions of web services (e.g. SAWDSL) enable us to identify, invoke and dynamically compose applications across the web using distributed services.

Similarly, ontologies were already used to describe users (e.g. FOAF) and they are now extending to the description of the users' context (e.g. geo-location, current activity, food preferences, access rights, devices at hand, past interactions, etc.) giving the applications what is called an awareness of context. Linking these profiles, ontologies also enable us to describe social networks (e.g. FOAF, SIOC), communities of interest (e.g. DOAP, SCOT), communities of practice, etc.

Finally, while ontologies are often presented as a means to facilitate access to information and applications, we can also see how it can be used in describing and applying, rights (e.g. Creative Common), security and privacy rules at high levels of abstraction, allowing us to control accesses with great flexibility. In an information system based on ontologies, privacy and its rules are also based on the semantics of ontologies and the inferences they allow us to control access to information and the accuracy of information disclosed.

Our information systems are increasingly complex. This complexity, even if it is artificial, raises difficult scientific challenges that must be addressed to see the technological expansion continue. The ability to design systems that reconfigure, adapt to the context, identify their mistakes, and even correct them to some extent, is a key factor in scaling the technological growth. Making explicit the conceptualizations of the world on which we base our software architecture, data structures, and the design choices, is a way to participate in this evolution of software and programming.

The challenge is now to move ontologies to software engineering practices, so that current conceptualizations which are often only watermarks in the code, in its comments or its documentation in the best cases, are more and more often made explicit and captured in formalisms. When exposed in ontologies, conceptualizations provide inference opportunities to information systems, reflexive processing on knowledge and procedures, introspection, dynamic alignment and interoperability, dynamic evolution and affordance in software components, *i.e.* their ability to describe how to interface with them and use them in order make interactions more dynamic and more automated, to make coupling more flexible and to move towards a more autonomous computing.

While web applications infiltrate all our information systems, the web is transforming from a document linking system to a universal virtual machine that combines resources of all kinds, made available by servers and web services. One can imagine a new programming paradigm, where data structures are representations based on shared ontologies, and applications would be obtained by composition of services (personal local software, online web services, grids, etc.). Get ready for ontology-oriented programming.

SEMANTIC WEB: THE WORLD-WIDE RISE OF ONTOLOGIES

A particularly promising course for the expansion of systems based on ontologies is known as the Semantic Web. It is an extension of the current web in which information is associated with a well-defined meaning, improving the ability of software to process the information available on the web. In this approach, the annotation of information resources of the Web is based on ontologies which are themselves available and exchanged on the web.

Ontology-Based Web Applications

A number of pioneer projects prepared the ground for the semantic web. SHOE (Heflin *et al.*, 1998) (Luke & Heflin, 2000) was one of the first languages to merge ontologies and Web markup languages. It stands for Simple HTML Ontology Extension and it was developed as an extension of HTML in order to incorporate semantic knowledge in web pages.

Ontobroker or On2broker (Fensel *et al.*, 1998) (Fensel *et al.*, 1999) was also one of the first projects intended to improve retrieval of information on the World Wide Web. It uses Flogic (Frame Logic) as the knowledge representation formalism to represent the ontology and to express the annotations of documents.

OntoSeek (Guarino et al., 1999) was dedicated to product catalogs and yellow pages on the web. It used structured content representations coupled with linguistic ontologies to increase both recall and precision of content-based retrieval. Queries and resource descriptions were represented in Lexical Conceptual Graphs a simplified variant of Conceptual Graphs

LogicWeb (Loke & Davison, 1998) divided the web into two layers: classic web page content at the first level, and the links between these pages at a logical layer for more flexibility. LogicWeb proposed the use of logic programming to improve both Web sites structuring and browsing and searching. Web pages were viewed as logic programs, consisting of facts and rules. The use of logic programming made it possible to define relationships between web pages and express complex information about pages.

WebKB (Martin & Eklund, 2000) succeeded to the earlier system CGKAT (Martin, 1996) and it was a web-based set of tools allowing its users to represent knowledge to annotate web resources and propose retrieval mechanisms using conceptual graphs. The inference engine exploited subsumption relations to compute specialization relations between a query graph and an annotation graph. WebKB annotations could be inserted anywhere in a web document by using <KR> ... </KR> tags and the language used was a hybrid of formalized English and conceptual graph linear form. WebKB allowed the use of undeclared types and automatically inserted them in the ontology according to the way they were used. WebKB was used to improve search, browsing and automatic document generation.

OSIRIX (Rabarijaona *et al.*, 1999 and 2000) stood for Ontology-guided Search for Information Retrieval In XML-documents. It was a tool proposing ontology-guided search in XML documents applied to corporate memory consultation. Taking into account the advantages of the World Wide Web and of ontologies for knowledge management, OSIRIX relied on XML for corporate knowledge management. The knowledge models that guided the search were CommonKADS expertise models, represented in standard CommonKADS Conceptual Modelling Language (CML) (Breuker & Van de Velde, 1994).

Semantic Web Frameworks

The Extensible Markup Language (XML) is a description language recommended by the World Wide Web Consortium (W3C) for creating and accessing structured data and documents in text format over internet-based networks. XML is a meta-language used to build languages to describe structured document and data; it can be considered as an SGML light for the Web. It comes with a set of tools and API to parse and process XML files; many of these tools are freely available on the net, but at the same time the format is being more and more available in commercial applications too (e.g. office tools).

RDF (Figure 9) is a data and metadata representation model based on graphs broken down into triples. RDF triples describe and connect resources *i.e.* objects anonymous or identified by a URI. The atom of knowledge in RDF consists therefore of triples of the form (subject, predicate, object). For example the assertion "Fabien has written a page doc.html about the Web" can be broken into two RDF triples (doc.html, author, Fabien) and (doc.html, subject, site). The triples can be seen as binary predicates in logics or as the arcs of a directed labeled graph. This model has an XML syntax allowing us to represent, store and exchange RDF graphs.

SPARQL provides a query language over RDF graphs, an XML language to represent the results of a query and a protocol for submitting a request to a remote server and for receiving the results.

RDFS is a lightweight language to declare and describe the resource types (called classes) and resource relationship types (called properties). RDFS allows us to name and define vocabularies

Figure 8. Semantic web recommendations

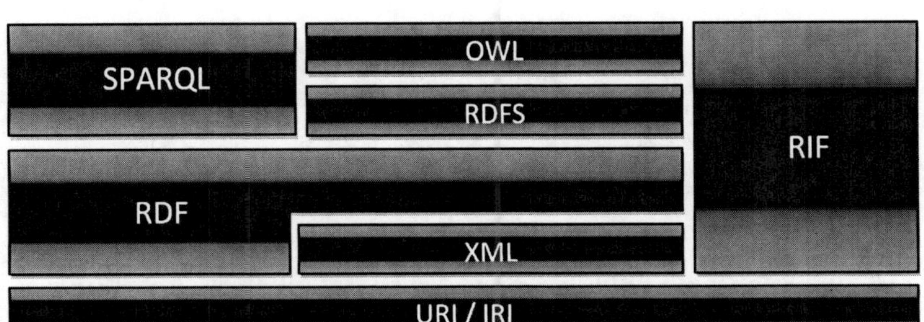

Figure 9. The RDF triple: the atom of knowledge on the semantic web inspired by "gödel, escher, bach: an eternal golden braid" (hofstadter, 1999)

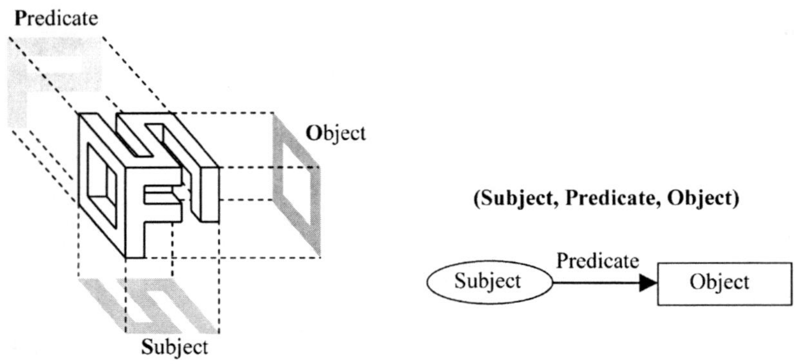

used in RDF graphs: naming the classes of existing resources; naming relation types existing between instances of these classes and giving their signatures, *i.e.*, the type of resources they connect. RDFS also allows inferences using these hierarchies of types and the signatures of properties. By allowing us to provide a URI for types, RDFS allows to declare the taxonomic skeleton of an ontology in a universal language and with universal identifiers.

OWL is a recommendation providing three layers of extension of the expressiveness of RDFS: OWL Lite, OWL DL and OWL Full. The first two layers of extension are based on description logics that allow additional inferences such as checking the consistency of a schema, the automatic classification of types to generate hierarchies, or the automatic identification of the type of a resource based on its properties. OWL allows the definition of classes by enumerating their contents or the union, intersection, complement and disconnection from other classes. OWL also allows the characterization of properties (restriction of their value or their cardinality) and their algebraic properties (symmetric, transitive, functional, inverse functional, inverse property). Finally OWL provides primitives for the management of equivalences between different ontologies and between different versions of an ontology. A second version of OWL is being reviewed at the time of writing this article.

With the Semantic Web, ontologies have found a world-wide standard and are being integrated in more and more web applications, without their users even knowing it. The integration these ontologies is increased tenfold by the scale of the web. As a result, more and more ontologies are available and used on the web.

ONTOLOGIES IN THE WILD

To conclude, let us imagine what could be the life of ontologies in the wild. Creating ontologies in the wild should be designed as a side effect of our daily tasks and not as an additional task. In such an approach the user is no longer simply the client of a service for which she provides inputs and expects outputs, but she becomes a computational resource of the software architecture; a computational resource that slips through classical computer science design techniques (Hutchins, 1995). In other words, the solution should be designed as comprehensive solutions including human and social elements along with technical artifacts. This requires from us to rethink the design phase too often confined to engineering technology, and to move toward anthropotechnic engineering.

The new practices introduced by Web 2.0 applications surprised everyone by their effectiveness. Social tagging produced folksonomies at a scale and a with speed we dream of in knowledge acquisition, in particular for building and populating ontology.

With folksonomies and ontologies we now have two kinds of cognitive artifacts that are more and more present in web applications. The contributions of a user to a folksonomy remain light and easy, and follow from its use: the tags used by users are collected and analyzed as background processing. On the opposite contributions to ontologies are very often seen as direct inputs and often require dedicated action: the concepts must be validated, organized, etc. defined.

However, a tag is a term used to mark a resource and it is not a concept or a class as in an ontology. It is closer to a new kind of "candidate term" (as in natural language approaches to build ontologies from text) and which specificities must be identified and exploited if we want to derive conceptualizations from folksonomies (Halpin et al., 2004) (Mika, 2005).

Ontologies are defined by the type of knowledge they contain. Folksonomies are defined by the way they are produced. Can we get ontologies from folksonomies, in particular domain ontologies, and envision folks-ontologies? (Van Damme *et al.*, 2007). In this new context, ontologies would no longer be the responsibility of ontologists who would subsequently become facilitators of a community federated around the use of applications of these ontologies. The community, by its activity, would feed the life cycle of these ontologies; the evolution of the ontology would be a side effect of the normal activity of the community.

From the perspective of ontological engineering, this is not about bringing the user into a loop of ontology design and maintenance but to assign basic tasks to a mass of users so that on a large-scale they handle problems notoriously difficult as the detection of a concept in a multimedia resource, the organization of concepts or the disambiguation. As a pioneering example, we can mention OntoGame (Siorpaes & Hepp, 2008).

Involving a community in the design of an ontology is a practice we already known and applied. But the idea of equipping a community with intelligent tools so that, in the normal course of its activity it maintains ontologies without knowing it, remains a perspective. The life cycles of folksonomies and ontologies are different and the management of their symbiosis seems a promising prospect. We can identify at least two approaches to combine them:

- To exploit folksonomies to build and populate ontologies (analysis of networks of tags, resources and users; use of linguistic resources or existing ontologies and alignment techniques; linguistic analysis; analysis of usage);
- To produce interoperable applications (e.g. tag images with Wikipedia-based disambiguation mechanisms) to capture as much as possible knowledge when it is explicit.

In the field of knowledge engineering and cognitive sciences, we witness a shift from centralized cognitive engineering to distributed cognitive engineering. The cognitive systems to be built are now no longer sole computer system, but larger systems composed of artifacts and people; complete solutions have to be designed and operated. "An important aspect of the larger unit is that it contains computational elements (persons) who cannot be described entirely in computational terms." (Hutchins, 1995). The coupling of knowledge engineering with this new participatory Web or Web 2.0 will accelerate this trend.

REFERENCES

Aussenac, N. (1989). *Conception d'une méthodologie et d'un outil d'acquisition des connaissances expertes*. Ph.D Thesis, University P. Sabatier, Toulouse, France.

Baader, F., Calvanese, D., McGuinness, D., Nardi, D., & Patel-Schneider, P. (2003). *The Description Logic Handbook: Theory, Implementation and Applications*. Cambridge, UK: Cambridge University Press.

Bachimont, B. (2000). Engagement sémantique et engagement ontologique: conception et réalisation d'ontologies en ingénierie des connaissances, In J. Charlet, M. Zacklad, G. Kassel, D. Bourigault, (Eds.), Ingénierie des connaissances Evolutions récentes et nouveaux défis. Eyrolles.

Bachimont, B. (2001), Modélisation linguistique et modélisation logique des ontologies: l'apport de l'ontologie formelle. In *Proceedings of IC 2001, Plate-forme AFIA,* (pp. 349-368), Grenoble, France.

Breuker, J., & Van de Velde, W. (1994). *CommonKADS Library for Expertise Modeling: reusable problem solving components*. Amsterdam, Tokyo: IOS-Press/Ohmsha.

Caroll, J.-M. (1997). Scenario-Based Design . In Helander, M., Landauer, T. K., & Prabhu, P. (Eds.), *Handbook of Human-Computer Interaction* (2nd ed.). Amsterdam: Elsevier Science B.V.

Corcho, O., & Gómez-Pérez, A. (2000). A Roadmap to Ontology Specification Languages. In R. Dieng et O. Corby (éd.), Proceedings of EKAW2000, Knowlege Engineering an Knowledge Management Methods, Models, and Tools, (pp.80-96).

Dieng, R. (1990). *Méthodes et outils d'acquisition des connaissances*. Technical report from INRIA n° 1319, Novembre 1990.

Dieng, R. (1993). Méthodes et outils d'acquisition des connaissances . In Spérandio, J.-C. (Ed.), *L'ergonomie dans la conception des projets informatiques* (pp. 335–411). Toulouse: Octares.

Dieng, R., Giboin, A., Amergé, C., Corby, O., Desprès, S., & Alpay, L. (1998, Winter). Building of a Corporate Memory for Traffic Accident Analysis . *AI Magazine*, *19*(4), 80–100.

Ducournau, R., Euzenat, J., Masini, G., & Napoli, A. (1998). *INRIA Langages et modèles à Objets Etat des recherches et perspectives*. Collection Didactique.

Fensel, D., Angele, J., Decker, S., Erdmann, M., Schnurr, H., Staab, S., et al. (1999). On2broker: Semantic-Based Access to Information Sources at the WWW. In *Proceedings of the World Conference on the WWW and Internet WebNet 99*, Honolulu, HI.

Fensel, D., Decker, S., Erdmann, M., & Studer, R. (1998). Ontobroker: Or How to Enable Intelligent Access to the WWW. In B. Gaines & M. Musen (eds.), *Proc of the 11th Workshop on Knowledge Acquisition, Modeling and Management KAW'98,* Banff, Canada, April 18-23. Retrieved from http://ksi.cpsc.ucalgary.ca/KAW/KAW98/KAW98Proc.html

Fernandez, M., Gomez-Perez, A., & Juristo, N. (1997 March). METHONTOLOGY: From Ontological Arts Towards Ontological Engineering. In *Proceedings of the AAAI97 Spring Symposium Series on Ontological Engineering*, Stanford, CA, (pp. 33-40).

Giboin, A., Gandon, F., Corby, O., & Dieng, R. (2002). Assessment of Ontology-based Tools: Systemizing the Scenario Approach. In J. Angele, & Y. Sure, (Eds.) *Proceedings of EON2002: Evaluation of Ontology-based Tools Workshop, 13th International Conference on Knowledge Engineering and Knowledge Management EKAW 2002*, Siguenza (Spain), September 30th, (pp. 63-73).

Gómez-Pérez, A., Fernandez, M., & De Vicente, A. (1996). Towards a Method to Conceptualize Domain Ontologies. In *Workshop on Ontological Engineering at ECAI'96*, (pp. 41-51).

Gruber, T. (1993). A Translation Approach to Portable Ontologies. *Knowledge Acquisition*, *5*(2), 199–220. doi:10.1006/knac.1993.1008

Guarino, N. (1992). Concepts, Attributes, and Arbitrary Relations: Some Linguistic and Ontological Criteria for Structuring Knowledge Bases. *Data & Knowledge Engineering*, *8*, 249–261. doi:10.1016/0169-023X(92)90025-7

Guarino, N. (1997). Understanding, Building and Using Ontologies. A Commentary to Using Explicit Ontologies in KBS Development, by van Heijst, Schreiber, and Wielinga. *International Journal of Human-Computer Studies*, *46*(2/3), 293–310. doi:10.1006/ijhc.1996.0091

Guarino, N., & Giaretta, P. (1995). Ontologies and Knowledge Bases: Towards a Terminological Clarification . In Mars, N. J. I. (Ed.), *Towards Very Large Knowledge Bases*. Amsterdam: IOS Press.

Guarino, N., Masolo, C., & Vetere, G. (1999). Ontoseek: Content-based access to the web. *IEEE Intelligent Systems*, *14*(3), 70–80. doi:10.1109/5254.769887

Guarino, N., & Welty, C. (2000). Towards a methodology for ontology-based model engineering. In *Proceedings of ECOOP-2000 Workshop on Model Engineering*, Cannes, France.

Halpin, H., Robu, V., & Shepherd, H. (2007). The Complex Dynamics of Collaborative Tagging . In *WWW 2007*. New York: ACM Press.

Heflin, J., Hendler, J., & Luke, S. (1998). Reading between the lines: Using SHOE to discover implicit knowledge from the Web. In *proc. of the AAAI Workshop on Artificial Intelligence and Information Integration*, WS-98-14, (pp. 51-57). Menlo Park, CA: AAAI Press.

Hofstadter, D. R. (1999, January). Gödel, Escher, Bach: An Eternal Golden Braid, (20th anniversary ed.). New York: Basic Books.

Hutchins, E. (1995). *Cognition in the Wild*. Cambridge, MA: MIT Press.

Kassel, G. (2002). OntoSpec: une méthode de spécification semi-informelle d'ontologies. In Actes des Journées Francophones d'Ingénierie des Connaissances: IC'2002, Rouen, France, (pp. 75-87).

Kayser, D. (1997). *La représentation des connaissances*.

Korfhage, R. (1997). *Information Storage and Retrieval*. Chichester, UK: John Wiley & Sons.

La France, M. (1992). Questioning Knowledge Acquisition . In Lauer, T. W., Peacock, E., & Graesser, A. C. (Eds.), *Questions And Information Systems*. Mahwah, NJ: L. Erlbaum associates.

Loke, S. W., & Davison, A. (1998). LogicWeb: Enhancing the Web with Logic Programming. *The Journal of Logic Programming*, *36*, 195–240. doi:10.1016/S0743-1066(98)00002-8

Luke, S., & Heflin, J. (2000, February). *SHOE 1.01 Proposed specification*. SHOE Project. Retrieved from http://www.cs.umd.edu/projects/plus/SHOE/

Martin, P. (1996). *Exploitation de Graphes Conceptuels et de documents structurés et Hypertextes pour l'acquisition de connaissances et la recherche d'informations*. Ph.D. Thesis, University Nice - Sophia Antipolis, October 14, 1996.

Martin, P., & Eklund, P. (2000). Knowledge retrieval and the world wide web . In Dieng, R. (Ed.), *IEEE Intelligent Systems* (pp. 18–25). Special Issue on Knowledge Management and Knowledge Distribution Over the Internet.

McCarthy, J. (1980). Circumscription—A Form of Nonmonotonic Reasoning. *Artificial Intelligence*, *13*, 27–39. doi:10.1016/0004-3702(80)90011-9

Mika, P. (2005). Ontologies are Us: a Unified Model of Social Networks and Semantics. In ISWC, (LNCS 3729, pp. 522-536). Berlin: Springer.

Mizoguchi, R., Ikeda, M., & Sinitsa, K. (1997). Roles of Shared Ontology in AI-ED Research, Intelligence, Conceptualization, Standardization, and Reusability. In *Proc. of AIED-97*, (pp.537-544), also as Technical Report AI-TR-97-4, I.S.I.R., Osaka University

Peirce, C. S. (1867), On a New List of Categories. In *Proceedings of the American Academy of Arts and Sciences, 7*(1868), 287–298.

Quillian, M. R. (1966). *Semantic Memory.* Report AD-641671, Clearinghouse for Federal Scientific and Technical Information.

Rabarijaona, A., Dieng, R., & Corby, O. (1999). Building a XML-based Corporate Memory. In *Proceedings of the IJCAI'99 Workshop on Knowledge Management and Organizational Memories*, Stockholm, Sweden.

Rabarijaona, A., Dieng, R., Corby, O., & Ouaddari, R. (2000). *Building a XML-based Corporate Memory, IEEE Intelligent Systems, Special Issue on Knowledge Management and Internet* (pp. 56–64). May-June.

Roberts, D. D. (1973). *The Existential Graphs of Charles S. Peirce*. The Hague, The Netherlands: Mouton.

Salvat, E., & Mugnier, M. L. (1996) Sound and complete forward and backward chainings of graph rules. In *Proc. of the 4th ICCS'96*, Sydney, Australia, (LNCS 1115, pp. 248-262). Berlin: Springer-Verlag.

Sebillotte, S. (1991). *Décrire des tâches selon les objectifs des opérateurs de l'Interview à la Formalisation.* Technical report from INRIA n°125.

Siorpaes, K., & Hepp, M. (2008). Games with a Purpose for the Semantic Web. *IEEE Intelligent Systems, 23*(3), 50–60. doi:10.1109/MIS.2008.45

Sowa, J. F. (1984). *Conceptual Structures: Information Processing in Mind and Machine*. Reading, MA: Addison–Wesley.

Sowa, J.-F. (2000a). Ontology, Metadata, and Semiotics Proceedings of ICCS'2000 in Darmstadt, Germany, on August 14, 2000. In B. Ganter & G. W. Mineau, (eds.), *Conceptual Structures: Logical, Linguistic, and Computational Issues*, (LNAI 1867, pp. 55-81). Berlin: Springer-Verlag.

Sowa, J.-F. (2000b). *Guided Tour of Ontology*. Retrieved from http://www.jfsowa.com/ontology/guided.htm

Sowa, J.-F. (2002). Conceptual Graphs Standard. *ISO/JTC1/SC 32/WG2N000.* Retrieved from http://users.bestweb.net/~sowa/cg/cgstand.htm

Uschold, M., & Gruninger, M. (1996). Ontologies: Principles, methods and applications. *The Knowledge Engineering Review, 11*(2), 93–136. doi:10.1017/S0269888900007797

Van Damme, C., Hepp, M., & Siorpaes, K. (2007). An integrated approach for turning folksonomies into ontologies . In *Bridging the Gap between Semantic Web and Web 2.0 (SemNet 2007)* (pp. 57–70). Folksontology.

Chapter 2
Ontology Theory, Management and Design: An Overview and Future Directions

Wassim Jaziri
MIRACL Laboratory, Tunisia

Faiez Gargouri
MIRACL Laboratory, Tunisia

ABSTRACT

Ontologies now play an important role in providing a commonly agreed understanding of a domain and in developing knowledge-based systems. They intend to capture the intrinsic conceptual and semantic structure of a specific domain. Many methodologies, tools and languages are already available to help anthologies' designers and users: However, a number of questions remain open: what ontology development methodology provides the best guidance to model a given problem, what steps to be performed in order to develop an ontology? which techniques are appropriate for each step? how ontology' lifecycle steps are upheld by the software tools? how to maintain an ontology and to evolve it in a consistent way? how to adapt an ontology to a given context? To provide answers to these questions, the authors review in this chapter the main methodologies, tools and languages for building, updating and representing ontologies that have been reported in literature.

INTRODUCTION

Nowadays, we can easily notice a mass information sources, accompanied by a proliferation of users' requirements, which became more complex and demanding. In fact, new information systems have to handle a variety of information sources, from proprietary ones to those available in web services worldwide.

DOI: 10.4018/978-1-61520-859-3.ch002

Since the information systems are imperative for the survival of any organization, they must guarantee a good circulation, coherence and an assistance to make appropriate decisions. However, currently, the new information systems are increasingly complex, requiring an enormous work of modeling. Designers are often confronted with a set of difficulties related mainly to the complexity of the domain of study and to the multitude of terms used to express the domain concepts. These problems are due to the lack of a consensus on the vocabulary used for a given domain. So, designers can, in some cases, make syntactic, structural and/or semantic errors. These errors will affect the coherence of the conceptual schema and consequently the quality of their implementation.

In this context, ontologies could play an important role, as they do in other disciplines, since they provide a source of precisely defined terms that can be communicated across people, organisations and applications. They offer a consensual shared understanding concerning a domain of knowledge to support communication among humans, computers and softwares (Gruninger et al., 2002). Ontologies are also used to share a common understanding of information structure and allow analyzing knowledge based on the terms specification of the domain area. The formal analysis of terms is extremely valuable for reuse of existing ontologies and for its extension (Bachimont, 2000).

In this chapter, we are interested in work conducted in the domain of ontology engineering and particularly in the approaches, languages and tools for ontology building, contextualization and evolution. It is intended to give an intuitive view, not an exhaustive account. In fact, regardless of the complexity of the ontology engineering setting, what is currently lacking is a unified overview of the wide variety of models and mechanisms that can be used to support all steps of ontology lifecycle.

The rest of chapter is structured as follows. In Section 2, we discuss some problems confronted during the information system modeling and the importance of ontology as support for the modeling of information systems. Then, the theoretical foundations of the ontological engineering field will be presented while commenting on the ontology utility, use and definitions. This will be followed by a presentation of related work regarding the ontology building and design. An overview of the notions of context and multi-representation problems in the domain of ontology and information systems is also proposed in section 5. Section 6 reviews approaches and works which focus on ontology evolution. Finally, we conclude in section 7.

FROM INFORMATION SYSTEMS DESIGN TO ONTOLOGY MODELING

Conceptual modeling is one of the most important tasks in the development of information systems in terms of both organizational understanding and systems development. It requires a determination of the domain entities and their relationships as well as different static and dynamic views of the expected system. The domain entities are not always simple nor organized since we must consider, when modeling, all the field's concepts belonging to the universe of discourse as well as their pertinent relationships.

The result of a design step is usually expressed using a model (Sánchez et al., 2005). This model may contain some ambiguities and errors due to the incomprehension of the domain of study and the difficulty to determine its concepts and relationships.

Conflicts in Information Systems

The information system modeling requires a perfect knowledge of the studied domain and a deep analysis of the user's requirements. This task becomes very difficult because the current applications become increasingly complex and use an enormous quantity of concepts coming from heterogeneous sources. For example, in the case of cooperative applications, the design step requires the extraction of an enormous quantity of data concerning the various intervening actors (e. g. customers, suppliers, produced). Modeling such data requires an analysis step allowing the determination, the distinction and the classification of the domain concepts. This step must be based on the designers' knowledge and expertise, helped by some domain's specialists. However,

Figure 1. Example of synonymy

Figure 2. Example of homonymy

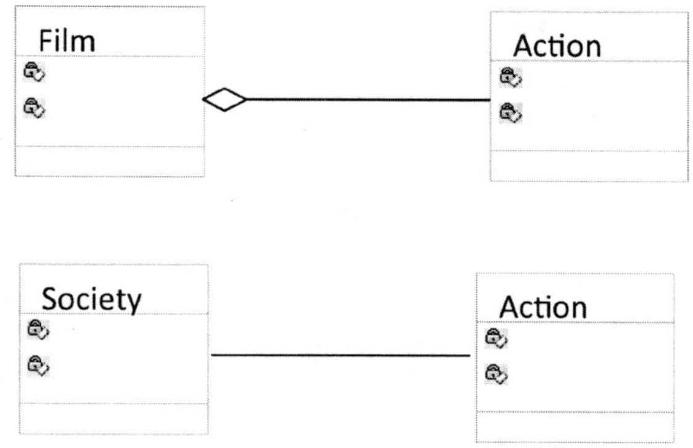

the traditional tools do not present any methodological help to synthesize this field's expertise. Moreover, designers are also confronted with conflicts at different levels of abstraction when modeling information systems. In addition, the design and development of a software product are carried out, more and more, between various geographically and temporally distributed teams. Those teams use a diversity of methods and languages to analyze and design information systems (e.g. OMT, UML). This diversity causes inevitably various types of conflicts when integrating the conceptual representations produced by the teams. These problems may generate various types of conflicts (syntactic, semantic and structural) and generally cause semantic inconsistencies of the resulted conceptual representations. The following sections present examples of syntactic, structural and semantic conflicts that may be generated when designing an information system.

Syntactic Conflicts

The syntactic[1] conflicts result from the differences between the used terminologies, at design time, by the various designers and teams working on the same application. They occur when naming schemes of information differ significantly. The syntactic conflicts may be avoided both by replacing the simple word that denotes a concept by an absolute and language independent identifier, and by replacing the textual definition of the concept by a complete model that describes it by means of a set of relationships and of meta attributes. This model makes explicit the definition context in which the corresponding concept is unambiguous and meaningful (Goh et al., 1999).

Various types of terminological conflicts exist, such as:

- **Synonymy**: two different concepts can have the same meaning (figure 1).
- **Homonymy**: the same concept can have different meanings (figure 2).

One of the reasons of these conflicts may be the cultural differences between the members of various groups implied in the design of the same application (even between the members of the same group). To solve this kind of conflicts, a vocabulary consensus, relating to the studied field, must be established between the various teams.

Structural Conflicts

The structural[2] conflicts are related to the adoption of various levels of abstraction, by several designers, to a same concept (e. g. class/attribute, attribute/method). They occur when different reference systems are used to measure the value of some properties. The structural conflicts may be avoided, either by associating explicitly at the schema level a computer-interpretable representation of the unit that shall be used for any value of a property, or by associating explicitly with each value its own unit (Goh et al., 1999).

Figure 3 shows that the attribute "Author" in the class Book can be considered as a class in another conceptual representation.

The design step is, generally, considered as a nondeterministic intellectual process. Indeed, the same data can be modeled in various manners by several designers. To solve this kind of conflicts, it is essential to define clearly and to specify the abstraction level of each domain concept.

Semantic Conflicts

The semantic[3] conflicts are related to the ambiguities which can be generated by the relationships between the concepts. They occur when information items seem to have the same meaning, but differ in reality, e.g. due to different temporal contexts (Goh et al., 1999). These conflicts can, in certain cases, cause structural and semantic errors which are not detected by current CASE.

Goh et al. (1999) identified three main types of conflicts: *Naming conflicts*, *Scaling conflicts* and *Confounding conflicts*.

Another typology of conflicts is given by Spaccapietra et al. (1992) that distinguish semantic, descriptive, heterogeneity and structural conflicts:

Figure 3. Example of a structural conflict

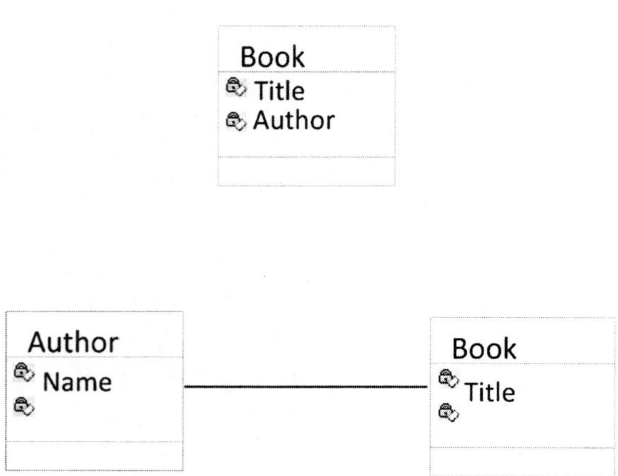

1. **Semantic conflicts**: two designers do not perceive exactly the same set of real world objects, but instead they visualize overlapping sets (included or intersecting sets). For example, a "Student" object class may appear in one schema, while a more restrictive "AI-Student" object class (grouping students majoring in Artificial Intelligence) is in another schema.
2. **Descriptive conflicts**: when describing the given related sets of the real-world objects, two designers do not perceive exactly the same set of properties. For example, let us assume two classes, C1 and C2, describing the same information. They can be perceived differently as:
   - **C1**: Customer (Id_customer, name, age, address)
   - **C2**: Client (Id_Client, name_client)

   Descriptive conflicts include naming conflicts due to homonyms and synonyms (Navathe et al., 1982) (Batini et al., 1986), attributes, constraints and operations (Larson et al., 1989).
3. **Heterogeneity conflicts**: designers use different data models, such as relational or object-oriented.
4. **Structural conflicts**: even if they use the same data model, designers can choose different constructs to represent common real-world objects. For example, in object-oriented models when a designer describes a component of an object type O1, (s) he has to choose between creating a new object type O2 or adding an attribute to O1.

Approaches for Solving Conflicts

The design of complex applications requires knowledge related to the domain of study, in particular to the concepts used as well as the relationships between them. Knowledge can be extracted, when analyzing user's requirements, from the existing applications, and using the anterior expertise in a domain of study. However, the conceptual representations obtained generally contain some ambiguities. They can cause semantic and/or structural errors related to the complexity of the modeled field and the heterogeneity of the obtained conceptual representations. The current information systems design methods and languages CASE are very limited in the detection and the resolution of these ambiguities.

Several approaches have been proposed in the literature to solve the conflicts that can occur and to assist the designers in representing and modeling knowledge during the design step. For example, the use of keywords represents a rapid way to find useful information. However, this approach can quickly be exceeded when information becomes complex (Bachimont, 2000). As an extension to this approach, some authors proposed in the literature another one based on the use of dictionaries. A dictionary represents a more elaborate structure than keywords (Huhns et al., 1997). However, it does not consider semantic relationships between the concepts of a given field. The use of taxonomies[4] constitutes another approach that provides a classification structures more complete than the dictionaries by adding the power of the inheritance relationship. However, it limits the various possible relationships between concepts to the only inheritance one.

Analysis of the types of conflicts studied above shows that the lack of common background calls for explicit guidance in understanding the exact meaning of the data.

As evolution and generalization of the various approaches, previously presented, the concept of ontology seems to be most complete and adequate for the resolution of conflicts. In particular, ontology allows a larger variety of structural and nonstructural relationships between concepts, producing so precise and complete models of the studied field. Then, it is an emerging mechanism for dealing with semantic interoperability. Semantic interoperability is a knowledge-level concept that provides the ability to bridge semantic conflicts arising from differences in implicit meanings, perspectives, and assumptions, thus creating a semantically compatible information environment based on the agreed concepts between different business entities.

The ontology can be used as a solution to represent all the concepts and the relationships characterizing a specific field. It allows the identification and the representation of concepts and their relationships, allowing a semantic verification at the specification step. It will be possible to couple this ontology to a given CASE, such as Rational Rose, to check, when designing applications, the semantic coherence of the specified representation. Thus, an ontology can be used as a tool to the semantic validation of the various CR as it contributes to represent the semantic rules related to the field, the concepts and the relationships between these concepts.

The issue is that, in mainstream academic and commercial work, practitioners typically regard their data models as representations of the 'real world', rather than 'a reification of an agreement on knowledge'.

For this reason, using a common set of concepts and terms to refer to these concepts is crucial for the development of high-quality software. It can be argued that fewer misunderstandings and misinterpretations will arise in any communication process when the involved parties use an agreed-upon, well-defined conceptual base. For the sake of common sense, this common conceptual base, or ontology, must have the following properties (González-Pérez et al., 2006):

- It must be complete, so that no area of software development lacks coverage.
- It must be unambiguous, so that misinterpretations are avoided.
- It must be taken from the appropriate domain, so that concepts are familiar and intuitive to their users.

- It must be as generic as possible, so that different usages in different contexts are possible.
- It must be extensible, so that new concepts can be added to it without breaking the existing ones.

Most of the causes of semantic conflicts in data integration result from implicit contexts, either in schema definition or in value evaluation. They may be solved if both the *modeling context* and the *value context* are made explicit (Pierra, 2008).

Among various other classification schemes and structures, including keywords, thesauri, and taxonomies, ontologies are often viewed as allowing more complete and precise domain models. Ontologies increasingly appear as the solution to the problems of conflicts. They are the most sophisticated form of semantics repository. From a database perspective, they may be intuitively understood as the most recent form of data dictionaries, i.e., a knowledge repository the purpose of which is to explain how concepts and terms relevant to a given domain should be understood.

The use of ontologies allows the study of the conceptualization independently of the programming language, the platform and the communications protocols. Their use was largely justified and useful for several fields: semantic Web, knowledge engineering, information retrieval, natural language processing, etc.

Ontologies have been proposed to overcome the difficulties raised by *monolithic, isolated knowledge systems* (Gruber, 1991), by specifying content specific agreements to facilitate knowledge sharing and reuse among systems that submit to the same ontology by means of ontological commitments (Gruber, 1995) (Spyns et al., 2002). Ontologies provide a way of building *external coupling interfaces that would enable the developer to reuse software tools and knowledge bases as modular components* (Gruber, 1991).

Ontology-Driven Information Systems

An information system uses ontology to represent explicit specification of a domain and to serve as a support for providing and searching knowledge sources. Even if not explicitly stated, any information system has its own ontology. Generally, an information system consists of three components: application programs, information resources (e.g. databases and/or knowledge bases) and user interfaces (Chira, 2003).

Guarino (1998) discussed the role an *explicit* ontology can play within an information system and argued in favor of an architectural perspective in which the ontology plays a central role at system development or run time, calling the resulting system an *ontology- driven information system.*

Conceptually modeling the universe of discourse with an ontology allows taking advantage of the ontology capacity to automatically check the model consistency and subsumption, i.e., detecting all the classes that are feasible and have at least one instance according to their restrictions, and detecting subclass relations which are not explicitly stated by the modeler but are inferable from those explicitly stated (Borgida, 1995).

The information system community (that includes the database community), uses the same components as those composing the ontology to build domain models, i.e., using concepts, relationships, properties, etc., but most of times the information system community imposes less semantic constraints than those imposed in heavyweight ontologies.

As simple as it would be, any software application, database component, or user interface requires specification languages (e.g. it might be an OO language for application, relational databases representation language for databases, and GUI for user interfaces) and agreements on the semantics intended for the objects of the domain the system is concerned with. In the traditional approaches, ontologies play a kind of fuzzy role consisting of a set of implicit agreements among the developers, even if not explicitly stated. Much can be won if ontology-based techniques will play a central role in the development of information systems (especially complex information systems), "driving all aspect and all components of an information system, so that we can speak of *ontology-driven information systems"* (Guarino, 1998). Guarino indicates that there are two orthogonal dimensions concerning the impact of the use of ontologies upon an information system (Guarino, 1998): the temporal dimension and the structural dimension.

The main benefits of using ontologies at development time are as follows (Guarino, 1998):

- helps the designer to increase the quality of conceptual analysis;
- increases the use and maintainability of an information system (any information system component can be directly translated into and from ontologies library);
- higher level of reuse (knowledge reuse instead of traditional software reuse);
- increases the reuse and sharing of application domain knowledge using a common vocabulary across heterogeneous software platforms;

When using ontologies at run-time, the ontology is not just a separate part, accessible to an information system (i.e. ontology-aware information systems), but is a component of the information systems that cooperates with all the other components for achieving the overall goal (Guarino, 1998). At this level the ontologies are used to enable the communication among software agents by providing a shared meaning for the agents that commit to them.

Concerning the structural dimension, each information system component (e.g. database component, user interface component and application program component) can use ontologies for its specific goals and in its specific ways. Depending on the moment ontologies are employed, it can be further distinguished between development and run-time aspects (Guarino, 1998).

Usually at development time, the use of ontology plays an important role in requirement analysis for the specific information system component, while at run-time the ontology enables processes such as sharing and reuse at higher levels than those reached by traditional means.

Spaccapietra et al. (2004) show the benefits of using conceptual data models for modeling ontologies. They assume that conceptual modeling and the database approach provide better readability of the content of an ontology, and more efficient management for large ontologies and associated knowledge bases.

Ontology vs. Database Schema

From a functional point of view, an ontology can be seen as an equivalent of a database schema. In general, a data model such as database schema represents the structure and the integrity of the data elements (Spyns et al., 2002). Hence, a data model usually implements some kind of informal agreement between the developers and the users of that specific data model regarding the semantics of data. But this agreement starts and ends with the above mentioned developers and users and it is not intended to be shared with other communities. An ontology, such as any database schema is a partial account of a conceptualization (Guarino et al., 1995) (Spyns et al., 2002), so they both have the same functions (e.g. establishing agreements, albeit in varying degrees). The difference is the domain they cover: database schemas are task-specific and implementation oriented (Spyns, Meersman et al., 2002), while ontologies are as generic and as task-independent as possible. These result in some key differences between ontologies and databases, as follows (Fensel, 2000):

- A language for defining ontologies is syntactically and semantically richer than common approaches for databases.
- The information that is described by an ontology consists of semi-structured natural language text and not tabular information.
- An ontology must be shared and consensual terminology because it is used for information sharing and exchange.
- An ontology provides domain theory and not the structure of a data container.

Therefore, an ontology lays somewhere between a knowledge base and a database schema. As any knowledge base incorporates explicitly or implicitly an ontology, a shared ontology will declaratively specify the *ground rules* (Neches et al., 1991) for modeling a domain in the form of top-level interconnected abstraction. In this way, an ontological commitment is viewed as a guaranteed assurance for providing specific services to systems (e.g. humans, agents, conventional software) that adopt a specific ontology or a specific library of ontologies.

ONTOLOGY BASIS AND THEORY

An ontology provides a source of shared and precisely defined terms. It typically consists of a hierarchical description of important concepts in a domain, along with descriptions of the properties of each concept. The degree of formality employed in capturing these descriptions can be quite variable, ranging from natural language to logical formalisms, but increased formality and regularity clearly facilitates machine understanding.

What is Ontology?

Ontologies aim at capturing domain knowledge in a generic way and provide a commonly agreed understanding of a domain, which may be reused and shared across applications and groups (Chandrasekaran et al., 1999). They define with different levels of formality the meaning of the terms and the relationships between them. They are usually organized in taxonomies and typically contain modeling primitives such as classes, relationships, functions, axioms and instances (Staab et al., 2001).

The origin of the ontology is Greek. It has emerged from Philosophy as a "branch of metaphysics concerned with identifying, in the most general terms, the kinds of things that actually exist. Thus, the "ontological commitments" of a philosophical position include both its explicit assertions and its implicit presuppositions about the existence of entities, substances, or beings of particular kinds" (Chira, 2003).

Since Aristotle's time, there has been an interest to represent the existing knowledge of the world with a methodology that identifies classes of objects with common properties in a hierarchical structure where some classes are specializations of others (Muñoz, 2006). This way to represent knowledge called *Ontology.*

Once used outside philosophy, several interpretations of the concept of ontology are possible. In fact, when a researcher would employ ontology-based meanings, he would depict his own understanding of the concept in his particular field for his particular needs. This is because the term ontology has an unquestionable defined meaning in the science of philosophy. Once imported to other domains loses some of its characteristics and gains others (specific to the borrowing domain) without any of these phenomena to be explained and defined by the borrowers (Chira, 2003).

In the last decade, the word ontology became a relevant word for the Knowledge Engineering community. It has appeared with the aim of sharing and reusing knowledge. Many definitions of ontology have been proposed and evolved over the time. Although ontology as a science comes from philosophy, it has mainly been developed by the artificial intelligence community. This community has focused on developing reasoning mechanisms that would alleviate the task of enriching an ontology by addition of new concepts. Typically, an ontological reasoner is expected to be able to check the consistency of new concepts with already known ones and to determine their most accurate placement within the structure of the ontology (Spaccapietra et al., 2004).

However, the artificial intelligence interpretation of an ontology differs from the philosophical understanding. While ontology for a philosopher is a *particular system of categories accounting for a certain visions of the world* (Guarino, 1998), independent of a particular language, for the artificial intelligence researcher, it refers to a *particular artifact constituted by a specific vocabulary* that describes a certain domain by explicitly constraining the intended meaning of the vocabulary words. Usually, the constraints are implemented in respect to the First Order Logic form and the vocabulary words are unary (concepts) or binary (relationships) predicate names (Guarino, 1998) (Gruber, 1995). Therefore, a commitment to a certain language should be undergone in order to develop an ontology.

A body of formally represented domain knowledge generally consists of objects and relationships between objects (i.e. universe of discourse) based on the *conceptualization* of that domain (Genesereth et al., 1987) (Gruber, 1993).

Ontologies in Computer Science evolved from semantic networks (Quillian, 1967) and have been used in Configuration Systems, Software Engineering, Information Retrieval, Conceptual Modeling, Interoperability, Enterprise Modeling, Electronic Commerce, and many other fields in the research and production areas.

In computer science, one of the first definitions of ontology was given by Neches and colleagues (Neches et al., 1991) "an ontology defines the basic terms and relations comprising the vocabulary of a topic area as well as the rules for combining terms and relations to define extensions to the vocabulary". This definition identifies basic terms, relationships between terms and rules to combine them. It also provides the definitions of such terms and relationships. Thus, an ontology includes not only the terms that are explicitly defined in it, but also the knowledge from which it can be inferred.

A few years later, Gruber (Gruber, 1993) proposes a new definition of ontology, by establishing its relationships with the concept of formal knowledge: *An ontology is an explicit specification of a conceptualization.* A conceptualization is viewed as an *abstract, simplified view of the world* to be formally represented (Gruber, 1993). This definition became the most quoted in literature and by the ontology community but needs further clarifications of the terms it uses, especially the distinction ontology-conceptualization.

The main advantage of Gruber's definition is that it requires the ontology to be explicit that is to be *publicly available*, not implicitly incorporated in some knowledge base. Gruber's definition is based on the assumption that every system incorporating formally and representing knowledge is explicitly or implicitly committed to a conceptualization.

In order to avoid possible confusions, Guarino and Giaretta (1995) suggest a terminological clarification of the three possible classes as follows:

1. To use *Ontology* with capital "*o*" as the term to identify the philosophical discipline.
2. To use the term *conceptualization* to identify a conceptual semantic entity.
3. To use the term *ontological theory* to identify a specific syntactic object-intended to represent knowledge.

Therefore, while ontological theories are a special kind of *artifacts*, conceptualizations are their semantic counterpart, with the specification that *the same ontological theory may commit to different conceptualizations, as well as the same conceptualization may underline different ontological theories* (Guarino et al., 1995).

Borst (1997) has given an elaboration of Gruber's definition, as follows: *Ontologies are defined as formal specification of a shared conceptualization.*

Based on Gruber's definition, many other definitions of ontology were proposed. In 1995, Guarino et al. (1995) collected and analyzed seven definitions of ontologies and provided their corresponding syntactic and semantic interpretations. They proposed to consider an ontology as "*a logical theory which gives an explicit, partial account of a conceptualization*", where conceptualization is basically an idea of the world that a person or a group of people can have. Although on the surface the notion of conceptualization is quite similar to Studer et al. notion (Studer et al., 1998), we can say that Guarino et al., (1995) went a step forward because they established how to build the ontology by making a logical theory. Hence, strictly speaking, this definition would be only applicable to ontologies developed in logic.

Based on the process followed to build the ontology, some other definitions have been proposed. According to Bernaras et al. (1996), an ontology "provides the means for describing explicitly the conceptualization behind the knowledge represented in a knowledge base".

Ontology is also defined as "a hierarchically structured set of terms for describing a domain that can be used as a skeletal foundation for a knowledge base" (Swartout et al., 1997). According to this definition, the same ontology can be used for building several Knowledge bases, which would share the same skeleton.

In 1998, Gruber's and Borst's definitions have been merged and explained by Studer and colleagues (Studer et al., 1998): "*Ontologies are explicit formal specification of a shared conceptualization*". According to the same authors, *Conceptualization* refers to an abstract model of some phenomenon in the world by having identified the relevant concepts of that phenomenon. *Explicit* means that the type of concepts used, and the constraints on their use are explicitly defined. *Formal* refers to the fact that the ontology should be machine readable, which excludes natural language. *Shared* reflects the notion that an ontology captures consensual knowledge, that is, it is not private for some individuals, but accepted by a group.

Guarino proposes a refined definition of an ontology that clarifies the difference between an ontology and a conceptualization: *An ontology is a logical theory accounting for the intended meaning of a formal vocabulary, i.e., its ontological commitment to a particular conceptualization of the world. The intended models of a logical language using such a vocabulary are constrained by its ontological commitment. An ontology indirectly reflects this commitment (and the underlying conceptualization) by approximating these intended models* (Guarino, 1998).

Uschold and Jasper (1999) provided a new definition of the ontology: "An ontology may take a variety of forms, but it will necessarily include a vocabulary of terms and some specification of their meaning. This includes definitions and an indication of how concepts are inter-related which collectively impose a structure on the domain and constrain the possible interpretations of terms. An ontology is virtually always the manifestation of a shared understanding of a domain that is agreed between a number of agents. Such agreement facilitates accurate and effective communication of meaning, which in turn leads to other benefits such as inter-operability, reuse and sharing".

Thus, ontologies represent a consensual, shared description of the pertinent objects considered as existing in a certain domain of knowledge (the domain of discourse). They constitute a special kind of software artifact conveying a certain view of the world (conceptualization), specifically designed with the purpose of explicitly expressing the intended meaning of a set of agreed existing objects.

In 1999, Fikes and Farquhar have given the following definition: "We consider ontologies to be domain theories that specify a domain-specific vocabulary of entities, classes, properties, predicates, and functions, and a set of relationships that necessarily hold among those vocabulary items. Ontologies provide a vocabulary for representing knowledge and describing specific situations in a domain" (Fikes et al., 1999).

Sowa's definition (Sowa, 2000) takes a philosophical perspective:

"The subject of ontology is the study of the categories of things that exist or may exist in some domain. The product of such a study, called an ontology, is a catalog of the types of things that are assumed to exist in a domain of interest D from the perspective of a person who uses a language L for the purpose of talking about D".

Mentzas (2002) defines ontologies as *a consensual agreement on the concepts and relations characterizing the way in which knowledge in a domain is expressed.*

In 2003, a pragmatic alternative, based on their experience in building ontologies, for defining an ontology, has been given by Noy and McGuinness as follows: "an ontology is a formal explicit description of concepts in a domain of discourse, properties of each concept describing various features and attributes of the concept, and restriction on slots" (Noy et al., 2003).

The common understanding of all the definitions and interpretations of an ontology orbit around two main characteristics: formality and consensus (Chira, 2003). All of the ontology definitions accentuate the importance of representing the knowledge in a consensual manner, at least among a specified group so knowledge sharing to be possible and implementable. Not the same thing can be said about formality requirement. Uschold allows ontologies to be expressed in a restricted and structured form of natural language, while Gruber's line of definition enforces a well-defined logical model for ontologies. Nevertheless, the general vision is that ontologies should be machine-enabled and, if not directly human-readable, they should at least contain plain text notices or explanations of concepts and relations for the human user (Borst, 1997) (Guarino, 1998) (Studer et al., 1998) (Uschold, 1998) (Fikes et al., 1999) (Sowa, 2000) (Noy et al., 2003).

While analyzing the above most relevant definitions of ontology, we can say that there is consensus among the ontology community and so there is not confusion about its usage. Different definitions provide different and complementary points of view of the same reality. Some authors provide definitions that are independent of the processes followed to build the ontology and also independent of its use in applications, while other definitions are influenced by the ontology development process. As a main conclusion to this section, we can say that ontologies aim to capture consensual knowledge in a generic and formal way, and that they may be reused and shared across applications and by groups of people. Ontologies are usually built cooperatively by a group of people in different locations (Corcho et al., 2003).

Moreover, the notion of ontology is sometimes diluted, in the sense that taxonomies are considered full ontologies (Studer et al., 1998). The ontology community distinguishes ontologies that are mainly taxonomies from ontologies that model the domain in a deeper way and provide more restrictions on domain semantics (Corcho et al., 2003).

The community calls them *lightweight* and *heavyweight* ontologies respectively. Lightweight ontologies include concepts, concept taxonomies, relationships between concepts and properties that describe concepts. Heavyweight ontologies add axioms and constraints to lightweight ontologies (Corcho et al., 2003).

Ontology Components

Despite the representation language being used, ontologies share a common set of characteristics and components in order to make knowledge representation and inference tasks possible (Muñoz,

2006). We distinguish the following components: concepts, slots, relationships, axioms, instances and operations.

Concepts (General Things)

A concept (also called class or frame) is the description of the common features that a set of individuals/objects have. Concepts are general, abstract or concrete notions within a domain of discourse (Uschold et al., 1995) (Gomez-Perez, 1999) (Noy et al., 2003). They are similar to the classes in the object-oriented modeling paradigm. From a logical point of view, a concept is a unary predicate which denotes a set of individuals (Muñoz, 2006).

Each concept has an associated term, generally a description in natural language, which expresses its name and a set of properties that characterize it. Concepts can be defined by extension, i.e., enumerating their instances, or by intension, i.e., giving restrictions that their instances must verify. A concept can have sub-concepts[5] and sub-concepts related using inheritance relationships.

Slots

Slots (also called properties, attributes or roles) describe the various features and attributes of a concept (and its instances) (Noy et al., 2003). They contribute to identify concepts by characterizing them and can be used in intensional definitions of concepts, to relate individuals or to give attribute values. Facets (sometimes called role restrictions) describe restrictions on slots (Noy et al., 2003).

Slots allow expressing relationships among concepts into a domain, such as hierarchy and consequently they are the basis for the hierarchical structure of the ontology. The values a property could have can be restricted to elements of a given class or can have cardinality constraints restricting their minimum or maximum number of possible values.

Relationships

A relationship represents a type of interaction between concepts of the domain (Gomez-Perez, 1999). Several types of relationships can be distinguished according to the number of concepts: reflexive relationship (links only one concept), binary relationship (links two concepts) and n-ary relationship (links more than two concepts).

Axioms

Axioms (constraints) are formal sentences that are always true (Guarino, 1998) (Gomez-Perez, 1999). They specify some constraints on the ontology elements (they constrain their interpretation). We distinguish[6] *axioms intra-element and inter-elements*. An inter-elements axiom specifies conditions relating more than an ontological element. For example "A disjoint axiom between two concepts A and B means that no instance can be at the same time an instance of concept A and also an instance of concept B. An *intra-element axiom is* local to a single ontology element and constrains its interpretation or the possible values.

Instances (Particular Things)

Ontology instances are individuals holding concepts definitions and facts representing relationships between individuals. An ontology together with a set of individual instances of classes constitutes a knowledge base (Noy et al., 2003) (Gomez-Perez, 1999).

Operations/Functions/Rules

Rules are generally used to infer knowledge in the ontology, such as attribute values, relation instances, etc. Ontological representation languages enable the execution of a certain basic set of operations to cover updating and querying tasks on ontologies. Likewise, new concepts can be defined, properties related to concepts and values changed or added during the entire life of the ontology (Muñoz, 2006).

In practical terms, developing an ontology includes (Noy et al., 2003):

- Defining classes in the ontology,
- Arranging the classes in a taxonomic (subclass–superclass) hierarchy,
- Defining slots and describing allowed values for these slots,
- Filling in the values for slots for instances.

We can then create a knowledge base by defining individual instances of these classes, filling in specific slot value information and additional slot restrictions.

Ontology Life-Cycle

The ontology life cycle identifies the *set of stages* through which the ontology moves during its lifetime, describes what activities are performed during each stage and how the stages are related. (Fernandez-Lopez et al., 1997) (Fernandez-Lopez, 2001).

The construction of each prototype starts with the specification process, sustained by the knowledge acquisition activity. Once the first prototype has been specified, the ontology building continues with the development of the conceptual model, formalization and implementation. During all these phases, the knowledge acquisition activity supplies the ontology development activities with needed knowledge. Also control, quality assurance, integration, evaluation documentation and configuration management activities are carried out simultaneously to development-oriented activities. Some of them have different degrees of intensity during specific life-cycle stages (e.g. evaluation and knowledge acquisition are more intense during conceptualization, while integration activity is more important at specification time (Chira, 2003).

The lifecycle of an ontology comprises several activities, such as: detection of needs, management and planning, design, evolution, dissemination, use and evaluation.

Typology

Ontologies can be of different types depending on several criteria as those described in (Guarino, 1997):

1. *The level of detail* allows distinguishing 2 types of ontology:
    - *Reference ontologies* (off-line) and
    - *Shareable ontologies* (on-line).

An ontology approximates the intended models of a logical language (Guarino, 1997) (Guarino, 1998) but how accurately should an ontology specify a conceptualization? Moreover how close should an ontology get to the underlying conceptualization? There is no formula to calculate an *optimal distance* between the intended models of a logical language (according to an ontological commitment) and the underlined ontology. This distance depends on the practical needs that ontology should fulfill. Nevertheless, there are tradeoffs between a detailed approach and coarse approach to designing ontologies. While a fine-grained ontology will specify more precisely the intended meaning of a vocabulary (and therefore it can be used off-line for reference purposes), it would be difficult to be assembled and reasoned on it (Guarino, 1998). On the other hand, a coarse (shareable) ontology would be much more easily shared among its clients that. *already agree on the underlying conceptualization*. (Guarino, 1998), and therefore it can be used on-line to support the system's services (Guarino, 1997).

2. *The level of dependence* on a particular task or point of view allows to distinguish:

- **Top-level ontologies**: specify very general concepts, *which are independent of a particular problem or domain* (Guarino, 1997) (Guarino, 1998) (e.g. engineering, person, agent);
- **Domain ontologies**: *domain ontologies* specialises the general concepts (of *top-level ontologies*), referring to a generic domain (e.g. mechanical-engineering, software-engineering, car disposal);
- **Task ontologies**: *domain ontologies* and *task ontologies* specialise the concepts (of *top-level ontologies*), referring to a generic task or activity (Guarino, 1997) (Guarino, 1998) (e.g. requirements analysis, disassembly);
- **Application ontologies**: level further specialization is involved by describing concepts *depending on a particular domain or task* and are often *roles* of domain or task entities performed during a certain activity (Guarino, 1997) (Guarino, 1998).

3. **Representation ontologies**: constitutes a special kind of meta-level ontologies describing a classification of the primitives used by a knowledge representation language (like concepts, attributes, relationships) (Guarino, 1997). An example of representation ontology is the Frame Ontology (Gruber, 1993) introduced within the Ontolingua system for capturing common knowledge-organization convention with the purpose to enable translations among different knowledge representation languages.

While agreeing that any ontology development, after all, depends on particular circumstances and needs (so any typology should depend on the practical use of ontologies), Uschold (1996) sets out three main dimensions by which ontologies may be classified, as follows:

1. **Formality:** the degree of formality by which a vocabulary is created and meaning specified.
2. **Purpose:** the intended use of the ontology.
3. **Subject Matter**: the nature of the subject matter that the ontology is characterizing.

While some authors request for an ontology language to be a formal one, Uschold adopts a *weak* position regarding the formality requirement. Uschold identifies four kinds of ontologies stretching from ontologies with no formality requirement at all to ontologies articulated in a meticulous formal language, as follows:

- Firstly, there are ontologies expressed in loosely natural language (Uschold, 1996) in the case of *highly informal* ontologies;
- Secondly, adding some degree of structure to the natural language results in *structure informal* ontologies that are expressed in a restricted and structured form of natural language, greatly increasing clarity by reducing ambiguity (Uschold, 1996);
- Thirdly, an artificial language has to be developed for expressing *semi-formal* ontologies;
- Fourthly, *rigorously formal* ontologies are expressed by meticulously defined terms with formal semantics, theorems and proofs of such properties as soundness and completeness (Uschold, 1996).

In the case of *communication ontologies*, normally intended for the use among people and not machines, informal and unambiguous ontologies may suffice. The *inter-operability ontologies* act as interchange formats for translations between different modeling methods, paradigms, languages and software tools (Uschold, 1996). The *system engineering* benefits when using ontology-based development can be summarized as follows (Jasper et al., 1999) (Uschold, 1996):

- **Reusability**: the ontology encodes domain information (including software components) in such way that sharing and reuse are possible;
- **Search**: an ontology may be used as metadata serving as an index into a repository of information.
- **Knowledge acquisition**: the ontology guides knowledge acquisition;
- **Reliability**: the ontology allows the automation of consistency checking resulting in more reliable software;
- **Specification**: the ontology can assist the process of identifying requirements and defining specification for an IT system (knowledge based, or otherwise);
- **Maintenance**: use of ontologies in system development, or as part of an end application, can render maintenance easier in a number of ways. Systems which are built using explicit ontologies serve to improve software documentation which reduces maintenance costs. Maintenance is also an important benefit if an ontology is used as a neutral authoring language with multiple target languages - it only has to be maintained in one place.

The main categories of the purpose dimension can be further granulated in categories and subcategories (e.g. within communication between people one may want to specify who the intended users are) (Uschold, 1996). A special kind of category related to *purpose dimension* is *genericity*, which is the extent to which an ontology can or is intended to be reused in a range of different situations (Uschold, 1996). These sorts of ontologies cover efforts that range from organizing human knowledge (in case of upper-level ontologies) to particular knowledge systems for specific applications (in case of application ontologies).

Noy and Hafner have identified four classes of ontologies relative to their genericity dimension, as follows: natural language applications, theoretical investigations, knowledge sharing and reuse, simulation and modeling ontologies (Noy et al., 1997).

The subject matter dimension covers all the topics an ontology may represent, that is anything conceivable, under the following headings (Uschold, 1996):

- The subjects as geography, medicine, etc.
- The subject matter of problem solving
- The subject matter of knowledge representation languages

Ontologies concerned with different areas of science (e.g. geography, medicine) are usually called *domain* ontologies; ontologies involved in problem solving are named *task*, *method* or *problem solving* ontologies; ontologies used for knowledge representation languages are labelled *representation ontologies* or *metaontologies* (Uschold, 1996).

According to (Sowa, 2000), ontologies can be classified as *formals* and *terminologicals* by the degree of axiomatization in their definitions.

- **Formal ontologies** have their categories and individuals distinguished by axioms and definitions stated in logic or in some computer-oriented language that can automatically be translated into logic. In formal ontologies, it is possible to do complex inferences supported by their logical fundamentals to check consistency in their building time and to infer new facts on query time. An example of a formal ontology is the ontology *Cyc* (http://www.Cyc.com).
- **Terminological ontologies** need not to have axioms restricting the use of their concepts. Examples of terminological ontologies are *Wordnet*[7] used for natural language processing and *The Electronic Dictionary – EDR*[8].

Theoretically, the difference between terminological and formal ontologies is one of degree. As more axioms are added to a terminological ontology, it may evolve into a formal one. The amount of time and human knowledge that formal ontologies need makes their cost prohibitive for certain applications, and they are generally restricted to a reduced number of terms. On the other hand, terminological ontologies are structurally simpler, cheaper, and possibly larger.

According to (Van Heijst et al., 1997), ontologies can be classified into 4 categories depending on the subject of their conceptualization.

- **Domain ontologies** contain conceptualizations that are specific for particular domains. They can be reused in various applications of the same domain (e.g., electronic, medical and mechanic domain).
- **Generic ontologies** are similar to domain ontologies, but the concepts that they define are considered to be generic across many fields. Typically, generic ontologies define concepts like state, event, process, action, component, etc. The concepts in domain ontologies are often defined as specializations of concepts in generic ontologies. Generic ontologies are constructed to be reused and extended. An issue is that there is no consensus in the research community as to what is the best way to express the general knowledge about the world that they are intended to represent. Examples of these kinds of ontologies are the proposal of Sowa (Sowa, 2000), The *Sensus ontology* constructed by the Information Sciences Institute – ISI of the University of Southern California, *Microkosmos ontology* developed by the Computer Research Laboratory of the New Mexico State University, *CYC*[9], *Wordnet* and *The Generalized Upper Model (GUM*[10]*)*.
- **Representation ontologies** explain the conceptualizations that underlay knowledge representation formalisms. That is, they provide a representational framework without making claims about the world. Domain ontologies and generic ontologies are described using the primitives provided by representation ontologies. An example in this category is the *Frame Ontology*, which is used in Ontolingua (Gruber, 1993). This ontology defines terms like relation, function, class, and the other primitives used in modeling ontologies in an object-oriented or frame-based way.
- **Application ontologies** contain all the definitions that are needed to model the knowledge required for a particular application. Typically, application ontologies are a mix of concepts taken from domain and generic ontologies. Application ontologies are not constructed to be reused. According to the characteristics of each group viewed, it can be observed that the more an ontology is suitable to be used in a defined context of a domain, the less suitable it is to be reutilized in other contexts of the same domain.

Ontologies Application Domains

Ontologies are currently very popular mainly within fields that require a knowledge-intensive approach to their methodologies and system development, such as knowledge engineering (Gruber, 1993) (Uschold et al., 1996) (Gaines, 1997) (Guarino, 1998) (Gomez-Perez, 1999), knowledge representation (Artala et al., 1996), qualitative modeling, language engineering, database design (Van de Riet, 1998), information modeling (Weber, 1997), information integration (Bergamaschi et al., 1998) (Mena et al., 1998), knowledge management and organization etc.

Ontologies can serve as a fundamental aid in the Software Engineering field, supporting the software developer by relating knowledge about the domain of the application being developed with existing code components to facilitate their reuse (Devambu et al., 1991).

Ontologies can play a central role in *Configuration Systems.* According to (McGuinness, 2002) a configuration system addresses the problem to assembly a complex artifact from its components. Potentially, the components have subcomponents, thus the artifact may be modular or hierarchical in nature. Likewise, each of the components typically has a number of properties and connections to other components. A domain model can be defined with concepts containing descriptions of parts, and interactions between properties that can be defined to condition the values of some properties to the values given to others. The input description for the configuration problems should be able to be given incrementally by human or automatic agents. The input could also be incomplete and inconsistent. Ontologies can be used to complete the input with the knowledge the system has on the domain and also to detect inconsistencies. Configuration Systems need not only to achieve an output with the result of the configuration process, but also to give the user the explanation about the line of reasoning followed which justifies the parts used in the final product.

Considering the Information Retrieval field, search engines such as Google (http://www.google.com), AltaVista (http://www.altavista.com) or Lycos (http://www.lycos.com) search in a full text way from the huge quantity of documents they have indexed. They recognize the relevance of an indexed document with respect to a posed question by how many times the words forming the question appear in the document. If some documents contain only synonyms of the words of the query, these documents will not be recognized as relevant, and if the words are used in indexed documents with different semantics than in the query, these indexed documents will be retrieved as relevant (although actually they are not relevant). On the other hand, with search engines powered by ontologies, the semantic context in which a word is used could be inferred many times and then the irrelevant documents could be omitted from the result set. Likewise, the query could be answered not only with documents containing the words in the query, but also with documents containing words that the ontology states are synonymous or related. As examples (Desmontils et al., 2001) presented a Web site indexed with the aid of a terminological ontology, and (Freitas et al., 2002) presented a method of cross-language information retrieval (CLIR) to classify both the information of the documents in a given collection and the user queries according to the concepts of the terminological ontology MeSH (Medical Subject Headings). These concepts are used as semantic units that minimize the linguistic problems like polysemy.

Ontologies are used in Intelligent Integration of information as metadata explaining the information content of data repositories enabling semantic queries into these subjacent sources with no need to consider the underlying structure of the sources in the query formulation. An example of this kind of information system is OBSERVER (Mena et al., 2000).

In Natural Language Processing, ontologies can help the semantic analysis of text by representing grammatical structures as related concepts in order to reduce the existent gap in the interpretation of the semantic ambiguity of the natural language. Since then, ontologies can be useful in text mining and machine translation. Wordnet is an ontology used in Natural Language Processing.

Ontologies play a main role in Enterprise Modeling by creating and maintaining an organizational memory that lets the different enterprise areas interoperate in a common language and with unified rules, for example modeling *Business Process*. They can also be the basis for the agents interoperation language in automated manufacturing processes. The TOVE[11] ontologies and the Enterprise ontology[12], are examples of this kind of ontologies.

In the field of Knowledge Engineering, ontologies and *Problem Solving Methods (PSMs)* are intended to enable the reuse of the domain knowledge across different intelligent applications. While ontologies are the repositories of the declarative knowledge and rules of the domain, PSMs specify the reasoning to solve concrete problems in a procedural way.

The Electronic Commerce area can take advantage of ontologies in several ways, for example, enabling a more intelligent access to online information and services, or providing structure to interoperability. For example, Yahoo! (http://www.yahoo.com) introduced metadata structures resembling ontologies early on by means of human tagging to help its users navigate into the content of the site. The idea was to have a small number of top level categories allowing drill-down in each of them to specialize the search. Another example is The Universal Standard Products and Services Classification Code – UN/SPSC[13]. The UN/SPSC is a freely available class taxonomy classifying products and services. Many B2B sites are currently using and extending it to better achieve their particular purposes.

In the medical domain, we can find several taxonomies, as MeSH, which enables the search in digital medical collections as MEDLINE[14] providing, among others, semantic expansion of queries so that searching for a more general topic can retrieve results of more specific topics in the taxonomy and vice-versa.

Before using an ontology, we have to build it, beginning from existing applications or from scratch. To build ontologies, several basic questions arise related to the methodologies, languages and tools to be used in its development process (Corcho et al., 2003):

- Which methods and methodologies can be used for building ontologies?
- Which tools support the ontology life-cycle stages?
- Which language should be used to formalize and implement an ontology?

In the following section, we will present the main characteristics of methodologies, tools and languages, which can help practitioners and researchers in this field to obtain answers to the previous questions.

ONTOLOGY DESIGN AND BUILDING

The ontology development process identifies what activities need to be carried out when building an ontology. It is crucial to identify these activities if agreement is to be reached on ontologies that are to be built by geographically distant cooperative teams with some assurance of correctness and completeness. If this is the case, it is advisable to perform the three categories of activities presented below and steer clear of anarchic constructions. Generally, the practice of building ontologies highly depends on the context and goals of a specific project. There are a growing number of methodologies that specifically address the issue of the development and maintenance of ontologies.

Van der Vet (1998) distinguishes two broad methods for creating ontologies:

- A bottom-up (data driven) approach that takes vast amount of diverse data sets and structures them into a set of top- level concepts,
- A top-down (manual, expert-driven) approach that relies primarily on a person's deep domain knowledge to create an overarching conceptual representation.

It is recognized that there is no unique and "correct" way or methodology for developing ontologies and there is not yet a general methodology which is adopted by all researchers. In fact, there is no one correct way to model a domain but there are always viable alternatives. Among several viable alternatives, we will need to determine which one would work better for the projected task, be more intuitive, more extensible, and more maintainable.

METHODOLOGIES FOR BUILDING ONTOLOGIES

An ontology is a model of reality of the world and the concepts in the ontology must reflect this reality. Ontology development is an iterative process, the best solution generally depends on the application we model and the designers' viewpoint. After defining an initial version of the ontology, we can evaluate and debug it by using it in applications or problem-solving methods or by discussing it with experts in the field. As a result, we will almost certainly need to revise the initial ontology. This process of iterative design will likely continue through the entire lifecycle of the ontology.

An ontology can be built either from scratch, through the re-engineering of other existing ontologies or by a process of ontology merging or ontology learning approach.

WordNet defines the term *methodology* as "the branch of philosophy that analyzes the principles and procedures of inquiry in a particular discipline". It is also defined as "the system of methods followed in a particular discipline". According to WordNet, A *method* is defined as a way of doing something, especially a systematic one and implies an orderly logical arrangement (usually in steps). A methodology can be seen as a systematic approach to conduct an engineering project that suggests activities to be performed at certain stages of the ontology development process (Chira, 2003).

According to Noy et al. (2003), there is no absolutely one correct way or methodology for developing ontologies. However there are some fundamental rules in ontology design that can help to make wise design decisions. These are given as follows (Noy et al., 2003):

- There is no one correct way to model a domain- there are always viable alternatives. The best solution almost always depends on the application that one has in mind and the extensions that are anticipated.
- Ontology development is necessarily an iterative process.
- Concepts in the ontology should be close to objects (physical or logical) and relationships in the domain of interest. These are most likely to be nouns (objects) or verbs (relationships) in sentences that describe the domain.

Series of approaches and methods have been reported in literature for building ontologies (Corcho et al., 2003). Although the general steps for building large knowledge-based systems have been initially reported by Lenat and Guha in 1990, in the setting of the Cyc Project (Lenat et al., 1990), the first methodologies have been proposed some years later (in 1995) by Gruninger (Gruninger et al., 1995) and Uschold (Uschold et al., 1995), based on their experience in the Enterprise Ontology and the TOVE (TOronto Virtual Enterprise) projects. The methodologies guidelines were later refined in (Uschold, 1996) (Uschold et al., 1996).

In 1996, the methodology METHONTOLOGY was developed within the Laboratory of Artificial Intelligence at the Polytechnic University of Madrid to enable the construction of ontologies at the knowledge level. This methodology developed by Gomez-Perez et al., (1996) was extended later (Fernandez-Lopez et al., 1997) (Fernandez-Lopez et al., 1999) (Gomez-Perez, 1998). At the same period, Bernaras et al. (Bernaras et al., 1996) presented a method used to build an ontology in the domain of electrical networks as part of the KACTUS project (Schreiber et al., 1995). In 1997, a new method was proposed for building ontologies based on the SENSUS ontology (Swartout et al., 1997). In 1998, Hearst (1998) developed the *hyponymy pattern approach*, for automatically learning relationships between concepts in an ontology.

Some years later, Assadi (2000) proposes an approach for textual ontology construction, to transform a natural language specification to formal language. Agirre et al. (2000) also propose a methodology to enrich the concepts in existing large ontologies using text retrieved from the World Wide Web. The approach is based on the use of topic signatures, used in text summarization that have been described in (Hovy et al., 1999) and (Lin et al., 2000).

Aussenac-Gilles et al., propose a method for ontology learning based on knowledge elicitation from technical documents (Aussenac-Gilles et al., 2000a) (Aussenac-Gilles et al., 2000b). This method allows the creation of a domain model by means of a corpus analysis using natural language processing (NLP) tools and linguistics techniques.

In 2001, the On-To-Knowledge methodology (Staab et al., 2001) (Sure et al., 2004) was developed at the ISI (Information Sciences Institute) natural language group, based on an analysis of usage scenarios.

In 2002, Leclère et al. (2002) propose a consensus for an ontology design process, based on three steps (i) conceptualization, (ii) ontologization and (iii) operationalization. It allows passing from the untreated data to operational ontology.

The Alfonseca and Manandhar's method was developed in 2002 (Alfonseca et al., 2002) as part of the Ensenada CICYT project (2002-2005), funded by the Spanish Ministry. This project includes knowledge acquisition from free texts for automatic generation of e-learning materials.

Bachimont et al. (2002) proposed a method for building ontologies, taking into account linguistic techniques that have come from Differential Semantics. In this method the construction of ontologies follows 3 steps: (1) *Semantic Normalization* (2) *Knowledge Formalization* and (3) *Operationalization.*

In (Benslimane et al., 2003), two approaches for ontology construction are proposed: from scratch or starting from the existing databases. These approaches allow capturing the semantic of the concepts and to clarify the contents of the information sources.

At the same times, Noy et al. (2003) propose an approach based on the following steps:

- Defining classes in the ontology,
- Arranging the classes in a taxonomic hierarchy (subclass, super-class),
- Defining slots and describing allowed values for these slots,
- Filling in the values for slots for instances.

Mhiri and Gargouri (2009) propose an ontology building approach based on 4 steps: (1) construction of the initial ontology; (2) ontology enrichment; (3) ontology representation and (4) ontology maintenance.

Another methodology is proposed by Châabane and Jaziri (2009) in the context of spatial modeling. This methodology extends the ontology design process proposed by Leclère et al. (2002) by integrating a new step of spatialization to allow to take into account the spatial characteristics and relationships between the concepts of the domain.

We will summarize in the following, the most relevant methods and approaches used for ontology building.

Uschold and King' Methodology

The methodology proposed by Uschold and King (1995) is based on 4 steps:

1. Identify purpose: why the ontology is being built and what its intended uses are.
2. Building the ontology: this step consists in (1) Ontology capture; (1) Coding and (1) Integrating existing ontologies.

3. Evaluation.
4. Documentation

The Enterprise Ontology is an example of project developed using this methodology by the Artificial Intelligence Applications Institute at the University of Edinburgh[15].

Grüninger and Fox' Methodology

The methodology proposed by Gruninger and Fox (1995) during the development of the TOVE project ontology (Gruninger et al., 1995) within the domain of business processes and activities modeling, consists of the followings steps:

1. Capture of motivating scenarios.
2. Formulation of informal competency questions.
3. Specification of the terminology of the ontology within a formal language.
4. Formulation of formal competency questions using the terminology of the ontology.
5. Specification of axioms and definitions for the terms in the ontology within the formal language.
6. Establish conditions for characterizing the completeness of the ontology.

METHONTOLOGY

This methodology (Bernaras et al., 1996) (Fernandez-Lopez et al., 1997) was developed within the Laboratory of Artificial Intelligence at the Polytechnic University of Madrid. The METHONTOLOGY framework (Fernandez-Lopez et al., 1997) (Fernandez-Lopez et al., 1999) enables the construction of ontologies at the knowledge level and includes (Bláz98): 1) identification of the ontology development process; 2) a life cycle that is based on evolving prototypes; and 3) particular techniques for carrying out each activity in the management, development-oriented and support activities.

The methodology has adequate support tool to aid the ontology development process. ODE, WebODE, Protégé, OntoEdit are among the available tools that give automation support to METHONTOLOGY. The METHONTOLOGY approach allows the stepwise refinement of the components of an ontology as new versions or prototypes evolve which makes the ontology to be very dynamic and open to change and growth.

The Methontology methodology enables a dynamic control of interconnected ontologies (e.g. different activities performed when building an ontology may require performing other activities on already build or under construction ontologies). It also supports the process of ontological reengineering, that is, the process of retrieving and mapping a conceptual model of an implemented ontology to another, more suitable conceptual model, which is re-implemented (see.Ontological reengineering an reuse.), by the means of reverse engineering, restructuring and forward engineering activities (Chikofsky et al., 1990).

The METHONTOLOGY framework is supported by ODE (Fernandez-Lopez et al., 1999), WebODE, Protégé, OntoEdit.

This methodology has been proposed for ontology construction by the Foundation for Intelligent Physical Agents (FIPA[16]), which promotes interoperability across agent-based applications.

Examples of ontologies built according to METHONTOLOGY: CHEMICALS (Fernandez-Lopez et al., 1999) (Gomez-Perez et al., 1996) Environmental pollutants ontologies (Gomez-Perez et al., 1999), the Reference-Ontology (Arpirez-Vega et al., 1998), the restructured version of the (KA) ontology (Blazquez et al., 1998).

On-To-Knowledge

The On-To-Knowledge methodology[17] (Staab et al., 2001) (Sure et al., 2004) includes the identification of goals that should be achieved by knowledge management tools and is based on an analysis of usage scenarios. The steps proposed by the methodology are: kick-off, where ontology requirements are captured and specified, competency questions are identified, potentially reusable ontologies are studied and a first draft version of the ontology is built; refinement, where a mature and application-oriented ontology is produced; evaluation, where the requirements and competency questions are checked, and the ontology is tested in the application environment; and ontology maintenance.

Aussenac-Gilles and Colleagues' Approach

This method for ontology learning allows the creation of domain model by means of corpus analysis using natural language processing (NLP) tools and linguistics techniques (Aussenac-Gilles et al., 2000a) and (Aussenac-Gilles et al., 2000b). It combines knowledge acquisition tools based on linguistic with modeling techniques that allow keeping links between models and texts. The method uses texts and may use other existing ontologies or terminological resources to build the ontology. This method, proposes to perform ontology learning in three levels: linguistic, normalization and formal level. The *linguistic level* is composed of terms and lexical relations extracted from texts by means of a linguistic analysis. These elements are used to create lexical clusters and convert them into concepts and conceptual relations at the *normalization level*. The process goes from terminological analysis to conceptual analysis, this means from terms to concepts and from lexical relations to semantic ones. Finally, concepts and relations are formalized by means of a *formal language*.

Mhiri and Gargouri' Approach

This approach consists in representing the set of the concepts, characterizing a specific field, and their relationships. This ontology allows the identification of all the studied domain's concepts and

Figure 4. Ontology building steps (mhiri et al., 2009)

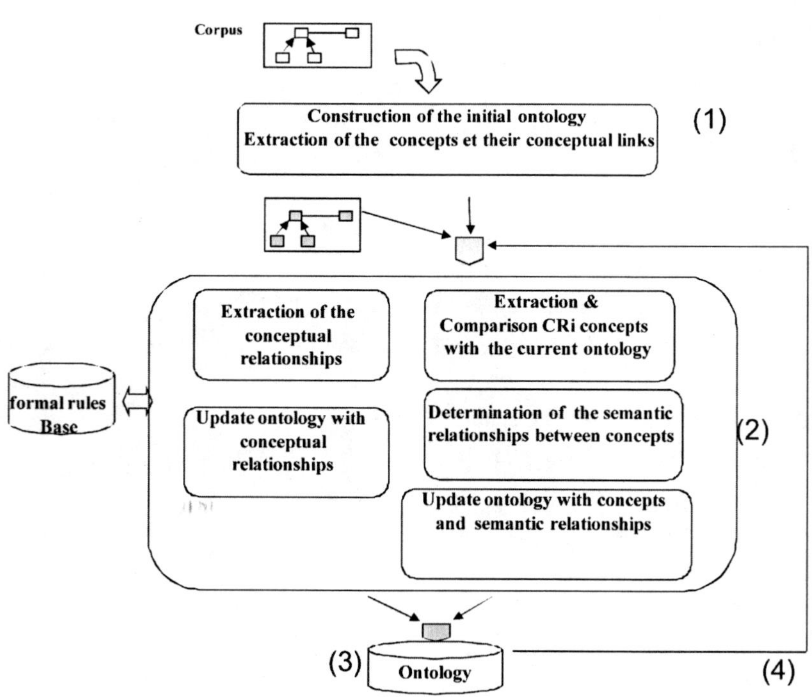

their relationships. It will be possible to couple this checking to the CASE, when designing an application (Mhiri et al., 2009). The main steps are (figure 4):

- *Construction of the initial ontology*: this step is based on the choice of the conceptual representation representing the greatest number of concepts. This choice is justified by the fact that this conceptual representation contains the greatest set of concepts and their possible relationships, making it possible to represent the semantic of the studied field. This construction consists in the extraction of the syntax of class names, their properties (attributes and operations) and their conceptual relationships. Conceptual relationships concern all the links supported by UML.
- *Ontology enrichment*: consists in adding new concepts coming from conceptual representations elaborated by the designers. For each conceptual representation:
- Extract the set of concepts (classes, attributes, operations and links).
- Compare the extracted concepts and the current ontology concepts, in order to determine the semantic relationships.
- Update the current ontology with the set of CR concepts and their semantic relationships.
- Update the current ontology with the set of conceptual relationships of CR.
- *Ontology representation*: this step requires the use of a comprehensible and semi-formal language to represent the concepts and their relationships. For this reason, an extension of the UML language has been proposed (Mhiri et al., 2005). This extension adds new stereotypes characterizing the proposed ontology. It consists in the concept of class, the concept of class association, the semantic relationships between the concepts (synonymy, homonymy, equivalence, antonymy). These stereotypes are modeled as a diagram called concepts diagram.
- *Ontology maintenance*: an ontology must be adapted to the increasing evolution of the users' needs. For that, the obtained ontology must be progressive, supporting the changes of the real word and according to the several contexts.

Several others methodologies and methods for building ontologies have been reported in literature, such as KACTUS approach (Bernaras et al., 1996), Cyc method (Lenat et al., 1990), Sensus method (Swartout et al., 1997), ONTOSPEC (Kassel, 2002), Biebow (Biebow et al., 1999), (Faatz et al., 2002), (Gupta et al., 2002), (Hahn et al., 1998) (Hahn et al., 2001), (Hwang, 1999), (Khan et al., 2002), (Kietz et al., 2000), (Lonsdale et al., 2002), (Missikoff et al., 2002), (Moldovan et al., 2001), (Nobécourt, 2000), (Roux et al., 2000), (Wagner, 2000), (Xu et al., 2002), (Gandon et al., 2001) (Leclere et al., 2002).

Moreover, many other methods and methodologies have been proposed for other tasks, such as ontology reengineering (Gomez-Perez et al., 1999), ontology learning (Aussenac-Gilles et al., 2000b) (Kietz et al., 2000), ontology evaluation (Gomez-Perez, 1996) (Gomez-Perez, 2001) (Guarino et al., 2000a) (Guarino et al., 2000b) (Kalfoglou et al., 1999a) (Kalfoglou et al., 1999b), ontology evolution (Klein et al., 2001) (Klein et al., 2002) (Noy et al., 2004) (Stojanovic et al., 2002), ontology merging (Noy et al., 2000), etc.

In (Fernández López, 1999), the most representative methodologies used in ontology development are presented. An analysis of these methodologies is proposed based on 9 criteria:

1. Inheritance from Knowledge Engineering.
2. Detail of the methodology. Consideration of whether the activities and techniques proposed by the methodology are exactly specified.

3. Recommendations for knowledge formalization. Consideration of the formalism or formalisms proposed for representing knowledge (logic, frames, etc.).
4. Strategy for building ontologies. Discussion of which of the following strategies are used to develop ontologies:
    a. Application-dependent: the ontology is built on the basis of an application knowledge base, by means of a process of abstraction.
    b. Application-semi dependent: possible scenarios of ontology use are identified in the specification stage.
    c. Application-independent: the process is totally independent of the uses to which the ontology will be put in knowledge-based systems, agents, etc.
5. Strategy for identifying concepts. The possible strategies are: from the most concrete to the most abstract (bottom-up), from the most abstract to the most concrete (top-down), or from the most relevant to the most abstract and most concrete (middle-out).
6. Recommended life cycle. Analysis of whether the methodology implicitly or explicitly proposes a life cycle.
7. Differences between the methodology and IEEE 1074-1995.
8. Recommended techniques: specification of whether particular techniques are proposed for performing the different activities of which the methodology is composed.
9. What ontologies have been developed using the methodology and what systems have been built using these ontologies.

The ontology development process involves many activities that can present a high level of complexity, depending on the intended scope, size and level of detail of the ontology under construction (Uschold et al., 1996) (Jones et al., 1999) (Mizoguchi, 2004).

As a consequence, the construction of an ontology cannot be conducted in an improvised manner. The complexity of activities like conceptualisation, knowledge structuring (ontologisation), ontology evaluation, etc., require the use of management processes, in order to control cost, risks, schedules and to ensure that the artifacts produced are of the intended quality (Mendes et al., 2004).

An important number of methodologies are presently described in the literature and offer guidance to different portions of the ontology development cycle. However, until now, there is no consensus about the best practices to adopt concerning the construction of an ontology. The ontology development process did not have any technical standard to guide the development process, despite major efforts in this direction. As a consequence, a number of questions remain open: which ontology building methodology provides the best guidance to develop an ontology? which lifecycle model (cascade, incremental prototyping, evolutionary prototyping, etc.) is best suited to the planned ontology development?

To help the developer in selecting the proper techniques and tools for building ontologies, Uschold has created a general framework (Uschold, 1996). The framework identifies the common steps and techniques applicable in all the cases and the conditions that require specific steps and techniques. It consists of a set of five sequential steps containing techniques to apply and guidelines to follow (Uschold, 1996):

1. Identify the *purpose* of the ontology
2. Decide the *level of formality*
3. Identify the *scope*
4. The *building* of the ontology
5. Evaluation/Revision cycle

The ontology development process should not be started before clearly identifying the *purpose* for which the ontology is built. This step includes activities such as the detection of the target users, the specification of the purpose relative to the range of purposes already identified, the elaboration of motivation scenarios and competency questions (Uschold, 1996). After the completion of the first step, enough information should be available so that the developer to be able to decide the *level(s) of formality* required for its ontology. Within the third step, the *scope* (e.g. what *is* and what *is not* in the ontology) has to be identified and the terminology (i.e. set of terms) has to be determined. This step can be completed through the use of one of the following techniques: motivating scenarios and informal competency questions or brainstorming and trimming (Uschold, 1996). Which technique to be used, depends on the specific circumstances of every ontology development process. The third step concludes with a document containing the terms and concepts (structured or not) the ontology has to define. The next step consists of *building* the ontology itself. This can be done in a few ways.

The analysis of related work concerning ontology construction allows defining a consensus for an ontology design process based on three main steps: (i) conceptualization, (ii) ontologization and (iii) operationalization (Leclere et al., 2002) (Noy et al., 2003):

- **Conceptualization**: this step provides, from a corpus, an informal (or a semi-formal) model, generally expressed in a natural language. This model aims at defining a conceptual vocabulary based on terms describing the domain of interest and a description of their semantics.
- **Ontologization**: this step aims to model, in a formal language, the formal properties of the domain. It provides, from the conceptual model, an ontology which formally represents the domain knowledge.
- **Operationalization/Implementation**: transforms the formal model into a computational model. (Blazquez et al., 1998) proposed a similar construction process with an additional step: the *specification* that defines why the ontology is built and who the final users are. The process of building an ontology is a collaboration that involves experts in the domain of knowledge, knowledge engineers and the future users of the ontology (Farquhar et al., 1997). In any case, the development of an ontology is based on linguistic and cognitive resources that constitute a corpus.

As explained in (Blazquez et al., 1998), ontology development process has to be done in four distinct phases.

- **Specification**: states why the ontology is built and who are the final users.
- **Conceptualization**: leads to a structured domain knowledge.
- **Formalization**: transforms the conceptual model into a formal model.
- **Implementation**: transforms the formal model into a computational model.

The majority of methodologies and methods provide general guidelines and do not guide in a detailed manner the ontology building and design. In addition, the proposals are not unified: at present each group applies its own methodology. Therefore, efforts are required along the lines of unifying methodologies to reach a situation resembling Knowledge and Software Engineering.

KNOWLEDGE REPRESENTATION

One important step of the ontology life-cycle is the formalization. When formalizing an ontology, it is important to find a formalism which provides adequate primitives to capture the aspects of the ontology. A formalism provides a symbolic system (e.g. syntax, axioms, inference rules) and the semantics attached to it. This section gives an overview of the existing knowledge representation languages which can be used for ontology formalization. A view of knowledge representation languages can be found in (Corcho et al., 2000).

Logic of Propositions

Propositional logic originates from philosophy and is the foundation knowledge formalization language. Propositional logic expressiveness is limited: it only considers relations between propositions without considering the structure and the nature of the proposition. For instance, it does not enable to represent the difference between individuals and categories.

First Order Logic

First order logic (FOL) includes propositional logic. The addition of existential and universal quantifiers enables to differentiate individuals from categories. The Cyc ontology (which aims at being an ontology of common sense) is formalized in FOL.

Semantic Networks

Semantic networks were introduced by Quillian (1985). Semantic networks are graphs which represent concepts and their relationships to each other. Graph nodes represent concept and graph arcs represent relations between concepts. When using semantic nets, several problems can be encountered.

The unpredictability of the inference process makes reasoning difficult to debug. Semantic nets are relatively under-constrained. This entails a large number of possible representations of the same situation. Moreover, knowledge bases represented in the semantic nets formalism often seem to be disorganized. The reasoning methods for a semantic net system have to be specified for each possible interaction of arcs. Whereas a logic-based representation usually has a very small number of powerful inference techniques available, a semantic net system usually has a large number of special purpose inference methods.

To answer queries, especially queries with negative answers, it is often necessary to search most or all of the semantic net. Heuristic methods to reduce the size of the search have been proposed, but have not been particularly successful.

Conceptual Graphs

Conceptual Graphs are inspired by semantic networks (Sowa, 1984). The main improvement on semantic networks is that they rely on a formal logical layer. Conceptual graphs enable friendly presentation of logic to human and have several representations:

- **Display Form**: a graphical representation;
- **Linear Form**: a textual representation equivalent to the Display Form;
- **Conceptual Graph Interchange Format**: for transmission between systems.

A conceptual graph is a bipartite oriented graph. There are two types of nodes in the graph: concept nodes and relation nodes. The arcs are oriented and always link a concept node to a relation node. Conceptual graphs are existential and conjunctive statements. The arity of relations is an integer n which represents the number of concepts they can be linked to. Concepts and primitives have a type which can be a primitive type or a defined type. The ontological knowledge upon which conceptual graphs are built is represented by the support, made of two subsumption hierarchies structuring concept types and relation types. This is the terminological level. Relations have a fixed arity and a signature (i.e. the types of the concepts linked by the relation).

The core reasoning operator of Conceptual Graphs is the computation of subsumption relations between graphs.

Frames

The notion of Frame was introduced by Minsky (1975). In this formalism, frames are organized hierarchically based on a specialization relation: a-kind-of. This relation is a partial order and is reflexive, transitive and anti-symmetric. From a structural point of view, this means that all the descendants of a class own all the attributes and methods of the inherited class. From a conceptual point of view, an instance of a subclass C is also an instance of super-class of C. A frame is composed of attributes and facets. Facets can be declarative or procedural to specify the nature (type, domain, cardinality, value) or the behavior of an attribute (default value, procedures to calculate the value, filters). Point of views can be defined to build different hierarchies of classes capturing different conceptualization while enabling an object to take into account all the aspects of its class in the different views.

Description Logics

Description Logics (Baader et al., 2007) are based on predicate logics, semantic networks and frame languages. Two types of knowledge are distinguished: terminological knowledge where concepts and roles are represented and manipulated and assertional knowledge where assertions and manipulations about individuals are made. The assertional level is usually called the A-Box and the terminological knowledge is called the T-Box. A T-Box is composed of a set of concepts being either primitive or defined by a term, a set of roles, or a set of individuals. A concept is a generic entity of an application domain representing a set of individuals. An individual is a particular entity, an instance of a concept. A role is a binary relation between individuals. A role can be primitive or defined. Concepts are organized in a subsumption hierarchy. There are several description logics, associated with a set of predefined constructors.

Today, many applications of information systems have adopted ontologies as conceptual infrastructure. The ontologies play a key role in the analysis, modeling and implementation in the field of knowledge (Studer et al., 1998). Moreover, ontology constitutes an integral and important part of the information systems design. For that information systems' ontology must be continually opened on its environment by taking account the several contexts. The next section, presents the necessity of using the context in ontology.

ONTOLOGY AND MULTI-CONTEXT

Usually, an ontology is mono-contextual (Rifaieh et al., 2004). But in some cases, a given domain can be modeled by several ontologies, where each one is related to a particular context. The problem of multitude of contexts is commonly known in the information system modeling. A concept is defined according to several contexts. This semantic dynamism is encountered by the possibility of adding other concept definitions according to another context in the same ontology.

Ontologies are becoming an essential component of many domains and area of interests such as the Web, e-commerce, medical applications, and government agencies. Ontologies, as an explicit specification of conceptualization, play an essential role in information integration and in the software engineering (3). Combining several context values may generate a more powerful understanding in the knowledge engineering. When a given field allows only one context, it also allows only one ontology. Such ontology is called, according to (Rifaieh et al., 2004), a mono-representation or a mono-contextual ontology.

The multitude of contexts occupies a particular place for the following factors:

- Considerable increase of data volumes represented in various forms and generated from users when using varied contexts.

- Contextual sharing and data exchange between applications treating the different elements of the same system.
- Complexity of the data which reflects the extreme complexity of the real world.
- Multiple facets of the data which translate users' diversity.
- Dynamic requirement evolution.

DEFINITION

Contexts appear in many disciplines as meta-information to characterize the specific situation of an entity, to describe a group of conceptual entities or to share knowledge.

A context can be seen from different perspectives. For instance, it could be considered from an abstraction level, granularity scale, interest of users' communities, or perception. Therefore, the same domain can have several contexts, where each concept is described in a particular context.

The use of the word "context" tends to be unclear because everything in the world happens in a given context. According to Brézillon (1999; 2002), a context is *a collection of relevant conditions and surrounding influences that make a unique and comprehensible situation*. It can be defined as *the interrelated conditions in which something exists or occurs* (Schmidt et al., 1999). The same authors define the context as *knowledge about the user's and its device's state, including surroundings, situation, and to a less extent, location*. Dey et al. (1999; 2006) provide the following definition: it is *any information that can be used to characterize the situation of an entity*. Thus, combining several context values may generate a more powerful understanding in the knowledge engineering. But, in information systems, several problems can occur such as lack of coherence, ambiguity and difficulty when users admit different contexts.

According to Rifaieh (2004), a contextual ontology provides local and global semantics. It allows a global view without losing original representation. It adapts the models to coexist through the contexts relationships. The contexts are related to an interpretation with a predefined structure. Thus, contextual ontology provides a dynamic consensus, rather than a static consensus, which is offered by integrated ontology (Rifaieh et al., 2004). However, an integrated ontology is a representation of global semantics. It defines a global understanding based on an established consensus (Staab, 2004). The consensus is needed for the common integrated ontology, which needs to be renewed each time an update occurs. It suffers from the loss of information in the profit of the unified representation.

IMPORTANCE OF CONTEXT

Context plays an important role since a long time in domains where reasoning intervenes in understanding, interpretation, diagnosis, etc. The reason is that these activities of reasoning rely heavily on a background or experience that is generally not made explicit and gives a contextual dimension to knowledge. Indeed, everybody uses context in his daily life. However, the lack of a clear definition and an explicit representation of context is one of the reasons of failures in many systems.

Importance of context explication for data integration was identified by several researchers in the field of multi-database system in the 90's. Kashyap et al. (1996) proposed to represent definition context at the schema level as a set of property-value pairs, but value where only informally defined. Sciore et al. (1994) proposed to represent value context at the value level.

Pierra (2005) discussed the notion of context in order to investigate the role of ontologies for data integration and to present an ontology model that was developed to allow neutral exchange and automatic integration of industrial component catalogues and of technical data.

Several researches identify that ontologies are depending on context. So, the concepts interpretation depends on a particular context in which the concepts are used. Ontology should provide definitions and structures of contextual data to represent the diversity of perceptions. The context can be used to define a user view and thus allows choosing an ontology subset (Mtibaa et al., 2008).

In characterizing the similarity between objects based on the semantics associated with them, we have to consider the real-world semantics of an object. It is not possible to completely define what an object denotes or means in the model world. The *context* of an object is the primary vehicle to capture the real-world semantics of the object. Modeling and representing context can lead to several benefits (Kashyap et al., 1998):

- **Economy of representation**: in a manner akin to database views, contexts can act as a focusing mechanism when accessing the component databases or information sources.
- **Economy of reasoning**: instead of reasoning with the information present in the database as a whole, reasoning can be performed with the context associated with an information source.
- **Managing inconsistent information**: where information sources are designed and developed independently, it is not uncommon to have information in one source being inconsistent with information in another. As long as information is consistent within the context of the user' query, inconsistency in information from different databases may be allowed.
- **Flexible semantics**: An important consequence of associating abstractions or mappings with context is that the same two objects can be related to each other differently in two different contexts. Two objects might be semantically closer to each other in one context as compared to the other.

There are several proposals for representing context. An effective approach needs to bring together metadata, user profiles, information modeling abstractions, and ontologies, as well as to allow their dynamic construction to model application domain and user needs (Sheth, 1999). Besides their modeling and representation, a key challenge includes the ability to reason about or compare contexts (Kashyap et al., 1996) (Lee et al., 1996) (Ouksel et al., 1994). While there are many representations and associated reasoning techniques, practical application of context is expected to be a key research challenge for achieving semantic interoperability in information systems.

TOWARD AN OPEN ONTOLOGY

Usually, an ontology is defined for one field and a given problem. It means that we have to define, for this field, its functional and relational signature, a formal language of representation and its associated semantics (Bachimont, 2000). To ensure an open ontology, the context of the application must be taken into account. The multitude of contexts involves different semantics for the same term according to several contexts. The problem of multi-context is encountered in information systems design when a given concept can be seen in various prospects according to the context of use.

Ontology represents a given conceptualization of the real world from a given perspective and viewpoint. Semantics described on the ontology is dependent on the context.

Ontology defines a shared and common comprehension of a domain of study for a users' community. This definition neglects the particularity of each user and the context of use. The contexts are built to be locally maintained and represent interpretations of the not shared individuals or groups of individuals' schemas. These contexts are local what neglects, by consequence, collaboration work between the users. In fact, there is a complementarity relation between ontologies and contexts. Ontologies which take into account the context give the users views representing their contexts and a common and shareable view for all users.

With the multitude of contexts in the engineering information systems, several problems can emerge, such as:

- Data duplication: the same concept can be presented according to several contexts. Thus the classic problems raised by the data duplication emerge.
- Difficulty to maintain the global model's coherence: to maintain this coherence, we need to integrate all the specified multi-context concepts.

Using different contexts may cause many problems of incoherence, ambiguity and difficulty when users admit different contexts. Some concepts can be seen differently according to the user's context. These problems may lead to a poor and inadequate future system.

Domain ontologies are developed by capturing a set of concepts and their links according to a given context. Contextual ontologies characterize a concept by a set of properties that vary according to context. In some cases, a given domain can have more than one ontology, where each one is related to a particular context. Thus, a concept is defined according to several contexts and to several representations; this ontology is called a contextual ontology. An ontology described according to several contexts at the same time is called a multi-representation ontology. A multi-representation ontology is defined, according to (Benslimane et al., 2003) and (Rifaieh et al., 2004), as *an ontology characterizing the concepts by a variable set of properties (static and dynamic) or attributes in several contexts and several granularities.*

DYNAMIC CONTEXT AND STATIC ASPECT

According to (Brézillon et al., 2001), representing knowledge with its context implies to rethink knowledge representation. Context is composed of relationships between different knowledge elements. This gives a static view on knowledge and the way in which knowledge elements are related within their context of use. A dynamic view consists of considering context as a mechanism of contextualization for retrieving knowledge, and its links to the reasoning mechanism that associates the considered incident with incidents known by the system.

This is better explained with the example illustrated in figure 5 (Kuck, 2007), a user context comprises three categories: (1) the personal background of the mobile user. These are user's data about gender, age, domicile etc. (2) the environment category provides situational context information like the actual location and the environment. (3) the user's "information world" for example to read documents and to visit Web pages which reflect the user's interests. Owing to computational reasons, there is no voluminous

Figure 5. Context of a mobile user (kuck, 2007)

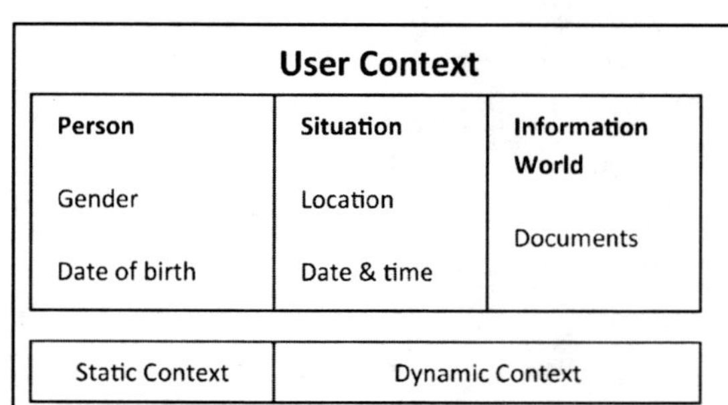

user context model on the mobile client. Only the essential user's data reside there.

ONTOLOGY AND CONEXT

Mtibaa and Jaziri (2009) developed a framework to manage contextual ontology characterized by:

- **Information visibility**: the information should be seen from one context or a set of contexts.
- **Information zooming**: to attain this feature, the notion of granularity or levels of information detail should enable the user to zoom in or zoom out by going from coarse-grain level to fine-grain one and vice versa.
- **Navigation features** are essentially important to view the structure of any concept and its relationships with other concepts. In contextual ontologies, it should be noted that a concept is allowed to occur in different nodes of taxonomies depending on the context to which it refers.

The same authors proposed the use of a multi-representation ontology, which will assist users to information systems' designing. They justified their proposition based on the following points:

- Possibility to have more than one context when specifying the user's system. Consequently, a concept has one or more definitions according to each context.
- Dynamic aspect of multi-representation ontology: we can add or modify, when requested, a given ontology concept according to another context.

Multi-representation ontology is used as a reference for users having different contexts in order to assist them in system modeling. Multi-representation ontology allows guiding future users in specifying their system and in solving conflicts resulting from the multitude of contexts.

The ontology contextualization process is composed of three steps (Mtibaa et al., 2008). It starts with pretreatment and acquisition step to clarify certain useful information about the used concepts in the field of information system design. The next step compares between concepts according to different contexts while referring sometimes on the expert or the designer. The last step extends the field ontology by a contextual ontological layer (figure 6).

Figure 6. The ontology contextualization process (mtibaa et al., 2008)

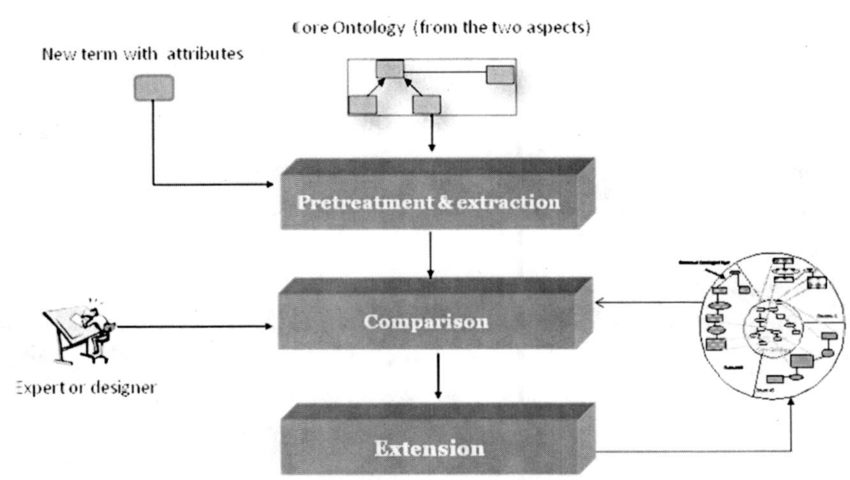

In the ontology engineering, mainly when using ontologies in dynamic environments, supporting ontology's evolution becomes essential and extremely important. It enables users to integrate changes and to treat different ontology versions. In changing environments, ontology over a period of time needs to be modified to reflect changes in the real world, changes in the user's requirements, and drawbacks in the initial design. A core aspect in the evolution process is to guarantee consistency of the ontology when changes occur. Ontology evolution concerns different facets: the needs to update and to evaluate data, the changes to apply in conformity with these needs, the management of inconsistencies in all parts of the ontology.

Several works (Corcho et al., 2006) (Fonseca, 2007) (Pierra, 2008) demonstrate that the ontology is important to analyze the knowledge in a given field, by modeling the relevant concepts for one or more applications of this field. For that, ontologies have become essential tools for information representation and processing at the semantic level. But, dynamically changing environments imply changes in the conceptualization of a domain that are reflected on the underlying domain ontology. The latter represents a structure capturing semantic knowledge about a given domain by describing relevant concepts and relations between them. The following section focus on the ontology evolution and consistency.

ONTOLOGY EVOLUTION AND VERSIONING

An important characteristic of today's systems is their ability to adapt themselves efficiently to the changes in their environment, as well as to the changes in their internal structures and processes. However, building and maintaining long-living applications that will be "open for changes" is still a challenge for the entire software engineering community (Stojanovic, 2004).

Ontology-based applications are subject to a continual change. Thus, to improve the speed and to reduce costs of their modification, the changes have to be reflected on the underlying ontology. Moreover, as ontologies grow in size, the complexity of change management increases significantly. If the underlying ontology is not up-to-date, then the reliability, accuracy and effectiveness of the system decrease significantly (Klein et al., 2001).

Particularly, there are three challenges for the efficient realization of the ontology evolution (Stojanovic, 2004):

- **Complexity***:* an ontology model is rich and, therefore, an ontology has an interwoven structure. Each change leads to a change specific workaround. Even when the effects of a change are minor, the cumulative effect of all changes realizing a user's request can be enormous;
- **Dependencies***:* ontologies often reuse and extend other ontologies. Changes in an ontology may affect the ontologies that are based on it. Therefore, changes between dependent ontologies are interrelated, and the immediate synchronization between dependent ontologies is required. Obviously, the complexity of the ontology evolution increases with the number of dependent ontologies being evolved;
- **Physical distribution***: o*ntology development is a decentralized and collaborative process. Therefore, the physical distribution of the dependent ontologies has to be taken into account. The ontology evolution requires tracking the changes applied to ontology and broadcasting the group of changes when an explicit request arises.

FROM SCHEMA VERSIONING TO ONTOLOGY VERSIONING

In the setting of Software engineering, ontology evolution has many relationships and overlaps with software evolution research (De Leenheer et al., 2008). Madhavji et al. (2006) explore what software evolution is and why it is inevitable. They address the phenomenological and technological underpinnings of software evolution and explain the role of feedback in software maintenance. Mens et al., (2007) present the state of the art and emerging topics in software evolution research.

Temporal databases support time-varying information and maintain the history of the modeled data (Özsoyoğlu et al., 1995) (Tansel et al., 1993). They allow the maintenance of data histories though the support of time semantics at system level. Several temporal data models were proposed as extensions of the relational data model. Clifford et al. (Clifford et al., 1995) classified them into two main categories: temporally ungrouped and temporally grouped. In temporally ungrouped models, the temporal representation is realized at extensional level, by means of timestamps added to data values as additional attributes to represent their temporal pertinence. In temporally grouped models, the temporal dimension is implicit in the structure of data representation: attributes are represented as histories considered as a whole and without the introduction of distinguished attributes. Attribute histories can be regarded as functions which map time into attribute domains. The second representation is said to have more expressive power and to be more natural since it is history-oriented (Clifford et al., 1995).

In information systems, ontologies are large and complex structures holding information. They are like database schema and need to change every time the modeled real world has changed. Thus, schema versioning in databases can be useful in order to propose an approach for ontology versioning which can benefit from principles and tools developed for schema versioning in temporal databases.

Schema versioning is a powerful technique not only to ensure reuse of data and continued support of legacy applications after schema changes, but also to add a new degree of freedom to database designers and users. In fact, different schema versions actually allow representing different points of view over the modeled application reality.

According to Roddick (1995), a *schema evolution* is the ability to change a schema of a populated database without loss of data, the latter which means providing access to both old and new data through the new schema. A *schema versioning* is the ability to access all the data through user-definable version interfaces. A version is a reference that labels a quiet point in the definition of a schema.

A schema describes the structure of data that are stored in a database. Whereas careful and accurate the initial design may have been, a database schema is likely to undergo changes and revisions after implementation. In order to avoid the loss of data after schema changes, many database management systems support schema evolution, which provides (partial) automatic recovery of the extent data by adapting them to the new schema. However, if only the updated schema is retained, all the applications compiled with the past schema may cease to work. In order to let applications work on multiple schemata, schema evolution is not sufficient and the maintenance of several schemas is required. This leads to the notion of schema versioning. A database management system supports schema versioning if it allows (1) the modification of the database schema without loss of existing data, and (2) the accessing of all data through user-definable version interfaces.

Most proposals deal with schema versions as natural extensions of studies on temporal data modeling and temporal databases. The same considerations on temporal dimensions applied for data can be applied at schema level. Schema versioning can then be done along one or two time dimensions. Thus, three temporal schema versioning mechanisms were defined in (De Castro et al., 1997): transaction-time schema versioning, valid-time schema versioning and bi-temporal schema versioning. Another temporal schema versioning mechanism is identified in (Brahmia et al., 2008): the application-time schema versioning.

The schema evolution in the relational databases is the starting point for all evolution issues. The standard SQL DDL (Data Definition Language) allows changes in the table definition (e.g. adding attributes). However, the related consistency problems have not been considered so far and are managed manually by the administrators, e.g. by using SQL UPDATE queries (Stojanovic, 2004).

The object-oriented database models, as an extension of the relational database evolution, provide a semantically richer model than the relational one. They are also more similar to the ontology models since they take into account the inheritance hierarchies. For these reasons, the object-oriented schema evolution seems more relevant for the ontology evolution.

However, there are many differences between ontology engineering and object-oriented modeling. According to (Stojanovic, 2004), an ontology reflects the structure of the world, it is often about the structure of concepts; besides, the actual physical representation is not an issue. On the other hand, an object-oriented structure reflects the structure of the data and code. It is usually about behavior since the integral part of an object-oriented model comprises methods. The physical representation of data (int, char, etc.) is also a part of a model.

Many approaches deal with the object-oriented schema evolution issue (Banerjee et al., 1987) (Ferrandina et al., 1996) (Huersch, 1997). They address two main questions: the effects of a schema change on the schema itself and on the underlying instances. The first problem is resolved either by defining rules that must be followed to maintain the schema consistency (Banerjee et al., 1987) (Zicari, 1991) or by introducing axioms, with an inference mechanism, that guarantee the consistency (Peters et al., 1997). The second problem is how to propagate the changes to the instances and can be solved based on (i) data migration for the immediate adaptation of the existing instances to the changed schema; (ii) mechanisms for the synchronization between data and schema. According to (Stojanovic, 2004), the synchronization mechanisms are realized as: (a) delayed conversion, when instances are only converted on demand; (b) screening, when changes are propagated via deferred object conversion; (c) versioning, when changes are never propagated and objects are indeed assigned to different schemas.

Although the issues in schema evolution are not entirely the same as in ontology evolution, the philosophy and results from schema evolution in generall have been fruitfully reconsidered for the treatment of the ontology evolution problem. The resemblances and differences between ontologies and data models are widely discussed in literature such as in Meersman (2001), Spyns et al. (2002), and Noy and Klein (2004). The basic argumentation behind comparing ontologies and data schemas is that (i) formally, all such kinds of formal artefacts are lexically represented by sets of predicates (data models); and (ii) they describe some domain by means of conceptual entities and relationships in a (not necessarily) shared formal language (Meersman, 2001).

ONTOLOGY EVOLUTION AND VERSIONING

The ontology can be defined as a formal representation based on the identification of conceptual entities (concepts or relationships) and their semantics. Since ontology has to be continually changed, it is important to take into account the ontology evolution. The task of the ontology evolution is to formally interpret all changes' requests coming from different sources (e.g. users, internal processes, business environment) and to perform them on the ontology, and its depending artifacts, while keeping consistency of all of them.

- Ontology is an explicit representation of knowledge related to a domain of study and a particular context. The application of changes in its conceptual entities is a modification of a subset of knowledge represented by the ontology. The application of changes requires defining the mechanisms specifying how knowledge can be changed and how to maintain the consistency of knowledge after each change.

- Ontology evolution is the process of adaptation of ontology to evolution changes and the consistent management of these changes to guarantee the consistency of ontology when changes occur (Klein et al., 2001). It encompasses the set of activities, both technical and managerial, which ensures that ontology continues to meet organizational objectives and users needs in an efficient and effective way (Stojanovic, 2004).
- The adaptation of ontology to evolution changes is a complex process from which several problems must be managed: identification of evolution changes, analysis of effects of changes, management of the ontology consistency, storage of ontology versions, etc.
- Main results in ontology evolution have been reported by Oliver et al. (1999), Heflin (2001), Klein et al. (2002), Stojanovic et al. (2002 ; 2003), Schlobach et al. (2003), Maedche et al. (2003), Rogozan et al. (2004), Haase et al. (2005), Flouris et al. (2005), Plessers (2006), Luong et al. (2007), Djedidi et al. (2008), Jaziri (2009) and Sassi et al. (2009a, 2009b).

ONTOLOGY VERSIONING

Klein et al. (2001) define ontology versioning as the ability to handle changes in ontologies by creating and managing different variants of it. In fact, ontology versioning implies the preservation of the old and new versions of ontology by giving transparent access to these various versions. It is thus necessary to identify the relationships between versions and their ontological entities (concepts, relationships and properties). By using these relationships, it becomes easy to identify the modifications between various versions.

Ontology versioning typically involves the storage of several ontology versions and takes into account identification issues (i.e., how to identify the different versions), the relationship between different versions (i.e., a tree of versions resulting from the various ontology modifications) as well as compatibility information (i.e., information regarding the compatibility of any pair of ontology versions).

Klein and Noy (Klein, 2002) (Noy et al., 2004) use the term versioning to describe their approach of ontology change. They define ontology versioning as the ability to manage ontology changes and their effects by creating and maintaining different variants of the ontology.

Adequate methods and tools must be used to distinguish and identify the versions, in combination with procedures of ontology change and a mechanism of interpretation of their effects on the ontology (Klein, 2002).

According to these same authors, a methodology of versioning must provide a mechanism to clear up the interpretation of the concepts for the users of the various versions of ontology, and must make explicit the effects of changes on the various tasks. Their methodology provides methods to distinguish and recognize versions, procedures for updates and changes in ontologies, and an interpretation mechanism for effects of change. The authors compared ontology evolution with database schema evolution. The framework they proposed contains a set of operators, on the form of ontology, useful for modifying another evolving ontology. Klein (2001) also proposes a change specification language based on the ontology of change operations.

A versioning methodology gives access to the data and concepts modeled via various versions of ontology. To control these versions, it is necessary to control the links of derivation between them. The derivation links allow defining and checking the compatibility between versions and following the transformations of data from a version to another one (Haase et al., 2004).

TOOLS TO SUPPORT THE ONOTOLOGY EVOLUTION

Ontology evolution can be defined as the timely adaptation of ontology and a consistent management of changes. The complexity of ontology evolution increases as ontologies grow in size (Haase et al., 2005).

There exist numerous scientific and commercial tools for creating and managing ontologies which have been used to build applications in several domains such as knowledge engineering, semantic web, information retrieval, database design, data warehousing etc.

KAON[18] (Oberle et al., 2004) (Psyché et al., 2003) is an open source suite of tools for ontology management. KAON uses logs of changes at the granularity of single knowledge-base operations as versioning information. It provides a comprehensive implementation allowing easy ontology management and application. Its ontology API (KAON API) consists of a set of interfaces for access to ontology entities (concept, property and instance). KAON consists of a number of different modules providing a broad bandwidth of functionalities centered on creation, storage, retrieval, maintenance and application of ontologies. The Core of KAON supports programmatic access to ontologies by including both *APIs* and *implementations* for managing local and remote ontology repositories. KAON saves in a journal all changes made during the evolution of ontology. In the journal, changes are described using the concept ChangeLog which specifies only basic types of changes: AddEnitity or DeleteEntity or ModifyEntity. Therefore, the track changes provided by the journal may not reflect the complex changes, for example merge or separate ontological entities, which is an important limitation. Moreover, KAON records changes in the order of their execution: this requires additional analysis that is, for example, a group changes depending on their type or function of the entities in which they operate. A more detailed technical description of the KAON components can be found in (Gabel et al., 2004).

OntoView (Klein et al., 2002) is a web-based system that helps users to manage changes in ontologies. It allows keeping different versions of web-based ontologies interoperable, by maintaining transformations between ontology versions and conceptual relationships between their concepts. OntoView compares versions of ontology at the structural level, i.e. the level of concepts and definitions of properties to find definitions of a modified version to another.

Unlike KAON, OntoView requires no additional information about the evolutionary process of ontology. This represents a great advantage in the context of the Semantic Web, where it is difficult to predict a log of changes for each version of ontology. OntoView allows users to specify that a change has altered the meaning of an ontological entity, if the entity is conceptually different from an old version, or if a change has only enriched the significance of an ontological entity which remains identical with that of the old version. Nevertheless, it has no feature to characterize the relationship between semantic ontological entities of two versions in terms of equivalence, inclusion or specification / generalization.

OntoManager (Stojanovic et al., 2003) is a tool for guiding ontology managers through the modification of ontology with respect to users' needs. It is based on the analysis of users' interactions with the ontology-based applications, which are tracked in a usage-log. OntoManager has been designed to provide the methods and tools that support the ontology managers in managing and optimizing the ontology according to the users' needs. The system incorporates mechanisms that assess how the ontology (and by extension the application) is performing based on different criteria, and then enable to take action to optimize it. One of the key tasks is to check how the ontology fulfils the perceived needs of the users. In that way, an in-depth view of the users' perspective on the ontology and the ontology-based application is obtained, since on the top of this ontology the application is going to be conducted. The technique that can be used to evaluate/estimate the user needs depends on the information source. By tracking users' interactions with the application in a log file, it is possible to collect useful information that can be used to assess what the main interests of the users are. In this way, it is avoided to ask the users explicitly, since they tend to be reluctant to provide the feedback via filling questionnaires or forms.

TextToOnto (Maedche et al., 2001) (Maedche et al., 2003) is a tool suite built upon KAON which is based on text mining techniques to support the ontology engineering process by text mining techniques. TextToOnto provides a collection of independent tools for both automatic and semi-automatic ontology extraction and assists users in creating and extending OIModels. Moreover, efficient support for ontology maintenance is given by modules for ontology pruning and comparison. TextToOnto does not allow mapping textual changes to the ontology. The current distribution of TextToOnto comprises the following tools:

- TaxoBuilder for building concept hierarchies;
- TermExtraction for adding concepts to an ontology;
- InstanceExtraction for adding instances to an ontology;
- RelationExtraction for semi-automatic learning of conceptual relations;
- RelationLearning for automatic *and* semi-automatic relation learning;
- OntologyComparison for comparing two ontologies;
- OntologyPruner for adapting an ontology to a domain-specific corpus.

SHOE (Heflin et al., 2000) (Haase et al., 2004) extends HTML with a set of knowledge oriented tags. Unlike HTML tags, SHOE provides a structure for knowledge acquisition as opposed to information presentation. SHOE associates meaning with the content by making each web page commit to one or more ontologies. These ontologies allow the discovery of implicit knowledge based on taxonomies and inference rules. The syntax of SHOE is defined as an application of SGML, a language that defines tag-based languages and was the influence for HTML's syntax. A slight variant of the syntax exists for compatibility with XML.

PromptDiff (Noy et al., 2004) is an ontology versioning tool that serves as an important role for maintaining ontology views and mappings between ontologies. It automatically performs structural comparisons of ontology versions without needing a trace of the evolutionary process based on an ontology comparison API which can be used by other applications. It identifies both simple and complex changes, presents the comparison results to the user in an intuitive way, and enables the user to accept or reject the changes between versions. However, this tool provides no information about the ontological changes. It only indicates whether the entities have been changed or not.

Protégé is an open source tool based on a graphical environment for developing ontology. It has evolved since its first version (Protégé-2000) to integrate from 2003, the semantic web standards including OWL[19](Ontology Web Language). This tool has many optional components such as graphical interfaces. The knowledge model of Protégé-2000 is based on the model of frames and contains classes (concepts), slots (properties) and facets (property values and constraints), as well as instances of classes and properties. Many plug-in are available and can be added by users. The three types of objects in Protégé: classes, attributes, properties and the graphical interface allow us to express the ontology properties without using a formal language. The ontology specification is made automatically with the OWL syntax. Protégé can also test uniformities of new data with ontology, reuse other existing ontologies etc.

Moreover, the field of system engineering offers a number of techniques and tools for versioning, merging and evolving software artefacts, and many of these techniques can be reused in an ontology engineering setting (De Leenheer et al., 2008). For managing the evolution of ontologies, established techniques from data schema evolution have been successfully adopted, and consensus on a general ontology evolution process model seems to emerge.

Ontology technology is nowadays mature enough: many methodologies, tools and languages are already available. However, most of the existing systems for the ontology development provide only one possibility for realizing a change, and this is usually the simplest one. For example, the deletion of a concept always causes the deletion of all its sub- concepts. It means that users are not able to control the way changes are performed (supervision).

The analysis of the developed approaches shows that no complete framework for managing ontology coherence is proposed. Except for Sassi et al. (2009) who propose an anticipatory approach acting before the appearance of inconsistencies, the majority of the proposed approaches are based on the correction of inconsistencies after they occur.

CONCLUSION

Until recently, most data interoperability techniques involved central components, e.g., global schemas or ontologies, to overcome semantic heterogeneity for enabling transparent access to heterogeneous data sources.

The research in knowledge representation and knowledge engineering, although successful, failed to provide by themselves cost-effective and time-effective shareable knowledge, so that different knowledge-based systems to be able to communicate between each other (Neches et al., 1991) (Gruber, 1993) (Gruber, 1995) (Guarino et al., 1997) (Guarino, 1998). A solution to this deadlock is to use ontologies for describing concepts and relations assumed to be always true independent from a particular domain by a community of humans and/or agents that commit to that view of the world (Guarino, 1997). In this way, ontologies may be viewed as knowledge bases integration mechanisms with the condition of an agreed upon vocabulary used. Therefore, while ontology is a kind of a shallow knowledge base from the specific domain integration point of view, a generic knowledge base may also contain background information explicitly and implicitly within its structure, describing a particular instantiation or state of affairs (Guarino, 1997).

A large body of research is being moving around ontologies, and contributions have been produced regarding methods and tools for covering the entire ontology life cycle, from design to deployment and reuse.

Ontologies are a keystone of knowledge representation and sharing and play an important role in the new generation of information systems. Because of the encouraging results as well as the potential positive outcomes, ontologies have widened beyond the boundaries of Artificial Intelligence, in domains such as Database Theory and Computational Linguistics (Guarino, 1998).

The role of ontologies is to represent knowledge in such a way so as to make possible the communication between different machines and between machines and humans to a knowledge level contrasting to communication at data level (as happen these days).

This study treats many ontology aspects like design, evolution and contextualization. The main methodologies, tools and languages for building, updating and representing ontologies that have been reported in the literature have been presented. However, other aspects will be proposed to ensure a good ontology quality by defining quality measures. This study shows that developing ontologies is expensive, but evolving them is even more expensive. Ontologies must be able to evolve because application domains and users' needs are changing as well as the developed system can be improved.

The future work in this field should be driven towards the creation of a common integrated workbench for ontology developers to facilitate ontology development, exchange, evaluation, evolution and management, to provide methodological support for these tasks, and translations to and from different ontology languages. This workbench should not be created from scratch, but instead integrating the technology components that are currently available.

Because of some particular terminological and semantic issues raised by the concept of ontology, a special attention has to be given to its interpretation regarding multi-contexts situations.

In addition, more powerful supports for designing, versioning and merging are required in order to allow designers, users and experts to collaboratively build and manage ontologies in increasingly complex applications requiring knowledge sharing and reuse. It turns out that much can be learned from other domains where formal artifacts are being collaboratively engineered. In particular, the fields of system and software engineering as well as information systems modeling offer a splush of techniques and tools for versioning, merging and evolving software artifacts (Mens, 2002), and many of these techniques can be reused in an ontology engineering setting.

REFERENCES

Agirre, E., Ansa, O., Hovy, E., & Martinez, D. (2000). Enriching very large ontologies using the WWW. In *Proceedings of the Workshop on Ontology Construction of the European Conference of AI (ECAI-00)*.

Alfonseca, E., & Manandhar, S. (2002). An unsupervised method for general named entity recognition and automated concept discovery. In *Proceedings of the 1st International Conference on General WordNet*, Mysore, India.

Arpirez-Vega, J. C., Gomez-Perez, A., Lozano-Tello, A., & Sofia Pinto, H. (1998). (ONTO)2Agent: an ontology-based WWW broker to select ontologies. In *Proceedings of the workshop on Applications of Ontologies and Problem-Solving Methods*, (pp. 16–24).

Artala, A., Franconi, E., Guarino, N., & Pazzi, L. (1996). Part-Whole Relations in Object-Centered Systems: an Overview. *Data & Knowledge Engineering, 20*(3), 347–383. doi:10.1016/S0169-023X(96)00013-4

Assadi, H., & Bourigault, D. (2000). Analyses syntaxique et statistique pour la construction d'ontologies à partir de textes, Ingénierie des connaissances, évolution récentes et nouveaux défis, chapter 15, Eyrolles, collection technique et scientifique des Télécommunications, (pp. 243-255).

Aussenac-Gilles, N., Biébow, B., & Szulman, S. (2000). Corpus Analysis For Conceptual Modelling, Workshop on Ontologies and Text, Knowledge Engineering and Knowledge Management: Methods, Models and Tools. In *12th International Conference EKAW'2000*, Juan-les-pins, France. Berlin: Springer- Verlag.

Aussenac-Gilles, N., Biébow, B., & Szulman, S. (2000). Revisiting Ontology Design: A Methodology Based on Corpus Analysis (LNCS, pp.172–188). Berlin: Springer-Verlag.

Baader, F., Calvanese, D., McGuinness, D. L., Nardi, D., & Patel Schneider, P. F. (Eds.). (2007). *The Description Logic Handbook: Theory, Implementation, and Applications*. Cambridge, UK: Cambridge University Press.

Bachimont, B. (2000). Engagement sémantique et engagement ontologique: conception et réalisation d'ontologies en ingénierie des connaissances. In J. Charlet, M. Zacklad, G. Kassel, D. Bourigault (Ed.), Ingénierie des connaissances: évolutions récentes et nouveaux défis, Eyrolles, (pp. 305-323).

Bachimont, B., Isaac, A., & Troncy, R. (2002). Semantic commitment for designing ontologies: a proposal. In A. Gomez-Perez & V.R. Benjamins (Eds.), EKAW 2002, (LNAI 2473, pp. 114–121).

Banerjee, J., Kim, W., Kim, H. J., & Korth, H. (1987). Semantics and implementation of schema evolution in object-oriented databases. In *Proceedings of the Annual Conference on Management of Data (ACM SIGMOD 16*(3)), San Francisco, (pp. 311-322).

Batini, C., Lenzerini, M., & Navathe, S. B. (1986). A comparative analysis of methodologies for database schema integration. *ACM Computing Surveys*, *15*, 323–363. doi:10.1145/27633.27634

Benslimane, D., Arara, A., Yetongnon, K., Gargouri, F., & Ben Abdallah, H. (2003). Two approaches for ontologies building: From-scratch and From existing data sources. In *the 2003 International Conference on Information Systems and Engineering (ISE 2003)*, Montreal, Canada.

Bergamaschi, S., Castano, S., Vimercati, S., & Vincini, M. (1998). *An Intelligent Approach to Information Integration, Formal Ontology in Information System*. Amsterdam: IOS Press.

Bernaras, A., Laresgoiti, I., & Corera, J. (1996). Building and reusing ontologies for electrical network applications. In *Proceedings of the European Conference on Artificial Intelligence (ECAI_96)*, Budapest, Hungary, (pp. 298–302).

Biébow, B., & Szulman, S. (1999). TERMINAE: a linguistic-based tool for the building of a domain ontology. In *Proceedings of the 11th European Workshop on Knowledge Acquisition, Modelling and Management*, (pp. 49-66), Germany.

Blazquez, M., Fernandez, M., Garcia-Pinar, J. M., & Goméz-Pérez, A. (1998). Building ontologies at the knowledge level using the ontology design environment. In *Proceedings of the Eleventh Workshop on Knowledge Acquisition, Modeling and Management*, Banff, Canada, Borgida A., (1995). Description Logics in Data Management. *IEEE Transactions, Knowledge and Data Engineering, 7*(5), 671-682.

Borst, W. N. (1997). *Construction of Engineering Ontologies*. PhD Thesis, University of Tweenty, Enschede, The Netherlands, Centre for Telematica and Information Technology.

Brahmia, Z., & Bouaziz, R. (2008). Schema Versioning in Multi-Temporal XML Databases. In *Proceedings of the 7th IEE/ACIS International Conference on Computer and Information Science (IEEE/ACIS ICIS 2008)*, (pp. 158-164), Oregon, OR.

Brézillon, P. (1999). Context in problem solving: A survey. *The Knowledge Engineering Review, 14*(1), 1–34. doi:10.1017/S0269888999141018

Brézillon, P., & Pomerol, J.-C. (2001). *Some comments about knowledge and context.* Research Report 2001-022, LIP6, University of Paris VI, Paris, France, Brézillon, P. (2002). *Modeling and using context: Past, present and future.* Research Report, LIP6 2002/010, University of Paris 6, France, Châabane, S. & Jaziri, W. (2009). Méta-modélisation du processus de construction d'ontologies géographiques: Application au domaine routier. In *Proceedings of ICWIT '2009*, Kerkennah, Tunisia, (pp. 653-661).

Chandrasekaran, B., Josephson, J. R., & Benjamins, R. V. (1999). What are ontologies, and why do we need them? *IEEE Intelligent Systems and Their Applications*, *14*(1), 20–26.

Chikofsky, E. J., & Cross, J. H. (1990). Reverse engineering and design recovery: a taxonomy. *IEEE Software*, 13–17. doi:10.1109/52.43044

Chira, O. (2003). Ontologies. *IDIMS Report.*

Corcho, O., Fernández-López, M., & Gómez, P. A. (2006). *Ontological Engineering: Principles, Methods, Tools and Languages, Ontologies for Software Engineering and Software Technology* (pp. 1–48). Berlin: Springer.

Corcho, O., Fernandez-Lopez, M., & Gomez-Perez, A. (2003). Methodologies, tools and languages for building ontologies: Where is their meeting point? *Data & Knowledge Engineering*, *46*, 41–64. doi:10.1016/S0169-023X(02)00195-7

Corcho, O., & Gomez-Perez, A. (2000). A roadmap to ontology specification languages. In *Knowledge engineering and knowledge Management Methods* (pp. 80–96). Models and Tools. doi:10.1007/3-540-39967-4_7

De Leenheer, P., & Mens, T. (2008). Ontology Evolution: State of the Art and Future Directions. In Hepp, M., De Leenheer, P., de Moor, A., & Sure, Y. (Eds.), *Ontology Management for the Semantic Web, Semantic Web Services, and Business Applications, from Semantic Web and Beyond: Computing for Human Experience*. Berlin: Springer.

Desmontils, E., & Jacquin, C. (2001). Indexing a Web Site with a Terminology Oriented Ontology. In *International Semantic Web Working Symposium (SWWS'2001),* Stanford, CA, (pp. 549-565).

Devambu, P., Brachman, R. J., Selfridge, P. J., & Ballard, B. W. (1991). LASSIE: A knowledge based software information system. *Communications of the ACM, 34*(5), 36–49.

Dey, A. K., & Abowd, G. D. (1999). *Towards a Better Understanding of Context and Context-Awareness. Technical report git-gvu-99-22.* Georgia: Institute of Technology.

Dey, N., Boucher, A., & Thonnat, M. (2002). Image formation model of 3D translucent object observed in light microscopy. In *Proceedings of ICIP'02*, (Vol. 2, pp. 469-472).

Djedidi, R. & Aufaure, M-A. (2008). Enrichissement d'ontologies: maintenance de la consistance et évaluation de la qualité. *Journées Francophones sur l'Ingénierie des connaissances (IC'2008).*

Faatz, A., & Steinmetz, R. (2002). Ontology enrichment with texts from the WWW. In *Semantic Web Mining*. Helsinki, Finland: Second Workshop at ECML/PKDD.

Farquhar, A., Fikes, R., & Rice, J. (1997). Tools for Assembling Modular Ontologies in Ontolingua. In [Menlo Park, CA: AAAI Press.]. *Proceedings of AAAI, 97*, 436–441.

Fensel, D. (2000). *Ontologies: A Silver Bullet for Knowledge Management and Electronic Commerce*. Berlin: Springer.

Fernandez-Lopez, M. (1999). Overview of Methodologies for Building Ontologies. In *IJCAI'99 Workshop on Ontologies and Problem-Solving Methods: Lessons Learned and Future Trends*, Stockholm, Sweden, Fernandez-Lopez, M., Gomez-Perez, A., Pazos-Sierra, A. & Pazos-Sierra, J. (1999). Building a chemical ontology using METHONTOLOGY and the ontology design environment. *IEEE Intelligent Systems & their applications, 4* (1), 37–46.

Fernandez-Lopez, M. (2001). Overview of Methodologies for Building Ontologies. *Intelligent Systems*, *16*(1), 26–34.

Fernandez-Lopez, M., Gomez-Perez, A., & Juristo, N. (1997). Methontology: From Ontological Art Toward Ontological Engineering. In *Spring Symposium Series on Ontological Engineering, AAAI97*, Stanford, CA.

Ferrandina, F., & Lautemann, S. E. (1996). An integrated approach to schema evolution for object databases. In *Proceedings of the Internation Conference on Object Oriented Information Systems (OOIS 1996)*, (pp. 280-294), London.

Fikes, R., & Farquhar, A. (1999). Distributed Repositories of Highly Expressive Reusable Ontologies. *IEEE Intelligent Systems*, *14*(2), 73–79. doi:10.1109/5254.757634

Flouris, G. & Plexousakis, D. (2005). *Handling ontology change: survey and proposal for a future research direction.* Rapport technique FORTH-ICS/TR-362.

Fonseca, F. (2007). The Double Role of Ontologies in Information Science Research. *Journal of the American Society for Information Science and Technology*, *58*(6), 786–793. doi:10.1002/asi.20565

Freitas, V. (2000). *Autoria Adaptativa de Hipermídia Educacional.* Master's thesis, Instituto de Informática, UFRGS, Porto Alegre, National Library Of Medicine.

Gabel, T., Sure, Y., & Voelker, J. (2004). *KAON – ontology management infrastructure.* SEKT informal deliverable 3.1.1.a, Institute AIFB, University of Karlsruhe, Gaines, B. (1997). Editorial: Using Explicit Ontologies in Knowledge-based System Development. *International Journal of Human-Computer Systems*, *46*(2-3), 181.

Gandon, F., & Dieng-Kuntz, R. (2001). Ontologie pour un système multi-agents dédié à une mémoire d'entreprise. In *Proceedings of the journées francophones d'Ingénierie des Connaissances IC'2001*, (pp. 1-20). Grenoble, France: Presses Universitaires de Grenoble.

Genesereth, M. R., & Nilsson, N. J. (1987). *Logical Foundations of Artificial Inteligence.* San Francisco: Morgan Kaufmann Publishers, Goh, C.H., Bressan, S., Madnick, S. & Siegel, M. (1999). Context Interchange: New Features and Formalisms for the Intelligent Integration of Information. *ACM Transactions on Information Systems*, *17*(3), 270–293.

Gomez-Perez, A. (1996). A framework to verify knowledge sharing technology. *Expert Systems with Applications*, *11*(4), 519–529. doi:10.1016/S0957-4174(96)00067-X

Gomez-Perez, A. (1998). Knowledge sharing and reuse. In Liebowitz, J. (Ed.), *Handbook of Expert Systems*. New York: CRC.

Gomez-Perez, A. (2001). Evaluation of ontologies. *International Journal of Intelligent Systems*, *16*(3), 1–10. doi:10.1002/1098-111X(200103)16:3<391::AID-INT1014>3.0.CO;2-2

Gomez-Perez, A., Fernandez-Lopez, M., & De Vicente, A. (1996). Towards a Method to Conceptualize Domain Ontologies. In *ECAI96 Workshop on Ontological Engineering*, (pp. 41–51), Budapest, Hungary.

Gomez-Perez, A., & Rojas, M. D. (1999). Ontological reengineering and reuse. In D. Fensel, R. Studer (Eds.), *11th European Workshop on Knowledge Acquisition, Modeling and Management* (EKAW_99, LNAI, Vol. 1621, pp. 139–156). Berlin: Springer.

González-Pérez, C., & Henderson-Sellers, B. (2006). An Ontology for Software Development Methodologies and Endeavours. In *Ontologies for Software Engineering and Software Technology* (pp. 123–151). Berlin: Springer. doi:10.1007/3-540-34518-3_4

Gruber, T. (1991). The role of a common ontology in achieving sharable, reusable knowledge bases. In *Proceedings of the 2nd International Conference on Principles of Knowledge Representation and Reasoning*, Cambridge, UK, (pp. 601-602).

Gruber, T. (1993). A translation approach to portable ontology specifications. *Knowledge Acquisition, 5*, 99–220. doi:10.1006/knac.1993.1008

Gruber, T. (1995). A translation approach to portable ontology specification. *Knowledge Acquisition, 5*(2), 199–220. doi:10.1006/knac.1993.1008

Gruninger, M., & Fox, M. S. (1995). Methodology for the design and evaluation of ontologies. In *Workshop on Basic Ontological Issues in Knowledge Sharing*, Montreal, Canada.

Gruninger, M., & Lee, J. (2002). Ontology Applications and Design. *Communications of the ACM, 45*(2), 1–2.

Guarino, N. (1997). Understanding, Building and Using Ontologies: A Commentary to Using Explicit Ontologies in KBS Development. *International Journal of Human-Computer Studies, 46*, 293–310. doi:10.1006/ijhc.1996.0091

Guarino, N. (1998). Formal Ontology and Information Systems. In Guarino, N. (Ed.), *Formal Ontology and Information Systems*. Amsterdam: IOS Press.

Guarino, N., Carrara, M., & Giaretta, P. (1995). Ontologies and knowledge bases: towards a terminological clarification. In Mars, N. (Ed.), *Towards Very Large Knowledge Bases, Knowledge Building and Knowledge Sharing* (pp. 25–32). Amsterdam: IOS Press.

Guarino, N., & Welty, C. (2000). Ontological analysis of taxonomic relationships. In *19th International Conference on Conceptual Modeling (ER_00),* (LNCS, Vol. 1920, pp. 210–224). Berlin: Springer.

Guarino, N., & Welty, C. (2000). A formal ontology of properties. In *12th International Conference in Knowledge Engineering and Knowledge Management (EKAW_00)*, (LNAI, Vol. 1937, pp. 97–112). Berlin: Springer.

Gupta, K. M., Aha, D. W., Marsh, E., & Maney, T. (2002). An architecture for engineering sublanguage WordNets. In *Proceedings of the First International Conference On Global WordNet*, (pp. 207-215). Mysore, India: Central Institute of Indian Languages.

Haase, P., & Stojanovic, L. (2005). Consistent Evolution of OWL Ontologies, (LNCS, Vol. 3532, pp. 182-197). Berlin: Springer.

Haase, P., & Sure, Y. (2004). *State of the Art on Ontology Evolution*. Retrieved from http://www.aifb.uni-karlsruhe.de/WBS/ysupublications/SEKT-D3.1.1.b.pdf

Hahn, U., & Markó, K. (2001). Joint knowledge capture for grammars and ontologies. In *Proceedings of the First International Conference on Knowledge Capture K-CAP 2001*, Victoria, Canada.

Hahn, U., & Schnattinger, K. (1998). Towards text knowledge engineering. In *Proceedings of the 15th National Conference on Artificial Intelligence*, Madison, WI, (pp. 524-531). Menlo Park, Ca: AAAI Press / MIT Press.

Hearst, M. A. (1998). Automated Discovery of WordNet Relations. In Fellbaum, C. (Ed.), *WordNet: An Electronic Lexical Database* (pp. 132–152). Cambridge, MA: MIT Press.

Heflin, J. (n.d.). Towards the Semantic Web: Knowledge and representation in a dynamic, distributed environment. *Faculty of the Graduate School of the University of Maryland.*

Heflin, J., & Hendler, J. (2000). Dynamic Ontology on the Web. In *Proceedings of the 17th National Conference on Artificial Intelligence*, (pp. 443-449). Menlo Park, CA: AAAI/MIT.

Hovy, E. H., & Lin, C.-Y. (1999). Automated Text Summarization in SUMMARIST. In Maybury, M., & Mani, I. (Eds.), *Advances in Automatic Text Summarization*. Cambridge, MA: MIT Press.

Huersch, W. (1997). Maintaining consistency and behaviour of object-oriented systems during evolution. In *Proceedings of the ACM Conference on Object-Oriented Programming, Systems, Languages and Applications (OOPSLA '97), ACM SIGPLAN Notices, 32*(10), 1-21.

Huhns, M. N., & Singh, M. (1997). Ontologies for Agents. *IEEE Internet Computing, 1*, 81–83. doi:10.1109/4236.643942

Hwang, C. H. (1999). Incompletely and imprecisely speaking: Using dynamic ontologies for representing and retrieving information. In *Proceedings of the 6th International Workshop on Knowledge Representation meets Databases (KRDB'99)*, Linköping, Sweden, July 29-30.

Jasper, R. (1999). Uschold M., A Framework for Understanding and Classifying Ontology Applications. In *IJCAI-99 Workshop on Ontologies and Problem-Solving Methods (KRR5)*.

Jaziri, W. (2009). A methodology for ontology evolution and versioning. *The Third International Conference on Advances in Semantic Processing (SEMAPRO 2009)*, (pp. 15-21), Sliema, Malta.

Jones, D., Bench-Capon, T., & Visser, P. (1999). Methodologies for Ontology Development. In *Proceedings of the IJCAI-99 workshop on Ontologies and Problem -Solving Methods*.

Kalfoglou, Y., & Robertson, D. (1999). Managing Ontological Constraints. In *Proceedings of the IJCAI99 Workshop on Ontologies and Problem-Solving Methods: Lessons Learned and Future Trends*, Stockholm, Sweden.

Kalfoglou, Y., & Robertson, D. (1999). Use of formal ontologies to support error checking in specifications. In D. Fensel, R. Studer (Eds.), *11th European Workshop on Knowledge Acquisition, Modeling and Management (EKAW_99)*, (LNAI, Vol. 1621, pp. 207–224). Berlin: Springer.

Kashyap, V., & Sheth, A. (1996). Semantic and schematic similarities between database objects: a context-based approach. *The VLDB Journal, 5*, 276–304. doi:10.1007/s007780050029

Kashyap, V., & Sheth, A. (1998). Semantic heterogeneity in global information systems: the role of metadata, context and ontologies. In Papazoglou, M., & Schlageter, G. (Eds.), *Cooperative Information Systems: Current Trends and Directions* (pp. 139–178). New York: Academic Press.

Kassel, G. (2002). Ontospec: une méthode de spécification semi-informelle d'ontologies. In *Proceedings of the journées francophones d'Ingénierie des Connaissances (IC'2002)*, (pp. 75–87).

Khan, L., & Luo, F. (2002). Ontology Construction for Information Selection. In *Proceedings of the 14th IEEE International Conference on Tools with Artificial Intelligence*, (pp. 122-127). Washington DC.

Kietz, J. U., Maedche, A., & Volz, R. (2000). A method for semi-automatic ontology acquisition from a corporate intranet. In *EKAW_00 Workshop on Ontologies and Texts, CEUR Workshop Proceedings*, Juan-Les-Pins, (Vol. 51).

Klein, M. (2001). Combining and relating ontologies: an analysis of problems and solutions. In *Proceedings of the IJCAI-20001 Workshop on Ontologies and Information Sharing*, Seattle, WA.

Klein, M. (2002). *Versioning of distributed ontologies*. EU/IST Project WonderWeb.

Klein, M., & Fensel, D. (2001). Ontology versioning on the Semantic Web. In the First International Semantic Web Workshop (SWWS01), Stanford, CT.

Klein, M., Fensel, D., Kiryakov, A., & Ognyanov, D. (2002). Ontology versioning and change detection on the Web. In A. Gomez-Perez & V.R. Benjamins (Eds.), *13th International Conference on Knowledge Engineering and Knowledge Management (EKAW02),* (LNAI 2473). Berlin: Springer.

Kuck, J., & Reichartz, F. (2007). A collaborative and feature-based approach to context-sensitive service discovery. In *16th International World Wide Web Conference, Workshop on Emerging Applications for Wireless and Mobile Access*, Banff, Alberta, Canada.

Larson, J., Navathe, S., & Elmrasri, R. (1989). A theory of attribute equivalence in databases with application to schema integration. *IEEE Transactions on Software Engineering, 15*, 449–463. doi:10.1109/32.16605

Leclère, L., Trichet, T., & Frûst, F. (2002). Construction of an ontology related to the projective geometry. In RFIA 13th congrès des Reconnaissance des Frames et Intelligence Artificielle, France.

Lee, J., Madnick, S., & Siegel, M. (1996). Conceptualizing semantic interoperability: a perspective from the knowledge level. *International Journal of Cooperative Information Systems, 5*(4), 367–393. doi:10.1142/S0218843096000142

Lenat, D. B., & Guha, R. V. (1990). *Building Large Knowledge-Based Systems: Representation and Inference in the Cyc Project*. Boston: Addison-Wesley.

Lin, C.-Y., & Hovy, E. H. (2000). The Automated Acquisition of Topic Signatures for Text Summarization. In *Proceedings of the COLING Conference,* Strasbourg, France.

Lonsdale, D., Ding, Y., Embley, D. W., & Melby, A. (2002). Peppering Knowledge Sources with SALT, Boosting Conceptual Content for Ontology Generation. In *Proceedings of the AAAI Workshop on Semantic Web Meets Language Resources*, Edmonton, Alberta, Canada.

Luong, P. (2007). *Gestion de l'évolution d'un web sémantique d'entreprise.* PhD Thesis, Ecole des Mines de Paris, Paris.

Madhavji, N. H., Fernandez-Ramil, J., & Perry, D. E. (2006). *Software evolution and feedback: Theory and practice*. New York: Wiley. doi:10.1002/0470871822

Maedche, A., Motik, B., & Stojanovic, L. (2003). Managing multiple and distributed ontologies in the semantic web. *The VLDB Journal, 12*(4), 286–302. doi:10.1007/s00778-003-0102-4

Maedche, A., & Staab, S. (2003). Ontology Learning. In S. Staab & R. Studer (eds.), *Handbook on Ontologies in Information Systems*. Berlin: Springer. Retrieved from http://www.aifb.unikarlsruhe.de/WBS/sst/Research/Publications/handbook-ontology-learning.pdf

Maedche, A., & Volz, R. (2001). The ontology extraction and maintenance framework text-toonto. In *Proceedings of the ICDM'01 Workshop on Integrating Data Mining and Knowledge Management*.

McGuinness, D. (2002). Description Logics for Configuration. In F. Baader, D. McGuinness, D. Nardi, P. Patel-Schneider, (Ed.), *The Description Logic Handbook: Theory, Implementation, and Applications*. Cambridge, UK: Cambridge University Press. Retrieved from http://www.ksl.stanford.edu/people/dlm/papers/dlhb-configuration.pdf

Meersman, R. (2001). *Ontologies and databases: More than a fleeting resemblance*. Rome: OES/SEO Workshop.

Mena, E., Illarramendi, A., Kashyap, V. & Sheth, A. (2000). OBSERVER: An Approach for Query Processing in Global Information Systems based on Interoperation across Pre-existing Ontologies. *International journal on Distributed and Parallel Databases (DAPD), 8*(2), 223-272.

Mena, E., Kashyap, V., Illarramendi, A., & Sheth, A. (1998). *Domain Specific Ontologies for Semantic Information Brokering on the Global Information Infrastructure, Formal Ontology in Information Systems*. Amsterdam: IOS Press.

Mendes, O., & Abran, A. (2004). Software Engineering Ontology: A Development Methodology. *Metrics News, 9*(1), 68–76.

Mens, T. (2002). A State-of-the-Art Survey on Software Merging. *Transactions on Software Engineering, 28*(5), 449–462. doi:10.1109/TSE.2002.1000449

Mens, T., & Demeyer, S. (2007). *Software Evolution*. Berlin: Springer.

Mentzas, G. (2002). *Knowledge asset management: beyond the process-centered and product-centered approaches*. Berlin: Springer.

Mhiri, M. & Gargouri, F. (2009). Méthodologie de construction des ontologies pour la résolution de conflits des Systèmes d'Information. *Revue Technique et Science Informatiques, 28*.

Mhiri, M., Mtibaa, A., & Gargouri, F. (2005). OntoUML: Towards a language for the specification of information systems' ontologies. In T*he Seventeenth International Conference on Software Engineering and Knowledge Engineering,* Taipei, Taiwan – China, (pp. 743-746).

Minsky, M. (1975). A framework for representing knowledge. In The Psychology of Computer Vision, (pp. 211-277)

Missikoff, M., Navigli, R., & Velardi, P. (2002). The Usable Ontology: An Environment for Building and Assessing a Domain Ontology. In *International Semantic Web Conference (ISWC'2002)*, Sardinia, Italia.

Moldovan, D. I., & Girju, R. C. (2001). An interactive tool for the rapid development of knowledge Bases. [IJAIT]. *International Journal of Artificial Intelligence Tools, 10*(1-2).

Mtibaa, A., & Jaziri, W. (2009). Contexte: vision d'un acteur de son système d'information à travers une ontologie. In *Second International Conference on Web and Information Technologies (ICWIT '09),* (pp. 17-31), Kerkennah, Tunisia.

Mtibaa, A., Jaziri, W., & Gargouri, F. (2008). Proposition d'une extension de l'ontologie de domaine pour supporter la multitude de contexte lors de la spécification des besoins. *International Conference on Web and Information Technologies (ICWIT '08),* Sidi Bel Abbes, Algeria.

Muñoz, L. S. (2004). *Ontology-based Metadata for e-learning Content*. Porto Alegre: Master's Degree in Computer Science.

Navathe, S., & Gadgil, S. (1982). A methodology for view integration in logical database design. In *Eighth International Conference on very Large Data Bases*, Mexico City, Mexico.

Neches, R., Fikes, R. E., Finin, T., Gruber, T. R., Senator, T., & Swartout, W. R. (1991). Enabling technology for knowledge sharing. *AI Magazine*, *12*(3), 36–56.

Nobécourt, J. (2000). A method to build formal ontologies from text. In *EKAW-2000 Workshop on ontologies and text*, France.

Noy, N., & Klein, M. (2004). Ontology evolution: Not the same as schema evolution. *Knowledge and Information Systems*, *6*(4), 428–440. doi:10.1007/s10115-003-0137-2

Noy, N. F., & Hafner, C. D. (1997). *The State of the Art in Ontology Design - A Survey and Comparative Review* (pp. 53–74). Menlo Park, CA: AAAI.

Noy, N. F., & McGuinness, D. L. (2003). *Ontology Development 101: A Guide to Creating Your First Ontology* (p. 94305). Stanford, CA: Stanford University.

Noy, N. F., & Musen, M. A. (2000). PROMPT: Algorithm and Tool for Automated Ontology Merging and Alignment. In *17th National Conference on Artificial Intelligence (AAAI_00)*, Austin, TX.

Oberle, D., Volz, R., Motik, B., & Staab, S. (2004). An extensible ontology software environment. In Staab, S., & Studer, R. (Eds.), *Handbook on Ontologies* (pp. 311–333). Berlin: Springer.

Oliver, D. E., Shahar, Y., Musen, M., & Shortliffe, E. H. (1999). Representation of change in controlled medical terminologies. *AI in Medicine*, *15*(1), 53–76.

Ouksel, A., & Naiman, C. (1994). Coordinating context building in heterogeneous information systems. *Journal of Intelligent Information Systems*, *3*(2), 151–183. doi:10.1007/BF00962977

Peters, R. J., & Oezsu, M. (1997). An axiomatic model of dynamic schema evolution in object base management systems. *ACM Transactions on Database Systems*, *22*(1), 75–114. doi:10.1145/244810.244813

Pierra, G. (2005). Context-explication in conceptual ontologies: PLIB ontologies and their use for industrial data. *Journal of Advanced Manufacturing Systems.*

Pierra, G. (2008). Context representation in domain ontologies and its use for semantic integration of data. *Journal on Data Semantics*, 174-211.

Plessers, P. (2006). *An Approach to Web-based Ontology Evolution*. PhD Thesis, Department of Computer Science, Vrije Universiteit Brussel, Brussel, Belgium.

Psyché, V., Bourdeau, J., & Mizoguchi, R. (2003). Ontology Development at the Conceptual Level for Theory-Aware ITS Authoring Systems. Journal of AI in Education, Shaping the future of Learning through Intelligent Technologies, 491-493.

Quillian, M. (1967). Word Concepts: A Theory and Simulation of some Basic Semantic Capabilities. *Behavioral Science*, *12*, 410–430. doi:10.1002/bs.3830120511

Quillian, M. R. (1985). Word concepts: A theory and simulation of some basic semantic capabilities. In Brachman, R. J., & Levesque, H. J. (Eds.), *Readings in Knowledge Representation* (pp. 97–118). Los Altos, CA: Kaufmann.

Rifaieh, R. (2004). *Utilisation des ontologies contextuelles pour le partage sémantique entre les systèmes d'information dans le entreprises.* PhD Thesis, INSA- Lyon, France.

Rifaieh, R., Arara, A., & Benharkat, N. (2004). A view of Enterprise Information Systems based on Contextual Ontologies. *IEEE International Con ference on Computational Cybernetics (ICCC'2004).*

Roddick, J. (1995). A Survey of Schema Versioning Issues for Database Systems. *Information and Software Technology, 37*(7), 83–393. doi:10.1016/0950-5849(95)91494-K

Rogozan, D., Paquette, G., & Rosca, I. (2004). Gestion de l'évolution de l'ontologie utilisée comme référentiel sémantique dans un environnement de téléapprentissage. In *TICE International Symposium on Technologies de l'Information et de la Communication dans les Enseignements d'ingénieurs et dans l'industrie*, France.

Roux, C., Proux, D., Rechermann, F., & Julliard, L. (2000). An ontology enrichment method for a pragmatic information extraction system gathering data on genetic interactions. In *Proceedings of the ECAI2000 Workshop on Ontology Learning (OL'2000)*, Berlin, Germany, Sánchez, D.M, Cavero, J.M. & Marcos, E. (2005). On models and ontologies. In *First International Workshop on Philosophical Foundations of Information Systems Engineering*, Porto, Portugal.

Sassi, N., Jaziri, W., & Gargouri, F. (2009). Anticipatory approach to maintain consistency in ontology versions. In *The Third International Conference on Advances in Semantic Processing (SEMAPRO 2009)*, (pp. 7-14), Sliema, Malta.

Sassi, N., Jaziri, W., & Gargouri, F. (2009). How to evolve ontology and maintain its coherence: A corrective operations-based approach. In *The International Conference on Knowledge Engineering and Ontology Development (IC3K-KEOD 2009)*, Madeira, Portugal.

Schlobach, S., & Cornet, R. (2003). Non-standard reasoning services for the debugging of description logic terminologies. In *Proceedings of the 18th International Joint Conference on Artificial Intelligence (IJCAI-03).*

Schmidt, A., Aidoo, K., Takaluoma, A., Tuomela, P., Van Laerhoven, K., & Van de Velde, W. (1999). Advanced interaction in context. In *Proceedings of First International Symposium on Handheld and Ubiquitous Computing, HUC'99*, (pp. 89-101). Karlsruhe, Germany: Springer Verlag.

Schreiber, A., Wielinga, B., & Jansweijer, W. (1995). *The KACTUS view on the 'O'word. Technical Report, ESPRIT Project 8145 KACTUS.* The Netherlands: University of Amsterdam.

Sciore, E., Siegel, M., & Rosenthal, A. (1994). Using Semantic values to Facilitate Interoperability Among Heterogeneous Information Systems. *ACM Transactions on Database Systems, 19*(2), 254–290. doi:10.1145/176567.176570

Sheth, A. P. (1999). Changing Focus on Interoperability in Information Systems: from System, Syntax, Structure to Semantics. In Goodchild, M. F., Egenhofer, M. J., Fegeas, R., & Koffman, C. A. (Eds.), *Interoperating Geographic Information Systems*. Amsterdam: Kluwer Academic Publishers.

Sowa, J. (2000). *Knowledge Representation: Logical, Philosophical, and Computational Foundations.* Pacific Grove, CA: Brooks Cole Publishing Co. Retrieved from http://www.jfsowa.com/ontology/index.htm

Sowa, J. F. (1984). *Conceptual structures: information processing in mind and machine*. Boston: Addison-Wesley Longman Publishing Co., Inc.

Spaccapietra S., Parent C. & Dupont Y. (1992). Model Independent Assertions for Integration of Heterogeneous Schemas. *VLDB Journat*, 81-126.

Spaccapietra, S., Parent, C., Vangenot, C., & Cullot, N. (2004). On Using Conceptual Modeling for Ontologies. In *WISE 2004 Workshops, LNCS* (pp. 22–33). Berlin: Springer.

Spyns, P., & Meersman, R. (2002). Jarrar M., Data modelling versus ontology engineering. *SIGMOD Record*, *31*(4), 12–17. doi:10.1145/637411.637413

Staab, S., Schnurr, H. P., Studer, R., & Sure, Y. (2001). Knowledge processes and ontologies. *IEEE Intelligent Systems*, *16*(1), 26–34. doi:10.1109/5254.912382

Stojanovic, L. (2004). *Methods and Tools for Ontology Evolution*. PhD Thesis, University of Karlsruhe, Germany.

Stojanovic, L., & Motik, B. (2002). Ontology evolution with ontology. In *EKAW02 Workshop on Evaluation of Ontology-based Tools (EON2002), CEUR Workshop Proceedings*, (Vol. 62, pp. 53–62), Sigüenza.

Stojanovic, L., Stojanovic, N., Gonzalez, J., & Studer, R. (2003). Ontomanager - a system for the usage-based ontology management. In [Berlin: Springer.]. *Proceedings of ODBASE, 2003*, 858–875.

Studer, R., Benjamins, V. R., & Fensel, D. (1998). Knowledge engineering: principles and methods. *Data & Knowledge Engineering*, *25*, 161–197. doi:10.1016/S0169-023X(97)00056-6

Sure, Y., Staab, S., & Studer, R. (2004). On-To-Knowledge Methodology (OTKM). In Staab, S., & Studer, R. (Eds.), *Handbook on Ontologies* (pp. 117–132). Berlin: Springer Verlag.

Swartout, B., Ramesh, P., Knight, K., & Russ, T. (1997). Toward Distributed Use of Large-Scale Ontologies. In *AAAI Symposium on Ontological Engineering*, Stanford, CA.

Uschold, M. (1996). Towards A Unified Methodology. In *Expert Systems*. Cambridge: Building Ontologies.

Uschold, M., & Gruninger, M. (1996). Ontologies: Principles methods and applications. *The Knowledge Engineering Review*, *11*(2), 93–155. doi:10.1017/S0269888900007797

Uschold, M., & Jasper, R. (1999). A framework for understanding and classifying ontology applications. In *Proceedings of the IJCAI99 Workshop on Ontologies and Problem-Solving Method*, Stockholm, Sweden.

Uschold, M., & King, M. (1995). Towards a Methodology for Building Ontologies. *IJCAI95 Workshop on Basic Ontological Issues in Knowledge Sharing*, Montreal.

Uschold, M. E. (1998). Knowledge level modelling: Concepts and terminology. *The Knowledge Engineering Review*, *13*(1), 5–29. doi:10.1017/S0269888998001040

Van de Riet, R., Burg, H., & Dehne, F. (1998). Linguistic Issues in Information System Design. In *Formal Ontology in Information System*. Amsterdam: IOS Press.

Van der Vet, P. E., & Mars, N. J. I. (1998). Bottom-Up Construction of Ontologies. *IEEE Transactions on Knowledge and Data Engineering*, *10*(4), 513–526. doi:10.1109/69.706054

Van Heijst, G., Schreiber, A., & Wielinga, B. (1997). Using Explicit Ontologies in KBS Development. *International Journal of Human-Computer Studies*, *46*(2/3), 183–292. doi:10.1006/ijhc.1996.0090

Wagner, A. (2000). Enriching a lexical semantic net with selectional preferences by means of statistical corpus analysis. In *Proceedings of the ECAI-2000 Workshop on Ontology Learning*, (pp. 37-42), Berlin.

Weber, R. (1997). *Ontological Foundations of Information Systems*. Coopers and Lybrand.

Xu, F., Kurz, D., Piskorski, J., & Schmeier, S. (2002). A Domain Adaptive Approach to Automatic Acquisition of Domain Relevant Terms and their Relations with Bootstrapping. In *Proceedings of LREC 2002, the third international conference on language resources and evaluation*, Las Palmas, Canary island, Spain.

Zicari, R. (1991). A framework for schema updates in an object-oriented database system. In *Proceedings of the Seventh International Conference on Data Engineering (ICDE'91)*, (pp. 2-13), Kobe, Japan.

ENDNOTES

1 Also called *Naming conflicts* in (Goh et al., 1999).

2 Also called *Scaling conflicts* in (Goh et al., 1999).

3 Also called *Confounding conflicts* in (Goh et al., 1999).

4 Taxonomy consists of a set of terms that alongside their definitions and relations among them form an ontology. Sometimes, a taxonomy is viewed as a simple ontology.

5 Sub-concepts are more specific than the super-concept.

6 Another classification is proposed in (Muñoz, 2006) who distinguish *Structural* and *non structural axioms. Structural axioms* constrain the structure of the ontology. *Non structural axioms* are local to a concept and constrain its interpretation stating conditions about its attributes.

7 http://wordnet.princeton.edu/

8 http://www2.nict.go.jp/r/r312/EDR/index.html

9 http://www.Cyc.com.

10 http://www.fb10.uni-bremen.de/anglistik/langpro/webspace/jb/gum/index.htm

11 http://www.ie.utoronto.ca/EIL.

12 http://www.aiai.ed.ac.uk/project/enterprise.

13 http://www.unspsc.org.

14 http://www.ncbi.nlm.nih.gov/PubMed.

15 http://www.aiai.ed.ac.uk/project/enterprise/enterprise/ontology.html.

16 http://www.fipa.org.

17 http://www.ontoknowledge.org/.

18 Karlsruhe Ontology and Semantic Web tool.

19 http://www.w3.org/TR/owl-guide.

Section 2
Theoretical Models and Aspects:
Formal Frameworks

Chapter 3
Exceptions in Ontologies:
A Theoretical Model for Deducing Properties from Topological Axioms

Christophe Jouis
Université Paris III, France

Julien Bourdaillet
Université de Montréal, Canada

Bassel Habib
LIP6, France

Jean-Gabriel Ganascia
LIP6, France

ABSTRACT

This chapter is a contribution to the study of formal ontologies. It addresses the problem of atypical entities in ontologies. The authors propose a new model of knowledge representation by combining ontologies and topology. In order to represent atypical entities in ontologies, the four topological operators of interior, exterior, border and closure are introduced. These operators allow to specify whether an entity, belonging to a class, is typical or not. The authors define a system of topological inclusion and membership relations into the ontology formalism, by adapting the four topological operators with the help of their mathematical properties. These properties are used as a set of axioms which allows to define the topological inclusion and membership relations. Further, the authors define combinations of the operators of interior, exterior, border and closure that allow the construction of an algebra. They model is implemented in AnsProlog, a recent logic programming language that allows negative predicates in inference rules.

INTRODUCTION

Some entities belong more or less to a class. In particular, some individual entities are attached to classes whereas they do not check all the properties of the class. To illustrate this phenomenon, let us consider the ontological network below (see Figure 1). This network corresponds to the seven following declarative statements:

DOI: 10.4018/978-1-61520-859-3.ch003

Figure 1. The element [paul] does not satisfy all the properties of the class [human-being]

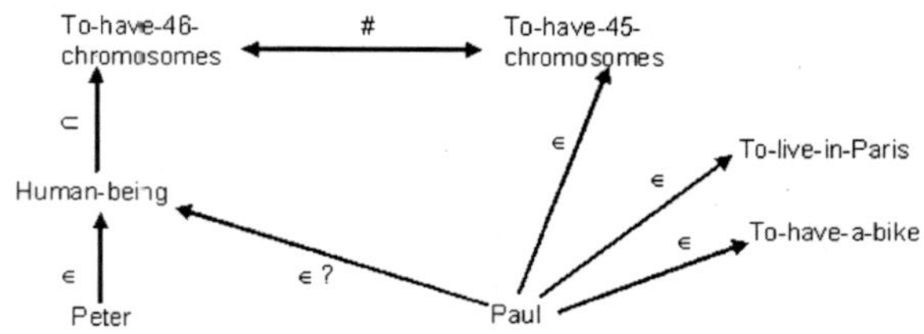

1. A human being has 46 chromosomes
2. Peter is a human being
3. Paul is a human being
4. Paul has 45 chromosomes
5. Paul lives in Paris
6. Paul has a bike
7. One thing can not have at the same time 46 chromosomes and 45 chromosomes

Because [Paul] is a [Human-being], he inherits all the typical properties of [Human-being], in particular [To-have-46-chromosomes]. A paradox is introduced by the statement (7) because "A human-being has 45 chromosomes" is a general fact but not a universal fact. The statement (1) means "In general, human beings have 46 chromosomes but there are some exceptions to this rule".

A similar phenomenon can be observed with distributive classes. Some subclasses are attached more or less to a general class because some of theirs elements may not check all the properties of this general class. To illustrate this phenomenon, let us consider the ontological network below (see Figure 2).

This network corresponds to the ten following declarative statements:

8. A thing which has an engine is a vehicle
9. A thing which has two wheels is a vehicle
10. A thing which has three wheels is a vehicle
11. A motorcycle has two wheels
12. A motorcycle has an engine
13. There are motorcycles with three wheels
14. A thing can not simultaneously have two wheels and three wheels
15. Paul's motorcycle is a motorcycle
16. Paul's motorcycle has three wheels
17. Paul's motorcycle is red

Figure 2. The individual entity [paul's motorcycle] does not satisfy all the properties of the class [motorcycle]. The subclass [motorcycle-with-3-wheels] does not satisfy all the properties of the class [motorcycle]

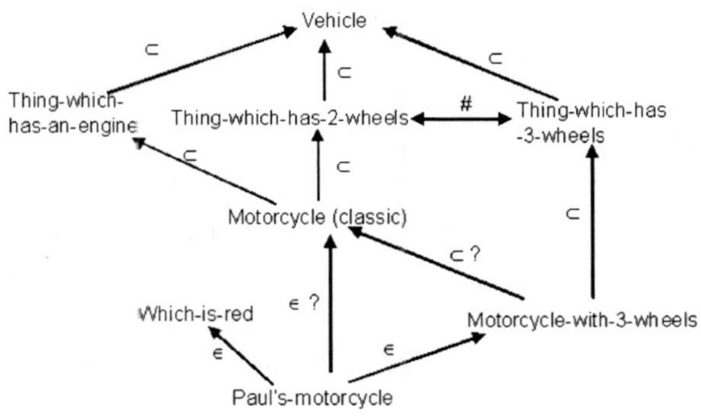

Because "Paul's motorcycle is a motorcycle", as introduced in the statement (15), it inherits all the typical properties of [Motorcycle] class, in particular, [Thing-which-has-2-wheels]. A paradox is introduced by the statement (16) because "A motorcycle has two wheels" is a general fact but not a universal one. The statement (11) "A motorcycle has two wheels" means that "In general, a motorcycle has two wheels but there are some exceptions to this law".

In Artificial Intelligence, a solution for this kind of problem is default reasoning: an individual A belonging to a concept F inherits concepts subsuming F except contrary indications. The technique of default reasoning led for example Reiter (Reiter 1981) to propose Default Logic (see Section 2).

We can consider that an individual entity is a typical element of a class if it checks all the properties of this class. Similarly, we can consider that a class is a typical subclass of another class if all its typical elements check the properties of the more general class. For example, the [Motorcycle] class is a typical subclass of [Thing-which-has-2-wheels] class.

An individual entity is an atypical element of a class if it does not check all the properties of the class. For example, [Paul's-motorcycle] is an atypical element of [Motorcycle] class. Similarly, we consider a class as an atypical subclass of a more general class if all its typical elements do not check all the properties of the more general class. For example, the [Motorcycle-with-3-wheels] class is an atypical subclass of the [Motorcycle] class.

It is important to distinguish between an atypical individual entity and an atypical class. When we consider [Motorcycle-with-3-wheels], we do not distinguish a particular object, but an undetermined number of objects that have typical properties.

According to (Desclés, 1990), we do not consider here a real object but a "typical object". This object does not exist in reality but checks all the properties of the concept by definition: using natural language, we express that by "a motorcycle with three wheels" (undetermined, i.e. without other determination). This corresponds to the definition of the concept in intention. The class associated with a concept includes all the objects, at one moment t, checking all the properties of this concept. This corresponds to the definition of the concept in extension. This set of objects may vary over time, while the "typical object" of a concept does not change. Using natural language, we express that by "those motorcycles with three wheels".

An individual entity is a particular object that has its own properties in addition to those of the class to which it is attached. For instance, the individual entity [Paul's-motorcycle] is an entity that inherits the properties of the [Motorcycle] class but is also assigned some other specific properties: [Which-is-red], [Motorcycle-which-has-3-wheels], etc.

The example of "Paul's motorcycle, which has three wheels" is a particular property of the individual entity [Paul's-motorcycle]. In the case of [Paul's-motorcycle], the property is not considered as a general property.

Thus, we must distinguish between two types of atypical properties, which are not similar: these on a concept, i.e. that are valid for all typical elements of the class associated with the concept; and those on a particular object, i.e. "Paul's motorcycle is red".

The chapter is organized as follows. In Section 2, we provide an overview of the key concepts used in this paper, including the main principles of some non-monotonic logics. We explain, in Section 3, the reason why we use topology and show two ways to define it. In Sections 4 and 5, we define the six relations of topological inclusion and membership and focus on their properties. Some interpretations, using the relations of topological inclusion and membership, are presented in Section 6. In this Section, we use the interpretations to illustrate the phenomenon of typicality presented in the introduction and show the possible inferences based on the six quoted relations. The Section 7 describes a new hybrid model of knowledge representation combining non-monotonic logic and topology that handles rules containing negative predicates. The model's implementation using ANSPROLOG is also described. Finally, the conclusion summarizes our work and presents some perspectives.

BACKGROUND ON NON-MONOTONIC LOGICS

Non-Monotonic Reasoning

The classical logic formalizes correct reasoning i.e. a type of reasoning that respects the monotony property. In an axiomatic formal system of deduction, this property can be formulated as following:

- Given {F1, F2, …, Fn} a set of logical formulas, F and G two logical formulas, if F can be deducted from {F1, F2, …, Fn} and F can be deducted from {F1, F2, …, Fn, G}, then the formal system is monotonic.

In other terms, if a set {F1, F2, …, Fn} of formulas considered as true permits to deduct that F is true, then the addition of a new formula in this set does not modify the truth of F. The monotony property is opposed to the notion of revisable reasoning (i.e. non-monotonic reasoning). In a formal system of revisable reasoning, it is possible (and desirable) to deduct from {F1, F2… Fn} that F is true and to deduct from {F1, F2… Fn, G} that F is false.

Default Logic

One major drawback of monotonic logic is that the knowledge base can not be updated by adding a fact which is not consistent with the base. In (Reiter, 1981), the author introduces default logic that allows revisable reasoning. It enables to deduct facts that are true in most of the cases, but which can also be invalidated for some exceptions. A default theory is formalized by a couple (P, V) where P is a set of first order closed formulas (all variables are bound) describing true facts and V is a set of inference rules called defaults. This means that default logic uses formulas of the first order logic, in particular inference rules that handle exceptions. Defaults are rules of the following form:

$$\frac{\mathrm{P(X)}\text{: } \mathrm{J}_1(X),\ \mathrm{J}_2(X)\ ,\ldots J_n(X)}{C(X)}$$

where P(X), J1(X), J2(X) … Jn(X) and C(X) are well-formed first order formulas in which all variables are free. They are interpreted in the following sense: if the fact P(X) is believed to be true and its justifications J1(X), J2(X), … Jn(X) are consistent with all the true facts in the theory (P,V), then the consequence C(X) can be inferred. Usually, only normal defaults are used; they are of the following form:

$$\frac{P(X) : C(X)}{C(X)}$$

where C(X) is both the justification and the consequence. In this case, only a modification of P, which involves an inconsistency of C(X) with V, can introduce non-monotony in the theory.

An example of normal default is:

$$\frac{\text{bird(x): fly(x) (birds typically fly)}}{\text{fly(x)}}$$

This kind of default rules makes the formalization of a system easier and supports the correspondence with other formalisms. Abnormal default rules are characterized by some differences between their justifications and their consequences. A special case concerns semi-normal default rules in which consequences are (only) included in their justifications, for example:

$$\frac{\text{bird}(x): \neg\,\text{penguin}(x), \text{fly}(x) \text{ (birds that are not penguins typically fly)}}{\text{fly}(x)}$$

The form of this kind of defaults prevents conflicts between rules. Defaults rules permit to construct extensions of P. All formulas that can be inferred by the successive applications of defaults, in which justifications are consistent, are included in P. This process is repeated until stabilisation in a fixed point. Every extension obtained by this process is a minimal and consistent set of closed formulas for classical deduction and containing P. A theory can have one, several mutually exclusive extensions, or no extension. One of the difficulties of this process is the possibility that an extension might be stabilized in a cycle instead of a fixed point: this problem is retrieved in other non-monotonic logics.

OVERVIEW OF THE CLOSED WORLD ASSUMPTION

The closed world assumption is a non-monotonic logic introduced by Ray Reiter in 1978 (Reiter, 1978; Morgenstern, 1999; Brewka et al, 1997; Sombé, 1989). It supposes that any given set of axioms is complete. If a formula is not included in this set and can not be inferred from this set, then it is supposed to be false until further revision of the facts. The addition of facts may require revising conclusions.

The closed world assumption is used by other logics such as default logic. It is often used in consultation systems of sub-domain databases: For example, if an air flight database does not have any information concerning a specific flight and if this flight can not be inferred from existing facts, then, under the closed world assumption, this flight is supposed not to exist.

Discussion on Default Logic

Difficulties with defaults logic can be enumerated as following:

- The hardness to define default theories that do not contain inference cycles.
- The hardness to decide if a formula of the language belongs or not to an extension of P. Excluding the case of closed theories containing a finite set of normal defaults, there does not exist a general decision process.
- The hardness to decide if there is an extension or not (even for normal defaults this can be decided).
- The hardness to choose between several extensions: a priori, all extensions can be justified for reasoning.

The two first points are directly correlated to implicit and non-enumerative principles of the approach. On the other hand, the absence of priorities between possible extensions led to numerous propositions using heuristic criteria, priorities or preferences. For example, the minimal extensions of a theory can be favored. These difficulties are retrieved in other approaches of non-monotonic reasoning.

Autoepistemic Logic (Moore, 1985)

Autoepistemic Logic formalizes non-monotonicity using sentences of a modal logic of belief with a belief operator called L. Autoepistemic Logic focuses on stable sets of sentences which can be viewed as the beliefs of a rational agent and the stable extensions of a premise set. Properties of stables sets include consistency and a version of negative introspection: if a sentence P does not belong to a belief set, then the sentence ¬ L P belongs to the belief set. This corresponds to the principle that if an agent does not believe a particular fact, he believes that he does not believe it.

Circumscription (McCarthy, 1980; 1986)

Circumscription formalizes non-monotonic reasoning within classical logic by limiting the extension of certain predicates, i.e. circumscribing them. For example, consider the theory containing the following assumptions: typical birds fly, atypical (usually called abnormal) birds do not fly, penguins are atypical, Opus is a penguin, and Tweety is a bird. Opus must be in the class of atypical non-flying birds, but there is no reason for Tweety to be in that class; thus we conclude that Tweety should fly. The circumscription of a theory is achieved by adding a second-order axiom (or, in a first-order theory, an axiom schema), limiting the extension of certain predicates to a set of axioms.

The system below describes a way of determining the non-monotonic consequences of a set of assumptions. entailment relations (Kraus et al., 1990) generalize these approaches by considering entailment operator |~, where P |~ Q means that Q is a non-monotonic consequence of P, and by formulating general principles characterizing the behaviour of |~. These principles specify how |~ relates to the standard entailment |- of classical logic, and how meta-statement referring to the entailment operator can be combined.

Belief Revision (Alchourron et al., 1985)

belief revision studies non-monotonic reasoning from a dynamic point of view. It focuses on how old beliefs are retracted while new beliefs are added to a knowledge base. There are four interconnected operators of interest: contraction, withdrawal, expansion and revision. In general, revising a knowledge base follows the principle of minimal change: as much information as possible is conserved.

Implementations

Applications of non-monotonic systems are rare. Programmers run into several difficulties. First, most non-monotonic logics explicitly refer to the notion of consistency of a set of sentences. Determining consistency is in general undecidable for first-order theories; thus non-monotonic predicate logic is undecidable. Determining inconsistency is decidable but intractable for propositional logic; thus performing propositional non-monotonic reasoning takes exponentional time. This precludes the development of general efficient non-monotonic reasoning systems (Selman & Levesque, 1993).

However, efficient systems have been developed for limited cases. Logic Programming, the technique of programming using a set of logical sentences in clausal form, uses a non-monotonic technique known as "negation as failure" (see Section 2.3).

Prolog (Colmerauer, A. and P. Roussel, 1992)

Prolog is the main logic programming language. It is based on first-order predicate calculus, but is restricted to Horn clauses in its initial version. Recent versions of this language accept more complex predicates, using the negation by failure.

AnsProlog* (Answer Set Prolog) (Baral 2003; Gelfond 2008)

AnsProlog* is a particular kind of logic programming. AnsProlog* is a language used to describe directly the facts (i.e. the initial network) and inference rules in the same formalism as the one used in PROLOG.

a. **Syntax:** An AnsProlog* program consists in a collection of rules of the form:
   - L_0 or … or L_k:- L_{k+1},…, L_m, not L_{m+1}, …, not L_n.
   - Where the L_{is} are literals in the sense of classical logic.
   - The rule can be read as: If $_{Lk+1}$… L_m are to be true, and if not L_{m+1}… not L_n can be safely assumed to be false then at least one of L_0 or … or L_k must be true.

b. **AnsProlog* vs. PROLOG AnsProlog*:** is a declarative alternative to PROLOG. Besides the fact that AnsProlog* allows disjunction in the head of rules, the following are the main differences between AnsProlog* and Prolog, according to (Baral, 2003, p.4):
   - *"The ordering of literals in the body of a rule matters in PROLOG as it processes them from left to right. Similarly, the positioning of a rule in the program matters in PROLOG as it processes them from start to end. The ordering of rules and positioning of literals in the body of a rule do not matter in AnsProlog*. From the perspective of AnsProlog*, a program is a set of AnsProlog* rules, and in each AnsProlog* rule, the body is a set of literals and literals preceded by not.*
   - *Query processing in PROLOG is top-down from query to facts. In AnsProlog* query processing methodology is not part of the semantics. Most sound and complete interpreters with respect to AnsProlog* do bottom-up query processing from facts to conclusions or queries.*
   - *Because of the top-down query processing, and start to end, and left to right processing of rules and literals in the body of a rule respectively, a PROLOG program may get into an infinite loop for even simple programs without negation as failure.*
   - *The cut operator in PROLOG is extra-logical, although there have been some recent attempts at characterizing it. This operator is not part of AnsProlog*.*
   - *There are certain problems, such as floundering and getting stuck in a loop, in the way PROLOG deals with negation as failure. In general, PROLOG has trouble with programs that have recursions through the negation as failure operator. AnsProlog* does not have these problems, and as its name indicates it uses the answer set semantics to characterize negation as failure."*

c. **AnsProlog vs. default logic AnsProlog:** is a subclass of AnsProlog* with only one atom in the head, and without classical negation in the body. The subclass AnsProlog can be considered as a particular subclass of default logic that leads to a more efficient implementation. Recall that default logic is a pair (P, V), where P is a first-order theory and V is a collection of defaults. AnsProlog can be considered as a special case of a default theory where $P = \emptyset$. Moreover, it has been shown that AnsProlog* and default logic have the same expressiveness. In summary, AnsProlog* is syntactically simpler than default logic and yet has the same expressiveness, thus making it more usable. (Baral, 2003, p.5). In this Section, we provide an overview of the main principles of non-monotonic logics. In the next Section, we will explain the raison why we use topology and how we will integrate non-monotonic logics and general topology to improve exceptions modeling in ontologies.

USING TOPOLOGY TO REPRESENT ATYPICAL ENTITIES IN ONTOLOGIES

Why Using Topology?

Networks of concepts and semantic relations between concepts can be represented on a plan. Instances are points of the plan, while classes are demarcated areas of the plan, which consist of: (a) an interior (the typical elements belonging to the class), (b) an exterior (the elements that are not in the class), (c) a border (atypical elements that do not check all the properties of the class, i.e. atypical elements that are neither within nor outside the class). Literally, topology means the study of the area and defines what an area (also called a space) is and what its properties are.

Different Ways to Define a Topology

A topology is based on the set theory and consequently on the first order logic. From the point of view of the first order logic, a concept F can be seen as a predicate (or a function) that applies itself to a variable X and returns a truth value defined on {True, False}. On the other hand, from the point of view of the set theory, this is equivalent to say that X belongs to the set F. Therefore, a concept F can be considered either as a set or as a predicate (Frege, 1893).

Two logically equivalent definitions of a topology are presented as follows: the first one, which is the most classical definition, is based on the set theory, and the second one adopts a functional approach. Because it is more adapted to logics of deduction, the functional approach will be used in the rest of this paper. Further, non-monotonic reasoning will be introduced to allow property inheritance of atypical entities in ontologies.

The Classical Definition of Topology Based on the Set Theory

Classically, a topology on a set E makes use of opens. Given O a family of parts of E such as:

1. $\varnothing \in O$
2. $E \in O$
3. If the $A_i \in O$, then $\cup A_i \in O$ (the number of A_i being finite or not)
4. If the $A_i \in O$, then $\cap A_i \in O$ (the number of A_i being <u>finite</u>)

An *open* of E is defined as any element of O. An element $a \in E$ is defined as interior to the part A of E if there exists an open P such as: $a \in P \subset A$.

The *interior* iA of A is the set of the elements which are interior to A: it is an open which is the union of the opens included in A. It is then possible to define the application $i: A \rightarrow iA$ in the set 2^E of the parts of E by:

$$iA = \bigcup_{\substack{o \subset A \\ o \in O}} o$$

If A is an open, iA=A, (because A is one of the open included in A), and reciprocally, if iA=A, because iA is still an open (a union of opens is an open from (3)), A is an open. This implies the following equivalence:

$$A \in O \Leftrightarrow iA = A.$$

Further, the following properties are true:

1. $iiA = iA$ (idempotence)
2. $iA \subset A$
3. a) $i(A \cap B) = iA \cap iB$ and b) $iA \cup iB \subset i(A \cup B)$
4. $A \subset B \Rightarrow iA \subset iB$ (monotony)

A closed set can then be defined as the complement of an open. Their properties are obtained using the laws of Morgan. They form a family F of parts of E such as:

a'. $E \in F$
b'. $\varnothing \in F$
c'. If the $A_i \in F$, then $\cap A_i \in F$ (the number of A_i being finite or not)
d'. If the $A_i \in F$, then $\cup A_i \in F$ (the number of A_i being finite)

The definition of the application "closure", f: A→fA is the dual of the definition of the « interior » i

$$fA = \bigcap_{\substack{A \subset X \\ X \in F}} X$$

This means that fA is the intersection of the closed sets containing A.

Its properties are the following

5. $A \subset B \Rightarrow fA \subset fB$ (monotony)
6. $A \subset fA$
7. $ffA = fA$ (idempotence)
8. a) $f(A \cup B) = fA \cup fB$ and b) $f(A \cap B) = fA \cap fB$.

Till now, we examined the behaviour of the application i with respect to union and intersection. It remains to examine the behaviour of i with respect to the third boolean operation, the complementary. In the following, the complement of a set A of E is noted cA.

In particular, icA is the set of the interior points to the complement of A. Such a set is called the exterior of A (and noted eA):

$eA = icA$.

With respect to the laws of Morgan on the complement, the properties (1), (2) and (4) of the application i have their correspondences for the application e:

9. $A \subset B \Rightarrow eA \supset eB$
10. $A \cap eA = \varnothing$
11. $e(A \cup B) = eA \cap eB$.

The property of idempotence does not have a simple correspondence.

The border of A, noted bA, is what is neither at the interior of A nor at the exterior of A. By definition, the border of A is the complement of $iA \cup eA$

12. $bA = c\,(iA \cup eA) = ciA \cap ceA$.

An Operational Definition of Topology: The Functional Approach

There exists a second way to define topology. It is less simple and classical than the previous approach, but it is more operational. This time, opens (or closed sets) are not given, but the application i (or the application f) with its properties is given.

Given a set E, an application f of the set 2^E of the parts of E in itself satisfying the following properties:

1. $f\varnothing = \varnothing$
2. $A \supset B \Rightarrow fA \supset fB$
3. $fA \supset A$
4. $ffA = fA$
5. $f(A \cup B) = fA \cup fB$

A set is called "closed" if it corresponds to the parts of E which are invariant in f

$F = \{ X: X \subset E \;\&\; fX = X\}$

This family F of the parts of E satisfies all the axioms of the closed sets given in Section 3.2.1. Consequently, giving f (or i) and its properties, is equivalent to give F (or O) and defines a topology on E. Once the closed sets are obtained, the remaining of the construction can use either the first approach, or the operational approach. If we use the second approach, we can define the application i as the transformed of f by its complement c as i = cfc. Then, the opens are the invariant parts of i. The relation ic = cf means that a part is closed if and only if its complement is open.

We presented two logically equivalent definitions of topology. The operational approach is more interesting to represent atypical entities in ontologies as we will exhibit in the following.

Integrating Non-Monotonic Logics

Reasoning systems using non-monotonic logics are useful only if they can be successfully integrated with other commonsense reasoning. For instance, we can integrate non-monotonic logics with temporal reasoning (Hanks & McDermott, 1987) or with multi-agent systems (Morgenstern & Guerreiro, 1993). In general, integration may require extending both the non-monotonic formalism and the particular theory of commonsense reasoning.

In this chapter, we will integrate non-monotonic logic and general topology to improve exceptions modelling in ontologies.

THE TOPOLOGICAL INCLUSION AND BELONGING RELATIONS

From Non-Monotonic Logic to General Topology

Non-monotonic logics can be seen as an illustration of the general topology exposed earlier, and they can be used to represent exceptions in ontologies. Given an element X and a predicate C (or a class, i.e. a part of E), the following assertions are true:

- If a formula C(X) is true except contrary indications in a non-monotonic logic, then it is similar to X being at the interior of C in topology. It is also similar to X being a typical element of C in ontologies.
- If a formula C(X) is partially true except contrary indications in a non-monotonic logic, i.e. it does not verifies all the properties of C, it is similar to X being at the border of C. It is also similar to X being an atypical element of C in ontologies.
- If a formula C(X) is wrong except contrary indications in a non-monotonic logic, then it is similar to X being at the exterior of C. It is also similar to X being an element which does not belong to C, i.e. which is neither typical nor atypical of C in ontologies.

Similar assertions are true for typical and atypical classes and subclasses.

In the following, X represents any given point and {X} a singleton (i.e. the smallest neighbourhood containing X). We note S as the set of singletons of E. A, B, C represent any parts of E that are not singletons.

Definitions of Topological Relations

a. Membership at the interior of a class (noted $\in_i$)
   - We define ($X \in_i A$) if and only if X inherits all the properties of A
   - $X \in_i A \Leftrightarrow \{X\} \subset iA$

b. Membership at the exterior of a class (noted $\in_e$)
   - We define ($X \in_e A$) if and only if X cannot belong neither to the interior nor to the border of A (and in the same way recursively for the subclasses of A):
   - $X \in_e A \Leftrightarrow \{X\} \subset eA$

c. Membership at the border of a class (noted $\in_b$)

Figure 3. $\subset_i$ is not symmetrical because in this counterexample $(x \subset_i y)$ and $(y \not\subset_i ix)$

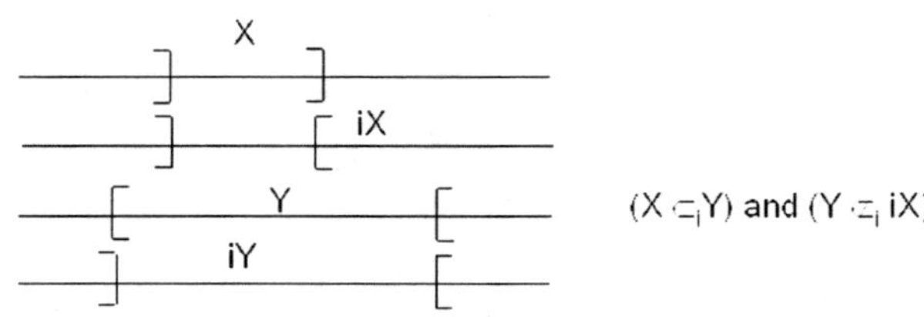

- We define $(X \in_b A)$ if and only if X is an atypical individual entity of A:
- $X \in_b A \Leftrightarrow \{X\} \subset bA$

d. Inclusion at the interior of a class (noted $\subset_i$)
- We define $(A \subset_i B)$ if and only if A is a typical subclass of B:
- $A \subset_i B \Leftrightarrow A \subset iB$

e. Inclusion at the exterior of a class (noted $\subset_e$)
- We define $(A \subset_e B)$ if and only if A cannot be a subclass neither at the interior nor at the border of B:
- $A \subset_e B \Leftrightarrow A \subset eB$

f. Inclusion at the border of a class (noted $\subset_b$)
- We define $(A \subset_b B)$ if and only if A is an atypical subclass of the class B:
- $A \subset_b B \Leftrightarrow A \subset bB$

Notice that $\in_I$ $\in_b$ et $\in_e$ are subsets of the Cartesian product S × D, while $\subset_i$, $\subset_b$ et $\subset_e$ are subsets of the Cartesian product D × D.

INFERENCE RULES FOR THE SIX TOPOLOGICAL RELATIONS OF INCLUSION AND MEMBERSHIP

Using properties of i, e, b and f operators, we can deduce inference rules for the relations of inclusion and membership. In order to demonstrate (or invalidate) these properties, two methods can be used:

1. The first method permits to demonstrate the truth of a property. It uses the properties of the operators i, e, b and f as axioms. For instance, the relation $\subset_i$ is transitive. In fact, it is equivalent to demonstrate that $(X \subset_i Y) \wedge (Y \subset_i Z) \Rightarrow (X \subset_i Z)$. However, $iY \subset Y$ (property (2) of i, see Section §3.2.1, this property is used as an axiom) thus we have, $X \subset iY \subset Y \subset iZ$ thus $X \subset iZ$.
2. The second method permits to invalidate a property. It consists in projecting the topological space E on the oriented straight line and to find a counter-example. For instance, the relation $\subset_i$ is not symmetrical because we suppose $(X \subset_i Y) \wedge (Y \subset_i X)$. We project this relation on the straight line (see Figure 3).

Reflexivity, Symmetry and Transitivity of Inclusion Relations

Note that it is inappropriate to consider these properties for relations of membership because they are define on SxE and not on ExE.

a. **Inclusion at the interior of a class $\subset_i$.** The relation $\subset_i$ is transitive, not symmetrical and not reflexive.

b. **Inclusion at the border of a class.** The relation $\subset_b$ is not reflexive, not symmetrical and not transitive.

c. **Inclusion at the exterior of a class.** The relation $\subset_e$ is not reflexive, not symmetrical but transitive. Notice that: $(F \wedge eF) = (G \wedge$

Figure 4. Spread of relations $\subset_e$ and $\in_e$ along the sub-hierarchies of interior inclusion of f and g

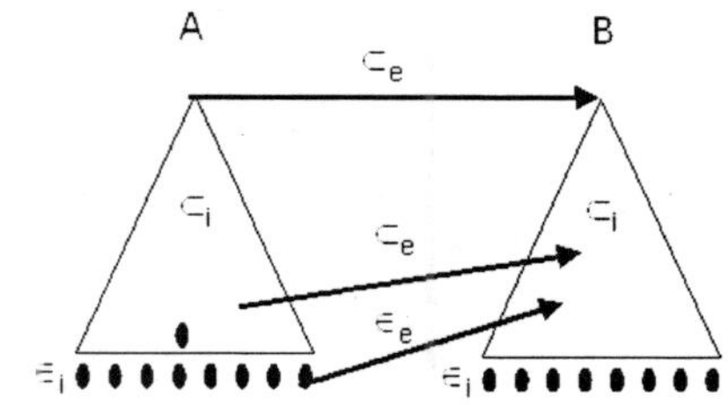

eG) = ∅ (axiom 8, Section 3.2.1). Thus, F $\subset_e$ G ⇒ G ∩ eF = ∅ because F ⊂ eG.

We have therefore the following rules of inference from G ⊂ eF:

F1: (G $\subset_e$ F) ∧ (H $\subset_i$ G) ⇒ (H $\subset_e$ F). In fact, H ⊂ iG ⊂ G ⊂ eF thus H ⊂ eF
F2: (G $\subset_e$ F) ∧ (X $\in_i$ G) ⇒ (X $\in_e$ F). In fact, {X} ⊂ iG ⊂ G ⊂ eF thus {X} ⊂ eF

These two rules spread in the sub-hierarchies of the interior inclusion of F and G. The same is true for the leaves of the hierarchies through the relations of interior membership (see Figure 4).

Properties of Combinations of the Topological Relations

In order to identify all the properties of combinations of the six topological relations, we consider all cases of a table [A … F] x [1 … 6] where an element Rz of the table represents the combination of an element Rx in lines and an element Ry in columns. For instance, A1 (if it exists) contains a result of the logical formula:

A1: (A ∈i B) ∧ (B ∈i C) ⇒ (A Rz? C)

we have figure 5.

In other words, we have the properties of the following relations:

A4: (X$\in_i$ B) ∧ (B $\subset_i$ C) ⇒ (X $\in_i$ C)

A5: (X$\in_i$ B) ∧ (B $\subset_b$ C) ⇒ (X $\in_b$ C)

A6: (X$\in_i$ B) ∧ (B $\subset_e$ C) ⇒ (X $\in_e$ C)

D4: (A $\subset_i$ B) ∧ (B $\subset_i$ C) ⇒ (A $\subset_i$ C)

D5: (A $\subset_i$ B) ∧ (B $\subset_b$ C) ⇒ (A $\subset_b$ C)

D6: (A $\subset_i$ B) ∧ (B $\subset_e$ C) ⇒ (A $\subset_e$ C)

E5: (A $\subset_b$ B) ∧ (B $\subset_b$ C) ⇒ (A $\subset_b$ C)

And also (see Section 5.1.3)

Figure 5. A table [A…F] x [1…6] where an element rz of the table represents the combination of an element rx in lines and an element ry in columns

		1	2	3	4	5	6
	(A Rx B) ∧ (B Ry C) ⇒ (A Rz C)	$\in_i$	$\in_b$	$\in_e$	$\subset_i$	$\subset_b$	$\subset_e$
A	$\in_i$	NIL	NIL	NIL	$\in_i$	$\in_b$	$\in_e$
B	$\in_b$	NIL	NIL	NIL	NIL	NIL	NIL
C	$\in_e$	NIL	NIL	NIL	NIL	NIL	NIL
D	$\subset_i$	NIL	NIL	NIL	$\subset_i$	$\subset_b$	$\subset_e$
E	$\subset_b$	NIL	NIL	NIL	NIL	$\subset_b$	NIL
F	$\subset_e$	NIL	NIL	NIL	NIL	NIL	NIL

Figure 6. (a) representation of a class c; (b) the starting point of an inclusion relation is situated within the interior of C1; (c) different end points of arcs representing relations

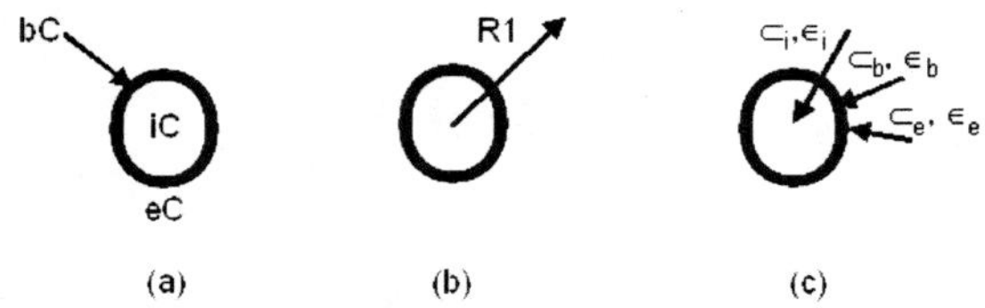

Figure 7. [Paul] is an atypical element of the class [human-being] (example 1)

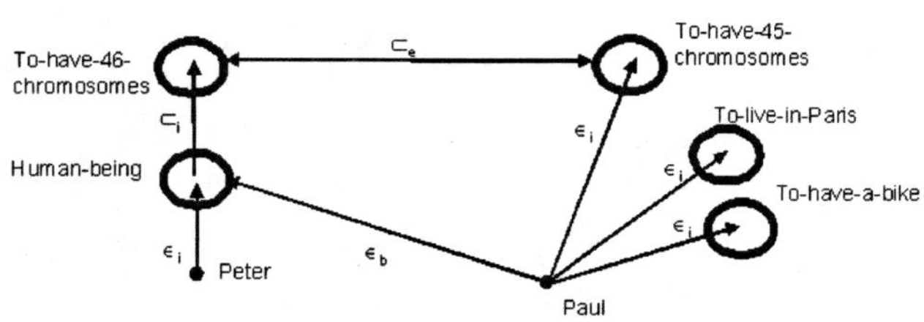

F1: $(A \subset_e B) \wedge (C \subset_i A) \Rightarrow (C \subset_e B)$

F2: $(A \subset_e B) \wedge (X \in_i A) \Rightarrow (X \in_e B)$

TOPOLOGICAL INTERPRETATION OF THE TWO EXAMPLES

An individual entity is represented by a point while a class C is represented in the form of a topological ball projected on a plan, with its interior, its border and its exterior as follows (see Figure 6, a):

The convention of representation is the following: the starting point of an inclusion relation between two classes R1 (C1 is the original class) is within the interior of C1, because the relation concerns all the typical elements of C1 and not necessarily the atypical elements located on the border of C1. Thus, we represent this differentiation as follows (Figure 5, b).

The end point of an arc representing a relation toward a class C2 depends on the relation. For $\subset_i$ and $\in_i$, the end point is within the interior of C2, because we have a typical class and a typical individual entity. For $\subset_b$ and $\in_b$, the end point is on the border of C2, because we have an atypical class and an atypical individual entity. For $\subset_e$ and $\in_e$, the end point is at the limit of the border of C2, because we have a class exterior to C2 and an individual entity exterior to C2 (Figure 5, c).

Figure 7 represents an interpretation of example 1 using our topological relations. In particular,

Figure 8. Some possible deductions in example 1

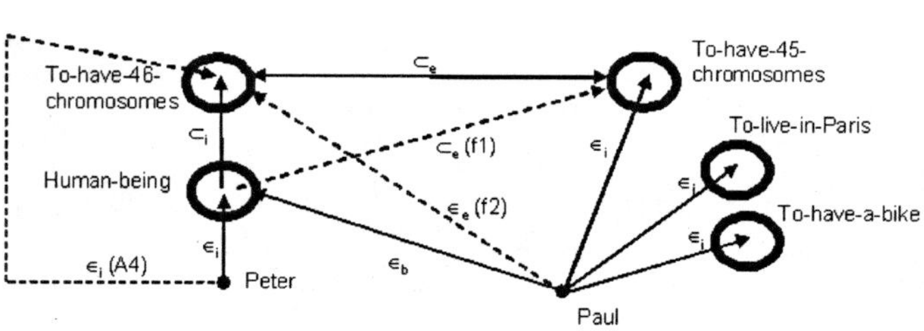

Figure 9. the class [motorcycle-with-3-wheels] is an atypical subclass of the class [motorcycle] (example 2)

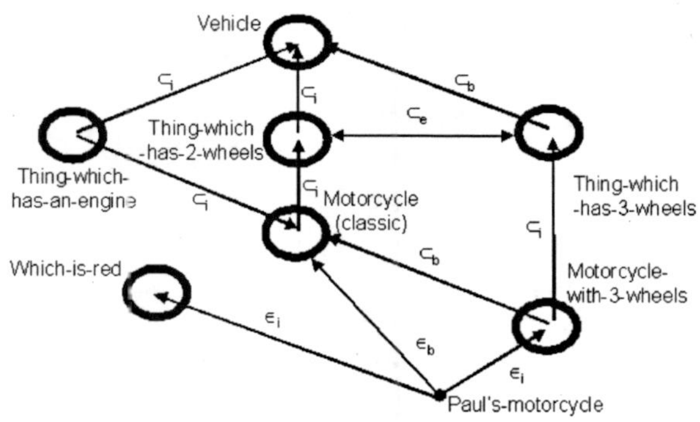

notice that [Paul] is an atypical element of the class [Human-being].

In Figure 8 dotted arrows represent some possible deductions in example 1 according to the rules of combination defined in Section 5. For example:

18. [Peter] is a typical element of [To-have-46-chromosomes] (statements 1 and 2 and rule A4)
19. [Paul] cannot belong neither the interior nor the border of [To-have-46-chromosomes] class (statements 4 and 7 and rule F2)
20. [Human-being] can be neither a typical subclass nor an atypical subclass of the [To-have-45-chromosomes] class (statements 1 and 7 and rule F1)

Figure 9 represents an interpretation of the example 2 using our topological relations. In particular, we notice that the [Paul's motorcycle] is an atypical element of the class [Motorcycle]. Furthermore, we notice that the class [Motorcycle-with-3-wheels] is an atypical subclass of the class [Motorcycle]

In Figure 10, dotted arrows represent some possible deductions in example 2 according to the rules of combination defined in Section 5.

For example:

21. The class [motorcycle] is a typical subclass of the class [Vehicle] (statements 8 and 12 and rule D4)
22. The class [Motorcycle-with-3-wheels] is an atypical subclass of the class [Vehicle] (statements 10, 13 and rule D5)
23. The class [Motorcycle-with-3-wheels] can be neither a typical subclass nor an atypical subclass of [Thing-which-has-2-wheels] (statements 13 and 14 and rule F1); etc.

Figure 10. Some possible deductions in example 2

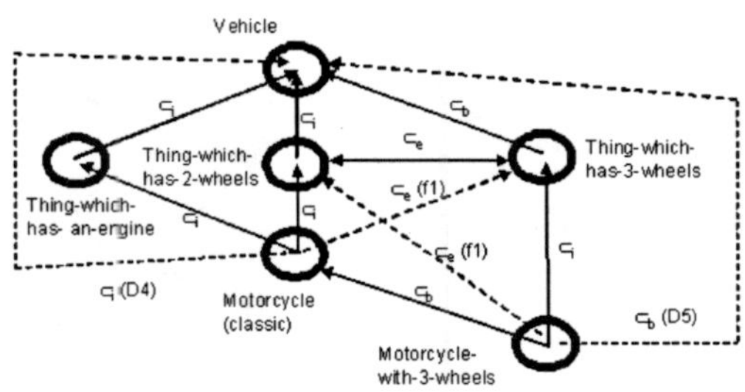

Figure 11. The class [motorcycle-with-3-wheels] should inherit the property of having an engine, by inheritance of property of the class [motorcycle]

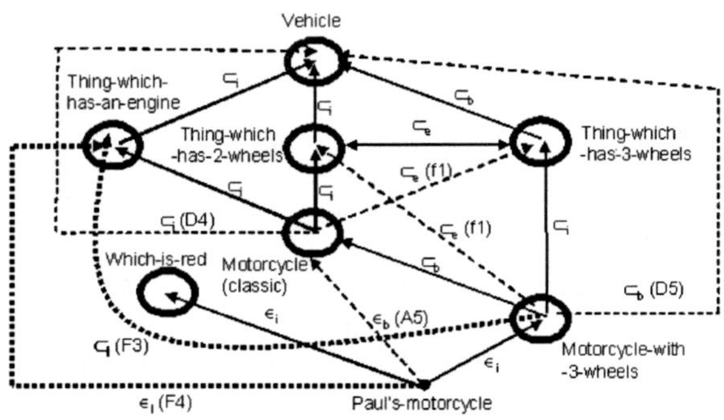

TOWARDS A NON-MONOTONIC SYSTEM: IMPLEMENTATION

Adding Non-Monotonic Rules

Let us consider the second example. In this example, the class [Motorcycle-with-3-wheels] is an atypical subclass of the class [Motorcycle], because [Motorcycle-with-3-wheels] is a typical subclass of the class [Thing-which-has-3-wheels]. This is incompatible with the typical property of motorcycles having two wheels. However, since we do not know why the class [Motorcycle-with-3-wheels] is an atypical subclass of the class [Motorcycle], all other inferences are blocked. In our example, we would like that the class [Motorcycle-with-3-wheels] inherits the property of having an engine (see Figure 11).

To do so, we simply add the rule F3 below:

F3: $(F \subset_b G) \wedge (G \subset_i H) \wedge \neg (F \subset_e H) \Rightarrow (F \subset_i H)$

We have the same phenomenon with the atypical individual entities. Let us consider the first example. Paul is an atypical human being because he has only 45 chromosomes instead of 46. But, knowing that all human beings are mammals, we would like Paul to inherit this property.

To do so, we simply add the rule F4 below:

F4: $(X \in_b F) \wedge (F \subset_i G) \wedge \neg (X \in_e G) \Rightarrow (X \in_i G)$

Note that we have in the right part of the two rules of inference F3 and F4 a negative predicate: ($\neg$ $(F \subset_e H)$ for F3 and $\neg$ $(X \in_e G)$ for F4). Both rules can be triggered only if both predicates are true at times of their applications. These are two non-monotonic rules. Indeed, we find ourselves in the case of « closed world assumption » (see Section 2.3). For example, the predicate $\neg$ $(F \subset_e H)$ is assumed false until a new modification of the ontology. The addition of new facts may require revision of the conclusions of F3 or F4.

Implementation in AnsProlog

Our model has been implemented in AnsProlog (Baral 2003) because this logic programming language has the ability to represent normative statements, exceptions and default statements, and is able to reason with them (see §2.8.2). In particular, AnsProlog allows negative predicates in inference rules.

The rules of inference are presented in AnsProlog directly as follows:

A4: $(A \in i\ B) \wedge (B \subset i\ C) \Rightarrow (A \in i\ C)$: membership_i(O, C):- membership_i(O, B), inclusion_i(B, C)

Figure 12. The results of inferences obtained from ansprolog on example 1

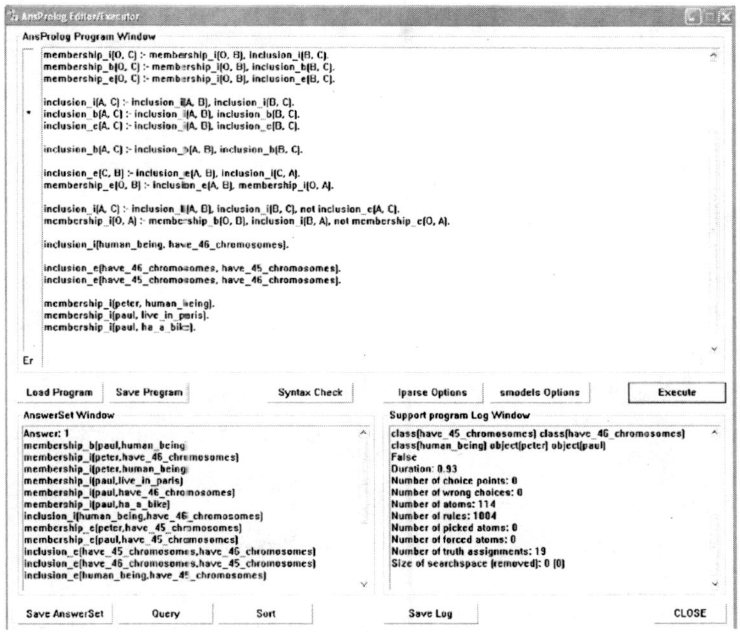

Figure 13. The results of inferences obtained from ansprolog on example 2

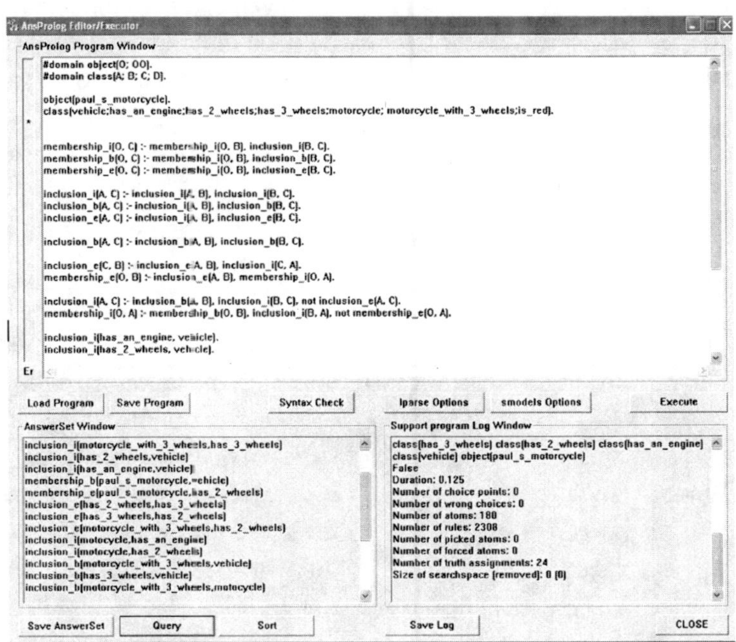

A5: $(A \in_i B) \wedge (B \subset_b C) \Rightarrow (A \in_b C)$: membership_b(O, C):- membership_i(O, B), inclusion_b(B, C)
A6: $(A \in_i B) \wedge (B \subset_e C) \Rightarrow (A \in_e C)$: membership_e(O, C):- membership_i(O, B), inclusion_e(B, C)
D4: $(A \subset_i B) \wedge (B \subset_i C) \Rightarrow (A \subset_i C)$: inclusion_i(A, C):- inclusion _i(A, B), inclusion_i(B, C)
D5: $(A \subset_i B) \wedge (B \subset_b C) \Rightarrow (A \subset_b C)$: inclusion_b(A, C):- inclusion _i(A, B), inclusion_b(B, C)
D6: $(A \subset_i B) \wedge (B \subset_e C) \Rightarrow (A \subset_e C)$: inclusion_e(A, C):- inclusion _i(A, B), inclusion_e(B, C)
E5: $(A \subset_b B) \wedge (B \subset_b C) \Rightarrow (A \subset_b C)$: inclusion_b(A, C):- inclusion _b(A, B), inclusion_b(B, C)
F1: $(A \subset_e B) \wedge (C \subset_i A) \Rightarrow (C \subset_e B)$: inclusion_e(A, C):- inclusion _e(A, B), inclusion_i(B, C)
F2: $(A \subset_e B) \wedge (C \in_i A) \Rightarrow (C \in_e B)$: membership_e(O, B):- inclusion_e(O, B), membership_i(O, A)
F3: $(F \subset_b G) \wedge (G \subset_i H) \wedge \neg (F \subset_e H) \Rightarrow (F \subset_i H)$: inclusion_i(A, C):- inclusion_b(A, B), inclusion_i(B, C), not inclusion_e(A, C)
F4: $(X \in_b F) \wedge (F \subset_i G) \wedge \neg (X \in_e G) \Rightarrow (X \in_i G)$: membership_i(O, A):- membership_b(O, B), inclusion_i(B,A), not membership_e(O, A).

The facts (i.e. the ontology), are presented in ANSPROLOG, by simple assertions such as, for example 1:

membership_i(peter, human_being).
membership_i (paul, live_in_paris).
membership_i (paul, have_a_bike).
membership_i (paul, to_have_45_chromosomes).
inclusion_e(to_have_45_chromosomes, to_have_46_chromosomes). Etc.

We obtain the necessary inferences by executing the program (see Figure 12).

On the example 2, the ontology is expressed in the following assertions:

inclusion_i(has_an_engine, vehicle).
inclusion_i(has_2_wheels, vehicle).
inclusion_i(motorcycle, has_2_wheels).
inclusion_b(has_3_wheels, vehicle).
inclusion_e(has_2_wheels, has_3_wheels).
membership_i(paul_s_motorcycle, motorcycle_with_3_wheels). Etc.

We obtain the necessary inferences by executing the program (see Figure 13).

CONCLUSION

In this chapter, we propose a new model of knowledge representation by combining ontologies, non-monotonic logic and topology. This model can be used by ontology builders during the modelisation process or the maintenance. When a certain size is achieved by an ontology and an atypical entity is discovered, the cost of ontology redesign may be too expensive. This model facilitates the ontology maintenance by avoiding redesign.

To specify whether an individual entity belonging to a class is typical or not, we borrow the topological concepts of interior, border, closure, and exterior. In order to represent atypical entities in the ontologies, we define a system of relations of inclusion and membership by adapting the topological operators. We propose to formalize the topological relations of inclusion and membership by using the mathematical properties of topological operators. However, there are properties of combining operators of interior, exterior, border and closure allowing the definition of an algebra. We propose to use these mathematical properties as a set of axioms. This set of axioms allows us to establish the properties of topological relations of inclusion and membership. Our model is implemented in ANSPROLOG, a recent logic programming

Figure 14. Thickness of the border of the "birds" class permits the introduction of typicality levels of "crow", "hen" and "ostrich"

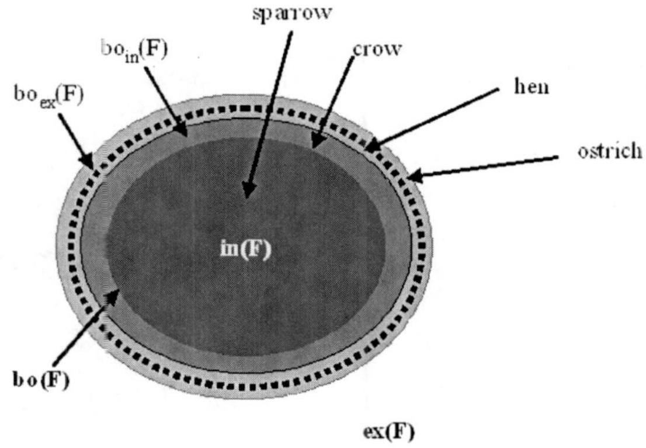

language that allows negative predicates in inference rules.

There are similarities between non-monotonic logics and topologies which seek to formalize atypical entities in the ontologies, such as *fuzzy logic* (Zadeh, et al. 1996) and *probabilistic reasoning*, especially *Bayesian networks* (Wong, et al. 2003). These fields are united in their attempt to represent and reason with incomplete knowledge. But our approach is different in that it uses a qualitative rather than a quantitative approach.

Future Works: A Scale of Typicality

It can be noticed there exists a third way to define a topology, using the notion of *neighbourhood* and an appropriate axiomatic. In general topology, a neighbourhood of an element X of E is any set containing an open whose X belongs to. In other words, a neighbourhood of X is any set whose X is interior.

As perspective let introduce a scale of typicality based on topology. It enables to define degrees of typicality where individual elements belonging to a class are more or less typical. The most typical elements are the elements where there exists no doubt on their class membership. Atypical elements miss some properties of class' typical elements which involve a reduction of their typicality degree. For example, in the "Bird" class a "Sparrow" is more typical than a "Crow" because for the common sense (at least in France) birds are small. A "Hen" which flies hardly is less more typical than a "Crow", but more than an "Ostrich" which does not. These differences involve a scale of typicality degrees. To model this scale of typicality, the thickness of the border of a class is introduced (see Figure 14).

The more elements with a low typicality degree are allowed to belong to a class, the more its border is thick. In our example, typicality degrees decrease with the loss of the flying property or in function of the common sense. Being given e, an integer constant arbitrarily set, which model the thickness of the border, it is then possible to define the interior in(F, e), the exterior ex(F, e), and the border bo(F, e) of a class F in function of e in the following way:

a) Interior of F in function of e, in(F, e):
   - if $e = 0$, $in(F, e) = in(F)$;
   - if $e > 0$, $in(F, e) \subset in(F)$;
   - $x \in in(F, e) \Leftrightarrow \exists\, n \in N(x): n \subset in(F, e\text{-}1)$, where N(X) represents the set of the neighbourhoods of X.

b) Exterior of F in function of e, ex(F, e):
   - if $e = 0$, $ex(F, e) = ex(F)$;
   - if $e > 0$, $ex(F) \subset ex(F, e)$;

Figure 15. A concept in which the border has a thickness

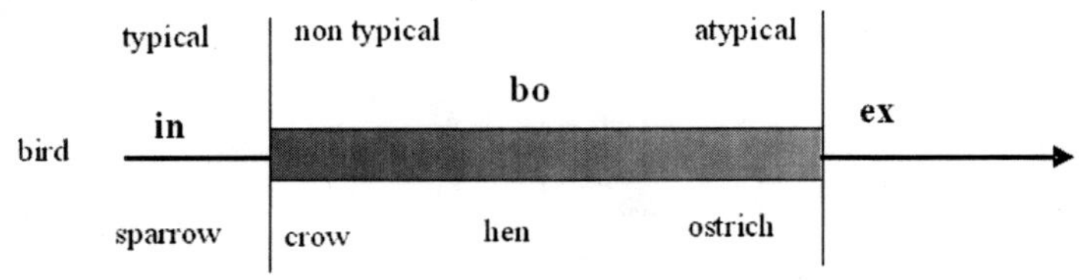

- ◦ $x \in ex(F, e) \Leftrightarrow \exists\, n \in N(x): n \cap F = \emptyset \wedge n \subset ex(F, e\text{-}1)$

c) Border of F in function of e, bo(F, e):
- ◦ if $e = 0$, $bo(F,e) = bo(F)$;
- ◦ if $e > 0$, $bo(F) \subset bo(F,e)$;
- ◦ $x \in bo(F, e) \Leftrightarrow \exists\, n \in N(x),\ n \cap in(F, e) = \emptyset \wedge n \cap ex(F, e) = \emptyset$.

These definitions remain compatible with the classical operators of topology by defining:

$ex(F) = \cap_{i=0..e}\, ex(F, i)$
$in(F) = \cap_{i=0..e}\, in(F, i)$
$bo(F) = \cap_{i=0..e}\, bo(F, i)$

Because the border has a thickness, it is then possible to define the interior border and the exterior border in the following way:

$x \in bo_{in}(F, e) \Leftrightarrow$
$\exists\, n_e \in N(x), \forall\, n \in N(F): n_e \subset n \wedge n_e \cap in(F, e) \neq \emptyset$.
$x \in bo_{ex}(F, e) \Leftrightarrow$
$\exists\, n_e \in N(x), \forall\, n \in N(F): n_e \subset n \wedge n_e \cap ex(F, e) \neq \emptyset$.

The interior border represents non-typical elements, i.e. elements with a lower typicality degree than typical elements of a class. The exterior border represents atypical elements which do not inherit all the properties of a class (see Figure 15).

ACKNOWLEDGMENT

The authors wish to thank Jie Liu for her contribution to the paper.

REFERENCES

Alchourron, C. E., Gardenfors, P., & Makinson, D. (1985). On the logic of theory change: partial meet functions for contraction and revision. *Journal of Symbolic Logic*, *50*, 510–530. doi:10.2307/2274239

Baral, C. (2003). *Knowledge Representation, Reasoning and Declarative Problem Solving*. Cambridge, UK: Cambridge University Press. doi:10.1017/CBO9780511543357

Brewka, G., Dix, J., & Konolidge, K. (1997). *Nonmonotonic Reasoning: an Overview. Center for the Study of Language and Information, Lecture Notes Number 73*. CA: Stanford University.

Colmerauer, A., & Roussel, P. (1992). The birth of Prolog. In The second ACM SIGPLAN conference on History of programming languages, (pp. 37-52).

Desclés, J.-P., & Pascu, A. (2005). Logic of Determination of Objects: the meaning of variable in quantification. In Flairs 2005, Florida, (pp. 610-616). Menlo Park, CA: AAAI Press.

Frege, G. (1893). *Logical Investigations* (Geach, P., Ed.). London: Blackwell.

Freund, M., Desclés, J.-P., Pascu, A., & Cardot, J. (2004). Typicality, contextual inferences and object determination logic. In Flairs 04, Miami, Florida, (pp. 491-495). Menlo Park, CA: AAAI Press.

Gelfond, M. (2008). Answer sets . In Harmelen, F., Lifschitz, V., & Porter, B. (Eds.), *Handbook of Knowledge Representation* (pp. 285–316). New York: Elsevier. doi:10.1016/S1574-6526(07)03007-6

Hanks, S., & McDermott, D. (1987). Nonmonotonic logic and temporal projection. *Artificial Intelligence*, *33*(3), 379–412. doi:10.1016/0004-3702(87)90043-9

Jouis, C. (2002). Logic of Relationships . In Green, R., Bean, C. A., & Myaeng, S. H. (Eds.), *The Semantics of Relationships: An Interdisciplinary Perspective* (pp. 127–140). Dordrecht: Kluwer Academic Publishers.

Kelley, J. L. (1975). *General Topology*. Berlin: Springer-Verlag.

Kraus, S., Lehmann, D., & Magidor, M. (1990). Nonmonotonic reasoning, preferential models, and cumulative logics. *Artificial Intelligence*, *44*, 167–207. doi:10.1016/0004-3702(90)90101-5

Kuratowski, C. (1958). *Topologie. Panstwowe Wydawnie two Naukowe*. Varsovie.

McCarthy, J. (1980). Circumscription –a form of nonmonotonic reasoning. *Artificial Intelligence*, *13*, 27–39. doi:10.1016/0004-3702(80)90011-9

McCarthy, J. (1986). Application of circumscription to formalizing common-sense knowledge. *Artificial Intelligence*, *28*, 86–116. doi:10.1016/0004-3702(86)90032-9

Moore, R. (1985). Semantical considerations on nonmonotonic reasoning. *Artificial Intelligence*, *25*(1), 75–94. doi:10.1016/0004-3702(85)90042-6

Morgensteirn, L., & Guerreiro, R. (1993). Epistemic logics for multiple agent nonmonotonic reasoning. In *Proceedings of the Second Symposium on logical Formalizations of Commonsense Reasoning (CS-93).*

Morgenstern, L. (1999). Nonmonotonic Logics. In R.A. Wilson & F. Keil (Ed.), *MIT Encyclopedia of the Cognitive Sciences*. Cambridge, MA: The MIT Press. Retrieved from http://www-formal.stanford.edu/leora/nonmon.pdf

Reiter, R. (1978). On closed world data bases . In Gallaire, H., & Minker, J. (Eds.), *Logic and Data Bases* (pp. 119–140). New York: Plenum.

Reiter, R. (1980). A logic for default reasoning. *Artificial Intelligence*, *13*, 81–132. doi:10.1016/0004-3702(80)90014-4

Selman, B., & Levesque, H. (1993). The complexity of path-based defeasible inheritance. *Artificial Intelligence*, *62*(2), 303–340. doi:10.1016/0004-3702(93)90081-L

Sombé, L. (1989). *Raisonnements sur des informations incomplètes en intelligence artificielle: comparaison de formalismes à partir d'un exemple*. Éditions Teknea.

Wong, S., Michael, K., Wu, D., & Butz, C. J. (2003). Probabilistic Reasoning in Bayesian Networks: A Relational Database Approach. In Advances in Artificial Intelligence, (LNCS Vol. 2671, pp. 583–590).

Zadeh, L. A. (1996). *Fuzzy Sets, Fuzzy Logic, Fuzzy Systems*. Amsterdam: World Scientific Press.

Chapter 4
An Algebra of Ontology Properties for Service Discovery and Composition in Semantic Web

Yann Pollet
CEDRIC Laboratory, France

ABSTRACT

The authors address in this chapter the problem of the automated discovery and composition of Web Services. Now, Service-oriented computing is emerging as a new and promising paradigm. However, selection and composition of Services to achieve an expected goal remain purely manual and time consuming tasks. Basing our approach on domain concept definitions thanks to an Ontology, the authors develop here an algebraic approach that enables to express formal definitions of Web Service semantics as well as user information needs. Both are captured by the means of algebraic expressions of ontology properties. They present an algorithm that generates efficient orchestration plans, with characteristics of optimality regarding Quality of Service. The approach has been validated by a prototype and an evaluation in the case of an Health Information System.

INTRODUCTION

The number of available Web data sources and services has exploded during the last years. This enables users to access rich information in many domains such as health, life sciences, law, geography, and many other domain of interest. Thanks to this wealth, users rely more on various digital tasks such as data retrieval from both public and corporate data sources and data analysis with Web tools or services organized in complex workflows [Gao, 2005, Kinsi,2007]. However, human users have to spend uncountable hours to explore and discover Web resources that meet their requirements. In addition, in many cases, users need to compose a specific set of Web resources in order to fulfill a complex question. This situation is mainly due to the inability of present standards in capturing Web Service semantics, i.e. the precise meaning of what a given Web Service exactly delivers regarding a specific user context.

DOI: 10.4018/978-1-61520-859-3.ch004

Meanwhile, Service-oriented computing (SoC) is emerging as a new and promising computing paradigm that centers on the notion of service as the fundamental element for accessing heterogeneous, rich and distributed resources in an interoperable way [Roman, 2005]. Web services are self-describing components that support a rapid and significant reuse of distributed applications. They are offered by service providers, which procure service implementation and maintenance, and supply service descriptions. Service descriptions are used to advertise service capabilities, behavior, Quality of Service, etc. (UDDI, WSDL, OWL-S). Service descriptions are meant to be used by other applications (and possibly other services), and not only by humans. WSDL and UDDI are the basic standards used for Web Service capabilities descriptions and advertising. However, they focus on the description of interfaces and syntactic considerations.

So, at present, the development of powerful applications on the Web is still facing two major problems. The first one is related to the increasing difficulties of identifying services that perform a specific task. The second one concerns the difficulty to orchestrate and compose services in a smooth, automated, and, if possible, optimal way, regarding the Quality of Service (QoS). This is still very challenging for many reasons. The main raison is the present limited ability of languages and models to describe the semantic of Web Services, despite tremendous efforts driven by the semantic Web Services community [Roman 2005, Kopecki 2007, Martin 2007].

In order to increase the benefits gained from rich Web resources, it would be of the highest importance to express formal semantic descriptions of Web Services. Such descriptions are in fact the absolute requisite condition to enable assisted or automated selection of relevant Web Services, and generate meaningful compositions of them. In addition, *non functional aspects* such as QoS (performance, availability, …) should be taken into account at Services selection and for the generation of a composition plan. This remains at present challenging and hard issue.

BACKGROUND

Emerging infrastructures such as the Semantic Web [Berners-Lee, 2001], the Semantic Grid [Goble, 2005] and Service Oriented architectures [Roman, 2005], support on-line access to a large number of resources from data sources and Web services to knowledge representation models such as taxonomies and ontologies. Ontologies play an important role in the Semantic Web and provide the basics for the definition of concepts and relationships that make information integration possible. OWL-S is proposed as a way to express more detailed descriptions of Web Services via a provided ontology of Web Services. But it remains limited and fails in expressing what a Service really provides, although services should ideally export also their semantics.

The new Semantic Service-Oriented Architecture (SSOA) leverages rich, machine-interpretable descriptions of data, services and processes to enable software agents to automatically interact and achieve collaborative goals. The SSOA integrates the principles of Service-Oriented Computing with semantics-based computing, Typically, a Semantic Service-Oriented Architecture (SSOA) includes four layers: the *data* layer, the *resource* layer, the *ontology* layer, and the *community* layer, as depicted in figure 1. The *data* layer represents data published by Web resources, and the hyperlinks that interconnect theses data objects, for example PubMed publications or medical records stored in Google Health. The *resource* layer is comprised of Web resources and their links, Resources can be either data source, e.g., SwissProt which is a protein database, or a Web service, e.g., BLAST which is a bioinformatics Web-based alignment tool. In the case of a data source, a resource implements some concepts and individuals of the ontology level, while in the case of Web Services, they

Figure 1. A layered web architecture

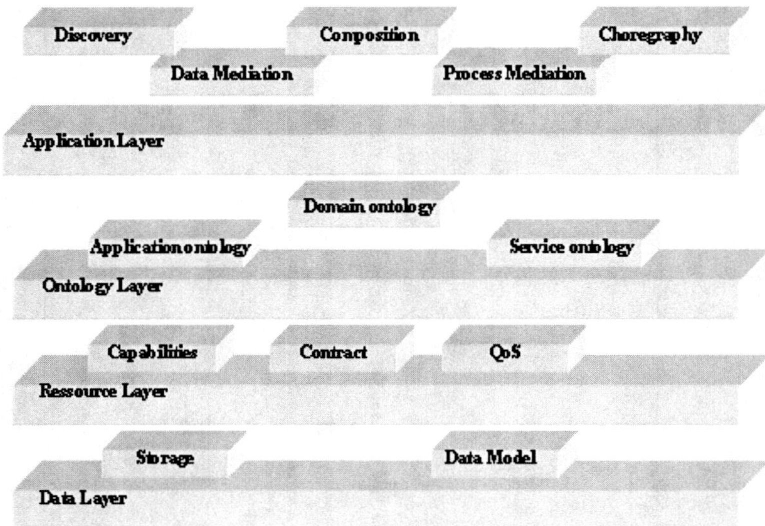

implement a semantic link that relates input and outputs parameters. Web Services infrastructure provides the syntactical basis for interoperability between resources thanks to standards such as WESL [Akkiraju, 2005], UDDI, and SOAP.

Semantic Web service (SWS) technology aims at providing richer semantic specifications of Web services in order to enable the flexible automation of service processes. The field includes substantial bodies of work to enhance resource descriptions with the use of an ontology including OWL for Services (OWL-S) [Martin, 2007], the Web Service Modeling Ontology (WSMO) [Roman, 2005], and SAWSDL [Kopecki, 2007]. Some approaches, such as [Ayadi, 2008] introduce a canonical set of semantic descriptions of Web services in order to extend SAWSDL standard and support automatic reasoning.

Regarding Service discovery, several techniques have been proposed to support service discovery using logical inference. Existing solutions, including those of Paolucci et al. [Paolucci, 2002, 2007] and Sycara et al. [Sycara, 2003, 2006] propose a method based on DAML-S descriptions for matching goals and capabilities of semantic Web services. Sycara et al. describe the implementation of the DAML-S/UDDI matchmaker that expands UDDI by providing semantic capability matching. OWLS-MX [Klusch, 2006] is a hybrid matchmaker that complements logic based reasoning with approximate matching based on similarity computations.

The proposed FUSION semantic registry [Kourtesis, 2008] relies on a combination of three standards: UDDI, for storing and retrieving syntactic and semantic information about services and service providers, SAWSDL for creating semantically annotated descriptions of service interfaces, and OWL, for modeling service characteristics and performing fine-grained service matchmaking via Description Logic reasoning. In contrast with prominent approaches, FUSION relies on functional and *non-functional* properties for matchmaking. However, it is not clear in [Kourtesis, 2008] how service discovery is realized in this System.

There is today a wide agreement on the fact that a flexible Web Service infrastructure, where resources can be discovered and smoothly composed into workflows, is strongly required. But, in spite of tremendous efforts around semantic Web Services, the automation of these tasks is still challenging and hard to achieve.

AN ALGEBRAIC APPROACH FOR SERVICE DISCOVERY AND COMPOSITION

We explore here an algebraic approach for Service Discovery and composition based on an ontology of a given application domain. In order to motivate the need for an automated selection and execution of relevant Services, we present at first a simplified case study in healthcare domain. Then we present the RCS algebra (Relationship Composition with Structured Expressions) that we propose as a new theoretical basis, and detail its mathematical foundations. We present then a mapping model based on the Algebra that enables to formally express the semantics of Services, referring to domain ontology. We present also an efficient algorithm for execution plans generation developed on the basis of our approach. At last, we give a short presentation of a developed software framework and of its application to the case study in order to validate our approach.

A Motivating Example

As a way to illustrate our problem, we present here a case study issued from the Healthcare domain. In a given regional area, we have several healthcare institutions (hospital and practitioner offices), each of them managing data about their patients. A medical file is a set of time labeled *medical events* from different types (diagnosis, treatment prescription, medical act), with standard codifications. An institution may have zero, one or several medical files for a given patient. Each practitioner from an institution has access rights regarding a patient file, depending of his /her role in the healthcare process of this patient (referent practitioner, consultant, etc.). Access rights may be limited in time, e.g. for some roles in relationship with specific acts. In addition to medical files, it may exist Identities Servers, at regional level, that deliver information about patient's administrative details, as well as links to existing medical files on the basis of a patient identity. The figure 2 below illustrates a fragment of a relevant simplified ontology (based on a consistent combination of different existing standards). For a concern of readability, datatype properties (attributes) and cardinalities don't appear.

A practitioner has access to a secured infrastructure where the different information servers that deliver data about patients are accessed via Web Services.

Figure 2. A simplified fragment of ontology in medical practice domain

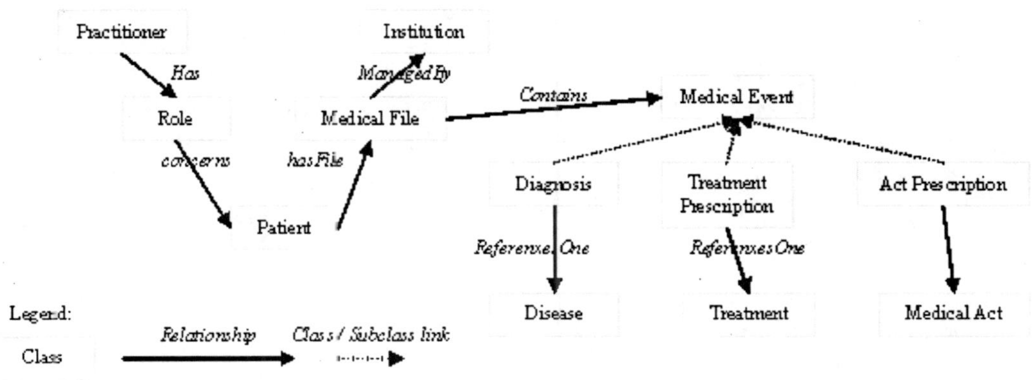

In healthcare domain, therapeutic decision, e.g. decision in oncology, requires access to various pieces of information scattered among various several institution servers. Therapeutic decision is the most convincing use case regarding the requirement for Services selection and composition, as there are many sources from which data have to be retrieved, with possible alternatives. But there exist many examples of support applications leading to the same requirements, such as epidemiologic studies, and access to anonymous patient files for medical students. Plans should be flexible in order to be automatically adapted when new sources are added to the community.

Problem and Basic Hypothesis

We assume here that the various information sources provide access to relevant data by the means of *Services*. We define here a *Service* just as a black box function that may be invoked by a distant software entity, with input data, and delivering output results. We make no assumption about technology, and these *Services* may be implemented as Web Services, or thanks to another technology. We consider here *stateless data access Services*, excluding *Services* having a side effect on internal data and/or on external world. A *Service* encapsulates all the details of operations executed to deliver a required piece of information. In particular, the user entity doesn't know whether the output is extracted from a local database, results from a calculation, or from a combination of both (e.g. rights determination from roles in our example).

We consider a domain Ontology O. This Ontology defines a set of classes $\{C_i,\}$, each of these having some *attributes* proprieties $\{V_{i,j}\}$ (*datatype properties*), and *directed relationships* $\{R_{i,k}\}$ to other classes (*object properties*).

By hypothesis, a *Service* will be such as $\{x'\} = S(x_1, \ldots, x_n)$, or $\{v\} = S(x_1, \ldots, x_n)$, where x' et x_i are *individuals* whose types directly correspond to Ontology classes, and where v is a type corresponding to a *datatype* in O. In the following, we shall also consider *Services* with more than a single output parameter. The development of relevant wrapping code, in the case of *Service* reuse, is out of the scope of the issue addressed by the chapter.

Principles

The domain Ontology provides a well defined formalisation of the various concepts of the domain, with meaningful properties and relationships. This enables to attach a precise meaning to a given piece of information, in particular when required by a user. However, the concern of a user may not exactly correspond to an Ontology concept. For example, the exams and the treatments that have been provided to a patient P have an interest for some users; nevertheless, this concept does not immediately correspond to a property of patient class. To define such a result, we have to consider at first all the medical files associated to the given individual patient, the union of all medical events from each file, and then the extraction of exams and treatments. There are in fact two problems. The first one is that we want to deal with an indirect access to medical events from a patient, with restriction condition on medical events. The second one is the fact that the type of results does not match with a concept known in the Ontology.

In order to define such information, we shall introduce the notion of *derived* property. A *derived* property will be formally defined by the means of an algebraic expression of the Ontology properties. A *derived* property will be attached to an Ontology class, or to new class defined on the basis of existing Ontology classes, and called here a *derived* class. A *native* or *derived* property will be called an *extended* property. So, a piece of information needed by a user will be always defined by both an individual, and an *extended* property to evaluate. This is the first principle of our approach.

The second principle concerns the capture of *Services* semantics. It consists in defining the semantic of a *Service*, i.e. the link between input and output by a correspondence with a property of the Ontology, such a *relationship* or an *attribute*. The most simple case is this where a *Service*, with an individual x as input, delivers as output the set of objects {x'} in correspondence with x via a given propriety R. As an example, if a *Service* delivers the content of a medical file starting from the reference to this file as input, the meaning of the *Service* is perfectly defined by the relationship "Contains" of the Ontology. Our principle is therefore this of defining the semantic of a *Service* by an Ontology property that this *Service* may *realize*. We shall call such a correspondence a *mapping* between a *Service* and a property. However, this is only a specific case, and there is no reason for a given *Service* to *realize* exactly a particular Ontology relationship. First of all, a *Service* may directly *realize* an indirect correspondence. For example this is the case if a *Service* delivers the overall content of a patient file given the patient's identity. This is the case in which a *Service realizes* a *derived* property. So, a *Service* may realize a *native* or a *derived* property.

Another case is this of a *partial realization.* Consider a *Service* provided by an institution delivering a set of medical events, for a given patient. The *Service* will be able, of course, to deliver results only for patients known in the institution, i.e., that have at least one event in this institution. In addition, it will be able to deliver only known events, i.e. those which have been performed in this institution. We shall present a general *mapping* model that enables to express sophisticated semantic correspondences between an available *Service* and properties. This model covers the more complex cases involving *Services* with more than one input parameter and/or more than one output parameter. The figure 3 synthesizes the two principles of our approach.

The RCS Algebra

We present here the RCS algebra that enables to express formal definition of new properties.

Derived Classes

Starting from the classes defined in the Ontology, called here *natives* classes, one can define new classes by application of the following OWL operators and their combinations. Such new classes will be called *derived* classes.

Intersection: $C = C_1.C_2$ ($C_1 \cap C_2$) defined by: $x \in C_1.C_2$ iff $x \in C_1$ AND $x \in C_2$, where C_1 and C_2 are *natives* classes, or already defined *derived* classes. Properties which are valid for $x \in C$ are those valid for $x \in C_1$ or for $x \in C_2$

Union: $C = C_1 + C_2$ (C_1 U C_2) defined by: $x \in C_1 + C_2$ iff $x \in C_1$ OR $x \in C_2$, where C_1 and C_2 are *natives* or already defined *derived* classes, Valid properties are those valid for $x \in C_1$ and $x \in C_2$

Property Restriction: $C' = C^P$ is defined by $x \in C^P$ iff $x \in C$ AND P(x), where C is a *native* or an already defined *derived* class, and P a predi-

Figure 3. Organization of the ontology based framework

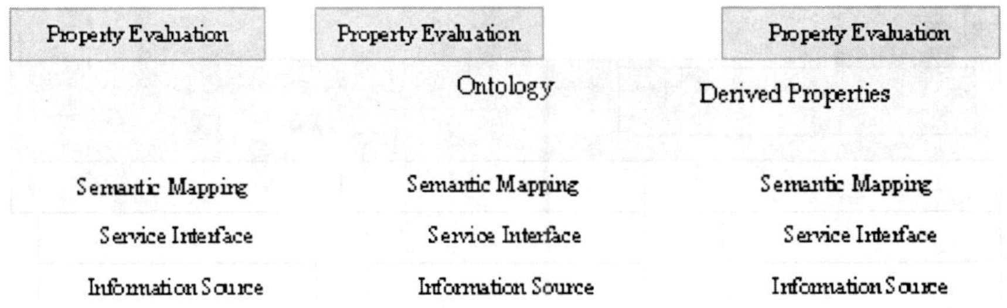

cate defined as a logical expression AND-OR of elementary predicates of the form $V_i \Theta v$, where V_i is a property defined on C, v a value of this property, and where Θ a comparison operator defined on the value domain. Valid properties for an $x \in C^P$ are those valid for $x \in C$

Property (Ontology Lattice). Let be an Ontology O, the set of native classes of O completed by the set of derived classes generated by intersection, union, and property restriction is a Lattice. The associated pseudo order is « $\leq$ », defined by: $C_1 \leq C_2$ iff $C_1 \subset C_2$ (set inclusion) and IsProperty (C_2) $\rightarrow$ IsProperty (C_1). This extends the Ontology relationship of specializations / generalization. A native or derived class will be called an extended class.

Algebraic Operators on Properties

We introduce here two binary operators on relationships: the composition and the union, plus a third operator called relationship restriction. These operators apply on *extended* (i.e. *native* of *derived*) relationships. In addition to relationships operators, we introduce the projection that enables to deal with attribute properties. We shall write <R, x, y> to express that the individual x and y are in relationship by R. In addition, dom(R) and range(R) will respectively denote the domain of a relationship R (i.e., the set of x), and its range (i.e. the set of y). At last, minCard(R) and maxCard(R) will respectively denote the minimum and maximum cardinalities of R (by default minCard(R) = 0 and maxCard(R) = ∞).

Composition

Definition (composition operator). Given two relationships R_1 and R_2, the composition $R_1 \,_* R_2$ is the relationship such as dom($R_1 \,_* R_2$) = dom(R_1), range($R_1 \,_* R_2$) = range(R_2), and <R, x, z> iff there exists $y \in$ range(R_1)$\cap$dom(R_2) such as <R_1, x,y> and <R_2, y,z>

The result is defined for any R_1 and R_2 and belongs to the set of relationships R. $_*$ is *associative* but not *commutative*. If range(R_1) $\cap$ dom(R_2) = Ø, then R is a *null* relationship, with no value for any individual. In addition, we have minCard($R_1 \,_* R_2$) = minCard(R_1) . minCard(R_2), and maxCard($R_1 \,_* R_2$) = maxCard(R_1).maxCard(R_2). The figure below illustrates the principle of the composition operator and an example of composition:

Union

Definition (union operator). Given two relationships R_1 and R_2, the union $R = R_1 + R_2$ is the relationship such as dom($R_1 + R_2$) = dom(R_1) $\cap$

Figure 4. Illustration of the composition operator

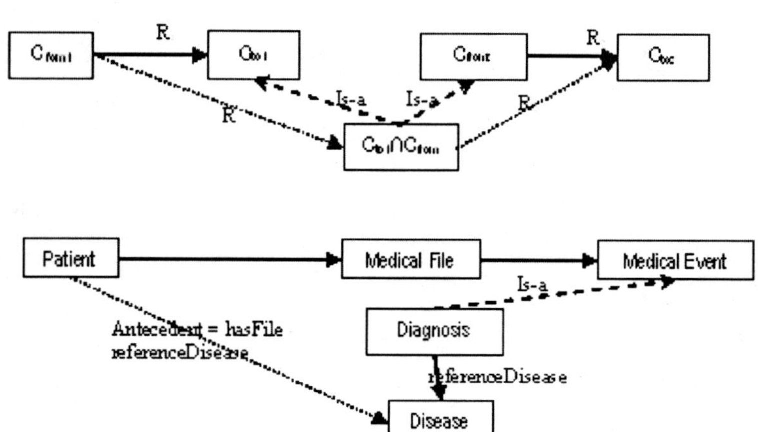

dom(R_2), range (R_1 + R_2) = range (R_1) U range (R_2), and <R, x, y> iff <R_1 x, y> or <R_2, x, y>

The set of individuals {y} associated to an individual x by R is so the union of the two sets of individuals respectively associated to x by R_1 and by R_2. $R_1 + R_2$ is always defined as a relationship, possibly empty (in particular in the case where dom(R_1) ∩ dom(R_2) = Ø). The union operator is *commutative* and *associative*. The composition operator $_*$ is *distributive* with respect to the union +. We have: minCard($R_1 + R_2$) = Max (minCard(R_1), minCard(R_2)) and maxCard(R_1 + R_2) = Min (maxCard(R_1), maxCard(R_2))

Restriction Filter and Relationship Restriction

Definition (restriction filter). Given two classes C_1 and C_2, the relationship [C_1, C_2], called restriction filter from C_1 to C_2, is the relationship such as dom([C_1, C_2]) = C_1, range([C_1, C_2]) = C_2, and < [C_1, C_2], x, y> iff (x=y) and (x ε $C_1 \cap C_2$)

Such a relationship is a *constant* as it is canonically defined from any ordered pair of classes, independently of the domain ontology relationships. [C_1,C_2] associates to any individual from C_1 either the individual itself, either an empty value set. An individual y from C_2 may be associated to an x from C_1 by [C_1, C_2] only if y ε C_1. If $C_1 = C_2$ = Universal, so [C_1, C_2] is the *Identity* relationship. If $C_2 = C_1^P$, where P is a predicate defined on C_1, [C_1, C_1^P] is a classical P-predicate restrictor. In particular, we have the following property:

$$[C, C^{PRE1}]_* [C, C^{PRE2}] = [C, C^{PRE1\ AND\ PRE2}]$$

So, we can define the *relationship restriction* operator in the following way:

Definition (restriction). Given a relationship R, two predicates PRE(x) and POST(y) respectively applying on individuals x from dom(R) and y from range (R), the relationship restriction of R by PRE and POST is defined by $R^{PRE, POST}$ = [C_1, C_1^{PRE}] $_*$ R $_*$ [C_2^{PRE}, C_2], where C_1 = dom(R) and C_2 = range(R)

So, the relationship $R^{PRE, POST}$ associates to any individual x from C_1 such as PRE its correspondent individual y by R if POST(y) is satisfied, and an empty set if not. The above figure 5 illustrates the relationship restriction operator:

Examples: One may derive from the relationship « contains » new relationships with the same domain (Medical File) and same range (Medical Event), but with a more specific meaning, e.g.:

- **R_1**, that associates to the medical files of a given institution H their content (with an empty value set for other ones): R_1 = Contains $^{ManagedByH,\ True}$, where ManagedByH (d) = d.managedBy = H
- **R_2**, that associates to any medical files the sets of its diagnosis: R_2 = Contains $_{True,\ Is(Diagnosis)}$

Canonical Decomposition of a Relationship

Let's consider a relationship R, two sets of predicates {PRE_i ; i=1, ..,n} and {$POST_j$; j=1, ..,m} respectively applying on the classes dom(R) and range(R), and such as:

PRE_1 OR .... OR PRE_n = True, and $POST_1$ OR .... OR $POST_m$ = True,

We have: $R = \sum_{i=1,\ldots,n\ OR\ PREi\ =\ True} R^{PREi,\ True} = \sum_{j=1,\ \ldots,m\ OR\ POSTj\ =\ True} R^{True,\ POSTj} = \sum_{i=1,\ldots,n,\ j=1,\ \ldots,m\ OR\ PREi\ =\ True,\ OR\ POSTj\ =\ True} R^{PREi,\ POSTj}$

Figure 5. Illustration of the relationship restriction operator

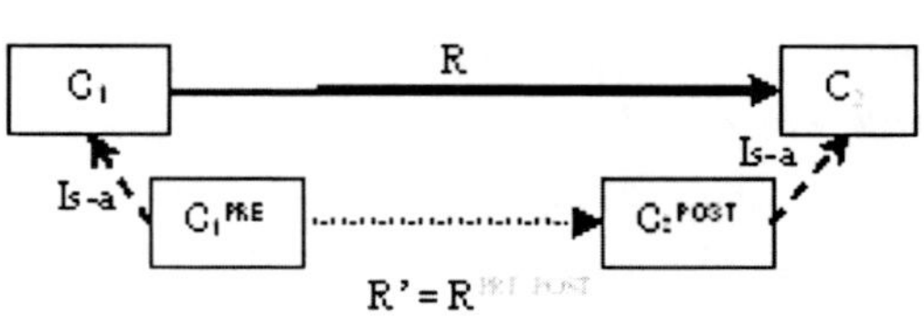

If R is a *native* relationship, and if the predicate sets PRE_i and $POST_j$ are composed of mutually exclusives conditions, the above property defines a *canonical way to decompose a relationship.*

Algebraic Expressions and Extended Relationships

The above operators enable to formulate algebraic expressions, where operands are *extended* relationships, i.e. *native* Ontology relationships, and/or already defined *derived* ones. Formulas define new *derived* relationships. Their domain and range are defined using the operator rules.

As an example, $R = R_1 + R_2 * R_3^{PRE1, POST1} + R_4^{PRE2, POST2} * R_5$ is a relationship with domain $dom(R) = dom(R_1).dom(R_2).dom(R_4)$ and $range(R) = dom(R_1) + dom(R_3) + dom(R_5)$

$R = has^{True, role=Referent} * concerns * hasFile^{True, managedBy=H} * contains^{True, Is(Diagnostic)}$ is a property of domain « Practitioner », and its range is the class « Diagnosis ».

It is possible to transform a given formula into equivalent expressions using the properties of operators (associativity of composition and union, commutativity of union, distributivity). If an operand is an *extended* relationship, another formula should have previously defined it, and we exclude in the present version of the theory cyclic definitions. The set of *derived* relationships associated to the Ontology O will be defined by a sequence of expressions of the form: $R_i = EXPR_i (R_{1,i}, \ldots, R_{ni,i})$; 1=1, …, N, where $EXPR_i$ is an expression whose operands $R_{k,i}$ k = 1, …, n_i are either *native* relationships, either *derived* relationship taken among the $R_1, \ldots, R_{i-1}$

Example 1: The relationship R_1 that associates to any patient registered in the institution H the set of his/her medical files is: $R_1 = hasFile * [MedicalFile, MedicalFile^{managedBy=H}] * contains$.

Example 2: The relationship R_2 that associates to any patient having hepatitis as diagnosis, the institution where his/her has a medical file is:

$R_2 = asFile * [MedicalFile, MedicalFile_{Hepatitis\ IN\ medicalFile.contains}] * managedBy$

R_1 and R_2 are *extended* properties of the class "Patient"

Extended Attributes

In order to support the definition of *derived* attribute properties, we define at first the operator of *projection* that gives access to the values of an attribute of a given individual.

Definition (projection). Given a native class C having an attribute property V. The projection C.V of the class C on the attribute V is the function that associates to any individual x from C the set $\{V_i\}$ of the values attached to x by the attribute V.

The projection allow to formulate expressions such as: $V' = EXPR(R_1, \ldots, R_n).V$, where $EXPR(R_1, \ldots, R_n)$ is an algebraic expression of the relationships $R_1, \ldots, R_n$, and where V is an attribute property attached to the class $C' = range(EXPR(R_1, \ldots, R_n))$.

This expression defines a new property attribute attached to the class $dom(EXPR(R_1, \ldots, R_n))$. It is such as $dom(V') = dom(EXPR(R_1, \ldots, R_n))$, and $range(V') = range(V)$. If $V = R.V$, we have $maxCard(V') = maxCard(R).maxCard(V)$ and $minCard(V') = minCard(R).minCard(V)$

Example: (hasFile * contains * [MedicalEvent, Diagnosis]).longName is the property attached to a patient giving his/her various diagnosis names in clear. Such a property will be said *derived attribute.* A *native* or *derived* attribute is said to be an *extended* attribute.

Semantic Services Mapping

Basic Principles

So far, we have at our disposal an Algebra enabling to manipulate and combine the properties of an Ontology in order to *define new properties with the help of rigorously defined operators*. In this section, we exploit this Algebra for the purpose of *Services* semantic definition. We define here a *mapping model* enabling to express the meaning of a given *Service* (i.e. the meaning of the transformation it performs from its inputs to its outputs) thanks to Cntology properties. We have indicated that the basic principle was this of capturing *Service* semantics by the means of relationships. In the simplest case, the operation performed by a 1-1 *Service* (i.e. a *Service* with one input and one output parameter) may exactly correspond to an Ontology relationship. In the case when there exists a *Service* with a patient identification as input and delivering as output the set of the references to his/her various medical files, the semantic of this *Service* corresponds exactly to the ontology relationship « hasFile». We call here *mapping*, such a semantic correspondence. We shall say that the Service *realizes* the relationship.

For a given practitioner, a *Service* may deliver the set of his/her patients (i.e. for who his/her is referent practitioner), although this relationship does not exist in O. In order to express such a link, we shall considerer more general *mappings*, linking a *Service* to an *extended* relationship. Such a correspondence may be *total*, or *partial* (case where a *Service* realize only a part of the relationship).

However, 1-1 Services are a particular case of a more general n-p *Services*, with n input parameters and p output parameter. So we need a *mapping* model more general that direct association. For that, we introduce *mapping* correspondences linking an output individual to n input individuals by the mean of algebraic expressions.

At last, if the notion of *mapping* expresses what the *Service* delivers regarding meaning, we should also be able to capture *non functional aspects*, such as e.g. performance or availability of Services. We introduce for that a model of *Quality of Service* (QoS) in order to have a complete definition of a *Service*.

Simple Mapping Assertions

We consider here a 1-1 *Service*, whose input parameter is an individual from a class C_{in} = In(S), and output parameter a set of individuals from class C_{out} = Out(S). The output set of S is so $2^{Out(S)} = 2^{Cout}$, i.e.: S: $C_{in} \rightarrow 2^{Cout}$

Definition (realization of a relationship by a service). Given a 1-1 Service, where In(S) and Out(S) are natives or derived classes, and R a native or derived relationship. We shall say that the Service realizes the relationship R iff y ε S(x) ⇔ <R, x, y>

We have so: In(S) = dom(R) and Out(S) = range (R). We shall write <MAP, S, R>. This expression being called a *mapping assertion.*

A service may also *realize* an attribute. Let's consider a 1-1 *Service* S, whose input set In(S) is a class C_{in}, and whose output set Out(S) is 2^D, where D is a datatype such as Interger, String, Date, etc.

Definition (realization of an attribute by a service). Given a 1-1 Service where In(S) is an extended class, and Out(S) a datatype D. Let be V a native or extended attribute of type D. S realizes V iff v ε S (x) ⇔ <V, x, v>. We write: <MAP, S, V>

We can therefore consider *individual-oriented Services*, which deliver sets of individuals from a given class, and *datatype-oriented Services*, which deliver as output sets of datatype values from a given Ontology datatype. In order to simplify the presentation, we describe here the mapping model with only *individual-oriented Services*, only a few extensions being necessary to integrate *datatype-oriented Services*.

Restricted Mappings

A existing *Service* may realize only a part of a relationship R, i.e. only apply to a sub domain of R, and only deliver a sub part of expected results regarding R. There are many reasons for that, the main one being that it is natural that an organization only delivers information in its perimeter of influence or knowledge. In our example, it is illustrated on one hand by the relationship "hasFile", and on the other hand by the *Services* provided by the various institutions, that have a view limited to their own patients and events. Such *Services* only realize parts of the "hasFile" relationship, with restrictions on inputs and output individuals.

In this case, we have so a weaker property, i.e.: $\{y_i\} = S(x) \rightarrow \langle R, x, y_i \rangle$, although the inverse may be false. S is said a *partial realization* of R. We consider here the case where the limitations in R realization follow criteria of rationality, in relationship with some known organizational rules. So, well defined criteria of limitation may be expressed (there exist cases where it is not true, e.g. in the case of a asynchronously lazy replicated databases, where limitation may depend on delay. This case will be captured in our approach by the means of QoS).

So, we consider a more general *mapping* model, based on a new form of *mapping assertions*.

Definition (general mapping assertion). Given a Service S, a relationship R, two predicates PRE and POST respectively applying on dom(R) and range(R), a mapping assertion states that S is a realization $R^{PRE, POST}$, i.e. $(y \in Y = S(x)) \Leftrightarrow (\langle R, x, y \rangle \text{ AND PRE}(x) \text{ AND POST}(y_i))$. We write: <MAP, S, R, PRE, POST>, that is equivalent to <MAP, S, $R^{PRE, POST}$>

It has to be noticed that this notion of *post condition* on output is not at all this of *effect*, related to the semantic capture of Web Service with side effect as it may be found in the literature.

The Algebra of Services

Let's consider a set of 1-1 *Services* $\{S_i\}$, a set of relationships $\{R_j\}$, and a set of *mapping assertions* <MAP, S_i, R_j, $PRE_{i,j}$, $POST_{i,j}$>. The relationship $R_j^{PREi,j, POSTi,j}$ defines the semantic of S_i. We define the operators of *Composition*, *Union* and *Restriction* applying on Services. These operators are symmetrical to those applying on relationships:

Definition (composition, union, and restrictions of services). Given three Services S_1, S_2, S, and two predicates PRE and POST respectively applying on In(S) and Out(S).

- $S_1 * S_2$ is defined by: $x'' \in (S_1 * S_2)(x)$ iff there exists x' such as $x' \in S_1(x)$ AND $x'' \in S_2(x')$
- $S_1 + S_2$ is defined by: $x' \in (S_1 + S_2)(x)$ iff $x' \in S_1(x)$ OR $x' \in S_2(x)$
- and $S^{PRE, POST}$ is defined by: $x' \in S^{PRE, POST}(x)$ iff $x' \in S(x)$ AND PRE (x) AND POST (x')

We have: $in(S_1 * S_2) = in(S_1)$, $out(S_1 * S_2) = out(S_2)$, $in(S_1 + S_2) = in(S_1).in(S_2)$, and $out(S_1 + S_2) = out(S_1) + out(S_2)$. The invocation of $(S_1 * S_2)(x)$ leads to an invocation of S_1, and a number of invocations of S_2 depending of result $\{S_2(o)\}$ cardinality, that may be 0, 1 or many. In addition to the Services provided by infrastructure, we consider *filters* $F_{[C1,c2]}$, that are predefined Services *realizing* $[C_1, C_2]$ relationships.

An algebraic expression of *Services* is equivalent to an execution graph, where elementary instructions are *Services* invocations, controlled by the means of *composition*, *union* and *restriction* operators that respectively stand for sequence, parallel activation (fork and join) and test conditions. The *Services* provided by the infrastructure are called *real Services*, as *Services* defined by algebraic expressions of *real* Services are *abstract Services*. As an example, consider three *Services* S_1, S_2 and S_3 that respectively realize the relationship « hasFile », the relationship "contains" for public institutions, and the same relationship « contains » other institutions. The complete medical file of a patient P is given by invocation of the abstract Service $S_1 * (S_2 + S_3)$.

If <MAP, S_1, R_1> AND <MAP, S_2, R_2>, we have <MAP, $S_1 * S_2$, $R_1 * R_2$> and <MAP, $S_1 + S_2$, $R_1 + R_2$>. If <MAP, S, R>, then we have <MAP, $S^{PRE, POST}$, $R^{PRE, POST}$>. This defines an algebraic *homomorphism* between the relationship Algebra and the 1-1 Services Algebra. An important result concerns the evaluation of *derived* relationships with several *Services* realizing *partial* mappings, thank to the following property:

Let be a relationship R, a set of Services S_i, i =1, …, n, such as <MAP, S_i, R, PRE_i, $POST_i$>. The Service $S = \sum_{i=1,\dots,n} S_i$ realizes the relationship R, i.e. <MAP, $\sum_{i=1,\dots,n} S_i$, R> iff $OR_{PREi, POSTi}$ (PRE_i AND $POST_i$) = True

Similar definitions and results may be developed for datatype oriented *Services*.

Complex Mappings

We consider here n-p *Services*, i.e. with n input parameters and p output parameters. We suppose that parameter types are still Ontology *extended* classes. We consider at first the n-1 *Services*, then the general case of n-p Services.

Let's consider three classes "Patient", "MedicalFile" and "Institution", linked by relationships "hasFile" (one or several files for a given patient) and "managedBy" (only one institution for a given file), as indicated on the figure 6.

Let's consider a *Service* provided by an identity server delivering references to medical files for each ordered pairs (patient, institution) given as input, i.e.: S: (Patient, Institution) → $2^{\text{MedicalFile}}$. This is a 2-1 Service, with $In_1(S)$ = Patient, $In_2(S)$ = Institution, and Out(S) = MedicalFile. The relationship that links the output set {Files D} to inputs Patient P and Institution I is simply:

$R_S = \text{hasFile}^{\text{True, PR (I)}}$, where PR is the predicate defined by PR(I) = P.Institution = E

This is a *restriction* of the Ontology relationship "hasFile" by the predicate POST, parameterized by the other input parameter I. This may be written:

<MAP, $S_{in2=I,}$ hasFile, True, P.Institution= $In_2(S)$>.

where $S_{in2=I}$ is the 1-1 *Service* that associates to a patient P the output S(P, I), and where $In_2(S)$ denotes the value of the second parameter of S, i.e. the current value of input I.

In order to express the semantic of such a n-1 Service S: $(C_{in\,1}, \dots, C_{in\,n}) \rightarrow 2^{Cout}$, one has to define an algebraic formula of the form: R_S = EXPR ($\{R_k\}$, $\{PR_l\}$) that gives the *extended* relationship R_S linking the output with an i_0^{th} input parameter. This relationship is expressed by the means of Ontology relationships, and predicates $\{PR_l\}$ involving the values of the other input parameters from $In_i(S)$, $i \neq i_0$, inside restriction predicates

The relationship R_S is the relationship *realized* by the Service S, i.e. such as <MAP, S, R_S>. It expresses the semantic of S on the basis of properties and predicates. This expression is not necessarily unique, in particular in the case where there exist in the Ontology *inverse* relationships of those used in the considered expression. The general method to define such an expression consists in: 1) determining a relationship path in the Ontology O linking one of the input parameter class to the output parameter class, and 2) adding predicates corresponding to the constraints involved by the data of other input parameters. The conditions of existence of such a mapping should be studied in detail.

Figure 6. Case of a mapping with (2, 1) service

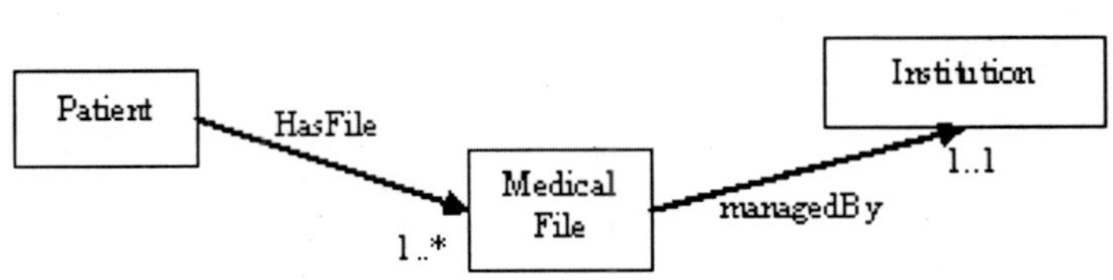

Now, let's consider the general case of a n-p Service, with n input and p output parameters, i.e. S: $(C_{in\,1}, \ldots, C_{in\,n}) \rightarrow (2^{Cout\,1}, \ldots, 2^{Cout\,p})$. The semantic of the *Service* S is perfectly defined by the semantic of the p partial *Services* S_j: $(C_{in\,1}, \ldots, C_{in\,n}) \rightarrow 2^{Cout\,j}$ (p projections of S on the p output sets $C_{out\,1}, \ldots, C_{out\,p}$), which are n-1 Services, and so, the previous results apply.

Quality of Service

It may exist several ways to *realize* a relationship, i.e. to evaluate a property by the means of *Services* invocations in response to a user query. E.g., in order to access to the complete medical file of a given patient, we may decide to address in parallel direct queries to each institution, via ad hoc *Services* provided by each of them. We may also decide to query first a relevant *Service* provided by the regional health server, that will deliver the set of institutions in which the given patient has a medical file, then to request only the relevant ones. This choice is influenced by many factors such as the number of institutions, the expected delay of execution of each individual *Service*, their average availability, and, may be, some additional factors such as the expected quality of data, factor that gives higher quality to fresh data against data with possible lack of recent pieces of information (e.g. data from mirror or cache sources). Each factor may be quantified with a magnitude relevant with its meaning (e.g. a time, a probability, etc.). We consider here a set of quality factors F_i, i=1, ..., p, with their associated metrics q_i. We consider the *quality function*:

$$q = [q_1, .., q_p] = Q(S_i),$$

that associates to each *Service* S_i, a p-dimension vector where the ith dimension is the quantification of the F_i factor. Let be a *function of preference*:

$$Pref = \Lambda(q) = \Lambda([q_1, .., q_p]) = \sum_{i=1, \ldots, p} \alpha_i . q_i$$

provided by a calling entity, and that aggregates the various quality dimensions is a single relevant value and enables to compare various *realisations* of a given evaluation. For each factor q_i, we should express rules that define how to aggregate values of q_i factors when *services* are combined by composition or by union:

$$q(S_1 * S_2) = F(q(S_1), q(S_2)) = [F_i (q_i(S_1), q_i(S_2))]$$

$$q(S_1 + S_2) = G(q(S_1), q(S_2)) = [G_i(q_i (S_1), q_i(S_2))]$$

where F_i and G_i functions depend on the semantic of the considered q_i factor. Depending of this semantic, each F_i or G_i function may be a sum, a maximum, a minimum, etc. In the case of composition, where the number of S_2 invocations depends on the cardinality of S_1 results, the definition of F_i should reflect the chosen strategy of optimization (e.g. minimax). So, the *Services* infrastructure should provide relevant meta information such as the maximum or the mean cardinal of results for each *Service*. As a simplification, we may write:

$$q(S_1 * S_2) = q(S_1) * q(S_2)$$

$$q(S_1 + S_2) = q(S_1) + q(S_2)$$

to denote the combination of QoS vectors by the $*$ and + operators.

Automated Execution Plans Generation

We study here the general issue of determining the set of *Services* that should be invoked in response to a request, as well as the way in which they have to be orchestrated to meet their objective. Firstly, we define in details the various elementary issues. Then we present the principles of new *mapping* assertion generation that may be followed to solve our problem. This enables to present at last an original algorithm providing solutions to our problem.

The Execution Plan Problem

So far, we have defined: 1) a way to express the semantic of the various *Services* provided by a given configuration, and 2) a way to express user queries under the form of *derived* properties evaluations. Such properties are expressed with the use of a combination of the various *native* elementary properties of the ontology.

Now, the main question is to determine the relevant execution plan of *Services* invocations that will deliver the expected result, i.e. the transformation of a given algebraic expression of properties into a plan of *services* invocations. Such a plan will be called here an *orchestration* plan. As the order of invocations is significant, and as some *Services* may be invoked in parallel, such an *orchestration* plan may be represented by an *execution graph*, where nodes represent the intermediate results, and where branches represent (sequential and/or parallel) tasks to execute.

In order to simplify the presentation, we focus here the presentation on the evaluation of *derived* object properties (relationships), the whole approach described in this section remaining valid for the evaluation of datatype properties (attributes), with a simple extension.

Having a configuration defined by an ontology, a set of defined *derived properties*, and a set of *Services*, with definitions of *mappings* relating *Services* to ontology properties, we consider a property R to evaluate, with a possible given *utility function*, and an individual x given as input. There are in fact three problems:

- A first issue is this of determining whether, in the given configuration, there exists or not a plan of *Service* invocations that may deliver the expected values.
- If we can be sure there exists a solution, a second issue concerns the construction of the solution graphs, and, if there are several solutions, the determination of the optimal plan regarding the criteria defined by the *function of preference*.
- If there is no solution, a third issue is this of determining possible restricting conditions regarding the input x, in case of which it would exist partial solutions

The second problem is this of the search for a *uniform optimal solution*, i.e. a unique solution which is optimal for any input x. If this solution exists, this will define an optimal *orchestration* plan that may be kept in memory for any further evaluation <R, x, ?y> to perform on this property R (static optimal execution plan). If it does not exist, there may exist solutions which are optimal for some $(dom(R))^P$ subclasses of dom(R).In this case, at runtime, the plan to execute should be the relevant one regarding the value of the input (dynamic execution plan).

In this the following, we present an approach that enables to deal with the three issues, thanks to one single algorithm. Such an algorithm generates a set of possible execution graphs. An execution graph determines a plan for *Service* invocations, with some *Service* composition (execution of two *Services* as a sequence), union (concurrent execution with join and result fusion), and restrictions (invocation with test condition). An execution graph G_i defines an algebraic combination of *Services*. During the execution of the algorithm, we shall consider plans that *realize* the property to evaluate, but also plans that *realize* only a subpart of the expression.

An execution graph G_i has an origin denoted $Orig(G_i)$, which stands for an input set of individuals, labeled by a $PRE(G_i)$ predicate, expressing the conditions to be satisfied by the input for the plan to be valid. The execution graph has also a end denoted $End(G_i)$, that stands for the output set of individuals, labeled with a $POST(G_i)$ predicate, that can express limitations in the delivered results. An execution graph is also labeled with a QoS value $Q(G_i)$.

An execution graph G_i which have the dom (R) class as origin, the range(R) class as end, and verifying POST(G_i) = True, will be called a *candidate partial solution*. If, in addition, PRE = True, then G_i will be a *candidate solution*. If we want to evaluate in advance an *orchestration* plan associated to a property, we shall apply at first the algorithm. Then, if there are candidate solutions, then the delivered solution (uniform optimal solution) will be the candidate solution G_0 maximizing QoS(G). This plan will enable to evaluate the property for any value of the input individual x. If there is no *candidate solution*, but if there exists some *candidate partial solutions* $\{G_j\}$, then it will be possible to evaluate the property iff the input given individual x satisfies the PRE(G_j) predicate. The delivered solutions will be the *partial candidate solution* G_1 maximizing QoS(G_j) among all partial candidate solutions such as PRE(G_j) (x) is satisfied. At the contrary of the previous case, there is here only a pseudo order on solution, as their associated valid input domains are not the same. This pseudo order become a total order as soon as the input individual x is specified. In this case, for a given input x_0, either there will be no solution; or there will be a solution for which the optimality is ensured, only for x_0.

Orchestration of Services

The basic idea on which our algorithm is based will be this of an iterative generation of new *mapping* assertions, derived from already defined ones. The problem is this of identifying the states in which it is actually possible to generate such new *mappings assertions*. Let be two relationships R_1 and R_2, and two *services* S_1 and S_2, such as:

<MAP, S_1, R_1, PRE_1, $POST_1$> and <MAP, S_2, R_2, PRE_2, $POST_2$>

In order to characterize $S_1 \ast S_2$ and $S_1 + S_2$ in terms of *mappings*, we use the following properties:

Property 1: It is possible to define a *mapping* for $S_1 \ast S_2$ iff: $POST_2$ = True AND PRE_1 = True, and this *mapping* is:

<MAP, $S_1 \ast S_2$, $R_1 * R_2$, PRE_1, $POST_2$>

As a particular case, we may notice that, if S_1 *realizes* R_1 (PRE_1 = $POST_1$ = True) and if S_2 *realizes* R_2 (PRE_2 = $POST_2$ = True), then $S_1 \ast S_2$ *realizes* $R_1 \ast R_2$

Property 2: the best *mapping* that may be defined for $S_1 + S_2$ is:

<MAP, ($S_1 + S_2$), ($R_1 + R_2$), PRE_1 OR PRE_2, $POST_1$ AND $POST_2$>.

So, the *mapping* exists iff $POST_1$ AND $POST_2$ $\neq$ False, i.e. iff out(S_1) $\cap$ out(S_2) $\neq \emptyset$.

Example: let be a class C_1 with a datatype property A, a class C_2 with a datatype property B, a relationship R_1 from C_1 to C_2, and a relationship R_2 from C_2 to C_3. If we have some *Services* $S_{1,1}$, $S_{1,2}$, $S_{2,1}$, $S_{2,2}$, such as:

<MAP, $S_{1,1}$, R_1, $PRE_{1,1}$ = (A=a_0), True> ; <MAP, $S_{1,2}$, R_1, $PRE_{1,2}$ = (A$\neq a_0$), True>

<MAP, $S_{2,1}$, R_2, $PRE_{2,1}$ = (B=b_0), True> ; <MAP, $S_{2,2}$, R_2, $PRE_{2,2}$ = (B=b_0), True>

At first, we may generate two *mappings*. The first one is: <MAP, $S_{1,1} + S_{1,2}$, R_1, True, True> and the second one is: <MAP, $S_{2,1} + S_{2,2}$, R_2, True, True>. Considering this new *abstract Services* S = $S_{1,1} + S_{1,2}$, and S' = $S_{2,1} + S_{2,2}$, it appears that we may generate a new *mapping* involving S, that is: <MAP, S, $\ast$ S', R_1, True, True>. At the contrary, in the case we would have at the beginning the following *mappings*:

<MAP, $S_{1,1}$, R_1, $PRE_{1,1}$ = (A=a_0), $POST_{1,1}$ = "B=b_0">

<MAP, $S_{1,2}$, R_1, $PRE_{1,2}$ = (A$\neq a_0$), $POST_{1,2}$ = "B$\neq b_0$">

the only *mapping* to be derived is: <MAP, $S_{2,1} + S_{2,2}$, R_2 >, and no other new *mapping* may be generated after that. This is symbolized on the Figure 7.

Before presenting the algorithm, we state the following definitions:

Definition (service equivalence). Given two Services S_1 and S_2, whose semantics are defined by the two mapping assertions <MAP, S_i, R_i, PRE_i, $POST_i$>, i = 1, 2. S_1 and S_2 are said to be equivalent ($S_1 \approx S_2$) iff $R_1 = R_2$, $PRE_1 = PRE_2$, and $POST_1 = POST_2$.

We shall say that $S_1 \geq S_2$ iff $R_1 = R_2$ AND ($PRE_1 \rightarrow PRE_2$) AND ($POST_1 \rightarrow POST_2$) (i.e. S_1 is a better realization than S_2 of the same relationship $R = R_1 = R_2$)

It has to be noticed that "≥" is not a pseudo order, because $S_1 \geq S_2$ AND $S_2 \geq S_1$ implies $S_1 \approx S_2$, but not necessarily $S_1 = S_2$. In an algebraic expression of *Services*, it will be possible, at first, to replace an operand *Service* S_i by an equivalent *Service* S_j or by a *Service* S_j such as $S_i \geq S_j$ if $q(S_j) \geq q(S_i)$. In order to *realize* an algebraic expression of properties, it is then necessary to find in the repository of available *Services*, all the *Services* that *realize* (totally or partially) a part of the expression. The *Services* found in the repository, that will "match" a given relationship, will be the possible building blocks of a future execution plan for the evaluation of R.

At last, we define the *matching* of *Service* with the following definition.

Definition (service matching). Given a relationship R defined as an algebraic expression, given a Service S, we shall say S matches R iff if exists a mapping assertion such as <MAP, S, R', PRE, POST>, where R' is a sub expression of R.

An Algorithm for Execution Plan Generation

We present here the algorithm enabling the construction of solution execution graphs in response to a given *derived* property evaluation. Depending on the case, the algorithm will provide: 1) A *uniform optimal plan* that will work for any individual x from the class dom (R), or 2) a set of plans with a associated constraints on input and QoS values for each plan.

The algorithm works in five main steps:

- **Step 1. Evaluate the input algebraic expression in terms of elementary *native* properties**. We replace, in a recursive way, each operand *derived* property by their definitions. So, for an expression such as: R = R' + R", where R' and R" are *derived* properties defined by: $R' = R_1 {}_* R_2$, $R'' = R_1 {}_* R'''$, and $R''' = R_3 {}_* R_4$, and where R_1, R_2, R_3 and R_4 are *native* properties. The expression of R will be transformed via the following iterations: $R = R_1 {}_* R_2 + R_1 {}_* R'''$, then $R = R_1 {}_* R_2 + R_1 {}_* R_3 {}_* R_4$
- **Step 2. Simplify the algebraic expression**, thank to algebraic properties of operators (factorization). So, the expression: $R = (R_1 {}_* R_2) + (R_1 {}_* R_3 {}_* R_4)$ becomes: $R = R_1 {}_* (R_2 + (R_3 {}_* R_4))$. This is done in order to minimize the number of *Service* invocations and the flow of intermediate results. The algebraic formula is stored under the form of an *execution tree*, where the leaves are operands R_i and the nodes are partial results. In the above example, we shall have the following nodes: (N_1):$R_1 {}_* (R_2 + R_3 {}_* R_4)$; (N_2): $R_2 + R_3 {}_* R_4$; (N_3): $R_3 {}_* R_4$

Figure 7. Example of mapping generation

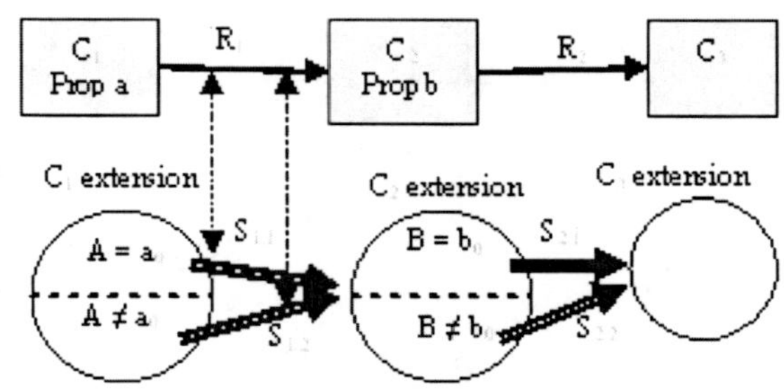

Figure 8. Example of flow – relationship graph

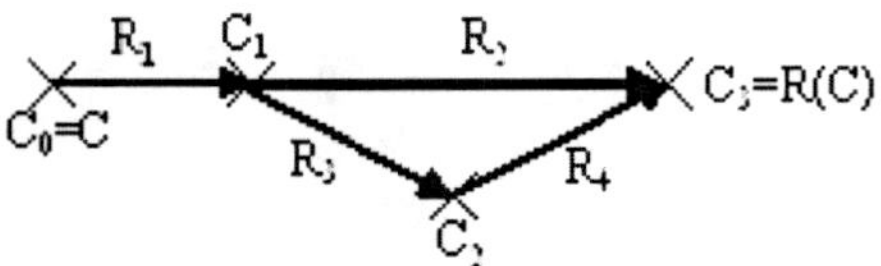

- **Step 3. Build the *flow - relationship graph* associated to the expression**. On the basis of the previous, maybe simplified, expression, one generates a directed acyclic graph, corresponding to the evaluation of the result property, where nodes $\{C_i\}$ stands for collections of values corresponding to the various intermediate levels of evaluation, and edges $\{R_j\}$ are instances of relationships, relating an input collection $C_{j,1}$ to an output one $C_{j,2}$, and labeled by the corresponding operand relationship.

The origin of the graph is the input x issued from the class $C_0 = \mathrm{dom}(R)$, the end of the graph is the expected collection of result values (it has to be noticed that the same relationship may have several instances as several distinct edges in the graph). As an example, the expression $R = R_1 * (R_2 + (R_3 * R_4))$, above considered, generates the following *flow - relationship graph*, which has four nodes and four edges.

- **Step 4. Find Services that matches parts of the graph**. This step consists in extracting from the repository of *Services*, all the *Services* that match a part of the *flow – relationship graph*, as it has been defined in the previous section (*match* operator). We get a subset $\{S_i\}$ of *Services* which will be the input *service* set of the following step of the algorithm.
- **Step 5. Generate the candidate execution graphs**. This is done by combining the selected *Services* in various manners, in order to construct a combination of *Services realizing* the relationship R, if this realization exists. There are two possible approaches for developing such an algorithm: the first one is this of a *descendant* algorithm that start from the relationship to evaluate, and tries to express it in function of the given input *Services*. The second one, that we have adopted here, is this of an *ascendant* algorithm. The algorithm takes the input *Services*, and combines them iteratively, in order to derive at each iteration new *mappings* that give *better* or *more complete* realizations (in the meanings of the QoS and of the above defined comparator $\geq$) than those already exhibited. The algorithm stops when there is neither new possible *matching*, neither new (better, or more complete) *mapping*. This stop is guaranteed due to the strict increasing of a function on a discrete set. When there exists a *mapping* involving both the graph origin and end, then there exists an execution plan which is at least a partial candidate solution.

If the *uniform optimal solution* does not exist, this algorithm stops with, as present state, the best partial solutions, solving de facto the third issue presented at the beginning of this section. With no global solution, these partial solutions will nevertheless permit to have a possible available solution for a given input x. In this case, the best solution will be selected at runtime.

The principle of the present step of the algorithm is so the following: we consider now the *realization graph*, that is a directed graph based on the previous flow – relationship graph, where nodes are those of the *flow – relationship graph*, but with possible additional edges. So, there are two types of edges in this graph:

- The *relationship edges*, that are the edges of the *flow - relationship graph*, standing for operand relationships,

- New edges iteratively generated by the algorithm, and standing for partial *realizations* of the *flow - relationship graph*. Such a *Service edge* may represent a *real Service*, as well as an *abstract Service*, i.e. an algebraic expression of some real Services. A *Service edge* (C_i, C_j) is a *realization* of the relationship relating C_i to C_j. It is labeled by a 3-uple (PRE, POST, q) where PRE and POST are *mapping* predicates, and q is the QoS vector associated to the realization.

The *relationship edges* are present at the beginning of the algorithm. The *Service edges* are incrementally added at each stage of the algorithm. The algorithm corresponding to this step may be expressed as a recursive procedure: at each stage, a new *Service* is considered. This *Service* may come from the input *Service* set or may have been generated at a previous stage. We integrate this *Service* as a new edge in the *realization graph*, labeled by the existing *mapping*. Among the already present *Services edges*, we consider those that may be combined with the new *Service* to generate at least a new *mapping*. In case it is possible, only one new *Service edge* is created, and a similar process is applies recursively. This recursive algorithm is so:

```
Algorithm
  for each S IN Input Service set
    Express mapping Mi = (R, PRE, POST)
    between S and a Relationship R
    Add a new Service edge in the graph
labelled
    with this mapping Mi and the QoS vec-
tor of S
    INTEGRATE (S)
  End for each
end Algorithm
Procedure INTEGRATE (in S: Service)
if There exists S' such as (S' ≥ S) OR
((S' Eq S AND q (S') ≥ q (S)
then delete S
else Associate S to possible Service
Edges Ei for each Ei
Determine the associated mappings (Ri,
PREi, POSTi) and the resulting quality of
service qi
Add a Service edge labelled with R, PRE,
POST and q
INTEGRATE (Ei)
end for each
end if
end Procedure
```

Redundant *Service edges* (i.e. that correspond to *Services inferior* to an already present *Service*, or *equivalent* with lower QoS) are removed in order to avoid the explosion of non significant *mappings*. At the termination of the algorithm, the partial solutions are the *Service edges* (if they exist) linking the origin with the end of the graph, and labelled with a (POST = True) condition. If one of such *Service edges* has a (PRE = True) condition, then it is an optimal solution. If not, the result of the algorithm is the set of 3-uples (S_i, PRE_i, q_i), where S_i is a partial realization, PRE_i the corresponding PRE validity condition, and q_i the associated QoS. In any case, a solution is an algebraic expression of the input *Services* that defines an orchestration of *Services*, i.e. an execution plan defining the *Service* invocation to execute with sequences and possibly concurrent branches, as well as test conditions to perform. The solution to the problem of a datatype property evaluation is based on the evaluation of a relationship, as seen above, with some additional specific operations not detailed here.

A Software Framework

On the basis of our approach, a software framework has been prototyped. This framework supports the definition and management of *derived* properties, issued from various user communities, and defined on the basis of a provided ontology defined via

PROTEGE. It supports the generation and run time execution of the relevant *Service* orchestration in response to a request for information. The framework also enables the management of Web *Services*, with all their relevant associated meta information. The framework includes: 1) A derived properties repository, where the description of derived object and datatype properties are stored, and queried, 2) a Web Service platform, that enables to develop and execute Web Services, with interface types conformant with classes and datatypes of the ontology, 3) a Web Service repository that stores the semantic descriptions of Services, and other required meta information, as defined in the present paper, In addition, we have two software components: 4) the Generator that generates execution plans, with an implementation of the algorithm we have proposed, and the Orchestrator, that executes such generated plans.

The framework has been tested in the context of the federation of several Health Information Servers, described in the Case Study section, and more specially in the context of a new application in oncology: a support to multidisciplinary meetings in which therapeutic decisions are taken. The results are very encouraging because the framework clearly adds important factors of openness and flexibility to the context. The experiment shows the approach constitutes an efficient way to easily integrate new information sources in an Information Server federation, and take into account new user needs, while avoiding huge amount of specific software coding.

FUTURE RESEARCH DIRECTIONS

To deal with the problem of Web Services automatic discovery and composition, we have presented in this chapter an Algebra allowing rigorous combinations of Ontology properties. This algebra enables to attach a precise meaning to any expected piece of information, as well as to confer to an existing Web Service a well defined semantic based on the Ontology concepts. On this basis, we show it is possible to develop an efficient algorithm generating optimal execution plans of Web Services.

The main hypothesis on which relies the applicability of the approach is this of a common agreement of user communities on an exhaustive and fine-grained ontology of their domain. Of course, this is at present a major limitation in the adoption of such an approach, but we do think that, on one hand, a capture of the application domain via an Ontology, and, on the other hand, a rigorous model, able to confer a well defined formal semantic to a Service, are the absolute requisite to achieve the expected objective. There are still many difficult issues to solve in the future in order to meet the complete objective of automated discovery and composition. To continue in the direction presented in this chapter, three main axes may be defined now:

A first axis consists in extending our approach to more general form of ontology properties. In particular, we may consider those that would be *deducted* by the means of inference rules. For example, such rules would be defined by permitting cyclic definitions of derived properties. This would introduce a reasoning aspect in the approach, and would lead to logical approaches for orchestration plan generation;

A second axis concerns the extension of the approach by considering Services having an *effect* on internal data (creation, update) and/or on external world. This would permit to address topics related to popular applications, such as these of electronic commerce, or construction of ad hoc processes in Information Systems of companies. Substantial works have already be done on these issues, and we think an Ontology-based algebraic approach would bring new developments;

The third axis is this of elaborating on the results in the framework of present standards and languages (OWL, OWL-S, …). In particular, declarative languages and user-friendly tools adapted to the problematic would be highly required. In addition, approach for the reuse and integration of existing Web Services in an Ontology-based approach is also a challenging issue.

CONCLUSION

The flexible integration of heterogeneous information sources, as well as the ways to query them by the mean of Web Services orchestrations, are not recent issues. But with the increasing importance of Ontologies, new approaches have to be developed. In the context of Semantic Web and widely spread decentralised architectures, a new challenge relative to the taking into account of the semantic dimension has now appeared with a very strong importance. This dimension concerns in particular the definition of semantic correspondences between on one hand Web Services, and, on the other hand, knowledge about domain, expressed via Ontologies.

In this paper, arguing that an important part of Services semantic may be captured by the means of Ontology relationships, we show that, on the basis of this hypothesis, it is possible to build a consistent and well formalized Algebra that enables to perceive any property definition, combination and evaluation of them as algebraic operations.

In this context, we propose (1) an Algebra of Ontology properties, that enables definitions of new properties on the basis of native ontological ones, (2) a model that enables to associate a formal semantic to a Service using mapping assertions using Ontology properties, and (3) a general algorithm that performs an automated generation of execution plans, and translates a property evaluation into a optimal orchestration of Services.

On the basis of our proposed approach, a software prototype has been developed and deployed in the context of an Health Information Systems, in order to provide new facilities. The evaluation shows that our approach provides to the application the properties of openness and flexibility, saving huge efforts that would have been spent in specific code development in the case of a classical software development approach.

REFERENCES

Akkiraju, R., Farrell, J., Miller, J., Nagarajan, M., Schmidt, M.-T., Sheth, A. & Verma K. (2005). *Web service semantics – wsdl-s*. W3C Member Submission, November 2005

Ayadi, N., & Lacroix, Z. (2008, November). Biomap: a deductive approach for resource discovery. In *UWAS'2008 - The Tenth International Conference on Information Integration and Web-based Applications Services*, 24-26 November 2008, Linz, Austria, pages 477-482

Berners-Lee, T., Hendler, J., & Lassila, O. (2001). The semantic web. *Scientific American*, (5): 35–43.

Gao, H. T., Hayes, J. H., & Cai, H. (2005). Integrating biological research through web services. *Computer*, *38*(3), 26–31. doi:10.1109/MC.2005.97

Goble, C., Kesselman, C., & Sure, Y. (Eds.). (2005). In Semantic Grid - Convergence of Technologies, number 05271 in Dagstuhl Seminar Proceedings.

Klusch, M., Pries, B., & Sycara, K. (2006). Automated semantic web service discovery with OWLS-MX. In *Proc. 5th International Joint Conference on Autonomous Agents and Multiagent Systems,* (pp. 915-922), Hakodate, Japan. New York: ACM.

Kopecky, J., Vitvar, T., Bournez, C., & Farrell, J. (2007). SAWSDL: Semantic Annotations for WSDL and XML Schema. *IEEE Internet Computing*, *11*(6), 60–67. doi:10.1109/MIC.2007.134

Kourtesis, D., & Paraskakis, I. (2008). Combining SAWSDL, OWL-DL and UDDI for Semantically Enhanced Web Service Discovery. In M. Hauswirth, M. Koubarakis, and S. Bechhofer, (Eds.), *Proc. 5th European Semantic Web Conference,* (LNCS 5021, pp. 614-628). Berlin: Springer.

Martin, D., Burstein, M., Mcdermott, D., Mcilraith, S., Paolucci, M., & Sycara, K. (2007). Semantics to Web Services with OWL-S. *World Wide Web (Bussum)*, *10*(3), 243–277. doi:10.1007/s11280-007-0033-x

Maximilien, E. M., & Singh, M. P. (2004). A framework and Ontology for Dynamic Web Services Selection. *IEEE Internet Computing*, 5.

Paolucci, M., Kawamura, T., Payne, T. R., & Sycara, K. P. (2002). Semantic Matching of Web Services Capabilities. In *Proc. 1st International Semantic Web Conference on The Semantic Web,* (LNCS 2342 pp. 333-347). London: Springer.

Papazoglou, M. P., Traverse, P., Dustdar, S., & Leymann, F. (2007). Service-oriented computing: State of the art and research challenges. *Computer*, *40*(2), 38–45. doi:10.1109/MC.2007.400

Roman, D., & Keller, U., H. Lausen H., De Bruijn, J., Lara R., Stollberg,M., Polleres, A., Feier, C., Bussler, C. & Fensel, D. (2005). Web Service Modeling Ontology. *Applied Ontology*, *1*(1), 77–106.

Sirin, E. (2003). Semi-automatic composition of Web services using semantic description . In *Proceedings of Web Services Modelling, Architectures and Infrastructures workshop in conjunction with ICEIS'2003, 12(8), 72-7*. Hendler J, Parsia B.

Sycara, K., Paolucci, M., Ankolekar, A., & Srinivasan, N. (2003, December). Automated discovery, interaction and composition of Semantic Web services. *Web Semantics: Science . Services and Agents on the World Wide Web*, *1*(1), 27–46. doi:10.1016/j.websem.2003.07.002

Chapter 5
Approaches for Semantic Association Mining and Hidden Entities Extraction in Knowledge Base

Thabet Slimani
ISG of Tunis, Tunisia

Boutheina Ben Yaghlane
IHEC of Carthage, Tunisia

Khaled Mellouli
IHEC of Carthage, Tunisia

ABSTRACT

Due to the rapidly increasing use of information and communications technology, Semantic Web technology is being increasingly applied in a large spectrum of applications in which domain knowledge is represented by means of an ontology in order to support reasoning performed by a machine. A semantic association (SA) is a set of relationships between two entities in knowledge base represented as graph paths consisting of a sequence of links. Because the number of relationships between entities in a knowledge base might be much greater than the number of entities, it is recommended to develop tools and invent methods to discover new unexpected links and relevant semantic associations in the large store of the preliminary extracted semantic association. Semantic association mining is a rapidly growing field of research, which studies these issues in order to create efficient methods and tools to help us filter the overwhelming flow of information and extract the knowledge that reflect the user need. The authors present, in this work, an approach which allows the extraction of association rules (SWARM: Semantic Web Association Rule Mining) from a structured semantic association store. Then, present a new method which allows the discovery of relevant semantic associations between a preliminary extracted SA and predefined features, specified by user, with the use of Hyperclique Pattern (HP) approach. In addition, the authors present an approach which allows the extraction of hidden entities in knowledge base. The experimental results applied to synthetic and real world data show the benefit of the proposed methods and demonstrate their promising effectiveness.

DOI: 10.4018/978-1-61520-859-3.ch005

INTRODUCTION

As described by Tim Berners-Lee, with the intense activity of Semantic Web (SW) in industry and academia (Berners-Lee, T. et al, 2001), it is reasonable to expect that increasingly more metadata describing domain information about resources on the Web will become available. The characteristic of distributed information, in the Semantic Web allows anyone to link anything to anything (linked data). The massive growth of the linked data stored in the Web is managed by a network of interrelated data sources (Semantic Link Network: SLN) which contains several types of entities (persons, companies, domain knowledge, general knowledge, scientific publications, books, etc.). SLN is a network containing semantic nodes and semantic links. A semantic node (Semantic Node: SN) can be a concept, an instance of concept, an URI, an entity, a particular form of resources in knowledge base (Zhuge H., 2004). A semantic link reflects a kind of relational knowledge represented as a pointer with a tag describing such semantic relations as *causeEffect*, *implication*, *subtype*, *similar*, *instance*, *sequence*, *reference* and *equal* (Zhuge H., 2007). Thus, SW is not a Web of documents, but a Web of semantic relations between entities denoting real world objects such as people, places and events.

The core idea of Semantic Web is to represent entities and relationships between them using ontologies for the purpose of the interoperability between machines themselves or the interoperability between machines and humans.

Consequently, the use of ontologies has been proven to be a good choice for knowledge and various information representations in a human understandable and machine-readable format consisting of entities, attributes and relationships. The direct relationship between two entities refers to us as semantic relations and indirect relationship as semantic association.

Semantic association (SA) discovery and mining is an important issue for applications including networked data. The approach of semantic association discovery initially appeared with the theories and the methods coming from the research of LSDIS laboratory at the University of Georgia[1]. A semantic association is essentially a graph-theoretic based approach that represents, discovers and interprets complex relationships between entities contained in RDF graph (Aleman-Meza et al, 2003) (Anyanwu K. et al, 2005) (Aleman-Meza B. et al, 2006) (Ning X. et al, 2006). A formal definition is presented in (Anyanwu & Sheth, 2002): "Semantic Associations capture complex relationships between entities involving sequences of predicates, and sets of predicate sequences that interact in complex ways". Since the predicates are semantic metadata extracted from various multi-source documents, this is an attempt to discover complex relationships between objects described in those documents. Detecting such associations is crucial for many research and analytical activities that are important to applications in national security, business intelligence and bioinformatics. The datasets that semantic associations operate over are RDF/RDFS graphs.

Thus, for a better exploitation of the extracted semantic association in several applications, it is essential to extract significant patterns and relevant information from preliminary extracted associations.

Semantic Web Mining aims at combining the two areas "Semantic Web" and "Web Mining". The Semantic Web addresses the first part of the new challenge posed by the great success of the current WWW aiming to make the data (also) machine-understandable, while Web Mining addresses the second part by (semi-) automatically extracting the useful knowledge hidden in these data, and making it available as an aggregation of manageable proportions. The term Semantic Web Mining can be interpreted as *Semantic* (*Web Mining*) and as (*Semantic Web*) *mining*.

In this chapter, we concentrate on mining approaches that refer to the explicit structure included in semantic association.

This chapter aims to give an overview of where the Semantic Web and Web Mining get together today. In our survey, we will first describe the current state of the two areas and then discuss their combination. We will provide references to typical approaches that have not been yet developed explicitly.

In the next section, we give brief overviews of the areas of Semantic Web (SW) and Web Mining (SM). Readers familiar with these areas can skip Section 2. We then go on, in the same section, to describe how these two areas cooperate together.

For a better semantic contribution, we propose to follow a process of association rules mining starting from a set of semantic associations already discovered. The process of semantic association mining is based on a generic parameterized request which allows the orientation of semantic association rule discovery. A survey of this method is contained in Section 3.

In relation to the cooperation between the two presented area (SW and SM), the set of semantic associations (Context of SW) returned to a given user can presents the ambiguity to select the appropriate semantic association which answer her/his needs. Consequently, we apply a hyperclique pattern approach (Context of Data Mining) to restrict the number of returned associations to the set which presents interests to user requirements. Section 4 presents our development of this approach.

Two connected semantic nodes could derive out a new semantic link if there is an applicable reasoning method. Semantic links between semantic nodes can be established in two ways: user definition and automatic discovery. The process of automatically generating semantic links can explain how this relation is established. In Section 5, we give a new method allowing the possible insertion of new semantic links based on the sequential characteristic contained in semantic association between a specified source node and a target semantic node not initially recognized. The future trends are described in Section 6. We conclude, in Section 7, that a flexible integration of these aspects will increase the comprehensibility of the Web for machines, and consequently will become the base for further generations of intelligent Semantic Web tools.

SEMANTIC WEB AND WEB MINING

In the first part of this Section, we briefly recall our understanding of the semantic web. In the second part, we give an overview of some data mining approaches which can be applied in Web Mining.

Semantic Web

The notion of Semantic Web (Berners-Lee, T. et al, 2001), proposed by Berners-Lee has become a research area with full expansion. Berners-Lee suggested a layer structure for the Semantic Web (Unicode/Unified Resource Identifiers, XML, RDF, ontologies, logic, proof, and trust). Each layer represents a step which will already provide added value, so that the Semantic Web can be realized in an incremental approach. In lower layer the standard way to refer to entities[2] is defined by URI. The URI (www.w3.org/Addressing/URL/URI_Overview.html) is used to uniquely identify resources.

The XML (Extensible Markup Language)[3], RDF[4](Resource Description Framework) and OWL[5] (Ontology Working Language) describe the semantics of resources at different levels. XML fixes a notation for describing labeled trees. Whereas XML is one step in the right direction, it only formalizes the structure of a document and not its content. RDF can be seen as the first layer where information becomes machine understandable. The components of each RDF document consist of three types of entities: Resources (subjects and objects), properties (predicates/relations). At the layer of ontology vocabulary it is possible to

Figure 1. Layers of semantic web

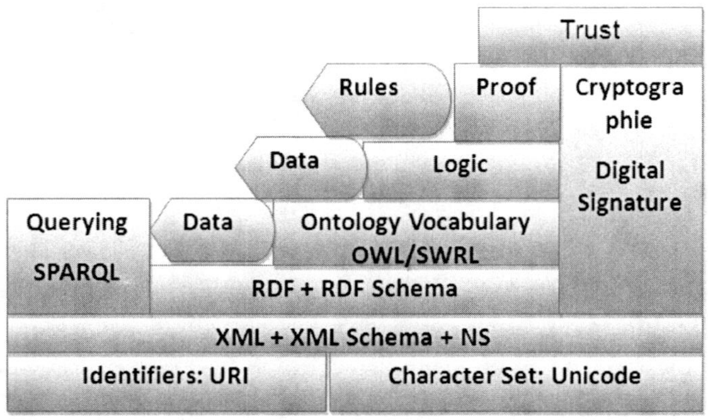

query any RDF database with the use of the latest RDF query language: SPARQL (www.w3.org/TR/rdf-sparql-query/). SPARQL serves as the query language of the Semantic Web.

The top layers of Semantic Web contain technologies that are not yet standardized which require the ability to check the validity of statements made in the (Semantic) Web, and trust for derived statements will be supported by (a) verifying that the premises come from trusted source and by (b) relying on formal logic during deriving new information. The snapshot of Figure 1 shows the layers of Semantic Web:

Our study is mainly focused on RDF, ontologies, and logic. We consider the content of the Semantic Web as being represented by ontologies and metadata. This approach is based on a formal definition of our understanding of what an ontology means. This definition constitutes a crucial structure that is quite straightforward, and may be derived from ontology representation languages. As example, the Karlsruhe Ontology framework KAON[6] reflects the approach represented by ontologies and metadata. The KAON definition can be extended by taking into account axioms, lexicons, and knowledge bases.

Ontology and Knowledge Base

Ontologies have been shown to be beneficial for representing domain knowledge, and are quickly becoming a fundamental layer of the Semantic Web. Ontologies can be seen as explicit specifications of conceptualizations (Gruber T, 1993). A common recognized source of confusion in ontology discussions is the (lack of) distinction between ontologies, knowledge representations and knowledge bases. The representation of terms (concepts/classes, relations, etc.) in some formal language constitutes the definition of knowledge base. The use of such formal language means the use of knowledge representation. The term ontology is normally used for specialized, domain-dependent formal languages, but not used for domain-independent knowledge representations (e.g., first-order logic).

In practice, we have found that in some works the ontologies are developed as knowledge bases, while in other works the ontologies are used as a knowledge representation or as a foundation for one. It can be argued that both are interchangeable, that is, given an ontology in knowledge base form (e.g., in KIF (Genesereth M.) or Ontolingua (Gruber, T, 1993) one could create a corresponding knowledge representation by creating guidelines to define each of the classes/concepts of things defined in that ontology.

We consider an ontology as special kinds of knowledge bases or representation of knowledge. In addition, we think that not all knowledge bases are ontologies.

Semantic Search from Knowledge Base

The quality of the underlying knowledge base is essential to ensure the effectiveness of the semantic search. The search methodologies discussed earlier either utilize explicit knowledge, which is asserted in the knowledge base, or implicit knowledge, which is derived using logical inference with rules.

In addition, we define another kind of knowledge, which is referred to us as "hidden knowledge", that cannot be directly discovered using techniques such as information extraction, logical inference, natural language processing and semantic analytics. Such knowledge can only be derived from large amount of data by using some sort of appropriate data analysis techniques. We refer to approaches that utilize techniques which infer hidden knowledge as mining-based semantic search.

Web/Data Mining

Web mining is the application of data mining techniques to the content, structure, and usage of Web resources. *Data Mining* (DM) is an application area appeared in the 1990s with the combination of several different research fields, especially Statistics, Machine Learning and Databases, as soon as developments in sensing, communications and storage technologies made it possible to collect and store large collections of scientific and commercial data (Fayyad U. et al, 1996).

The aim of DM is either *prediction* or *description*. The use of some variables or fields in the database to predict unknown or future values of other variables of interest is called Prediction. But, the description focuses on finding human-interpretable patterns describing data. As example of descriptive tasks, we can state data summarization which aims to extract compact patterns that describe subsets of data. We can distinguish two classes of methods which are horizontal and vertical approaches. In the horizontal approaches it is possible to produce summaries of subsets (e.g. producing sufficient statistics for subsets). But, in vertical approaches it is possible to describe relations between fields. The second class of methods is distinguished from the first in that rather than predicting the value of a specified field (e.g., classification) or grouping cases together (e.g. clustering) the goal is to find relations between fields. A regular output of this vertical data summarization is called *frequent (association) patterns*. These patterns state that certain combinations of values occur in a given database with a support greater than a user-defined threshold. Our system SWARM (Thabet S. et al. 2008) supports the DM task for Semantic Web Association Rule Discovery. It implements a framework to extract semantic association rule from RDF store.

Semantic Web Mining (Stumme et al., 2006) is a new application area which aims at combining the two areas of Semantic Web and Web Mining from a twofold viewpoint. The first perspective allows the new semantic structures in the Web to be exploited for improving the results of Web Mining. But, the second point of view allows the usage of the Web Mining results in fact to build Semantic Web. A good number of works in Semantic Web Mining simply extends previous work to the new application context. For example, in (Maedche & Staab, 2000) the authors discover conceptual relations from text by the application of a well-known algorithm for association rule mining. Certainly, we argue that Semantic Web Mining can be considered as DM for/from the Semantic Web. The existing DM systems could accomplish the purpose of Semantic Web Mining if they were combined with the standards of representation for ontologies and rules in the Semantic Web. In addition to the aptitude of interoperability with well-established tools for Ontological Engineering (OE) (Gomez-Perez A. et al, 2004), e.g. Protégé-2000 (Noy et al, 2000), that support these standards.

In this chapter we present an approach, SWARM that integrates Semantic Association

Rule Mining and data extracted from knowledge base in order to enable Semantic Web applications relation extraction. This solution suggests a methodology for building Semantic Web Mining systems.

Specification of Semantic Association

A core ontology O is a structure consisting of two disjoint sets C and R, whose elements are called classes/concepts and relations, respectively. At the top of the Figure 2, the set of classes C is defined by {Professor, University, Course, Project, Publication, Student}. The set *R* of relations is defined by *{Author_of, DegreeFrom, Offers, Related_to, Enrolled_in}*. A knowledge base is a structure of two disjoint sets including class instances and link instances. At the bottom of Figure 2, the set of class instances is defined by {P0, U0, C0, PR0, PU0, ST0}.

The relationship between the entity P0 of type "Professor" and the entity PR0 of type "Project", presented in Figure 2, constitute a Semantic Link (SL) presented by the following expression:

$$\mathrm{P0} \xrightarrow{DegreeFrom} \mathrm{U0} \xrightarrow{Offers} \mathrm{C0} \xrightarrow{Related-to} \mathrm{PR0}$$

Signature of Semantic Association

The signature of semantic relation is defined by the class source and class target pair. As example, the relation "Degree_From" has (Professor, University) as signature; the relation "Author_of" has two signatures (Student, Publication) and (Professor, Publication).

Based on the definition of semantic relation signature, we suggest that semantic association has three types of signatures defined as follows:

- **Relation Signature of Semantic Association (RSSA)**: Is defined by the properties which are contained in SA. Focus on the example of Figure 2, the Relation Signature of the semantic association reached from the source entity P0 and the target entity PR0 is defined by the set {Degree-From, Offers, Related-to}.
- **Instance Signature of Semantic Association (ISSA)**: Is defined by the entities contained in SA. As an example, the

Figure 2. An example of knowledge base describing social network

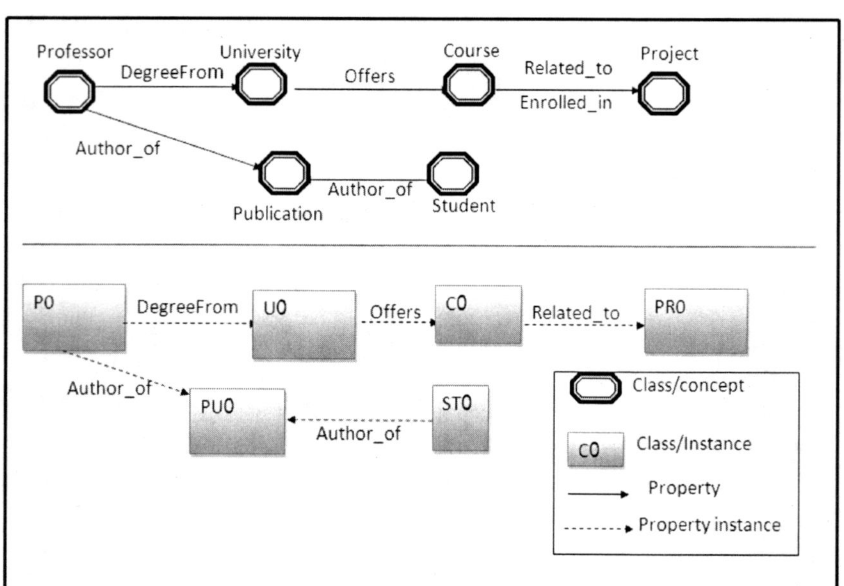

Instance Signature of the semantic association between P0 and PR0 is defined by the set {U0, C0}.

- **Class Signature of Semantic Association (CSSA)**: Is defined by the classes of the intermediate entities in a semantic association. As an example, the Class Signature of the SA between P0 and PR0 is defined by the set of classes {University, Course}.

Language of Semantic Association Querying

The process of semantic association querying is achieved by our PmSPARQL (Slimani et al, 2008a) language which doesn't only make possible to achieve the needs for semantic associations extraction, but also to allow, in some cases, the possibilities of ranking them to distinguish the most suitable association with the user's need.

The proposed PmSPARQL language provides a space of needs specification obtained by a request in such a way that the result will not be presented exclusively in the form of the assertion "entities A and B are connected through such and such a way", but also by the extraction of the best way according to some specified dimensions.

An example of simple semantic association query between the entities P0 and PR0 is given by this example:

```
SELECT @P WHERE
{<P0> @P <PR0>};
```

Where p0 indicates the source entity, pr0 indicates the target entity and @p indicates the path variable which returns all possible paths connecting p0 to pr0. A PmSPARQL query can includes a variety of constraints (further details are included in (Slimani et al, 2008a))

SWAR: A FRAMEWORK FOR SEMANTIC ASSOCIATION EXTRACTION FROM ASSOCIATION DATABASE

The discovered semantic associations are stored in a so-called association data base (ADB). Being given a large number of associations contained in ADB, the capacity to extract knowledge from this one for decision-making aims becomes increasingly important and desirable. However, a process of data mining applied to the vast quantities of semantically rich data (semantic associations) is useful to extract unexpected knowledge from it.

The discovery of semantic association rules is confronted with several challenges which are summarized as follows: firstly, the entities contained in semantic association have a sequential structure (sequence of entities connected via sequence of properties) having a presentation more complex than an attribute of a traditional database. Secondly, the elements of an association (entities and properties) have contextual positions which present semantic significance that is not present on the level of a simple attribute of ordinary database. Thirdly, the data of semantic association seem much richer in semantics than those of a traditional database.

To answer these challenges, we propose to adopt a process of data mining starting from these associations in order to extract some association rules whose evaluation will be based on the traditional confidence and support defined by (agrawal et al, 1993).

With an aim to make the semantic associations mining possible, we present a framework which helps the semantic association rules discovery (SWARM).

The works focused on associations rules showed the effectiveness of association rule mining based on request/constraint (lakshmanan et al, 1999) (Ng et al, 1998) (srikant et al, 1997) (baralis & psaila, 1997) (dehaspe & toivonen, 1999). These works are similar to our model SWAR.

Description of Semantic Association Rule

Formally, semantic association is a sequential structure, connecting two entities sources by a sequence of nodes/entities (instance signature) related to a sequence of properties (relation signature). A database of semantic associations is a database composed by the attributes: "attrsource", "attrtarget", "RSSA", "ISSA" and "assoc" indicating, respectively, the attribute of the source entity, the attribute of the target entity, the relation signature, the instance signature of semantic association and the expression of semantic association. For example, given a semantic association with the following expression:

$$P0 \xrightarrow{DegreeFrom} U0 \xrightarrow{Offers} C0 \xrightarrow{Related-to} PR0$$

If SA has inserted in database (ADB), then *attrsource*="P0", *attrtarget*="PR0", RSSA= {DegreeFrom, Offers, related-to}, ISSA= {U0, C0} and *assoc attribute* =Sa. A semantic association rule is defined as an implication of the form A→B (where, A and B are two associations variable) satisfying three conditions:

1. $A \subset \mu, B \subset \mu$ and $A \cap B = \phi$
2. For all semantic association sets E_A and E_B containing, respectively, the values of the association variables A and B, there exist respectively two set of associations E_{AR} and E_{BP} containing sources entities similar to the sets E_A and E_B. The variable μ indicates the set resulting from the union of E_{AR} and E_{BP} ($\mu = E_{AR} \cup E_{BP}$).
3. Association rules are generated based on the binary association between E_{AR} and E_{BP}. The rule support and confidence are defined as follows:

$$\textbf{Support (A→B)} = \frac{Ass_{ab}}{\mu} \qquad (1)$$

$$\textbf{Confidence (A→B)} = \frac{Ass_{ab}}{\mu_a} \qquad (2)$$

Where, the expression |μ| indicates the cardinality of the semantic associations in μ.

$|Ass_{ab}|$ = {entities| $\forall$ X $\in$ (A $\cup$ B)} represents the cardinality of the appearance of a source entity contained in the set E_{AR} in a similar manner to a source entity in the set E_{BR} with a well defined signature, μ_a={entities| $\forall$ X $\in$ (A)} represents the set of semantic associations contained in E_{AR} and $|\mu_a|$ represents the cardinality of μ_a.

A strong association rule is characterized by a support>MinSup and a confidence>MinConf. The thresholds of the minimum support (MinSup) and minimal confidence (MinConf) must be fixed in advance by the expert.

Constraints Specified on the Level of Semantic Rules

The generation of association rules from database (ADB) is a complex task which requires a paramount stage of modeling. We present, in the following section, the SWAR model (Semantic Web Association Rule) for semantic association rules mining. In this model, we present some definitions allowing the comprehension of it.

Being given the expression of the following form:

Let the following expression of the form: Δ = {Θ*A*: Θ*B*) →ΘX} where Θ*A* and ΘB indicate two associations variable sources and Θ*X indicates* a target entity variable allowing the direction of the associations rules mining process. In other words, the variable ϒ X=(ΔX, *VCT*, MinSup, MinConf) indicates a parameterized request which seeks all the strong associations rules (support>MinSup and confidence>MinConf). Where VCT represents a context variable (signature type) having as value "RSSA" (signature of relations), or (instance signature) or "*" (combination of all the elements of semantic association).

The ΘA or ΘB variable is presented by the following expression: (*<e1:p1>,...,<en:pn>*) which includes the generic semantic associations by classes and relations representations; *pi* indicates the property/relation connected to an entity *ei* and if *ei* is terminal (target entity of semantic association) then *pi* has the value "null".

Let Υ X= (ΔX, *VCT*, MinSup, MinConf) a request having as parameter ΔX and as search criterion VCT. The constraints of SWAR model is defined as follows: the first constraint is specific to the relation signature. In the case of the application of this constraint, the VCT variable is fixed to "RSSA". This type of constraints is adopted to obtain a set of rules which depends on the semantics of the operations performed to the relation signature contained in the semantic association variable (we interest only by the links in the semantic association). The second constraint is specific to the signature of instances. With this constraint, the VCT variable is fixed to "ISSA". This type of constraints is adopted to extract a set of rules concerning the entities contained in semantic association independently of the operations on which they are performed.

SWAR Model for Semantic Association Rules

The SWAR Model for semantic association rules discovery is based on the following steps.

Step 1: Transformation of the variables relatives to the request of association rule

To allow the feasibility of the association rule mining, it is essential to transform each association in ADB into a sequence of data stored in specific matrices, based on the specified VCT variable. As example, in the case of the variable ΘA, Transformation (ΘA) = <M_e [e_1,..., e_n]; M_p [p_1,...,p_n]> where e_i and p_i represent, respectively, an instance stored in the column i of the matrix Me and the properties stored in the columns i of the M_p matrix.

Request Example: Υ X = (ΔX, *, MinSup, MinConf): Detection of strong associations rules to discover the tendency of a student/authors who published an article in a given field to be coauthor with editors/professors (Figure 3).

Let ΔX = {ΘA: $\Theta B \rightarrow \Theta$X} the parameter of the request Υ X. The variables of the request are defined by the following expressions:

ΘA= (<Student|Author:"Author_of"><Publication:"have_subject_area"><Research_Area: "null">)

ΘB= (<Editor|Professor:"Advisee"><Student:"Enrolled_In"><Course:"Course_In"><Research_Area:"null">)

ΘX= (<PERSON>)=(<STudent|Author|Editor|Professor>) which represents the entity or the instance in the semantic association on which the matching between the semantic associations in the database is carried out.

In order to simplify the work of semantic association matching in ADB with the entities in M_e and the properties in M_p, the ΘA and ΘB variables are transformed as follows:

- Transformation (ΘA) = <M_e["Student", "Publication","Research_Area"]; M_p["Author_of","have-subject_area"]>,

Figure 3. An example of the formulation of request constraint

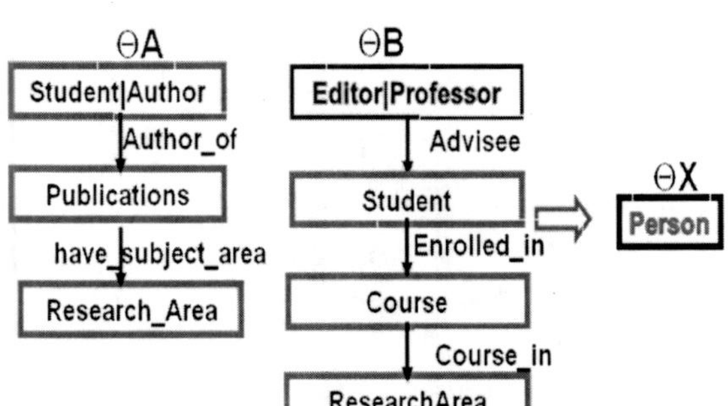

- Transformation (v*B*) =<M_ε["Professor", "Student", "Course ", "ResearchArea"] ; M_p["Advisee", "Enrolled_in", "Course_ in"]>.

These transformations allow the possibility to recover any semantic association in ADB having an exact matching with M_e and M_p *i*ndicating instance signature, relation signature or a combination between them (*).

Step 2: Extraction of the associations sequence relative to exact matching with the variables of association rules request

Let Θ*A* and Θ*B* be two variables concerned by a request of associations rules having as values E_A and E_B. The two sets E_A and E_B are built through a process of matching. Then, the research of association rule will be restrictive to the minimal association database ADB_{Min} (ADB_{Min} which includes the sets E_A and E_B) in order to reduce the extraction process. In table 1, we present an illustrative example of minimal association database ADB_{Min} (8 semantic associations) already extracted from the initial ADB.

Step 3: Conversion of the association sequence found in the step 2 under format of rough set methodology and generation of possible semantic rules

We have adopted the rough set methodology (Munkata, 1998) since it returns, from an optimal subset of the attributes sources, sufficient information concerning the target attributes already predefined. By adopting this methodology, the set of attributes Ω of a binary table (Table 3) is divided into two disjoint sets, a set which constitutes the condition attributes set Ω_C and a second set which constitutes the decision attributes Ω_d. The values of the condition attribute are contained in the set E_{AR}=

Table 1. An example of adb_{Min} after the step 2 execution

idAssoc	Assoc
A1	$S0\left(\xrightarrow{\text{Author - of}}\right)P0\left(\xrightarrow{has\_subject\_area}\right)RA0$
A2	$S1\left(\xrightarrow{Author-Of}\right)P1\left(\xrightarrow{has\_subject\_area}\right)RA0$
A3	$A0\left(\xrightarrow{Author-Of}\right)P1\left(\xrightarrow{has\_subject\_area}\right)RA0$
A4	$A1\left(\xrightarrow{Author\_Of}\right)P2\left(\xrightarrow{has\_subject\_area}\right)RA1$
A5	$E0\left(\xrightarrow{Advisee}\right)S0\left(\xrightarrow{Enrolled\_In}\right)(C0)\left(\xrightarrow{Course\_In}\right)RA1$
A6	$E1\left(\xrightarrow{Advisee}\right)S1\left(\xrightarrow{Enrolled\_In}\right)(C1)\left(\xrightarrow{Course\_In}\right)RA2$
A7	$P0\left(\xrightarrow{Advisee}\right)S2\left(\xrightarrow{Enrolled\_In}\right)(C2)\left(\xrightarrow{Course\_In}\right)RA1$
A8	$P1\left(\xrightarrow{Advissee}\right)S3\left(\xrightarrow{Enrolled\_In}\right)(C1)\left(\xrightarrow{Course\_In}\right)RA3$

{A1, A2, A3, A4} and the values of the decision attributes are contained in the set E_{BR}= {A5, A6, A7, A8} (Table 2). At the end of the phase 3, the minimal association database (ASB_{Min}) becomes as represented by the table 2.

To generate association rules it is critical to build a binary table (table 3) based on the new established ADB_{Min} . The binary table construction is based on the following principle: the associations contained in the set E_{AR} and E_{BR} must be compared together, in order to extract similar associations (comparison starting from the second entity of each association) according to a specified criterion.

However, the binary table must be structured as a binary matrix (Cartesian product of associations between the set E_{AR} and the set E_{BR}). The binary values to be found in the binary table are represented by 1 in the case of correct matching and 0 in opposite case.

The possible rules (table 4) corresponding to the example presented in figure 1 are: A1→A5, A1→A6, A1→A7, A1→A8, A2→A5, A2→A6, A2→A7, A2→A8, A3→A5, A3→A6, A3→A7, A3→A8, A4→A5, A4→A6, A4→A7, A4→A8,

An association rule X→Y has as support S if s% of the rows in the binary table contains X ∪Y (to count the values 1 of the binary table where X is associated with Y). A rule appears in the binary table with a confidence C if c% of

Table 3. Binary table construction: instance matching (vct=issa)

idAssoc	A5(E0)	A6(E1)	A7(P0)	A8(P1)
A1(S0)	0	0	1	1
A2(S0)	0	0	1	1
A3(A0)	0	0	1	1
A4(A1)	0	0	0	0

Table 2. ADB_{Min}: example of minimal association database containing the associations extracted after the execution of the phase 3.

idAssoc	Assoc
A1	$S0\left(\xrightarrow{Author\_Of}\right)Pu0\left(\xrightarrow{related\_to\_project}\right)PR1$
A2	$S0\left(\xrightarrow{Author\_Of}\right)Pu1\left(\xrightarrow{related\_to\_project}\right)PR1$
A3	$A0\left(\xrightarrow{Author\_Of}\right)Pu1\left(\xrightarrow{related\_to\_project}\right)PR1$
A4	$A1\left(\xrightarrow{Author-Of}\right)Pu0\left(\xrightarrow{Required\_Text}\right)C0\left(\xrightarrow{Taught\_By}\right)C0$
A5	$E0\left(\xrightarrow{Editor\_Of}\right)Pu0\left(\xrightarrow{Related\_to\_project}\right)PR0$
A6	$E1\left(\xrightarrow{Editor\_Of}\right)Pu0\left(\xrightarrow{Related\_to\_project}\right)PR1$
A7	$P0\left(\xrightarrow{Author\_Of}\right)Pu0\left(\xrightarrow{Related\_to\_project}\right)PR1$
A8	$P1\left(\xrightarrow{Editor\_Of}\right)Pu1\left(\xrightarrow{Related\_to\_project}\right)PR1$

Table 4. Possible semantic association rules

idRule	Supp	Conf
R1:A1→A5	0	0
R2:A1→A6	0	0
R3:A1→A7	0.5	0.5/0.5=1
R4:A1→A8	0.5	0.5/0.5=1
R5:A2→A5	0	0
R6:A2→A6	0	0
R7:A2→A7	0.5	1
R8:A2→A8	0.5	1
R9:A3→A5	0	0
R10:A3→A6	0	0
R11:A3→A7	0.5	1
R12:A3→A8	0.5	1
R13:A4→A5	0	0
R14:A4→A6	0	0
R15:A4→A7	0	0
R16:A4→A8	0	0

the rows in this table which contain X contains also Y. For example, the support of the rule R3: A1→A7 is obtained by the following expression: Supp (R3) = 0,5 (where 0,5=2/4 represents the value of the appearance of the entity source (S0) contained in A1 and A7 with value 1 compared to the total number of the rows in the binary table. The confidence of the rule R3 is obtained by the following expression: Conf (R3) =supp (R3)/ conf(A1)=0,5/0,5=1 (where the value conf(A1) =0,5 represents the support of the appearance of the entity source contained in the A1 association given the column of A7 association contained in the binary table).

In the case of the request ϒ X, if the user choices are MinConf=0.7 and MinSup=0.3, then the strong associations rules generated by our method are R7, R8, R11 and R12. For example, the R7 rule means that we have a certainty of 100% that the student "S0" is in relation to the professor "P0" and consequently, this student has a confident teamwork relation with professor.

SWARM Algorithm for Semantic Association Rules Mining (SWARM)

Being given a database of semantic associations (ADB) including N semantic associations, the main idea of our algorithm consists to traverse, in a first step, the ADB (according to the constraint of the request: RSSA or ISSA) in order to break up associations into matrices. Two matrices are devoted to contain the data of associations: properties matrix (Mp) and entities matrix (Me).

Each matrix has NxM dimensions where N indicates the number of semantic associations contained in ADB (having a correct matching with the parameter specified in the request) and M indicates the number of columns (properties or entities) which form the matrix Mp or Me). Then, the SWARM algorithm proceeds in the same way as the steps described in the previous section.

Experimental Evaluation

The purpose of this work is to propose a new method allowing the generation of semantic association rules. Indeed, we have developed a prototype to evaluate our work. In our experiments, we have used an ADB including some semantic associations generated from RDF graph. The architecture of the SWAR model is defined by the Figure 4. In this architecture, we present the different modules allowing the functionality of our application. To evaluate the performance of the proposed method, we perform a set of experiments on synthetic data.

Performance study: Three sets of experiments have performed on synthetically produced data.

We have performed several tests applied to ADB containing, respectively, 1700, 3000 and 5000 semantic associations. The minimum support and minimal confidence frequencies were fixed respectively to 20% and 80%. For example, the results applied to the test of 1700 associations give 234 strong association rules with variable condition formed by only one element, 22 rules

Figure 4. Design of swarm framework

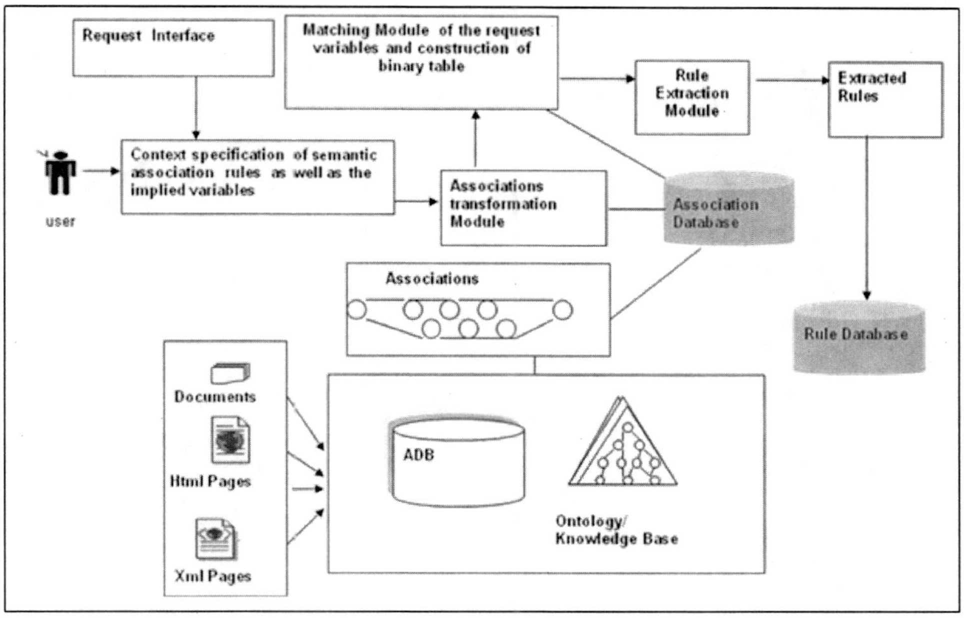

for attributes with variable condition composed by two elements and any rule for other combination (if the user interest is fixed to strong rules on the level of relation signature RSSA).

The performed experiments show that the execution time is higher. The growth of time execution results from the large size of the matrix at the time of semantic association transformation.

To accelerate the execution time, we have developed the algorithm BSWARM (Binary Semantic Web Association Rule Mining) inspired from the principle of our algorithm BARM (Binary Association Rule Mining) (Slimani et al, 2004). We recall that BARM is succeeded to decrease the computing time resulting from multiple database traversing, with this intention, it uses the Peano Tree structure (Ptree) making possible the database conversion into a file stored on disc which includes binary data allowing possible the binary operations over it (Ptrees ANDing). With these operations, the support of each candidate itemset can be obtained directly without traversing the database.

This same principle is adopted by our algorithm BSWARM where the associations contained in ASB are transformed into binary sequences of bits, according to the parameters of the request. Consequently this transformation helps their comparison with request parameters without performing expensive operations of the matrix construction. The graph of figure 5 shows a comparison of the results applied to the ADB of 1700 associations by adopting SWARM and BSWARM algorithm (varying the minimum support value in the interval [10%, 20%]). The obtained results show a remarkable acceleration of the execution time by adopting the BSWARM algorithm compared to the SWARM algorithm.

EXTRACTION OF RELEVANT SEMANTIC ASSOCIATIONS WITH THE USE OF HYPERCLIQUE PATTERN METHOD

In this section, we present a new method for discovering relevant patterns based on semantic and statistical metrics. Statistical metrics of SA are based on the incoming and outgoing links of

Figure 5. Comparative graph of swarm and bswarm algorithms applied to a support value in the 0.1 and 0.2 interval.

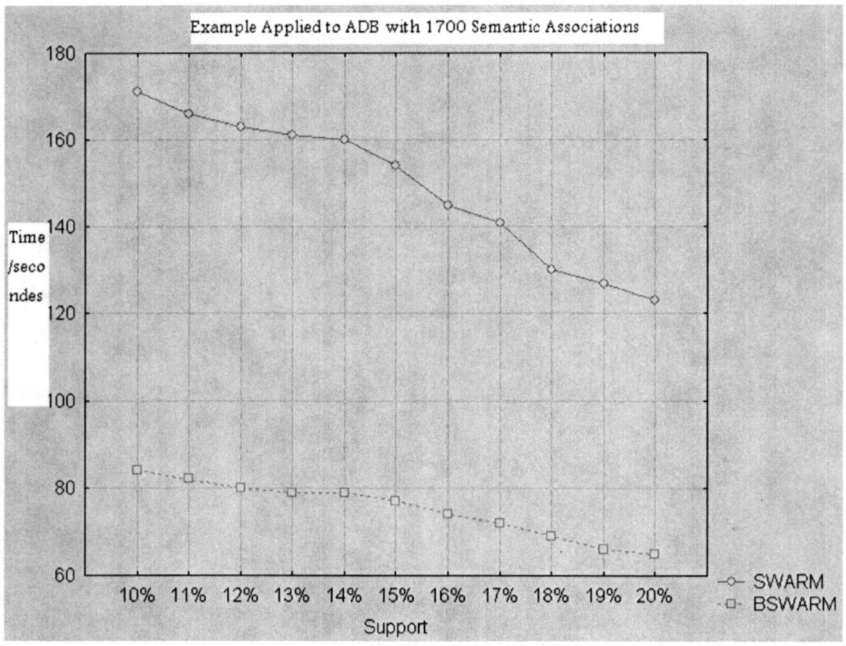

the instances contained in knowledge base. Semantic metrics of SA are based on the semantic characteristic of relevant ontology.

Our method considers each semantic association as a transaction in data base with two entries (source entity and target entity). Our approach is based on association pattern mining. The discovery of relevant pattern in SA is derived from the measure of hyperclique pattern (HP) discussed in (Xiang et al, 2003) in which the authors group the source entity and the target entity of SA into meaningful links.

A hyperclique pattern is a new type of association pattern includes items that are strongly affiliated with one another in (Xiang et al, 2003). Hyperclique pattern discovery can be adopted to capture frequently occurring entities in predefined characteristic with a set of SA (Useful for discovering association rule between predefined features and a set of grouped SA). Discovering such pattern is interesting for several analytical applications in the field of Semantic Web. The process of SA discovery with HP exploits semantic association ranking to establish semantic relevance.

Approach Description

In term of SA, a hyperclique pattern is a strong-kinship association pattern includes entities which are highly connected. More specifically, the presence of an item in one transaction (SA) implies the presence of every other item that belongs to the same HP.

The principle of HP is based on frequent itemsets described in (Agrawal et al, 1994). To measure the strength of association, Xiang et al introduce the h-confidence measure. The h-confidence measure is designed specifically for capturing such strong affinity relationships (Xiang et al, 2003), and for an itemset $P = \{i_1, i_2, ..., i_m\}$, h-confidence is defined as the minimum confidence of all associations rules of the itemset with a single source item in the left side, i.e., $hconf(P) = \min\{conf\{i_1 \rightarrow i_2, ..., i_m\}, conf\{i_2 \rightarrow i_1,$

Table 5. An example of possible extracted sas between two entities a and b

TID	Expression
SA1	$A \xrightarrow{Editor-Of} 12 \xrightarrow{Related-to-Project} 13 \xrightarrow{Project-In} B$
SA2	$A \xrightarrow{Author-Of} 14 \xrightarrow{Related-to-Project} 13 \xrightarrow{Project-In} B$
SA3	$A \xrightarrow{Author-Of} 15 \xrightarrow{Related-to-Project} 14 \xrightarrow{Project-In} B$
SA4	$A \xrightarrow{Advisee} 15 \xrightarrow{Author-Of} 16 \xrightarrow{Has-Subject-Area} B$
SA5	$A \xrightarrow{Advisee} 15 \xrightarrow{Enrolled-In} 12 \xrightarrow{Course-In} B$
SA6	$A \xrightarrow{Advisee} 15 \xrightarrow{Enrolled-In} 13 \xrightarrow{Course-In} B$
SA7	$A \xrightarrow{Author-Of} 17 \xrightarrow{Related-to-Project} 18 \xrightarrow{Project-In} B$

$i_3, ..., i_m\},...,$ conf$\{i_m \rightarrow i_1, i_2,..., i_{m-1}\}$, where conf follows the classic definition of association rule confidence (Agrawal et al, 1994).

An itemset P is a hyperclique pattern if the value of hconf(P) $\leq$ hc, where hc is the minimum h-confidence threshold. More specifically, the hconf for P has mathematically the upper bound function represented by the following expression:

$$upper\left(hConf\left(P\right)\right) = \frac{Min_{1<j<p} Supp\left(x_j\right)}{Max_{1\leq k\leq q} Supp\left(x_k\right)}$$

Where, Supp follows the classic definition of association rule support (Agrawal et al, 2004). This formula takes into account the set P divided into two subsets (a first subset with p items and a second subset with q items) and the item in the first set with min-supp is divided by the max-supp of the item in the second subset.

To be able to extract relevant pattern from SA it is recommended to follow these steps:

1. Discover all semantic associations in knowledge base,
2. Organize the discovered SA with table representation (Table 5).
3. Construct the co-occurrence matrix which stores the occurrence of each SA (Table 6) according to a predefined feature.
4. Calculate the h-confidence of items (SA) which presents a (support $\geq$ minsupp).

Example 1: To find all SA between the entity A of type "Professor" and the entity B of type "Research-Area" with the characteristic F={"Author-Of", "Has-subject-Area"}, the extraction of possible SA would present in the example of table 5.

Co-Occurrence Matrix Construction

The relevant information content of SA is then captured by creating the NxM (N represents the number of SA and M represents the number of defined features) of co-occurrence matrix, MA, defined by 0 or 1 values:

- MA [i, j] = 1 if the feature xj is contained in the item xi.
- 0 Otherwise.

A predefined feature can be a subset of properties (RSSA), a subset of instances (ISSA) or a subset of classes (CSSA).

In the Table 6, the set of features is defined by the properties subset {"Author-Of", "Has-subject-Area"}. We represent the presence of each feature which occurs in a specified SA by the value 1 and its absence by 0. The second row of the table 2 shows that SA2 (semantic association number 2) contains the property "Author-Of", but not includes the property "Has-Subject-Area".

Hconf Computation

For the running example in Table 5, suppose that the minimum support threshold (minsupp) is 10% and the minimum h-confidence threshold (minhconf) is 60%. The support of each SA is obtained by the set (0, 0.28, 0.42, 0.42, 0, 0, 0.28). The value of Supp (SAi) is obtained by the following formula:

$$Supp(SA_i) = Average_{k=1..p}\left(\left(\Sigma_{i=1...n}\ j\right) / \ n\right)$$

Where j represents each occurrence value equal to 1 in the column of the feature "k", p is the number of features in F and n is the number of SA or itemsets in P. The itemsets which have a support<minsupp will be pruned and the number of SA is reduced to the set P'= {SA2, SA3, SA4, SA7}.

hconf(P') is calculated by the following expression: hconf(P') = Min({Conf(SA2 →{SA3, SA4, SA7}), Conf(SA3→{SA2, SA4, SA7}), C onf(SA4→{SA2,SA3,SA7}),Conf(SA7→{ SA2, SA3, SA4, SA7})}= Min(0.28/0.28, 0.28/0.42, 0.28/0.42, 0.28/0.28) = Min{1, 0.66, 0.66, 1}=0.66>minhconf.

Table 6. Co-occurrence matrix of the sas with corresponding feature set

TID	Author-Of	Has-Subject_Area
SA1	0	0
SA2	1	0
SA3	1	1
SA4	1	1
SA5	0	0
SA6	0	0
SA7	1	0

Conf(SA2→SA3,SA4,SA7) is obtained by the following expression: Support (SA2, SA3, SA4, SA7)/Support(SA2). The support of a specified itemsets SA1,...,SAn is obtained by the following formula:

$$Supp\left(SA_1,\ldots,SA_n\right) = \Sigma_{i=1...n}\ j / n$$

Where j is equal to 1 for each SAi having the same values for all the corresponding features in F, and n is the number of SA in P. For example, support (SA2, SA3, SA4, SA7)= 2/7=0.28.

The hyperclique pattern method is an alternative to frequent itemsets. In the example presented earlier, the hypergraph model contains pattern (SA3, SA4) represented by a hyperedge with weight equal to the h-confidence of the hyperclique pattern (in our case equal to 0.66). We can derive another pattern with two source items (SA2, SA3→{SA4, SA7},...). The discovery of hyperclique pattern between itemsets having source entities>1 is useful in the case where the itemsets to be mined include different source entities or different target entities of SA.

The interpretation of the discovered relevant semantic associations has different meaning which depends on the fields of the items (SA) to mine and specifically the defined feature. In our example, the retained SA in P' is the relevant information required by the predefined features F. In other term, these SAs may present a pattern that is strongly supported in the proposed features with confidence=66%.

Example of SA Mining Based on the Instance Subset

We present here another example of HP, derived from Web service: In the field of Web service discovery the research of required service is characterized by: Inputs which include a set of necessary inputs that the requester should provide to invoke the service and Outputs which represent the results that the requester should expect after interaction when the service provider is completed. In our example, the inputs and the outputs are formulated as follows:

Inputs: booking information (string).

Outputs: passenger name (person), origin and destination cities, departure time-date (list-date-and-time, e.g. (20 33 16 5 1 2009))

- SA1:A1= "British Rail: CHRISTOPH", I00 = "the 66 going", I01= "Paris", C1= "London", T1= "16:49, 18, 9 2008"
- SA2: A2= "SNCF: SINUHE", I02= "on the 511 going", I03= "Paris", C2= "Lyon", T2= "5: 12, 18, 8 2008"
- SA3: A3= "Die Bahn: CHRISTOPH", I04= "on the 323 going", I05= "Paris", C3= "Lyon", T3= "17:11, 15, 9 2008"
- SA4:A4="Austrian Rail(OBB): SINUHE", I06= "on the 367 going", I07="Paris", C4= "London", T4= "16:47, 18, 9 2008"
- SA5: A5="European Student Rail Travel: JOHN", I08= "on the 916 going", I09= "London", C5= "Nice" T5="6:44, 18/08/2008"
- SA6: A6="Business Europe: LILIANA", I10="on the 461 going", I11= "PARIS", C6= "INNSBRUCK" T6= "6:12, 18, 9 2008".

The full description of the itemset given above is presented in Table 7.

Let the predefined features F = {"Paris", "London", "18/09/2008"} which take into account the origin, the destination and the departure values as parameters for mining HP pattern. Then, the co-occurrence matrix of the discovered SA is represented by the table 8:

We suppose the minimum support threshold (minsupp) is 20% and the min-hconf threshold is 35%. The support of each SA is obtained by the set: $\{0.55, 0.18, 0.18, 0.55, 0, 0.44\}$. According to the computed support of each item in the entry of table 8, the new reduced set P' will be SA1, SA4 and SA6. Then, the hconf(P') is computed by the following formula: Min(Conf(SA1→SA4,SA6), Conf (SA4→{SA1, SA6}), Conf(SA6→{SA1, SA4})= Min(0.6, 0.6, 0.37) = 0.37≥ minhconf.

In the relative example of Web service discovery, the hypergraph model includes pattern represented by two hyperedges whose weight is equal to the h-confidence of the hyperclique pattern (in our case equal to 0.37).

Table 7. An example of possible extracted sa between two entities: booking information and time-date

TID	Expression
SA1	$A1 \xrightarrow{is-booked-on} I00 \xrightarrow{going-from} I01 \xrightarrow{to} C1 \xrightarrow{at} T1$
SA2	$A2 \xrightarrow{is-booked-on} I02 \xrightarrow{going-from} I03 \xrightarrow{to} C3 \xrightarrow{at} T3$
SA3	$A3 \xrightarrow{is-booked-on} I04 \xrightarrow{going-from} I05 \xrightarrow{to} C3 \xrightarrow{at} T4$
SA4	$A4 \xrightarrow{is-booked-on} I06 \xrightarrow{going-from} I07 \xrightarrow{to} C4 \xrightarrow{at} T4$
SA5	$A5 \xrightarrow{is-booked-on} I08 \xrightarrow{going-from} I09 \xrightarrow{to} C5 \xrightarrow{at} T5$
SA6	$A6 \xrightarrow{is-booked-on} I10 \xrightarrow{going-from} I11 \xrightarrow{to} C6 \xrightarrow{at} T6$

Table 8. Co-occurrence matrix of the sas with corresponding feature set

TID	Paris	London	18/09/2009
SA1	1	1	1
SA2	1	0	0
SA3	1	0	0
SA4	1	1	1
SA5	0	0	0
SA6	1	0	1

Experimental Results

In this section we give an evaluation of the method described above. In the experimental evaluation, we mainly test the possibility of mining relevant SA with hyperclique pattern method. Our tests are performed on machine with Intel(R) Core(TM) 2 Duo 2.2GHZ CPUs, 2GB memory and Windows XP.

In this section, the application of hyperclique pattern to discover relevant SA in real world datasets has been presented.

The scope of this method is not to extract semantic associations, but to discover relevant SAs from preliminary extracted ones. The extraction process from knowledge base is obtained by the PmSPARQL language intended to produce SA with multi-paradigm path extraction (Slimani et al, 2008b).

To evaluate our approach, we apply it to a real world dataset "swetodblp-august-2007-part-1"[7] which maps visual patterns to feature spaces according to the hyperclique pattern method. The application of this method is similar to the principle of association rule mining, where each rule has two parts: the antecedent, which models low level feature subspaces, and the consequent, which represents a set of semantic associations. The result of a confident rule includes the relevant semantic associations set having a value of hconf ≥ minhconf proposed by user.

In our previous work (Slimani et al, 2008b), we have demonstrated the ability of our proposed

Figure 6. Number of discovered frequent sas with various value of minsupp and co-occurrence matrix construction time

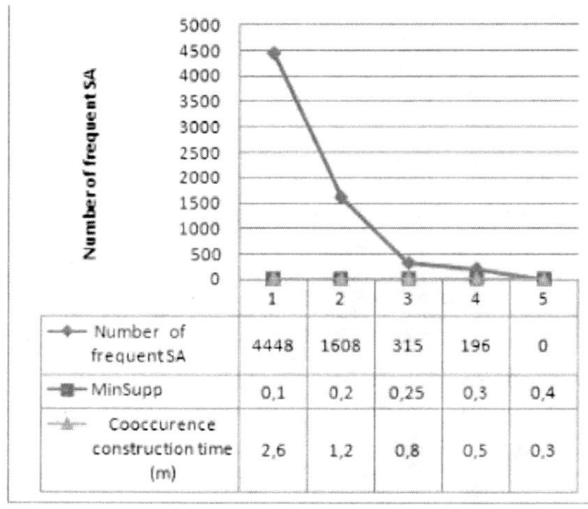

PmSPARQL language to produce long association. We exploit the ability to produce longer SA for mining relevant pattern from the enriched dataset. The executed PmSPARQL request to generate SA of length 6 hops has generated 10158 semantic associations. We have performed several tests with various values of min-supp and min-hconf applied to the generated SA. We have performed some tests which extract relevant patterns between the source features {"./FongWF04", "./Lamport2002", "./Spinellis03", "./TanSK2005"} and the consequent, which represents a set of extracted SA between the source entity (SE= "./Makoui2007") and all other publications after 2001 (SA with length=6).

In Figure 6, the number of the frequent semantic associations is drawn varying the min-supp value. In addition, each specified support value is presented with the amount of execution time. The obtained result shows the feasibility of the proposed method of co-occurrence matrix construction. If the value of min-supp increases, the value of the frequent SA decreases and the time of the matrix construction decreases.

Generally, the application of hyperclique pattern to itemsets (SA) with low support may

Figure 7. Number of relevant sas with various value of min-hconf and construction time of co-occurrence matrix

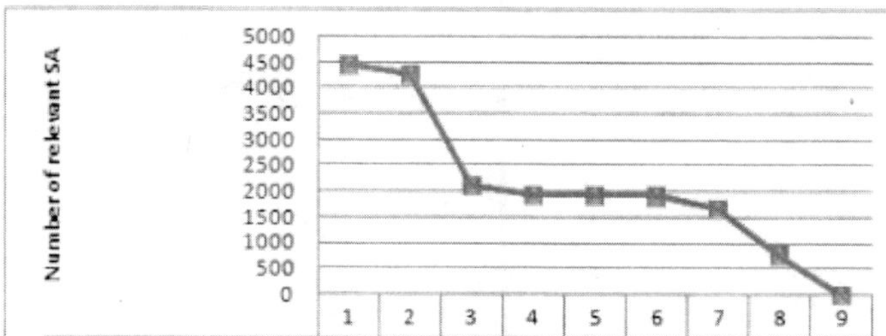

	1	2	3	4	5	6	7	8	9
MinhConf	0,1	0,2	0,3	0,4	0,5	0,6	0,7	0,8	0,9
Time (sec)	5,25	5	4,8	4,1	4	3,9	3,7	3,3	3
MinSupp	0,1	0,1	0,1	0,1	0,1	0,1	0,1	0,1	0,1
Number of relevant SA	4448	4264	2116	1932	1932	1915	1668	781	0

represent knowledge with good probabilistic evidence. The obtained results in Figure 7 show that the retrieved SAs with minhconf≥0.9 may present a pattern that is not convenient with the proposed features. In other term the frequent retained SA does not constitute a hyperclique pattern and consequently will not be considered as relevant pattern. The intuition with the use of the type of mining process in our test example is as follows: discover all relevant SAs which present supported publications (specified in the features set P) in the paths which lead from the source entity and any other publications after the year 2001. For example, in the plot of Figure 7, if min-supp=0.1 and min-hconf=0.7, we have 1668 SA which present a relevant pattern that is strongly supported in the features P with confidence=70%.

RELATION EXTRACTION IN KNOWLEDGE BASE

An object in OWL knowledge base is an instance of a specified concept/class. Let A and B two objects, we can extract new unexpected information through semantic link extraction between a given instance and another unrecognized instances.

Relation Extraction without Open World Assumption Links

For this purpose, we propose a measure which computes the maximum degree of the incoming relations which connects the instances entities through a specified relation R. The example presented in the Figure 8 describes a process of mining hidden entity which identifies the expertise field of the effective reviewer for a well defined conference through the use of statistical information defined in a semantic link. The results can be restricted to the identification of a new link (**Expert_in**) between the fixed source instance and an object not firstly recognized.

Let RE1 the specified source entity (instance of the entity Reviewer) and X the generic target entity initially not fixed. In addition to the fixed entities, the detection process of useful target entity is based on the specified relation "Has-Subject-Area".

The formula SAV (semantic association value) which identifies the relevant entity is defines as follows:

$$SAV = Max\left(Count_{Ir}\left(RE1, X\right)\right)$$

Figure 8. Example of rdf graph without open world assumption

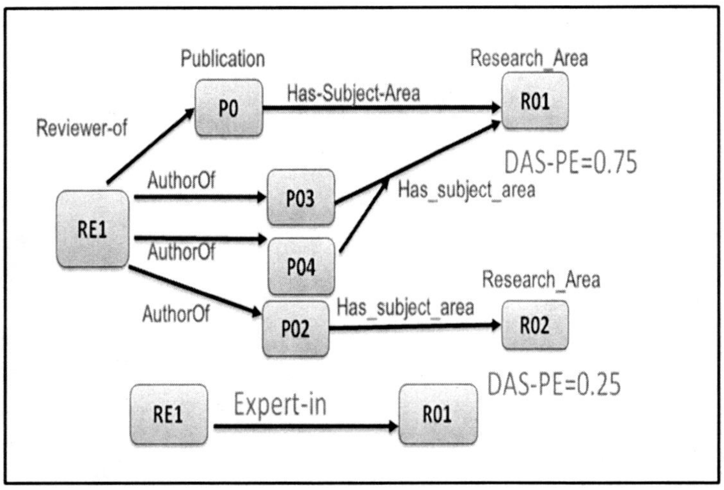

This formula, take as input the entity source, the relation R directly related to searched target entity (X) and turns as output the entity which presents a maximum number of the incoming relation I_r. In the example of the Figure 8, the entity R01 presents the maximum value of the incoming relation "Has-Subject-Area" and consequently will be related to RE1 by the new link "Expert_In".

Relation Extraction Based on Open World Assumption Links

If the knowledge base presents the special type of the relation "know" or "coauthor_Of", then the formula of entity identification will be changed.

If we apply the formula 6 to the example of Figure 9, we obtain two entities as results (R01 and R03), which does not make such sense. Indeed, the fact of knowing a person who published an article in a given Research_area does not mean his expertise in this field. Moreover, the fact to be coauthor of a person who published in a well defined Research_area does not mean his expertise in this Research_area.

To regulate this problem, we reformulate the SAV formula by adding a coefficient to the formula 6 which represents the weight of outgoing relations (O_r) from the source entity (RE1). This coefficient is specified only to some relations selected by the domain expert. The new formula is defined as follows

$$SAV = Max\left(W_{Or} * Count_{Ir}\left(RE1, X\right)\right)$$

Based on the formula 7, the weight of the relation "knows" is fixed to be zero in the figure 9 and for this reason the selected entity remains the same as the example of figure 8.

The discovered entity will be related to the source entity and will be included in the knowledge base for a best enrichment.

The intuition behind this process of entities extraction is important in the case where a kind of link doesn't have explicit existence in knowledge base. The new added link (expert_in) helps the knowledge base become richer, which in turn imports new unexpected links.

FUTURE TRENDS

The prototypes described in this chapter is a first step in applying semantic web mining approach applied to semantic associations data. A number of areas require further attention, principally due

Figure 9. Example of rdf graph with open world assumption

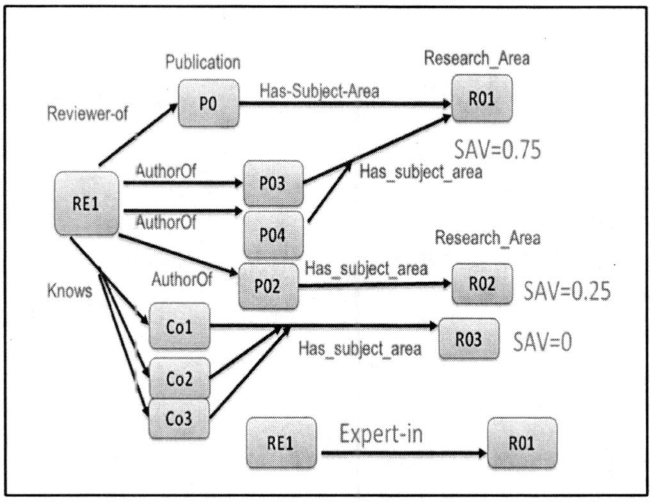

to the fact that the prototypes currently consider some specified methods (association rule, hyper-clique pattern method). Further work is required to address other requirements. For example, it is also possible to mine the ontologies, metadata and other databases so that the Web is made more intelligent. The ultimate goal is to make the Web easier to use. In addition, the objective is to arrive to mine the large quantities of information on the Web so that humans can better perform their tasks. Further evaluation of run-time performance is also required. Semantic Web mining can help to develop ontologies. With Semantic Web mining, it is possible to discover various resources, patterns, and trends. This in turn could help toward constructing various types of agents. In addition, we can use Semantic Web mining to support a variety of agents to carry out their jobs. These solutions will include locator agents, resource management agents, and agents for knowledge management and collaboration, among others.

CONCLUSION

The Semantic Web has become real. Currently, there are a huge number of RDF triples, Web pages, and ontology pages which have been published using RDF schema and OWL, with a growing level of industrial and academic support. This means that the information on Web pages may have to be mined so that the machine can understand the content. Essentially, we need to carry out machine learning on the Web.

So, in this chapter we have provided an outline of Semantic Web Mining. We examined the semantic web concepts and described the impact of web mining. We started with a discussion of semantic web mining concepts and discussed some concepts related to semantic association rule mining and the method of hyperclique pattern. We have detailed each method in a separate section and presented the experimental results for each method. Moreover, we have presented an approach for entity extraction given an entity source and a specified relation contained in the path between them. The developed prototypes show the feasibility of the Semantic Web Mining approach and its ability to extract hidden patterns and trends from large quantities of data.

REFERENCES

Agrawal, R., & Srikant, R. (1994). Fast algorithms for mining association rules. In *20th Int. Conf. Very Large Data Bases (VLDB).*

Aleman-Meza, B., Halaschek-Wiener, C., Arpinar, I. B., & Sheth, A. P. (2008). Context-aware semantic association ranking. In *Proceedings of SWDB'03, the first International Workshop on Semantic Web and Databases, Co-located with VLDB 2003*, Humboldt-Universit¨at, Berlin, Germany, (pp. 33-50).

Aleman-Meza, B., Nagarajan, M., Ramakrishnan, C., Ding, L., Kolari, P., Sheth, A. P., et al. (2006). Social networks: Semantic analytics on social networks: Experiences in addressing the problem of conflict of interest detection. In *Proceedings of the 15th international conference on World Wide Web, WWW 06,* Edinburgh, Scotland, (pp. 407-416). New York: ACM.

Anyanwu, K., Maduko, A., & Sheth, A. P. (2005). Sem-rank: Ranking complex relationship search results on the semantic web. In *Proceedings of International World Wide Web Conference*, Chiba, Japan, (Vol.14, pp. 117–127). New York: ACM.

Anyanwu, K., & Sheth, A. P. (2002). The rho Operator: Discovering and Ranking Associations on the Semantic Web. *SIGMOD Record, 31*(4), 42–47. doi:10.1145/637411.637418

Baralis, E., & Psaila, G. (1997). Designing templates for mining association rules. *Journal of Intelligent Information Systems*, (9): 7–32. doi:10.1023/A:1008637019359

Berners-Lee, T., Hendler, J., & Lassila, O. (2001). *The Semantic Web: A new form of Web content that is meaningful to computers will unleash a revolution of new possibilities.* Scientific American.

Dehaspe, L., & Toivonen, H. (1999). Discovery of frequent datalog patterns. In Data Mining and Knowledge Discovery.

Fayyad, U., Piatetsky-Shapiro, G., Smyth, P., & Uthurusamy, R. (Eds.). (1996). *Advances in Knowledge Discovery and Data Mining.* Menlo Park, CA: AAAI Press/The MIT Press.

Genesereth, M. (n.d.). *Knowledge Interchange Format (KIF).* Retrieved from http://www.ksl.stanford.edu/know- ledge-sharing/kif/

Gomez-Perez, A., Fernandez-Lopez, M., & Corcho, O. (2004). *Ontological Engineering.* Berlin: Springer.

Gruber, T. (1993). A translation approach to portable ontologies. *Knowledge Acquisition, 5*(2), 199–220. doi:10.1006/knac.1993.1008

Lakshmanan, L., Ng, R., Han, J., & Pang, A. (1999). Optimization of constrained frequent set queries with 2-variable constraints. In the ACM SIGMOD Intl. Conf. on Management of Data, (pp. 157–168).

Maedche, A., & Staab, S. (2000). Discovering Conceptual Relations from Text. In W. Horn, (Ed.), *Proceedings of the 14th European Conference on Artificial Intelligence*, (pp. 321–325). Amsterdam: IOS Press.

Munkata, T. (1998). Rough sets . In *Fundamentals of the New Artificial Intelligence* (pp. 140–182). New York: Springer-Verlag.

Ng, R., Lakshmanan, L., Han, J., & Pang, A. (1998). Exploratory mining and pruning optimizations of constrained association rules. In *Proc. of the ACM SIGMOD Intl. Conf. On Management of Data*, (pp. 13–24).

Ning, X., Jin, H., & Wu, H. *(2006)*. Semrex: Towards large-scale literature information retrieval and browsing with semantic association. In *Proceedings of IEEE International Conference on e-Business Engineering (ICEBE 06)*, (pp. 602–609). Washington, DC: IEEE Computer Society.

Noy, N. F., Fergerson, R., & Musen, M. (2000). The Knowledge Model of Protege-2000: Combining Interoperability and Flexibility. In R. Dieng & O. Corby, (Eds.), Knowledge Acquisition, Modeling and Management, (LNCS, Vol. 1937, pp. 17–32). New York: Springer.

Slimani, T., Ben-Yaghlane, B., & Mellouli, K. (2004). Approche binaire pour la génération de fortes règles d'associations. In EGC 2004, Revue des Nouvelles Technologies de l'Information, (pp. 329-340). Clermont Ferrand, France: Cépadues edition.

Slimani, T., Ben-Yaghlane, B., & Mellouli, K. (2008a). PmSPARQL: Extended SPARQL for multi-paradigm Path Extraction. *International Journal of Computer, Information, and Systems Science, and Engineering*, *2*(3), 179–190.

Slimani, T., Ben-Yaghlane, B., & Mellouli, K. (2008b). SWAR: Modèle de génération des règles d'associations sémantiques à partir d'une base d'association. In *Proceedings of EGC 08, Atelier de fouille de données complexes*, Sophia Antipolis, (pp. 83-94).

Srikant, R., Vu, Q., & Agrawal, R. (1997). Mining association rules with item constraints. In *Proc. of the 3rd Intl. Conf. on Knowledge Discovery and Data Mining*, (pp. 67–73).

Stumme, G., Hotho, A., & Berendt, B. (2006). Semantic Web Mining: State of the art and future directions. *Journal of Web Semantics*, *4*(2), 124–143. doi:10.1016/j.websem.2006.02.001

Xiong, H., Tan, P. N., & Kumar, V. (2003). Mining strong affinity association patterns in data sets with skewed support distribution. In *IEEE International Conference on Data Mining*, (pp. 387–394). Washington, DC: IEEE Computer Society.

Zhuge, H. (2004). *The knowledge grid*. Amsterdam: World Scientific.

Zhuge, H. (2007). *Autonomous semantic link network model for the Knowledge Grid, Concurrency and Computation: Practice and Experience* (pp. 1065-1085), *7*(19).

ENDNOTES

1. http://lsdis.cs.uga.edu/
2. URL *(uniform resource locator)* refers to a locatable URI, e.g., an http://... address. It is often used as a synonym, although strictly speaking URLs are a subclass of URIs.
3. XML:www.w3.org/XML/
4. http://www.w3.org/TR/REC-rdf-syntax-grammar-20040210/
5. www.w3.org/TR/owl-features/
6. http://kaon.semanticweb.org
7. http://lsdis.cs.uga.edu/projects/semdis/swetodblp/

Chapter 6
Reusing the Inter-Organizational Knowledge to Support Organizational Knowledge Management Process:
An Ontology-Based Knowledge Network

Nelson K. Y. Leung
RMIT International University Vietnam, Vietnam

Sim Kim Lau
University of Wollongong, Australia

Joshua Fan
University of Wollongong, Australia

ABSTRACT

Various types of Knowledge Management approaches have been developed that only focus on managing organizational knowledge. These approaches are inadequate because employees often need to access knowledge from external knowledge sources in order to complete their works. Therefore, a new inter-organizational Knowledge Management practice is required to enhance knowledge sharing across organizational boundaries in their business networks. In this chapter, an ontology-based Inter-organizational knowledge Network that incorporates ontology mediation is developed so that heterogeneity of knowledge semantic in the ontologies could be reconciled. The reconciled inter-organizational knowledge could be reused to support organizational Knowledge Management process semi- or automatically. The authors also investigate the application of ontology mediation that provides mechanisms of reconciling inter-organizational knowledge in the network.

DOI: 10.4018/978-1-61520-859-3.ch006

INTRODUCTION

Over the past two decades, a lot of efforts have been placed in order to integrate heterogeneous information systems. Although heterogeneity is an obstacle for system interoperation, the heterogeneity allows systems to be designed and developed according to the business requirements. This interoperation is essential because systems of different characteristics from organizations, companies or even individuals would be able to communicate, cooperate, exchange information as well as reuse knowledge and services with one another. Especially in the era of the Internet, a business transaction can hardly be completed without making use of others' data, information, knowledge and services. For instance, when customer is shopping in an online store, s/he may need to seek comments on the quality of a particular product from an external forum. Once s/he decides to purchase the product, the online store will have to contact related financial institutes for payment verification and confirmation. The online store is also required to arrange delivery service with shipping company. Such a simple online shopping transaction would involve interoperation of at least three heterogeneous information systems, the complexness could be imagined if it is a multi-million trade that involves the participation of more enterprises.

Artificial intelligence researchers first applied the concept of ontology in intelligence system development so that knowledge could be shared and reused among artificial intelligence systems. Ontology as a branch of philosophy is the science of what is, of the kinds and structures of objects, properties, events, processes and relations in every area of reality (Smith, 2003). Ontology can be further elaborated as a particular system of categories accounting for a certain vision of the world (Guarino, 1998). The term, ontology, was then borrowed by artificial intelligence community and Tom Gruber's definition was widely accepted within the community: an ontology is an explicit specification of a conceptualization while a conceptualization is an abstract, simplified view of the world that we wish to represent for some purpose (Gruber, 1993). Later on, Borst (1997) refines Gruber's definition by labelling an ontology as a formal specification of a shared conceptualization. Based on Gruber's and Borst's definitions, Studer et al. (1998) make the following conclusion: 1) an ontology is a machine-readable specification of a conceptualization in which the type of concepts used and the constraints on their use are explicitly defined, and 2) an ontology should only capture consensual knowledge accepted by large group of people rather than some individual. By representing knowledge with representational vocabulary in terms of objects and their interrelated describable relationships, inference engine and other application program from one intelligence system will be able to understand the semantic of knowledge in another knowledge base.

The popularity of the World Wide Web (WWW) further magnifies the importance of ontology. The Hypertext Markup Language (HTML)-based web content is solely designed for formatting and displaying information on the web and computers have no way to understand and process the semantics (Antoniou & Harmelen, 2004). The disadvantage of HTML-based web content is reflected completely when users attempt to retrieve information from the web using a search engine. It is very common for a search to return more than ten thousand results. The application of search operators may be able to narrow down the results to a few hundreds, but users still require extensive effort to locate the right information within a pool of results. It is due to the fact that application program resided in the search engine can only perform keyword search in the HTML-based document without understanding the actual semantic of the document, for example, searching the web with the keyword "bank" using Google search engine will return any webpages that contain "bank" or with "bank" as one of the indexes, regardless whether "bank" means a financial institute, river bank or cant on those webpages. This leads to the emergence of the Semantic Web. The Semantic Web is the extension of the current one, in which web content is represented in a structural form within ontologies by a finite list of vocabularies and their relationships (Berners-Lee et al., 2001). In this way, ontologies enable computer program, software agent and search engine to understand the semantics, thus making it possible for them to process the web content. Ontologies also provide a shared understanding of a domain which is necessary to overcome differences in terminology from various sources (Antoniou & Harmelen, 2004).

Unfortunately, it is unrealistic to expect all individuals and organizations will agree on using one or even a small set of ontologies (de Bruijn et al., 2006). The adoption of such an approach is problematic. On one hand, it is lengthy and non-trivial to define and maintain a large globally shared ontology, on the other hand, the globally shared ontology approach may hinder a system from reflecting its actual business requirements due to the fact that design of the system is restricted by terminologies defined in the ontology (Visser & Cui, 1998). Researchers such as Berners-Lee et al (2001) state that there would be a large number of small domain specific ontologies developed by communities, organizations, departments or even individuals. While multiple ontologies allow system to be designed according to its actual requirements without committing to a particular set of terminologies, data heterogeneity caused by multiple ontologies has become an obstacle for the interoperation of systems (Visser et al., 1998). Since vocabularies and their relationships defined in the ontologies are inconsistent, therefore it is impossible for one system to understand and reuse other ontologies unless the ontologies are reconciled in some form. The above inconsistent problem caused by multiple ontologies is commonly termed as ontology mismatches.

This research describes the three main meditation methods used to reconcile mismatches between heterogeneous ontologies. The research also investigates the application of ontologies and its mediation methods in the aspect of Knowledge Management (KM). The rest of the Chapter is organized as follows. Section 2 describes various approaches of ontology mediation. Section 3 discusses the application of ontology and its mediation methods in KM. This includes the development of a proposed mediation selection framework and ontology-based collaborative KM network. Finally, conclusion is given in Section 4.

APPROACHES OF ONTOLOGY MEDIATION

Based on the actual requirements, organizations and individuals are expected to develop their own ontologies of different languages, scopes, coverage and granularities, modelling styles, terminologies, concepts and encodings. To reuse other ontologies of different types, ontology mediation is required to reconcile mismatches between heterogeneous ontologies so that knowledge sharing and reuse among multiple data sources can be achieved (Predoiu et al., 2006). There are three major kinds of ontology mediations which include mapping, merging and integration. Ontology mapping is a process of relating similar concepts and relations from different ontologies to each other in which the correspondences between different entities of the two ontologies are formulated as axioms in specific mapping language (de Bruijn et al., 2006; Klein, 2001). Since the involved ontologies do not require any adaptation, ontology mapping often specifies just a part of the overlap between ontologies which is relevant for the mapping application (Scharffe et al., 2006). Two common approaches used to establish mapping between ontologies are listed as follows.

- The first approach is to relate all ontologies to a common top-level ontology so

Figure 1. Two mapping approaches

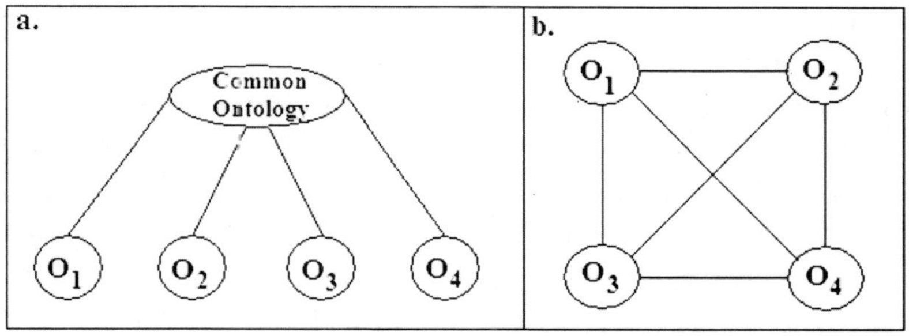

that different ontologies are mapped together indirectly by the top-level ontology as illustrated in Figure 1a (Wache et al., 2001). Consequently, conflicts and ambiguities can be resolved since concepts used in different ontologies are inherited from the common ontology. However, this approach has three major drawbacks. First, constructing a large-scale common top-level ontology from scratch is never a simple task. Even if we take a simpler path by merging various local ontologies together, the experiences of building the air campaign planning ontology and the Suggested Upper Merged Ontology (SUMO) tell us that the actual merging processes are trickier than expected, not only because there is inconsistency between chunks of theoretical content but also because there were structural differences between the local ontologies (Valente et al., 1999; Niles & Pease, 2001). Second, this approach can only be adopted in a relatively stable environment where maintenance is minimal because a substantial amount of resources and overheads are required to maintain a common top-level ontology. Third, established mappings between local ontologies and top-level ontology can easily be affected by the elimination and addition of local ontologies as well as changes in either local or common ontologies because local ontologies are related indirectly with each other through the common ontology.

- Rather than mapping all ontologies to a common top-level ontology, one-to-one mapping approach requires mappings to be created between each pair of ontologies as shown in Figure 1b (Predoiu et al., 2006). The lack of a common top-level ontology in this approach makes it possible to be adopted in a highly dynamic environment. This advantage may be offset by the lack of common terminologies, thus increasing the complexity of defining mapping between local ontologies. Another major drawback of this approach occurs when a large number of heterogeneous ontologies are involved in the interoperation. Such an interoperation will greatly increase the amount of mappings and extra effort is required to control and maintain the mappings.

The second type of ontology mediation is merging. Unlike mapping that links two separate ontologies together in a consistent and coherent form, ontology merging creates a new ontology (in one subject) by unifying two or more different ontologies on that subject and it is usually hard to identify regions of the source ontologies from the merged ontologies (Pinto & Martins, 2001). As compared with mapping that keeps the original ontologies unchanged, merging requires at least one of the original ontologies to be adapted so that the conceptualization and the vocabulary match in overlapping parts of the ontologies (Ding et al., 2002). While a majority of semantic web researchers foresee the main stream would switch to the approach of developing enormous amount of small domain specific ontologies, McGuinness et al. (2000) argues that some of the industries or organizations still require to develop very large and standardized ontology, for instance, SNOMED CT is a comprehensive clinical ontology developed by the College of American Pathologists that contains about 344,549 distinct concepts and 913,697 descriptions (Lussier & Li, 2004). Theoretically, it is more efficient and effective to merge existing ontologies than to build a large ontology from scratch. In practice, the process of ontology merging is more than just simple revisions, improvements or variations of the source ontologies since the involved ontologies are developed by different people for different purposes with different assumptions and using different vocabularies (Lambrix, et al., 2003; Pinto & Martins, 2001). McGuinness et al. (2000) specify the three major tasks that are required to merge two ontologies: 1) coalesce two semantically identical terms from different ontologies so that they can be referred to by the same name in the resulting ontology, 2) identify terms that should be related by subsumption, disjointness or instance relationships, 3) verify and validate correctness and consistency of the merged ontology. Chimaera, developed by the Stanford University Knowledge Systems Laboratory, is an example of a semi-automatic merging tool that supports the above three tasks.

One of the most important phrases in the process of ontology mapping and merging is ontology matching. In general, ontology matching can be defined as the process of discovering similarities between two ontologies with the purpose of establishing semantic relationships in between (Predoiu et al., 2006). It determines the relationships holding between two sets of entities that belong to two discrete ontologies (Shvaiko, 2004). In other words, it is the process of finding a corresponding entity in the second ontology for each entity (for example, concept, relation, attribute and so on) in the first ontology that has the same or the closest intended meaning. This can be achieved by analysing the similarity of the entities in the compared ontologies in accordance with a particular metric (Ehrig & Sure, 2004; INTEROP, 2004). Ontology matching (or similarity computation) can be processed exploiting a number of different techniques. To provide a common conceptual basis, researchers have started to identify different types of ontology matching techniques and propose classifications to distinguish them, for example, Abels et al. (2005) propose a classification that consists of nine matching techniques based on existing literature studies. Another example is the classification developed by Shvaiko & Euzenat (2005). Building on the foundation of Rahm & Bernstein (2001)'s schema matching techniques classification, Shvaiko and Euzenat develop a meticulous classification to categorize ten elementary ontology and schema matching techniques:

1. String-based technique is used to match names and name descriptions of ontology entities in terms of a sequence of alphabet letters.
2. Language-based technique uses natural language processing techniques such as tokenization, lemmatization and elimination to exploit morphological properties of the input words. The technique is usually applied before string-based technique in order to improve the results.
3. Constraint-based technique is used to match the definitions of properties in terms of their internal constraints such as datatypes and cardinality.
4. Linguistic resources technique utilizes common knowledge or domain specific thesauri such as WordNet to analyse linguistic relations in the word matching process.

5. Alignment reuse technique exploits the idea of reusing alignments of previously matched ontologies as many ontologies to be matched are similar to already matched ontologies with the same application domain.
6. Upper level formal ontologies technique uses external source of common knowledge in the form of ontology such as DOLCE (Gangemi et al., 2003) within the matching process.
7. Graph-based technique considers the input as labelled graphs containing terms and their inter-relationships. Basically, the similarity is obtained through the analysis of the positions of a pair of nodes (from two ontologies) on the graphs.
8. Taxonomy-based technique also considers the input as graphs but the technique concerns only the specialization relation (is-a links).
9. Repository of structures technique stores ontologies and their fragments together with pair-wise similarities between them. When new ontologies are to be matched, the stored similarities could first be checked to avoid the matching operation to be performed over the dissimilar fragments. The available similarities could help to identify fragments that are worth carrying out matching in more detail.
10. Model-based technique deals with input based on its semantic interpretation using well grounded deductive methods such as propositional satisfiability and description logics.

Finally, the third type of ontology mediation is integration. Pinto & Martins (2000, 2001) define ontology integration as a process of building an ontology in one subject reusing one or more ontologies in different subjects and it is always possible to identify regions of the source ontologies from the integrated ontologies. Source ontologies may need some sort of refinements before they can be aggregated, combined and assembled together to form the resulting ontology. It is also important to include ontology integration in the early stage of the ontology building process, preferable during conceptualization and formalization, so as to simplify the overall ontology building procedure.

APPLICATION OF ONTOLOGY IN KNOWLEDGE MANAGEMENT

The concept of ontology and its related mediation methods can also be applied to solve the interoperation problem in the distributed KM environment. At the very beginning, KM is emerged with the purpose of preserving and capitalizing on organizational knowledge for the future benefit of organizations. KM encourages organizations to create and use knowledge continuously for the innovation and enhancement of service, product and operation. Simultaneously, it also aims to improve the quality, content, value and transferability of individual and group knowledge within an organization (Mentzas et al., 2001). This is achieved by organizing formal, direct and systematic process to create, store, disseminate, use

Figure 2. Knowledge management process

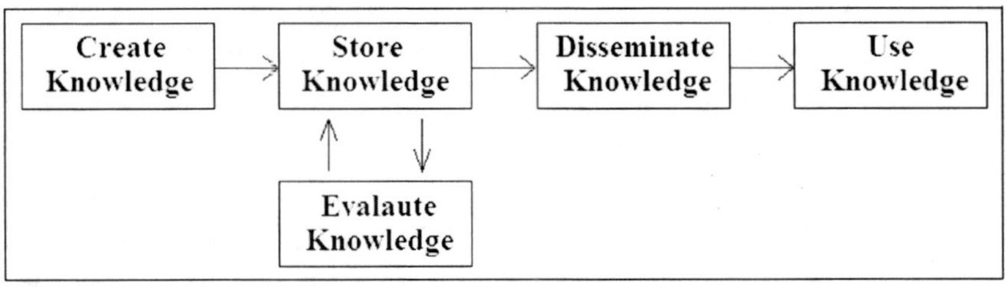

and evaluate organizational knowledge using the appropriate means and technologies (Leung & Lau, 2006) as illustrated in Figure 2.

Nonaka et al. (2001) suggest that there are four methods to create organizational knowledge by means of interaction between explicit and tacit knowledge. While tacit knowledge is personal, complex and hard to communicate and formalize because it is gained through individual insights overtime and is resided in human, mind and body, explicit knowledge is structured, relatively simple and can be captured, recorded, documented, codified and shared using formal and systematic language (Goh, 2002; Nonaka & Takeuchi, 1995). The first method to create knowledge is socialization. It is the process of developing new tacit knowledge from tacit knowledge embedded in human or organization through experience sharing, observation and traditional apprenticeship. The second method is called externalization. This is the process of turning tacit knowledge into new explicit knowledge simply by transforming tacit knowledge in the form of document such as manual and report. The third method is combination. This is the process of merging and editing "explicit knowledge from multiple sources" into a new set of more comprehensive and systematic explicit knowledge. The last one is called internalization. This is the process of embodying explicit knowledge as tacit knowledge by learning, absorbing and integrating explicit knowledge into individual's tacit knowledge base.

The second and third stages of KM, store and disseminate are often linked with technologies. Explicit knowledge created is collected and stored in some sort of database or knowledge base in which the users can access using "search and retrieve" tools, intranets, web access and applications, groupware and so on (Alavi & Leidner, 1999; Smith, 2001). The retrieved knowledge can then be used by knowledge workers to add value to current business process, implement and coordinate organizational strategy, predict trends in the uncertain future, deliver new market values, create new knowledge, solve existing problems and so on (Bailey & Clarke, 2001; Newman, 1997). The fifth stage of KM is knowledge evaluation. This phrase eliminates incorrect or outdated knowledge (Alavi & Leidner, 1999). In other words, organization must keep creating new knowledge to replace any knowledge that has become invalid.

Unfortunately, it is shown that some of the KM approaches, ranging from industrial specific, theoretical, to procedure-wise, are incompetent to cooperate with the current distributed knowledge environment, especially those that are designed to manage merely organizational knowledge, for example, the re-distributed KM framework is developed to manage organizational help desk knowledge (Leung & Lau, 2006). Those approaches are tailor-made according to different organizational KM strategies and business requirements without the concern of system interoperation. The lack of interoperability means that heterogeneous Knowledge Management Systems (KMSs) from different organizations are not able to communicate, cooperate, exchange as well as reuse knowledge with one another. Wagner & Buko (2005) argues that knowledge-sharing in an inter-organizational network allows a richer and more diverse body of knowledge to be created as compared with sharing in one organization.

The non-collaborative KMSs have several disadvantages for both knowledge workers and knowledge engineers. In terms of knowledge workers, they have to spend a lot of time and effort to look for relevant knowledge from different KMSs because they are often required to access knowledge from other knowledge sources in order to complete their works in the knowledge explosion era, for instance, an investment manager has to retrieve company's financial report, share performance report and regional economy reports from external sources if s/he wants to adjust the proportion of a particular share in the investment portfolio. In terms of knowledge engineers, they have to waste a lot of resources in creating and updating organizational knowledge even though the same knowledge is available in other KMSs. As external source of knowledge is essential for organizational performance, a new inter-organizational KM practice is required to enhance the interoperability among independent KMSs and to encourage the sharing of knowledge across organizational boundaries in their business networks (Oinas-Kukkonen, 2005).

Nevertheless, the absence of a common language or standardization has put up a barrier to prevent the collaboration of KMSs. (Sheth, 1999). Although the emergence of middleware technology has provided a way to enhance the interoperability of KMSs, the concept of middleware can hardly be accommodated in the era of the Internet as each pair of KMSs are required to implement a tailor-made middleware for interoperation (Leung et al., 2007). Since a single KMS is interconnected with a huge amount of systems via the Internet, it is impractical to customize and install a middleware for each connection. Another deficiency of middleware is that even if the involved systems only undergo minor modification, the middleware may require a complete re-construction.

ONTOLOGY-BASED COLLABORATION INTER-ORGANIZATION KNOWLEDGE MANAGEMENT NETWORK

As mentioned in the previous section, knowledge created from external source plays a very important role in supporting organizational activities because employees are often required to make use of the knowledge in their daily work. However, a majority of KM frameworks, KM practises and KMSs are merely designed to manage organizational knowledge. Let us consider a real life example: at the University of Wollongong (UOW), if an academic researcher is in need of IS related literatures (for instance, literatures that are related to "inter-organizational KM"), the first thing s/he can do is to access the website of the university library. S/he can then type in a set of keywords in the search interface of the website to see whether relevant literatures are available in library's collection either in the format of hardcopy or softcopy. If so, s/he can choose to pick up the literatures from the library or download them in virtual format. Otherwise, s/he has to search again in various literature knowledge bases that are subscribed by the library. Literature knowledge bases allow subscribers to retrieve literatures that include journals, conference papers, electronic books and thesis in forms of full text or abstracts and citations from their online knowledge repositories. Unfortunately, s/he has to search in every single knowledge base until s/he can find the required literatures because each knowledge base contains different sets of literatures based on publishers, disciplines and so on, for instance, IEEE Xplore (IEEE) and ACM Digital Library (ACM) mainly contain computer related journal and conference papers that are published by IEEE and ACM respectively whereas Australian Digital Theses Program stores thesis of any disciplines that are produced by the postgraduate research students at Australian universities. Finally, if s/he still cannot find any related literature, s/he may choose to search again using search engines such as Yahoo and Google.

This collaboration problem needs to be addressed in a technical way. In this research, we propose to use ontology and its related mediation methods to solve the collaboration problem of heterogeneous KMSs in the Internet environment. Ontology is incorporated to allow explicit knowledge to be annotated in the form of machine processable metadata. Although different organizations possess their own set of ontologies, the mediation methods are capable of reconciling the underlying heterogeneities of ontologies. In this way, the concept of ontology and mediation enables organizational KMSs to understand the incoming request and the return knowledge, thus making it possible for them to collaborate and communicate with each other. We argue that the knowledge reusability and mismatches reconcilability of ontology and its related mediation methods could be further contributed towards the reformation of the existing KM frameworks that only focus on managing organizational knowledge. Because of that, we propose to develop an ontology-based collaborative inter-organizational KM network that provides a platform for organizations to access and reuse inter-organizational knowledge with a similar domain. Here, inter-organizational knowledge is defined as a set of explicit knowledge formalized and created by other organizations. In the network, the formalized inter-organizational knowledge is reusable in a way that it can be retrieved by any organizations to support their own KM processes in terms of knowledge creating, storing, dissemination, using and evaluation.

Each network should only contain knowledge of a specific domain to ensure knowledge workers can retrieve relevant knowledge in an efficient manner, for example, an IS network should only provide knowledge in the aspect of IS. Once an organization recognizes the need for a certain type of knowledge, the organization can invite other organizations and knowledge providers that possess the knowledge of the same domain to establish a network together, for example, given that academic researchers in IS discipline have to access quite a number of knowledge bases so as to obtain the right knowledge, the library of the UOW decides to invite library at the University of New South Wales (UNSW), University of Sydney (SYDNEY) and Queensland University of Technology (QUT) as well as IEEE and ACM to establish a knowledge network that contains only IS knowledge. When the network for a particular knowledge is mature, organization in need may choose to join instead of establishing one. Within a network, each organization or knowledge provider must commit to a mutual agreement to allow other participants accessing an agreeable portion of ontology and the associated knowledge reposited in its knowledge base. Besides, a single organization or knowledge provider can commit to more than one network of different domains, for instance, the library of the UOW may choose to commit to networks of IS, economics, mechanism engineering, education and chemistry whereas IT help desk of the UOW may commit to network of Microsoft, Adobe and Dell knowledge.

Selection Framework for Ontology Mediation

Before we go any further in the description of the proposed network, the participant organization first need to make four important decisions related to ontology mediation. Figure 3 illustrates a selection framework for ontology mediation in form of a matrix. The first decision is whether to adopt top-level ontology or one-to-one as the network level mapping approach. As this decision is on network level rather than organizational one, organizations as a whole must compromise in order to select the most appropriate mapping approach for the benefit of the network. The decision process should include a thorough assessment and discussion in the aspects of resources, expertise and frequency of modification among all organizations in the network. The top-level ontology approach can only be applied to an environment where maintenance effort is minimal even though such an approach can provide a better mechanism to resolve conflicts and ambiguities. Whenever a minor modification is performed in one of the ontologies in the network, the shared ontology used in the top-level ontology approach may need a complete reconstruction. Organizations must also make sure that they have sufficient resources and expertise to build the shared ontology. If frequent maintenance is required or resources and expertise are insufficient, it is much more appropriate to exploit the one-to-one approach.

The second decision is whether to perform mediation automatically or semi-automatically. Mediation can be performed semi-automatically which requires the support of automatic tools

Figure 3. Selection matrix for ontology mediation

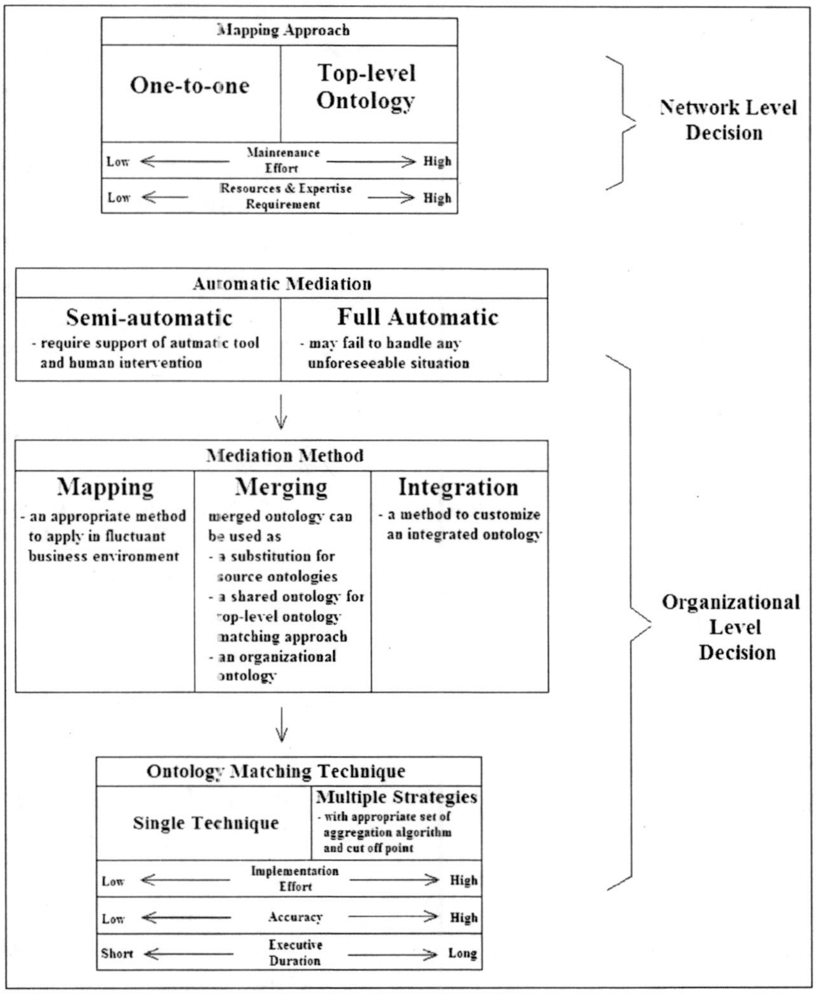

as well as human intervention, for instance, iPROMPT exploits simple heuristic to compare string similarity of concepts between two ontologies (Noy & Musen, 2000). Based on the computational results of similarity, iPROMPT will then be able to suggest a list of possible actions for ontology merging but the final decision on choosing the most suitable action will be left to users. Other forms of support provided by automatic tools include post-mediation verification, validation, critiquation as well as conflicts recognition and resolution. Although semi-automatic mediation could have a better performance than the manual one in terms of accuracy, it still substantially relies on human efforts and is rather time consuming. Without human intervention, process of semi-automatic mediation cannot be completed, thus compromising the accuracy of the mediation result. As semi-automatic tool is not capable of supporting mediation on-the-fly, it would be ideal to perform mediation automatically. Unfortunately automatic tools are unable to detect and interpret concepts that do not have a close correlation. Moreover, it may also fail to handle any unforeseeable situations since the tool is designed to perform mediation under certain pre-defined circumstances, for example, NOM may fail to find matching candidates if the matching condition is not described in the fifteen pre-defined similarity measures (Ehrig & Sure, 2004). Of course, there is not any tool that can guarantee full accuracy. However, if automatic mediation is done and inference builds on top of it, inaccurate results can bring down the value of the mediation process

The third decision is whether to adopt merging, mapping and/or integration as the chosen mediation method for each organization. Each organization can choose one or more methods in accordance with its own need. The concept of mapping enables ontology to be developed in response to its actual business requirement and is more suitable to be applied in a fluctuant business environment. Here, fluctuant business environment refers to an environment where organizations need to modify their ontologies in a frequent manner. Unless ontology has undergone a major modification, otherwise simple modification, such as adding or deleting a concept from an ontology, may merely require to update the mappings accordingly. Alternatively, merging is an appropriate method for creating an ontology that combines common views of multiple source ontologies. In other words, the merged ontology should include all possible correspondences and differences among the entire set of source ontologies. As a result, the merged ontology could act as 1) a single ontology used to substitute individual source ontology, 2) a shared ontology (reference point) used in top-level ontology mapping approach or 3) an organizational ontology that includes all possible views of other organizations' ontologies. Unlike merging, integration selects only apposite modules from individual source ontologies to form an integrated ontology. Thus, integration is a proper method for organizations to customize ontologies in accordance with their own needs, for instance, the library of the UOW can customize a KM based ontology by integrating portions of ontologies derived from the library at the UNSW, the IEEE, the ACM and so on.

The final thing needs to be considered is whether to adopt single or multiple matching techniques. In the decision process, organization must also take into consideration the executive duration, the level of matching accuracy that it can accept and the level of resources that it can afford for implementation. In general, multiple strategies are expected to generate more accurate result than single matching technique but it is not always the case. The choice of aggregation algorithm and cut off point also plays an important role in determining the level of matching accuracy. When choosing multiple strategies as its matching technique, organization must conduct a series of experiments with the purpose of finding a combination of multiple strategies, aggregation algorithm and cut off point that can produce the most accurate result. As compared with single matching technique, multiple strategies are relatively hard to design and implement as well as requires longer executive time.

Operation of the Inter-Organizational Knowledge Management Network

The reconcilability of ontology mediation allows the participant organizations to reuse inter-organizational knowledge within the network even though there are fundamental differences among organizations in terms of KMS structures and knowledge formats. Under a mutual agreement, organizations are permitted to retrieve inter-organization knowledge and the retrieved knowledge can be reused to support the five stages of KM process. Conventionally, technology has very limited contribution in knowledge creating stage especially in socialization, externalization and internalization where tacit knowledge is involved, for example, word processing tools can be used to record and visualise explicit knowledge in externalization and internalization whereas communication tools such as email and telephone provide a platform for the exchange of explicit knowledge in socialization.

However, ontology merging tool could provide a practical way to create knowledge by combining two or more ontologies together semi- or automatically in the network. This can be achieved on both network and organizational level. On the former level, merging tool is capable of creating a shared ontology for top-level mapping approach that contains common views of all organizational ontologies in the network. On the latter level, organization can create its own domain specific ontology by merging relevant ontologies from other organizations within the network. Other than that, ontology integration tool provides an alternative way to creating knowledge. Using integration, organization can create its own knowledge by integrating relevant parts of ontologies from other organizations in the network into its ontology building process. As a result, both merging and integration tool enable organizations to reuse not only the contents of other ontologies but also their associated inter-organizational knowledge that stored in knowledge bases of other organizations. While ontology merging and integration are never a trivial task even with the assistance of automatic tools, they are still less demanding than building from scratch.

Knowledge dissemination tool allows user to retrieve and use knowledge from organizational knowledge repository. If user cannot find suitable organization knowledge, s/he has to seek from other external sources. This can be achieved by creating mappings among ontologies of different organizations either semi- or automatically with the support of ontology mapping tool. The established mappings allow one KMS to access KMSs of other organizations in the same network so as to search for the relevant knowledge. Besides, it is also practical for mapping to be performed on-the-fly. In this case, automatic mapping tool is responsible to look for, select and establish mapping with the most relevant concepts and properties from other ontology in the network. Whenever required knowledge is not available in the organizational repository, the KMS is able to retrieve and deliver inter-organizational knowledge in a "black box" through the establishment of mappings. In addition, inter-organizational knowledge can be reused to support knowledge evaluation process in KM. This is accomplished by setting up dedicated mappings between two or more ontologies. Once a piece of inter-organizational knowledge is updated, this inter-organizational knowledge will be translated into a suitable format and delivered from source knowledge base to target one automatically via the pre-established mappings. To demonstrate the reconcilability of ontology mediation and reusability of inter-organizational knowledge in the network, let us take a look at the following example.

The UOW realizes that there is an increasing demand for information systems related knowledge but this demand can never be satisfied with the current collection of publications in the library. Consequently, the UOW decides to invite knowledge providers and libraries of other universities to establish a network that contains merely information systems related knowledge. Among all, library of the SYDNEY, UNSW and QUT, the IEEE and the ACM agree to join the network. Except for the QUT, they all possess ontologies that are used to describe the classification of literatures for each knowledge base. Figure 4a illustrates a partial view of the classification ontology adopted in the library of the UOW. In this ontology, concept publication has concept book, journal, proceeding and thesis as its subclasses and each subclass is described by a set of properties such as International Standard Book Number (ISBN), International Standard Serial Number (ISSN), and publisher. Concept category and its subclasses are used to distinguish publications into different subjects such as concept computer, medical, commerce, computer science and so on. Given that this network only supports information systems related knowledge, therefore the library of the UOW is willing to share publication that belongs to concept computer and its subclass information systems. As publication may contain chapters written by different authors in reality, the ontology reflects it by including concept book chapter, journal paper as well as conference paper and their related properties as an extension of concept book, journal and proceeding respectively. Figure 4b shows a partial view of the classification ontology exploited in the ACM. There are three major concepts in this ontology, that is, concept book, journal and proceeding. Each concept has a set of publication details (such as issue and edition), contains a set of literatures and belongs to one discipline (such as information systems). The above three components are represented by concept publication detail, literature and discipline respectively. Similar to the UOW, the ACM also agrees to share literatures that are classified under concept information systems.

After discreet consideration, the six organizations have reached a mutual agreement not to adopt top-level ontology as the network-wide mapping approach. This decision is based on the fact that there will be numerous organizations joining the newly established network, so the shared ontology built for the top-level ontology mapping approach may be required to undergo a series of reconstructions. Although they have sufficient

Figure 4. Partial view of the classification ontology adopted in the (a) library of the university of wollongong and (b) acm digital library (round rectangular nodes represent concepts) (rectangular nodes and labels on the arcs represent properties)

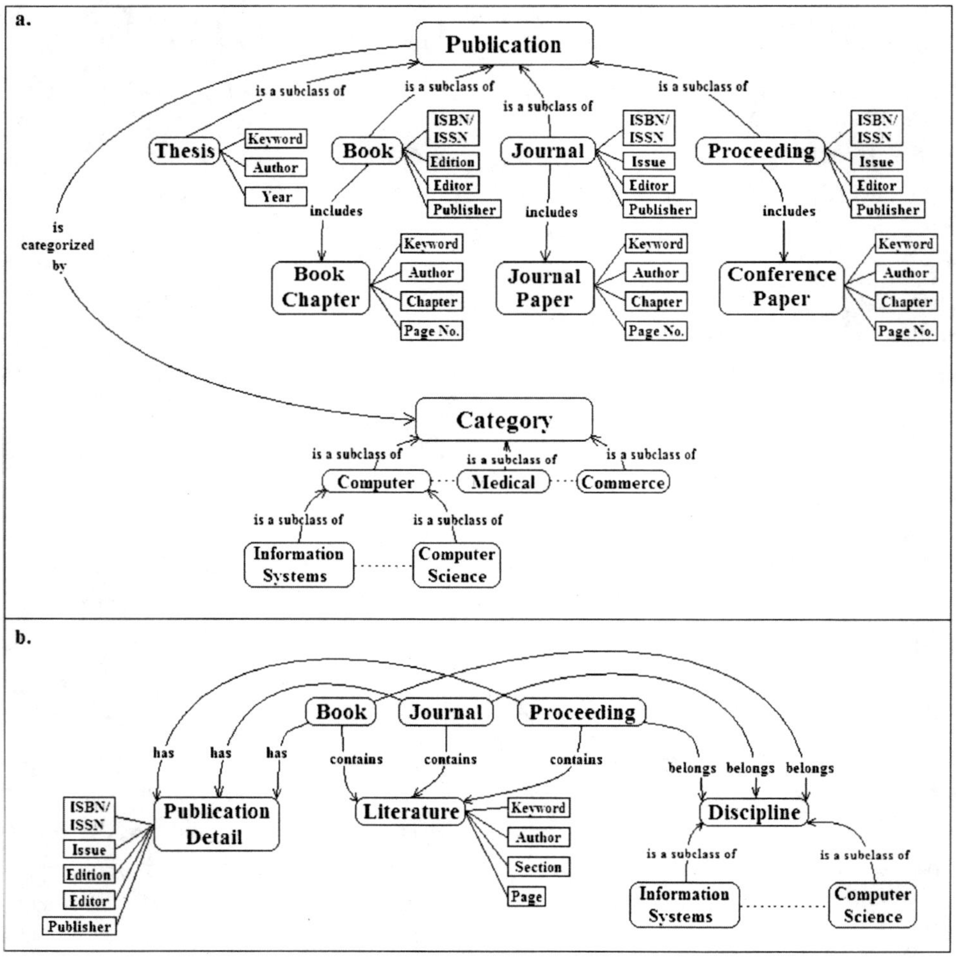

expertise and resources to build and reconstruct the shared ontology, it is not cost effective to do so. In addition, the reconstruction works will definitely affect the stability and performance of network-wide mediation because the shared ontology will be mapped by all other ontologies as a reference point. At this moment, the organizations prefer to use one-to-one mapping approach but they agree to review the mapping approach once the number of organization joining the network becomes steady.

As the library of the QUT does not possess an ontology, the library has to create one in order to fulfil the joining requirement. Instead of building from scratch, the library decides to reuse ontologies from other organizations and integrate them into its own development process using ontology integration method. However, the chosen ontologies must be similar to the library's actual classification in terms of publication and discipline so as to minimize the degree of modification, for instance, concept publication and its subclasses in the ontology of the UOW are more appropriate than those defined in the ACM as the subclass thesis, book, journal and proceeding defined in the ontology of the UOW are very similar to the actual classification used in the library of the QUT. Based on the criteria, the library reuses only a partition of two ontologies that include the concept publication and its subclasses derived from the ontology of the UOW as well as the concept discipline and its subclasses derived from the ontology of the ACM (see Figure 5). In the ontology development process, the library of the QUT could reuse not only the ontologies of other organizations, but also the inter-organizational knowledge associated with the instance of the integrated ontology. As illustrated in Figure 5, the softcopy of the thesis described by the instance of the integrated ontology, thesis "Turning User into First Level Support in Help Desk: Development of a Web-based User Self-help KM System" in

Figure 5. Process to develop qut's ontology using integration method

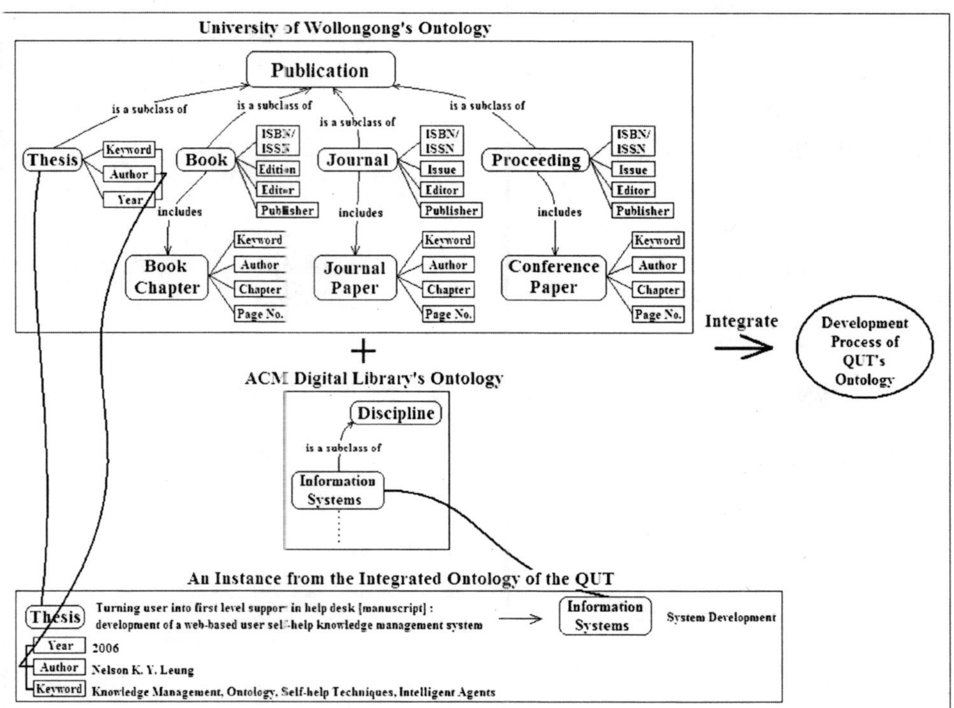

discipline information systems, could be captured from the knowledge base of the UOW and stored in the knowledge base of the QUT. This integrated ontology created by the library of the QUT has one more function. By establishing dedicated mappings between the integrated ontology and its ontology providers (that is, the ontologies of the UOW and ACM), the associated publication captured in the knowledge base of the QUT could be automatically updated as long as there is an revised version generated from the side of the ontology providers, for instance, when the thesis "Turning User into First Level Support in Help Desk: Development of a Web-based User Self-help KM System" has undergone a minor revision in the knowledge evaluation process, the revised thesis will not only be stored back in the knowledge base of the UOW, but will also be broadcasted to other KMS through the dedicated mappings that includes the knowledge base of the QUT.

To allow general user to retrieve and use inter-organizational knowledge, organizations are required to establish mappings between its own ontology and ontologies of other organizations in this network. As shown in Figure 7, each broken line represents a mapping between a pair of concepts or properties that belong to two different ontologies. Making use of string-based and linguistic resources matching techniques, two similar concepts from the ontologies of the UOW and ACM are mapped with each other, for instance, two identical concepts (such as journal) and two properties that are synonyms (such as section and chapter) from the ontologies of the UOW and ACM are mapped together. The mapping details of the two ontologies are summarized in the table in Figure 6.

In Figure 7, user is looking for suitable journal papers by filling in title, publisher and keyword fields on the "knowledge searcher" which is designed to be used as a searching interface for the KMS at the library of the UOW. Since the KMS cannot provide journal that satisfies this query, the system begins to turn its way to other KMSs including the one at the ACM. The mappings in between allow the KMS of the ACM to understand the incoming query, for instance, the details provided in the title, publisher and keyword fields on the knowledge searcher interface are actually referring to concept journal, property publisher and property keyword that belong to the ontology of the ACM. As long as the requested journal is available in the knowledge base of the ACM, it will be delivered to the knowledge searcher interface of the UOW. Subsequently, the journal will be displayed unnoticeably as is retrieved from its own knowledge base. In other words, the entire inter-organizational knowledge retrieval and displaying processes are performed in a "black box" automatically.

CONCLUSION

The organization based KM approaches have caused collaboration problem in which organiza-

Figure 6. Table of mapping summary of the ontology of the uow and acm

	ACM Ontology	UOW Ontology	Matching Reason
Concept	Journal	Journal	Identical
	Literature	Journal Paper	Synonym
	Discipline	Category	Synonym
	Information Systems	Information Systems	Identical
Property	ISBN/ISSN	ISBN/ISSN	Identical
	Issue	Issue	Identical
	Editor	Editor	Identical
	Publisher	Publisher	Identical
	Keyword	Keyword	Identical
	Author	Author	Identical
	Section	Chapter	Synonym
	Page	Page No.	Identical

Figure 7. Inter-organizational knowledge retrieval and reusing process

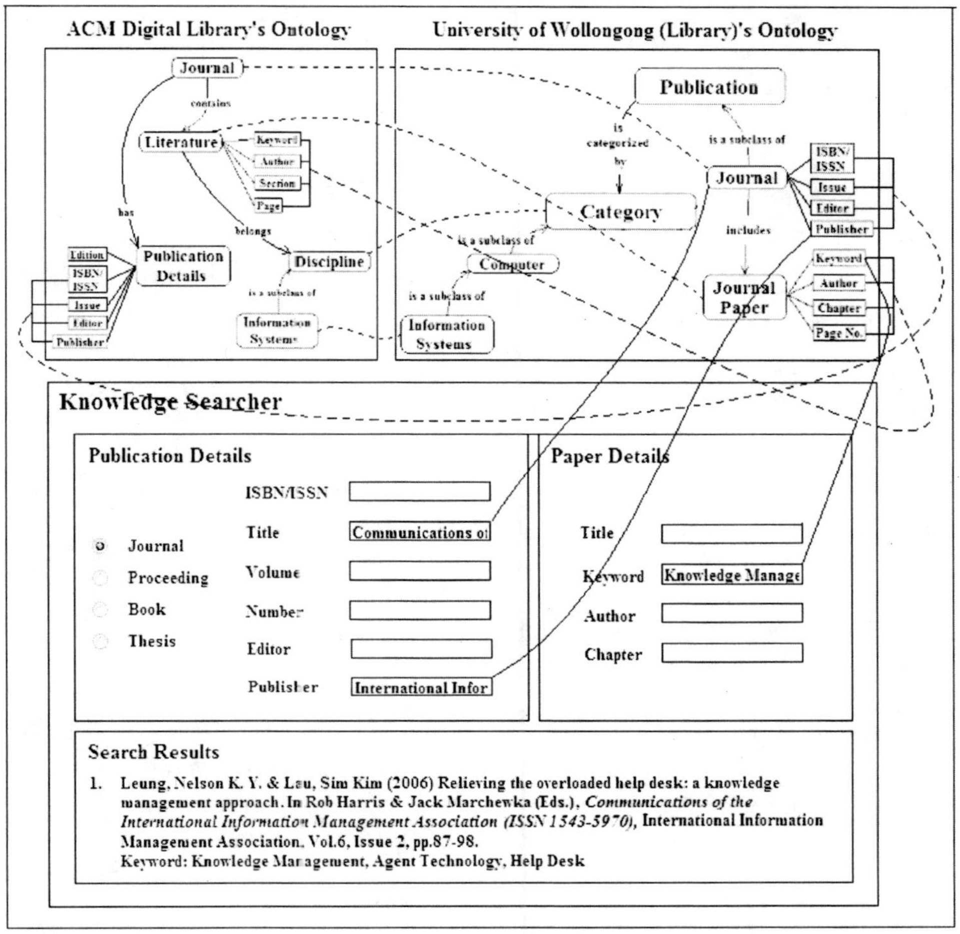

tion is not capable of reusing inter-organizational knowledge even though the required knowledge is available in knowledge bases of other organizations. An ontology-based collaborative inter-organizational KM network is proposed to solve the problems. To establish the network, a selection framework is proposed to assist organizations in choosing suitable ontology mediation approaches, ranging from mapping approaches, levels of automation, mediation methods to matching techniques. The knowledge reusability and mismatches reconcilability of ontology and its related mediation methods enable organizational KMSs to understand the incoming request and the return knowledge, thus making it possible for them to collaborate and communicate with each other. By annotating knowledge explicitly in the form of machine processable representation, organizations joining the network can access, retrieve and reuse domain specific inter-organizational knowledge to support the five stages of organizational KM process. While knowledge engineers could reuse inter-organizational knowledge to create and evaluate organizational knowledge, general users are benefit from the effectiveness and efficiency in searching for relevant inter-organizational knowledge within the network.

REFERENCES

Abels, S., Haak, L., & Hahn, A. (2005). Identification of Common Methods Used for Ontology Integration Tasks. In A. Hahn, S. Abels & L. Haak (Ed.), *Proceedings of the 1st International ACM Workshop on Interoperability of Heterogeneous Information Systems* (pp. 75-78). New York: ACM.

Alavi, M., & Leidner, D. E. (1999). Knowledge Management Systems: Issues, Challenges, and Benefits. *Communications of the Association for Information Systems*, *1*(7), 1–37.

Antoniou, G., & Harmelen, F. (Eds.). (2004). *A Semantic Web Primer*. Cambridge, MA: The MIT Press.

Bailey, C., & Clarke, M. (2001). Managing knowledge for personal and organisational benefit. *Journal of Knowledge Management*, *5*(1), 58–67. doi:10.1108/13673270110384400

Berners-Lee, T., Hendler, J., & Lassila, O. (2001). *The Semantic Web. Scientific American*. May Issue.

Borst, W. N. (1997). *Construction of Engineering Ontologies for Knowledge Sharing and Reuse*. Enschede, Netherlands: University of Tweenty, Centre for Telematica and Information Technology.

de Bruijn, J., Ehrig, M., Feier, C., Martin-Recuerda, F., Scharffe, F., & Weiten, M. (2006). Ontology Mediation, Merging and Aligning . In Davies, J., Studer, R., & Warren, P. (Eds.), *Semantic Web Technologies* (pp. 95–112). Chichester, UK: John Wiley & Sons. doi:10.1002/047003033X.ch6

Ding, Y., Fensel, D., Klein, M., & Omelayenko, B. (2002). The Semantic Web: Yet Another Hip. *Data & Knowledge Engineering*, *41*(3), 205–227. doi:10.1016/S0169-023X(02)00041-1

Ehrig, M., & Sure, Y. (2004). *Ontology Mapping – An Integrated Approach* (*Vol. 3053*). Lecture Notes in Computer Science.

Gangemi, A., Guarino, N., Masolo, C., & Oltramari, A. (2003). Sweetening WordNet with DOLCE. *AI Magazine*, *24*(3), 13–24.

Goh, S. C. (2002). Managing Effective Knowledge Transfer: An Integrative Framework and Some Practice Implications. *Journal of Knowledge Management*, *6*(1), 23–30. doi:10.1108/13673270210417664

Gruber, T. R. (1993). Toward Principles for the Design of Ontologies Used for Knowledge Sharing . In Guarino, N., & Poli, R. (Eds.), *Formal Ontology in Conceptual Analysis and Knowledge Representation*. Amsterdam: Kluwer Academic Representation.

Guarino, N. (1998). Formal Ontology and Information Systems. In N. Guarino (Ed.), *Proceedings of the International Conference on Formal Ontology in Information Systems* (pp. 3-17). Amsterdam: IOS Press.

INTEROP. (2004). *Ontology Interoperability*. State of the Art Report (SOA), WP8ST3 Deliverable, IST-508011.

Klein, M. (2001). Combining and Relating Ontologies: an Analysis of Problems and Solutions. In A. Gomez-Perez, M. Gruninger, H. Stuckenschmidt & M. Uschold (Ed.), *IJCAI-2001 Workshop on Ontologies and Information Sharing,* (pp. 53-62).

Lambrix, P., Habbouche, M., & Perez, M. (2003). Evaluation of Ontology Development Tools for Bioinformatics. *Bioinformatics (Oxford, England)*, *19*(12), 1564–1571. doi:10.1093/bioinformatics/btg194

Leung, N. K. Y., & Lau, S. K. (2006). Relieving the Overloaded Help Desk: A Knowledge Management Approach. *Communications of International Information Management Association*, *6*(2), 87–98.

Leung, N. K. Y., Lau, S. K., & Fan, J. P. (2007). An Ontology-based Knowledge Network to Reuse Inter-organization Knowledge. In M. Toleman, A. Cater-Steel, & D. Roberts (Ed.), *Proceedings of the 18th Australasian Conference on Information Systems*. The University of Southern Queensland.

Lussier, Y. A., & Li, J. (2004). Terminological Mapping for High Throughput Comparative Biology of Phenotypes. In R. Altman, A. Dunker, L. Hunter, T. Jung and T. Klein (Ed.), *Proceedings of the Pacific Symposium on Biocomputing* (pp. 202-213). World Scientific.

McGuinness, D. L., Fikes, R., Rice, J., & Widler, S. (2000). An Environment for Merging and Testing Large Ontologies . In Cohn, A., Giunchiglia, F., & Selman, B. (Eds.), *KR2000: Principles of Knowledge Representation and Reasoning* (pp. 483–493). San Francisco: Morgan Kaufmann Publishers.

Mentzas, G., Aposolou, D., Young, R., & Abecker, A. (2001). Knowledge Networking: a Holistic Solution for Leveraging Corporate Knowledge. *Journal of Knowledge Management*, *5*(1), 94–106. doi:10.1108/13673270110384446

Newman, V. (1997). Redefining Knowledge Management to Deliver Competitive Advantage. *Journal of Knowledge Management*, *1*(2), 123–128. doi:10.1108/EUM0000000004587

Niles, I., & Pease, A. (2001). Towards a Standard Upper Ontology. In C. Welty & B. Smith (Ed.), *Proceedings of the International Conference on Formal Ontology in Information Systems* (pp. 2-9). New York: ACM.

Nonaka, I., & Takeuchi, H. (Eds.). (1995). *The Knowledge-Creating Company: How Japanese Companies Create the Dynamics of Innovation*. Oxford, UK: Oxford University Press.

Nonaka, I., Toyama, R., & Konno, N. (2001). SECI, Ba and Leadership: a Unified Model of Dynamic Knowledge Creation. In S. Little, P. Quintas & T. Ray (Ed.), Managing Industrial Knowledge Creation, Transfer and Utilization (pp. 13-43). Milton Keynes, UK: The Open University.

Noy, N., & Musen, M. (2000). PROMPT: Algorithm and tool for automated ontology merging and alignment. In *Proceedings of the 17th National Conference on Artificial Intelligence*. Menlo Park, CA: AAAI Press.

Oinas-Kukkonen, H. (2005). Towards Evaluating Knowledge Management through the 7C Model. In *Proceedings of the European Conference on Information Technology Evaluation*. University of Oulu.

Pinto, H. S., & Martins, J. P. (2000). Reusing Ontologies. In *Proceedings of the AAAI Spring Symposium Series, Workshop on Bringing Knowledge to Business Processes* (pp. 77–84). Menlo Park, CA: AAAI Press.

Pinto, H. S., & Martins, J. P. (2001). A Methodology for Ontology Integration. In *Proceedings of the 1st International Conference on Knowledge Capture* (pp. 131-138). New York: ACM.

Predoiu, L., Feier, C., Scharffe, F., de Bruijn, J., Martin-Recuerda, F., Manov, D., & Ehrig, M. (2006). *State-of-the-art Survey on Ontology Merging and Aligning V2, EU-IST Integrated Project (IP) IST-2003-506826 SEKT: Semantically Enabled Knowledge Technologies*. Innsbruck, Austria: University of Innsbruck.

Rahm, E., & Bernstein, P. (2001). A Survey of Approaches to Automatic Schema Matching. *The International Journal on Very Large Data Bases*, *10*(1), 334–350. doi:10.1007/s007780100057

Scharffe, F., de Bruijn, J., & Foxvog, D. (2006). *Ontology Mediation Patterns Library V2. EU-IST Integrated Project (IP) IST-2003-506826 SEKT: Semantically Enabled Knowledge Technologies*. Innsbruck, Austria: University of Innsbruck.

Sheth, A. (1999). Changing Focus on Interoperability in Information Systems: from System, Syntax, Structure to Semantics . In Goodchild, M., Egenhofer, M., Fegeas, R., & Kottman, C. (Eds.), *Interoperating Geographic Information Systems* (pp. 5–30). Amsterdam: Kluwer Academic Publishers.

Shvaiko, P. (2004) A Classification of Schema-based Matching Approaches. In S. McIlraith, D. Plexousakis & F. van Harmelen, (Ed.), *Proceedings of the Meaning Coordination and Negotiation Workshop at the 1st International Semantic Web Conference*. Heidelberg, Germany: Springer.

Shvaiko, P., & Euzenat, J. (2005). A Survey of Schema-based Matching Approaches. *Journal on Data Semantics*, *IV*, 146–171.

Smith, B. (2003). Ontology . In Floridi, L. (Ed.), *Blackwell Guide to the Philosophy of Computing and Information* (pp. 155–166). London: Wiley Blackwell.

Smith, E. A. (2001). The Role of Tacit and Explicit Knowledge in the Workplace. *Journal of Knowledge Management*, *5*(4), 311–321. doi:10.1108/13673270110411733

Studer, R., Benjamins, V. R., & Fensel, D. (1998). Knowledge Engineering: Principles and Methods. *Data & Knowledge Engineering*, *25*, 161–197. doi:10.1016/S0169-023X(97)00056-6

Valente, A., Russ, T., Macgregor, R., & Swartout, W. (1999). Building and (Re)Using an Ontology of Air Campaign Planning. *IEEE Intelligent Systems*, *14*(1), 27–36. doi:10.1109/5254.747903

Visser, P. R. S., & Cui, Z. (1998). On Accepting Heterogeneous Ontologies in Distributed Architectures. In *Proceedings of the ECAI'98 workshop on Applications of Ontologies and Problem-solving methods* (pp. 112-119).

Visser, P. R. S., Jones, D. M., Bench-Capon, T. J. M., & Shave, M. J. R. (1998). Assessing Heterogeneity by Classifying Ontology Mismatches. In N. Guarino (Ed.), *Proceedings of the International Conference on Formal Ontology in Information Systems* (pp. 148-162). Amsterdam: IOS Press.

Wache, H., Vogele, T., Visser, U., Stuckenschmidt, H., Schuster, G., Neumann, H., & Hubner, S. (2001) Ontology-based Integration of Information – A Survey of Existing Approaches. In H. Stuckenschmidt (Ed.), *Proceedings of the IJCAI workshop on Ontologies and Information Sharing* (pp. 108–117).

Wagner, S. M., & Buko, C. (2005). An Empirical Investigation of Knowledge-sharing in Networks. *The Journal of Supply Chain Management*, *41*(4), 17–31. doi:10.1111/j.1745-493X.2005.04104003.x

KEY TERMS AND DEFINITIONS

Knowledge Management: A formal, direct and systematic process to create, store, disseminate, use and evaluate organizational knowledge using the appropriate means and technologies.

Inter-Organizational Knowledge: A set of explicit knowledge formalized and created by other organizations.

Ontology: A machine-readable specification of a conceptualization in which the type of concepts used and the constraints on their use are explicitly defined.

Ontology Mediation: A process used to reconcile mismatches between heterogeneous ontologies so that knowledge sharing and reuse among multiple data sources can be achieved. There are three major kinds of ontology mediations which include mapping, merging and integration.

Ontology Mapping: A process of relating similar concepts and relations from different ontologies to each other in which the correspondences between different entities of the two ontologies are formulated as axioms in specific mapping language.

Ontology Merging: Creates a new ontology (in one subject) by unifying two or more different ontologies on that subject and it is usually hard to identify regions of the source ontologies from the merged ontologies.

Ontology Integration: A process of building an ontology in one subject reusing one or more ontologies in different subjects and it is always possible to identify regions of the source ontologies from the integrated ontologies.

Chapter 7
Building and Use of a LOM Ontology

Ghebghoub Ouafia
University of Jijel, Algeria

Abel Marie-Hélène
Compiègne University of Technology, France

Moulin Claude
Compiègne University of Technology, France

Leblanc Adeline
Compiègne University of Technology, France

ABSTRACT

The increasing number of available resources that may be used during e-learning can raise problems of access, management and sharing. An e-learning application therefore shares the same problem of relevance to the Web concerning the access to learning resources. Semantic web technologies provide promising solutions to such problems. One main feature of this new web generation is the shared understanding based on ontologies. This chapter presents an approach to index learning resources semantically using a LOM ontology. This ontology was developed to clarify the concepts, and to describe the existing relations between elements of the LOM standard. The author present also our tool based on this ontology which allows to describe learning objects and helps retrieving them.

INTRODUCTION

The use of information and communication technology in the field of training gave birth to new forms of learning. The e-learning is a type of training whose context is based on the diffusion of resources through an electronic medium. It is defined as just-in-time education integrated with high velocity value chains. It is the delivery of individualized, comprehensive, dynamic learning content in real time, aiding the development of knowledge communities, linking learners and practitioners with experts" [Drucker, 2000].

With the expansion of the web, a new form of e-learning has appeared. Online learning or web-based training, offers many possibilities of collaboration and interactivity. Teachers involved in this type of

DOI: 10.4018/978-1-61520-859-3.ch007

learning do not have the total control of the material delivered to learners. Learners may discover different learning resources, use them as such, or combine them if necessary; these resources are called learning objects (LO).

%The use of simple metadata such as LOM to describe resources is an insufficient %solution.

The increasing number of available LOs leads to problems of management and sharing as well. One solution that facilitates research, exchange and data management teaching is to index learning objects with a set of shared metadata. This metadata can be information regarding the authors of learning materials, their fields of interest, their ideas, their learning objectives, and so on.

Currently, the most successful and most used standard for describing the LO is the LOM[1] (Learning Object Metadata). However, the semantic ambiguity of some of its elements and their subjectivities, especially in the educational category, makes difficult the use of its descriptors.

To overcome this difficulty and to allow the understanding of these elements, it is necessary to add semantics. The semantics include the definition of concepts, relationships between concepts, attributes and constraints. Using an ontology for modeling the LOM may facilitate the description of resources by defining the concepts and exploiting semantic relations between them. Each resource will be described by a concept rather than words that may be ambiguous.

In this chapter, we present the metadata approach for indexing learning resources and especially the use of the LOM standard. Then, we justify why a semantic web approach is relevant to overcome the metadata limits and present the LOM ontology we have constructed. Thereafter, we illustrate how to use our ontology by means of two examples. The first example concerns the tool we developed for indexing and retrieving the LOs. The second concerns MEMORAe2.0 [Abel & al., 2008] project. Finally we conclude by presenting the perspectives of this work.

LEARNING OBJECTS INDEXING

The basic idea behind the creation of learning objects is the ability to build components or small units that can be reused several times in different learning contexts.

Adopting this concept of small reusable units, Reigeluth and Nelson explain that often when teachers or learning content creators access for the first time to a learning material, they break it down into its components, and then assemble these components in order to build a material that supports their educational goals [Nelson & al. 1996].

In order to find these learning materials to use or reuse them, they must be described efficiently. A resource that is not indexed is an unexploitable resource and is difficult to retrieve. To develop and promote standards for learning technology, the IEEE[2] consortium created in 1996 the Learning Technology Standards Standing Committee (LTSC).

Learning Objects

The definition of a learning object gave rise to several debates. In the document representing the LOM (Learning Object Metadata) 1.0[3] standard, a learning object is defined as any digital or non-digital entity that can be used, reused or referenced during learning, education or training activities.

This definition is seen by Wiley as wide as it may include an object, a person or an idea [Wiley, 2002]. Polsani (2003) refines it by adding that a learning object is an " independent and self-standing unit of learning content that is predisposed to reuse in multiple instructional contexts" [Polsani, 2003].

In our work, we have adhered to the Wiley definition that considers that a learning object is a learning material that can be selected, combined with another according to the needs of teachers and learners. It is also a learning content that should exist as such, and can be searched and indexed easily.

Metadata and Standards

Metadata are used for retrieval and indexing of learning objects. The term metadata refers to data about data or information about information. It is defined as structured information describing a resource. It includes characteristics or properties such as file, format, size, location, etc. This information can be embedded into a resource or it can be stored separately.

In order to facilitate sharing and exchange of learning objects as well as their metadata, it is necessary to adopt a common metadata set to describe them. Several metadata schemas have been developed in a variety of user environments and disciplines. They are designed principally to standardize the description of resources. Some of them have been admitted and acknowledged by the organization for standardization such as ISO (The International Standardization Organization), they have become standards.

In March 1995, the first basis for a documents descriptors system was proposed. The Dublin Core[4] (DC) standard is a set of 15 simple elements such as title, creator, subject, language, etc. where each element is optional and repeatable. This schema was not designed specifically for the e-learning. It may be neither comprehensive nor suitable for the education domain. Therefore it has been adjusted to be applied to learning resources.

In August 1999, a DCMI (Initiative Metadata DC) Education Working Group[5] has been set up to think up how the DC standard could be adapted to educational resources. The DC Education helped enriching the initial elements.

Other metadata schemas have been designed to take into account the specificities of some professions or fields of applications. In the field of education and e-learning, working groups have attempted to define and specify suitable description elements. One of the most complete and successful standards for describing learning resources is the Learning Object Metadata.

LEARNING OBJECT METADATA

The standard developed by the IEEE LTSC, was approved by the IEEE-Standards Association on June 12, 2002. 1484.12.1 - 2002 Learning Object Metadata specifies the syntax of the elements composing it such as element names, definitions, vocabularies, occurrences number and data types.

LOM Model

The LOM data model is a hierarchy of elements (45 elements in LOM 1.0) grouped into nine categories:

1. *General category* groups the general information that describe the learning object as a whole (e.g. title, description, language, structure, aggregation level) ;
2. *Lifecycle category* groups the features related to the history and current state of a learning object and those who have affected this learning objects during its evolution (e.g. version, participants to its evolution) ;
3. *Meta-metadata category* groups information about the metadata instance itself ;
4. *Technical category* groups the technical requirements and technical characteristics of the learning object (e.g. format, size) ;
5. *Educational category* groups the educational and pedagogical characteristics of the learning object (e.g. interactivity type, difficulty level) ;
6. *Right category* groups the intellectual property rights and conditions of use for the learning object;

7. *Relation category* groups features that define the relationship between the learning objects and other related learning objects ;
8. *Annotation category* provides comments on the educational use of the learning object and provides information on when and by whom the comments were created (e.g. creator of the annotation) ;
9. *Classification category* describes this learning object in relation to a particular classification.

LOM Using

The LOM metadata instances are implemented using XML or RDF[6] binding [Nilsson & al., 2003]. Data in XML binding are considered as a tree. In RDF binding, data have semantics and can be viewed as objects having properties that relate them to other objects. In RDF, metadata have semantics, but the representation %remains insufficient to express the complex structure [Ghebghoub & al., 2008].

Based on the model and its bindings, several tools have been developed to create metadata, and various learning object repositories (LOR) have been created to store and index LO in common areas.

Some editors may be downloaded and used locally such as LomPad[7] others are used online. With the exception of SHAME[8] which is based on the LOM RDF binding, most of these editors generate files compatible with the XML binding. However, if these tools allow to describe the learning resources and to store these descriptions, they do not use them for learning objects retrieval.

LOR have their own editors and have more complex systems that allow to describe the LO. These repositories store these LO along with their descriptions. In addition, they allow the retrieval of these described LO's. Among these LORs, we find the ARIADNE[9] project of the Foundation for the European knowledge pool and the American collection of links towards sharable learning objects, MERLOT[10] (Multimedia Educational Resource for Learning and Online Teaching).

Although the standard is widely used, it is strongly criticized to be complex having a large number of elements. Also, many problems are detected during its use. For example in ARIADNE, the most used elements are the basic metadata such as title or author name. Elements belonging to the learning category (context, interactivity level, difficulty) are much less used, even when they are given average values similar to those by defaults [Najar & al., 2006].

These problems are due to an incorrect interpretation and misunderstanding of the elements as well as values. Indeed, the semantic ambiguity of some LOM elements, and especially their subjectivity in the educational category, makes the description of resources difficult [Motelet, 2007].

SEMANTIC WEB FOR E-LEARNING

As an e-learning application shares the same problem of relevance to the Web concerning the access to learning resources, adopting the same approaches can solve these problems.

The shared-understanding problem in e-learning appears especially when it comes to define the content of a learning resource during the process for accessing or searching a particular learning material [Stojanovic & al., 2001].

Among the efforts to solve this problem, we find the concept of "Semantic Web". Introduced by Tim Berners-Lee in 1998, the semantic web is defined as an extension of current web. This new generation of web is an environment in which humans and machine agents will communicate using a semantic base [Berner, 1998]. One of its main features is that it allows the shared understanding based on ontologies.

Gruber defined an ontology as "a specification of a conceptualization" [Gruber, 1993]. Thereby, an ontology provides a unifying framework that ensures shared understanding [Uschold & al., 1996]. It helps eliminating or reducing conceptual and terminological ambiguities.

So, several researchers have proposed to combine ontologies to certain elements of the LOM to resolve some of the problems associated with the use of LOM. Indeed, Mohan and Boorks determine the elements of the LOM for which they suggest the use of ontologies [Mohan & al., 2003]. These elements are the structure and learning strategies.

Phaedra and Permanand propose to mitigate the problem of lack of expressiveness of the LOM approach based not only on the use of a single ontology but on several ones [phaedra & al., 2006]. What they did is interesting but is still insufficient because they do not handle all LOM elements. We propose then to build an ontology of LOM standard.

Using an ontology of LOM for learning object indexing will allow a better understanding of the elements and values and thus facilitate their descriptions.

A LOM ONTOLOGY

Building a LOM ontology aims to clearly define the concepts and the relationships that will make the standard elements more explicit. Accordingly, it allows users to better interpret the proposed values and therefore to have less difficulty in filling the fields. The LOM ontology provides common and shared vocabulary and set of concepts that can facilitate the description and promote reuse and sharing of learning objects among users.

To construct the ontology of LOM, we began by studying the structure of the LOM 1.0 and XML files generated by an editor as LOMPad. We built the ontology with multilingual approach. Concepts, attributes and individuals have labels and comments in French and English. For a French use of our ontology, we have included definitions of terms from LOMFR[11] standard established by the French Association for Standardization AFNOR[12].

However for the sake of flexibility, we relied on the original structure of the LOM developed by the IEEE consortium. In this way, a user can switch from English to French easily.

Several methodologies have been proposed to develop ontologies [Jones & al., 1998]. For building our ontology we haven't used a particular methodology. We started from the RDF binding and have defined different rules:

- ***R1***: the learning object and LOM categories are represented by concepts that are *LearningObject* and *LOMCategory* ;
- ***R2***: each category of LOM is represented by a sub concept of *LOMCategory* ;
- ***R3***: the *LearningObject* concept is connected by relations with the sub-concepts representing the LOM categories ;

The LOM elements are in turn represented by a concept, an attribute or a relationship according to their type. Indeed, it lists three types of elements:

1. Simple elements (or atomic) with predefined values, such as the *Structure* element in the *General* category;
2. Simple elements without predefined values, such as the *Description* element in the *General* category ;
3. Complex elements such as the *Contribute* element in the *Life cycle* category.

The representation rules ***Ri*** for the element ***e*** are:

- ***R4***: if ***e*** is simple element with predefined values, then it is represented by a concept attached to its category, and its values ***v*** are represented by individuals ;
- ***R5***: if ***e*** is of simple element without predefined values, then it is represented by an

attribute associated with the concept representing its category or the concept representing the element to which it is attached;

- ***R6***: if ***e*** is complex element, then ***e*** is represented by a concept associated with its category.

Ontologies may be developed in RDF, or with other languages and concepts such as Topics Maps[13], and especially with the Web Ontology Language OWL[14]. The OWL syntax is in most cases handled by an editor.

To represent our ontology we chose OWL and used Protégé[15] editor. We use the formalism N3 to present examples because it is more readable than the formalism OWL.

In the following we present examples how we applied the defined rules to build our ontology.

By applying *R1*, we first created the *LearningObject* concept and the *LomCategory* concept. With *R2*, the *LomCategory* concept is specialized in nine sub-concepts representing the nine LOM categories (figure 1 shows a part of the LOM ontology structure). Figure 2 shows the super-class and two subclasses representing two categories of the LOM. *R3* allowed us to associate the *LearningObject* concept to the sub-concepts that represent the nine LOM categories using nine relations. This way, the *LearningObject* concept is attached to the *LomGeneralCategory* concept by the relationship *hasGeneralCategory*.

The LOM General category defines the notion of identifier for a learning object as " a globally unique label that identifies this learning object". The identifier is characterized by a catalog and an entry. The Identifier is a complex element, and so we apply *R6*. It is represented by a concept of the ontology. The Catalog and Entry elements are simple elements without predefined values, and so they are represented by attributes of the *Identifier* concept. A relationship *hasIdentifierElement* between the two concepts, *LomGeneralCategory* and *Identifier*, is linked the identifier to the LOM general category.

Figure 1. A part of the lom ontology structure

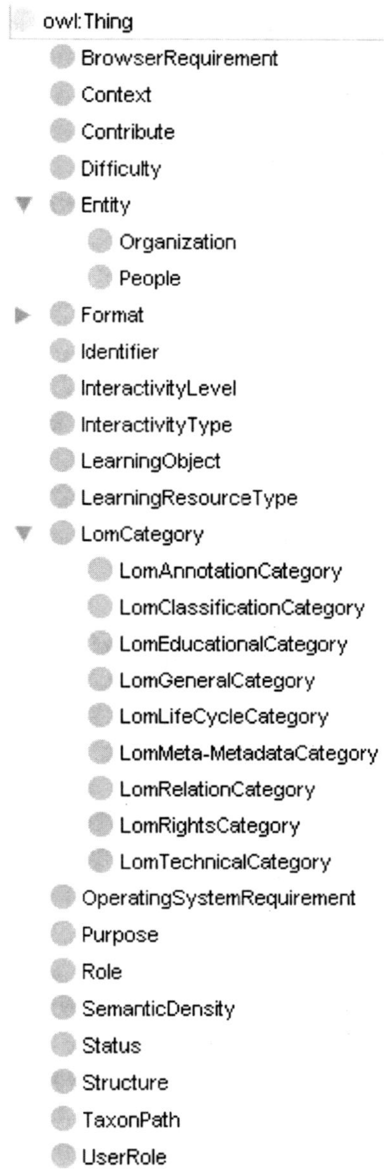

The *language*, D*escription*, *Keyword*, *Coverage* and *Title* are also are simple elements without predefined values, and they are therefore represented with attributes of the *LomGeneralCategory* concept.

The *Structure* and *Agregationlevel* elements are simple elements with predefined values, and they are therefore represented as concepts. Two relationships, *hasStructureElement* and *hasAg-*

Figure 2. Some lom categories formalized using owl

```
:LomCategory
  rdf:type owl:Class ;
  rdfs:comment "regroupement de métadonnées selon le type
d'informations qu'elles relatent."@fr ;
rdfs:comment " a group of related data elements."@en ;
   rdfs:label "catégories du LOM"@fr;
   rdfs:label "LOM category"@en .

:LomGeneralCategory
  rdf:type owl:Class ;
  rdfs:comment This category groups the general information
that describes this learning object as a whole "@en ;
  rdfs:comment "Cette catégorie présente les
caractéristiques générales pour
décrire une ressource."@fr ;
  rdfs:label "general category"@en ;
  rdfs:label "catégorie générale"@fr ;
  rdfs:subClassOf :LomCategory .

:LomEducationalCategory
  rdf:type owl:Class ;
  rdfs:comment "This category describes the key educational
or pedagogic characteristics of the learning object. "@en ;
  rdfs:comment "Cette catégorie décrit les caractéristiques
pédagogiques de l'objet pédagogique."@fr ;
  rdfs:label "educational category"@en ;
  rdfs:label "catégorie pédagogique"@fr ;
  rdfs:subClassOf :LomCategory .
```

gregationLevelElement, are created whose domain is the *LomGeneralCategory* concept, and ranges are the concepts *Structure* and *AgregationLevel* respectively.

The values for the *Structure* element (*Atomic, Collection, Networked, Hierarchical, Linear*) are represented by individuals. We apply our rules in the same way for the remaining categories.

In the next section, we will present the LOM ontology based tool that we developed for indexing and retrieving LOs.

TOOL ENVIROMENT

In order to overcome the problems associated with the use of the standard LOM in LORs, we developed a specific tool with two objectives: indexing a distributed set of resources referenced by their URL/URI, and helping the search for these resources. In addition to these objectives, the tool provides a support for a good understanding of the different items.

This tool can be used in LOR to facilitate the LO indexing by our ontology concepts, but it can be also used independently of any LOR. Even if it cannot store the LOs locally, it stores their links.

The LOIT Architecture

Our tool LOIT (learning objects indexing tool) is a component that can be added to any Java web application. It is built using the struts[16] open-source framework that allows developing web applications using MVC (Model-View-Controller) architecture. MVC has been widely implemented in user interfaces to separate the entity responsible of showing information (view) from the one responsible of storing it (model) and the one responsible of receiving user events (controller) [Gallego & al., 2005].

LOIT is made up of three main forms (see Figure 3):

- A form that allows entering information to describe learning objects.
- A second form used to enter search criteria according to the nine categories to perform an advanced search.
- A third form used to enter search criteria to perform a simple search.

The first two forms rely on nine java classes, each representing a category of LOM. The two actions, index and search, are defined in the struts configuration file (struts-config.xml). The user interaction is supported by JSP pages that present the ontology concepts and allow inserting values for describing or searching a resource.

The knowledge base is managed by the Jena[17] framework, whose last version extends the notion of property in order to access concepts using any keyword. The knowledge base exported in OWL also allows interoperability with systems using another technology. The possibility of finding and retrieving particular LO is a key issue for users in LOR.

Internally, descriptions of resources are transformed in an RDF knowledge base, compliant with our ontology, that can be exported as an OWL file. Discovery of resources is performed thanks to SPARQL[18] requests against this knowledge base.

Learning Object Indexing

When a user selects one of the nine LOM categories, all its editable elements are displayed (Figure 4). The user can edit any one of all these elements. However, some elements are mandatory such as URL/URI and title. The user can browse from one category to another in the left menu or by clicking on the button *continue*. Once the fields are filled,

Figure 3. Loit architecture

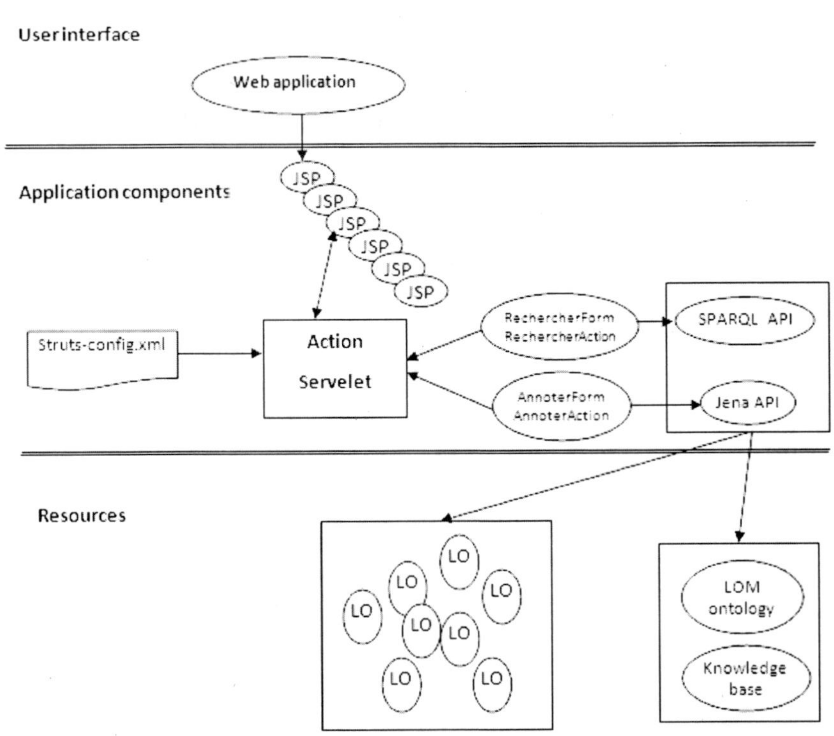

the user can click on a specific button to save the entered values. The values are then added to the knowledge base which is exported as an OWL file. To make more explicit the standard elements, each label is a link that refers to the page of the adequate LOM ontology documentation.

Search for Learning Objects

The simple search interface is intended for novice users who do not know the structure of the LOM and want to search without browsing the nine categories.

The query is written using the LARQ[19] technology, which is a combination of ARQ[20] and Lucene[21]. It is a library written in java allowing the creation of different kinds of indexes and providing a full-featured text search engine.

For example, the interface enables the user to search for resources whose keyword field contains the word "sql" and not the word "sparql", combining free text search from several fields (for more examples, see [Ghebghoub & al., 2008]).

To write queries, the wildcard characters are used, such as an asterisk (*) or a question mark (?) which respectively represent zero to several characters and one character. The symbols (+) or (-) are also used to represent respectively, the Boolean operators "and" and "not".

Advanced Search

The advanced search interface allows users to fill in the fields that will enable them to regain the learning objects they need. In the same way, a user can use one or more elements of the nine categories. When he runs his request, the URL/URI of learning objects that answers the search criteria are displayed in user interface.

For example, in order to search for the resources written in French, with an atomic structure, aggregation level equal to 1 and associated with the SQL word, the user inputs values for language, structure, aggregation level and language elements respectively of the General category. The tool then builds the request. A list of triples linking

Figure 4. Indexing interface - french interface

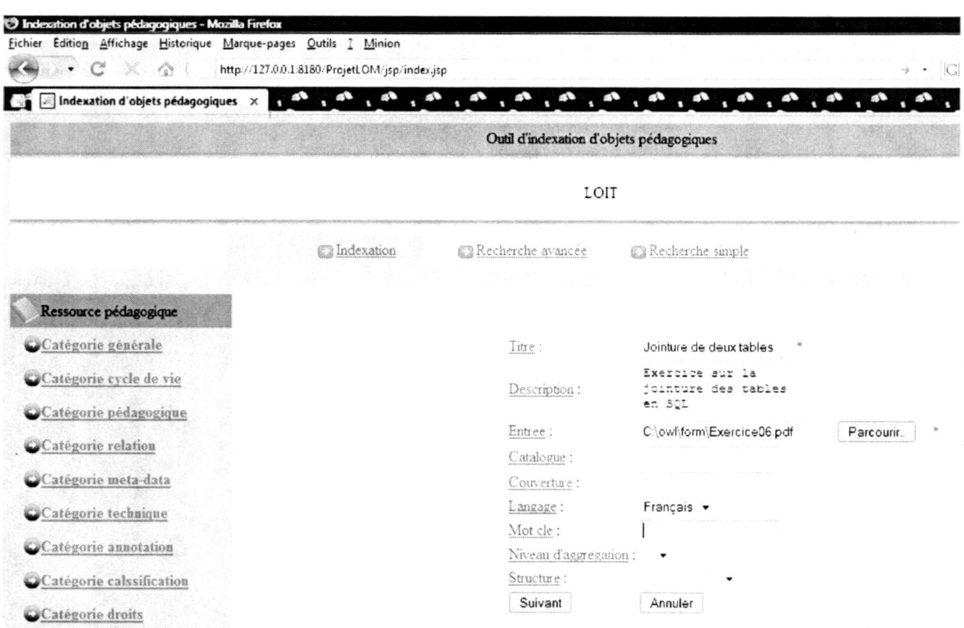

the variables to the user values is created and the URLs of the corresponding resources are returned in the entry variable. In the query interface, the response is displayed as links.

The constructed ontology allowed us to build a tool that facilitates the use of a standard such as LOM for indexing the LO, and allows the reduction of noise and silence in a search. This ontology can be also used to describe learning contents in training environments and e-learning applications. This is why we introduced this ontology in our project MEMORAe2.0.

USING THE LOM ONTOLOGY IN MEMORAE

Before explaining how we have integrated our ontology in E-MEMORAe2.0 environment, we will first clarify the approach MEMORAe. We also describe the MEMORAe model and we present subsequently the E-MEMORAe2.0 environment.

%Currently the information used to describe the learning resources are only %author name, description and date of the availability of the resource in %organizational memory.

The Approach MEMORAe

Our aim within the approach MEMORAe, is to make connections between e-learning and knowledge management operational [Leblanc & al., 2007a]. To that end, we chose to associate:

- Knowledge engineering and educational engineering,
- Semantic Web and Web 2.0 technologies,

To model and build learning collaborative platform as organizational learning support for SLO (Semantic Learning Organization). Let us note that SLO is a concept that extends the notion of learning organization in a semantic dimension. A SLO must be considered as a learning organization in which learning activities are mediated and enhanced through a shared knowledge representation of the organization [Sicilia & al., 2005].

In the e-learning side, we decided to follow the resources modeling approach based on the learning objects paradigm.

In the knowledge management side, we chose to adapt the concept of Organizational Memory. Dieng define such a concept as an " explicit, disembodied, persistent representation of knowledge and information in an organization, in order to facilitate its access and reuse by members of the organization, for their tasks" [Dieng & al., 1998]. Extending this definition, we propose the concept of Learning Organizational Memory for which users' task is learning. Thereby MEMORAe is more than a LOR.

In the Semantic Web side, we decided to structure and organize our memory content through ontologies.

Finally, in order to facilitate interactions and social processes we decided to use Web 2.0 technologies and to link them to semantic web in modeling and organizing micro resources by means of ontologies.

The MEMORAe Model

Within the project MEMORAe, we conceived the organizational learning memory around two types of sub-memory that constitute the final memory of the organization:

- **Group memory:** this kind of memory enables all the group members to access knowledge and resources shared by them. The group is at least made of two members. We distinguish three types of group memory corresponding to different communities of practice:
  1. **Team memory:** The team memory capitalizes knowledge, resources, communication relative to any object of interest for the group members.

2. **Project memory:** The project memory capitalizes knowledge, resources, communication relative to a project. All the information stored is shared by members working on the project.
3. **Organization memory:** this memory enables all members of the organization to access knowledge and resources without prior access right. These resources and knowledge are shared by all the organization members.

- **Individual memory:** this kind of memory is private. Each member of the organization has his own memory in which he can organize, and capitalize his knowledge, resources.

These memories offer a way to facilitate and to capitalize exchanges between organization members. The MEMORAe model relies on two ontologies [Abel & al., 2007].

- The first one (domain ontology) describes the concepts of the *organization* domain (see figure 5). They can be individual group (team, project, organization-wide), resource types (communication resource: book, course, ..., action resource: exercise or problem, exchange resource: chat, forum, ... or coordination resource: planning or engagement book), function (administrative, student, teacher or technician), etc.
- The second ontology (application ontology) specifies notions which are used by members of a particular organization. In order to assess our approach, we chose to build organizational memory for academics organization: a course B31.1 on applied mathematics at the University of Picardy (France) (see figure 6).

All the memories share these ontologies. These ontologies define and structure the organizational memory. Each sub-memory can have its own resources. However, they shared the same ontologies. Resources, even if they are private - stored in a particular memory - are indexed by the concepts of the ontology.

We used Topic Maps as a representation formalism facilitating navigation and access to the resources. The ontology structure is also used to navigate among the concepts as in a roadmap. The user has to access the appropriate resources.

Figure 5. Memorae2.0 domain ontology

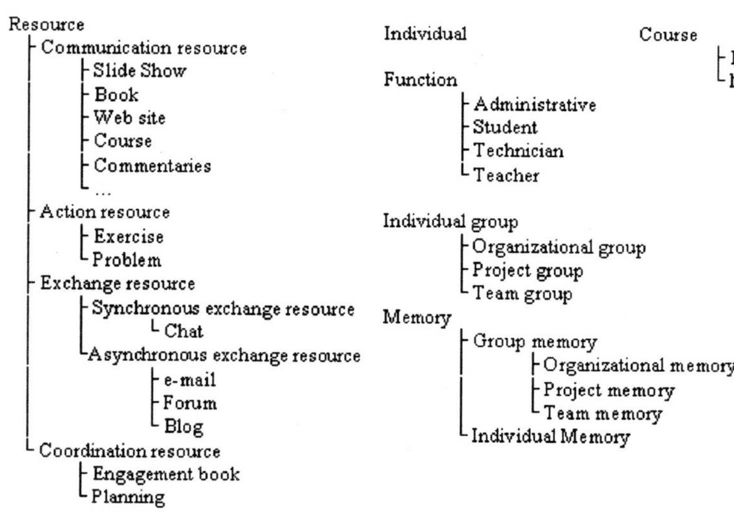

Figure 6. Part of the b31.1 application ontology

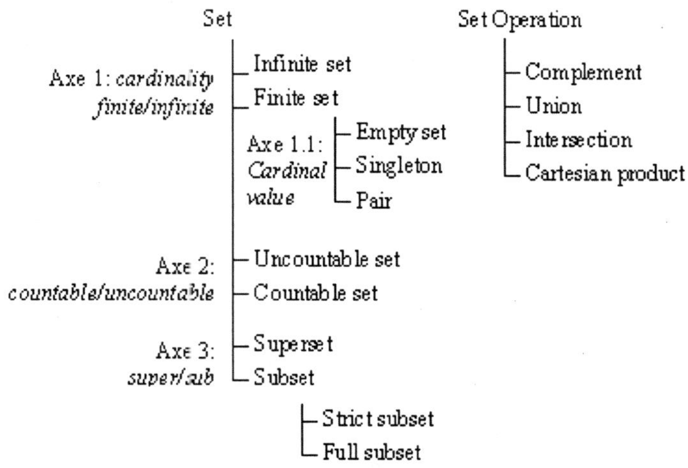

E-MEMORAe2.0 Environment

In order to put into practice this modeling we developed an environment called E-MEMORAe2.0 [Leblanc & al., 2007b]. It enables users to navigate through ontologies and gives the possibility for learners to have different memories.

All these spaces (memories) share the same ontology but store different resources and different entry points. We can see three spaces in Figure 7: one dedicated to organization members, one dedicated to group3 members and one to the connected individual. Let us note that by default, the user visualizes two memory levels: his private memory and the organization memory.

However he can choose spaces he wants to visualize by selecting them in the memories choice window. These choices are registered and will be considered for the next session.

Figure 8 illustrates the presentation of such spaces. We can see the space dedicated to the organization group according user wants to access entry point (entry point vertical tab is selected) or resources indexed by the concept selected in the ontology map (resources vertical tab is selected).

Thus, by means of this interface, users can explore the memory content. Vertical navigation allows to explore subsumption relations and to reach related concepts. For example, if the user wants to discover the *Finite Set* notion, the best entry point is *Set (population)*. By choosing this entry point, (s)he has access to the part of ontology associated to the notion of Set. Among the sub-concepts of *Set*, (s)he can find *Finite Set*. By clicking on this concept, a local taxonomy centered on this new concept is displayed. The iteration of this process allows the learner to browse the ontology.

Let us suppose now that the learner decides to temporarily stop the navigation and to focus on a particular concept. This concept is at first described by a short definition. If the user wants to learn more on the selected notion, (s)he has access to a list of resources ordered by type. For example, figure 7 shows that if the user wants to deepen the notion of *Set (population)*, (s)he can select among the associated resources. A description text is then displayed in a new window. When the resource is digital, it can be displayed or sent to someone by e-mail.

A concept can refer to concepts other than those which are displayed in the ontology. Access to these concepts is sometimes needed in order to understand some notions. Proximity relations (other than subsumption) are useful for that.

Figure 7. E-memorae environment (in french)

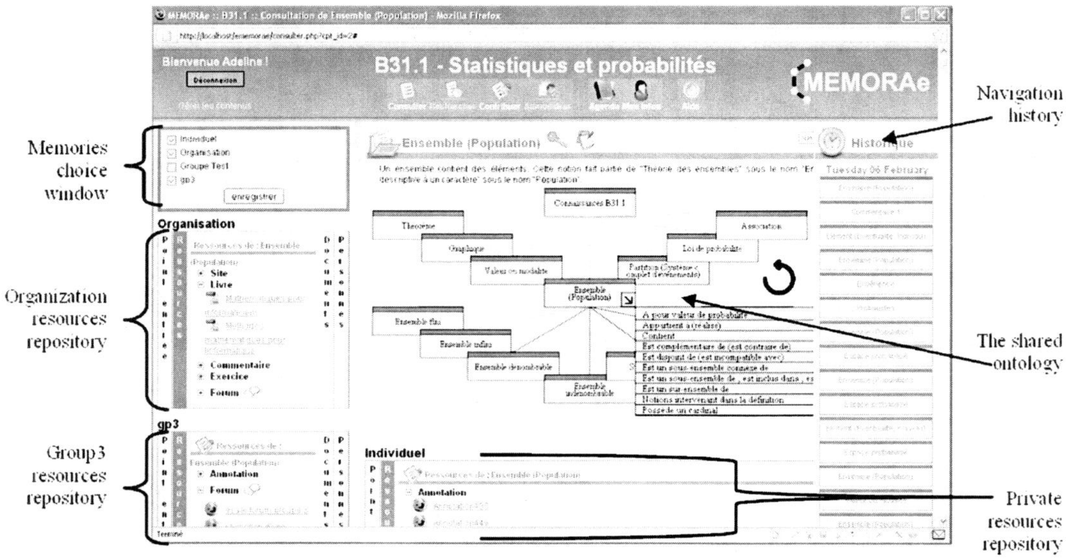

Figure 8. Visualization of different vertical tab of organizational memory (in french)

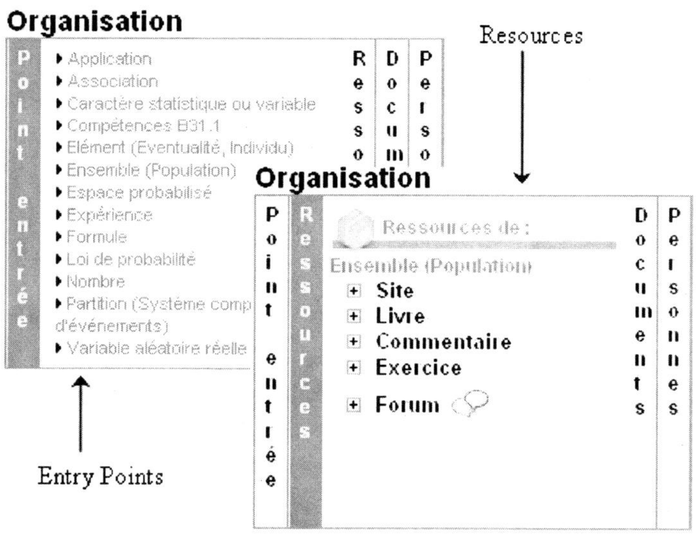

Examples of these relations are: prerequisite-of, in-the-definition-of, suggests, etc. We call this kind of navigation *horizontal navigation*, in comparison with the *vertical navigation* that we considered before. These relations are accessed by left-clicking in the arrow icon: a popup menu contextually displays the available relations starting from the concept.

LOM Ontology Integration in MEMORAe

A learning object is defined as learning material that can be selected, combined with another according to the needs of teachers and learners. It is also a learning content that should exist as such, and can be searched and indexed easily.

This definition fits perfectly the resource concept defined in the domain ontology which is presented in Figure 5. Therefore it was decided to replace the modeling of Resource concept by the one of learning object concept. This one will inherit the sub-concepts and relationships of the Resource concept, in addition to concepts and relationships defined in LOM ontology.

The ontology domain extended, thereby allows a better description. Previously the information used was only the author name, the description and the date of availability of the resources in the organizational memory.

The link between the ontology domain extended and the application ontology presented in Figure 6 allows describing the contents of LO and their form with LOM standard. The concept *keyword* associated with the concept representing *general category* becomes then redundant and less effective.

This integration also allows users to navigate through LOM ontology concepts by means of interface presented in figure 9. For example, if the user wants to discover the *General Category* concept, the best entry point is *LOM Category*. By choosing this entry point, (s)he has access to the part of ontology centered on the concept of LOM Category. Among the sub-concepts of *LOM Category*, (s)he can find *General Category*. By clicking on this concept, (s)he has access to the part of ontology centered on this new concept is displayed. (s)he can access to relations hasIdentifierElement and hasStructureElement thanks to the arrow icon.

To integrate our ontology developed in OWL, we used the Protégé plug-in developed in [Leblanc & al., 20008], which enables exporting an OWL ontology developed with Protégé into an XTM[22] compatible with the format used in the E-MEMORAe environment. Consequently, this adds the possibility to integrate ontologies developed in OWL into the E-MEMORAe2.0 environment.

CONCLUSION

In order to solve the problems of management, sharing and access to resources we have presented our approach to semantically indexing the learning objects through the concepts of LOM ontology that we constructed. We have explained the construction methodology of this ontology, and then we presented the tool developed for indexing and retrieval of learning objects. We also explained

Figure 9. LOM ontology in memorae

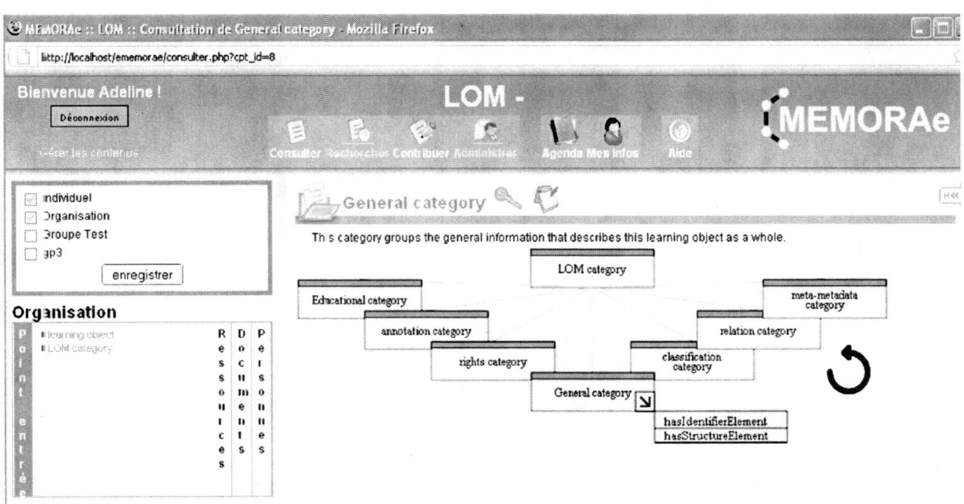

how to integrate this ontology in a training environment such as MEMORAe.

Our ontology based on the LOM mainly describes the form of resources and not directly their content. For example, it is possible to say that a resource is a very difficult exercise, but it is not possible to say that it is a math exercise about the Pythagora's theorem. However thanks to links with application ontology, these limits are overcome in the MEMORAe2.0 project.

So, we plan to change free text in keyword by concepts identifier among application ontologies concepts. These ontologies application can be downloaded by the users.

Concerning the LOM ontology use in the MEMORAe2.0 project, the specialization of the Resource concept is redundant with some LOM elements. So, we have decided to modify these sub-concepts.

REFERENCES

Abel, M.-H., & Leblanc, A. (2008). An operationnalization of the connections between e-learning and knowledge management: the memorae approach. In *Proceedings of the 6th IEEE Internationa l Conferences on Human System Learning*, Toulouse, France (pp. 93--99).

Abel, M.-H., Lenne, D., & Leblanc, A. (2007). Organizational learning at university. In *Second European Conference on Technology Enhanced Learning, EC-TEL2007*, (pp. 408—413).

Berners-Lee, T. (1998). *What the Semantic Web can represent*. Retrieved from http://www. w3. org/DesignIssues/RDFnot. Html

Dieng, R., Corby, O., Giboin, A., & Ribière, M. (1998). Methods and tools for corporate knowledge management. In 11th workshop on Knowledge Acquisition, Modeling and Management (KAW' 98), Banff Canada.

Drucker, P. (2000). *Need to Know: Integrating e-Learning with High Velocity Value Chains*. A Delphi Group White.

Gallego-Carrillo, M., Garcia-Alcaide, I., & Montalvo-Herranz, S. (2005). Applying Hierarchical MVC Architecture to High Interactive Web Applications. In *Proceedings of the Third International Conference I. TECH*, (pp. 110--115).

Ghebghoub, O., Abel, M., & Moulin, C. (2008). Semantic Indexing of e-Learning Resources. In *3rd International Conference Information and Communication Technologies: From Theory to Applications. ICTTA 2008*, (pp. 1--6), Damascus, Syria.

Gruber, T. (1993). A translation approach to portable ontology specifications. *Knowledge Acquisition, 5*, 199–199. doi:10.1006/knac.1993.1008

Jones, D., Bench-Capon, T., & Visser, P. (1998). *Methodologies for ontology development. In 1998 IFIP world computer congress N°15* (pp. 62–75). Budapest: HONGRIE.

Leblanc, A., & Abel, M.-H. (2007a). Using organizational memory and forum in an organizational learning. In *IEEE/ACM ICDIM'07, The Second International Conference on Digital Information Management*, (pp. 266--271).

Leblanc, A., & Abel, M.-H. (2007b). E-memorae2.0: An e-learning environment as learners communities support. *IJCSA, 5*(1), 108–123.

Leblanc, A., Abou Assali, A., Abel, M.-H., & Lenne, D. (2008). Use of topic maps to support learning organisational memory. In *Proceedings of the Fourth International Conference on Topic Maps Research and Applications, TMRA 2008 Subject-centric Computing*, Leipzig, Germany, (15--17).

Mohan, P., & Brooks, C. (2003). Learning objects on the semantic Web. In *Advanced Learning Technologies, Proceedings. The 3rd IEEE International Conference on*, (pp. 195--199).

Motelet, O. (2007). *Improving Learning-Object Metadata Usage during Lesson Authoring*. PhD thesis, Chile university, October 2007.

Najjar, J. & Duval, E. (2006). Actual use of learning objects and metadata: An empirical analysis. *Technical Committee on Digital Libraries Bulletin (TCDL), 2*(2).

Nilsson, M., Palmer, M., & Brase, J. (2003). The LOM RDF Binding-Principles and Implementation. In *Proceedings of the Third Annual ARIADNE conference, 140*.

Phaedra, M., & Permanand, M. (2006). *Incorporating multiple ontologies into the ieee learning object metadata standard* (pp. 143–154). CSWWS.

Polsani, P. (2003). Use and Abuse of Reusable Learning Objects. *Journal of Digital Information*, (4): 2003–02.

Reigeluth, C. (1996). A New Paradigm of ISD? *Educational Technology, 36*, 13–20.

Sicilia, M. & Lytras, M. (2005). The semantic learning organization. *The Learning Organization: An International Journal*, 12.

Stojanovic, L., Staab, S., & Studer, R. (2001). eLearning based on the Semantic Web. In *WebNet2001-World Conference on the WWW and Internet.*

Uschold, M., & Gruninger, M. (1996). Ontologies: Principles, Methods and Applications. *The Knowledge Engineering Review, 11*(2). doi:10.1017/S0269888900007797

Wiley, D. (2002). *Connecting learning objects to instructional design theory: A definition, a metaphor, and a taxonomy* (pp. 571–577). The Instructional Use of Learning Objects.

ENDNOTES

1 http://ltsc.ieee.org/wg12/20020612\-Final\-LOM\-Draft.html

2 Institute for Electrical and Electronic Engineers, http://ltsc.ieee.org/wg12

3 http://ltsc.ieee.org/wg12/files/LOM_1484_12_1_v1_Final_Draft.pdf

4 http://dublincore.org/documents/usageguide/elements.shtml

5 http://dublincore.org/groups/education

6 http://www.w3.org/RDF/

7 http://demo.licef.teluq.uquebec.ca/LomPad/

8 http://kmr.nada.kth.se/shame/

9 http://www.ariadne-eu.org/

10 http://www.merlot.org/home.po/

11 http://www.cura.fr/unr-ra/page18/files/LOM-FR\%20experimentale.pdf

12 http://www.afnor.fr

13 http://www.topicmaps.org/

14 Ontology Web Language: http://www.w3.org/2004/OWL/

15 http://protege.stanford.edu

16 http://struts.apache.org/

17 Jena, a Semantic Web Framework for Java: http://jena.sourceforge.net/

18 SPARQL: http://www.w3.org/TR/rdf-sparql-query

19 http://jena.sourceforge.net/ARQ/lucene-arq.html/

20 http://jena.sourceforge.net/ARQ/

21 http://lucene.apache.org/java/docs/index.html

22 http://www.topicmaps.org/xtm/

Section 3
Ontology Management:
Construction, Evolution and Alignment

Chapter 8
Ontology Evolution:
State of the Art and Future Directions

Rim Djedidi
Supélec – Campus de Gif, France

Marie-Aude Aufaure
MAS Laboratory, France

ABSTRACT

Ontologies evolve continuously throughout their lifecycle to respond to different change requirements. Several problems emanate from ontology evolution: capturing change requirements, change representation, change impact analysis and resolution, change validation, change traceability, change propagation to dependant artifacts, versioning, etc. The purpose of this chapter is to gather research and current developments to manage ontology evolution. The authors highlight ontology evolution issues and present a state-of-the-art of ontology evolution approach by describing issues raised and the ontology model considered (ontology representation language), and also the ontology engineering tools supporting ontology evolution and maintenance. Furthermore, they sum up the state-of-the-art review by a comparative study based on general characteristics, evolution functionalities supported, and specificities of the existing ontology evolution approaches. At the end of the chapter, the authors discuss future and emerging trends.

INTRODUCTION

Today, ontologies are finding their way into a wide variety of applications. In addition to the Semantic Web, they are also applied to knowledge management, content and document management, information and model integration, etc. They offer rich and explicit semantic conceptualizations and reasoning capabilities and facilitate query exploitation and system interoperability. However, ontological knowledge cannot be considered as being fixed and static. Just like any structure holding knowledge, it needs to be updated as well. Ontology development is a dynamic process starting with an initial rough ontology, which is later revised, refined and filled in with the details (Noy & McGuinness, 2001). Even during the usage of the ontology, the knowledge of the modeled domain can change

DOI: 10.4018/978-1-61520-859-3.ch008

and develop. To remain useful, the ontology has to cope with frequent changes in the application environment.

The purpose of this chapter is to gather research and current developments to manage ontology evolution and to discuss future direction. The chapter is composed of three main parts. In the first part, we outline ontology evolution requirements, we present a comparative study of ontology, database schema, and knowledge-based system evolution; and we detail ontology change management issues.

In the second part, devoted to a state-of-the-art review, we present an overview of existing ontology evolution approaches and highlight the functionalities supported by their process. The study also takes into account the ontology representation language and its consistency constraints. We also describe ontology engineering tools supporting a part of/or a complete ontology evolution process. At the end of this part, we sum up the review by a comparative study based on general characteristics, evolution functionalities supported, and specificities of the existing ontology evolution approaches.

The third part focuses on future research directions by gathering the latest research in progress on ontology evolution and giving perspectives on open issues.

ONTOLOGY EVOLUTION ISSUES

In this section, we try to give a better understanding of the ontology evolution problem by analyzing the context of the problem, comparing it with problems and solutions in related areas and outlining its issues.

The increasing number of ontologies in use and their costly adaptation to change requirements make ontology evolution very important. Ontology evolution regards the capability of managing the modification of an ontology in a consistent way. It is defined as being "the timely adaptation of an ontology and consistent propagation of changes to dependent artifacts" (Maedche, Motik, & Stojanovic, 2003, pp.287).

Ontology Evolution Requirements

Ontology evolution requirements have been discussed in (Blundell & Pettifer, 2004; Noy & Klein, 2004; Stojanovic & Motik, 2002; Stojanovic, 2004; Klein, 2004). Ontology evolution is a complex problem (figure 1): Besides identifying change requirements from several sources (modeled domain, usage environment, internal conceptualization, etc.), the management of a change –from a request to the final validation and application– needs to formally specify the required change, to analyze and resolve change effects on ontology, to implement the change, and

Figure 1. Ontology evolution requirements

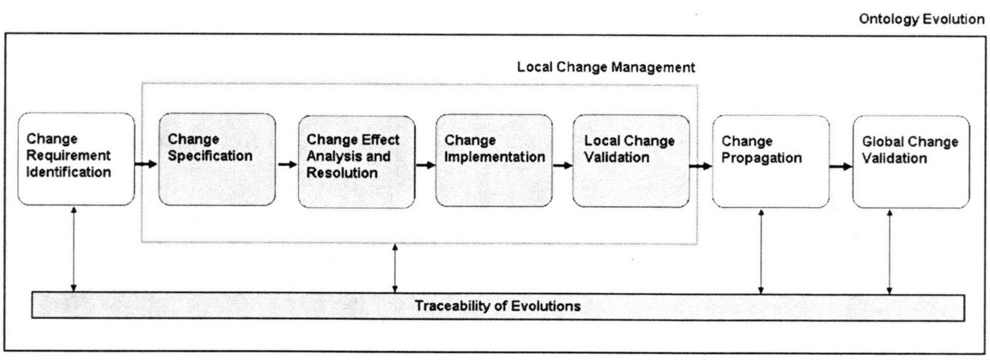

to validate its final application. In a collaborative or distributed context, it is also necessary to propagate local changes to dependent artifacts and to globally validate changes. Moreover, to justify, explain or cancel a change, and manage ontology versions, change traceability has to be kept.

Before going into depth with ontology change management issues, we compare existing strategies to handle changes in database schemas and knowledge bases with the field of ontology evolution.

Comparison with Database Schema and Knowledge Base Evolution

Change management is a well-known topic from research on database schema evolution (Roddick, 1996; Breche & Wörner, 1995; Banerjee, Kim, Kim, & Korth, 1987) and knowledge-based system maintenance (Menzis, 1999). Several issues from these fields are relevant to ontology evolution. In the following sub-sections, we summarize comparative studies of respectively database schema vs. ontology evolution and knowledge-based system vs. ontology evolution.

Database Schema vs. Ontology Evolution

Ontology evolution is closely related to the area of evolving database schemas (Roddick, 1996) and especially object-oriented databases (Banerjee et al., 1987; Franconi, Grandi, & Mandreoli, 2000). General ideas from version modeling in database evolution are relevant for ontology evolution. However, the scale of the issues around ontology changes is more extended and has to be considered on its own. A comparative study of database schema vs. ontology evolution is summarized in Table 1.

Knowledge-Based System vs. Ontology Evolution

To study the problem of ontology evolution, it is interesting to look at existing approaches to handling changes in formal knowledge structures (Coenen & Bench-Capon, 1993) (Menzis, 1999) and compare the evolution of these structures with ontology evolution. The maintenance of a knowledge-based system focuses on the maintenance of its knowledge base (Menzis, 1999). A comparative study of knowledge-based system vs. ontology evolution is summarized in the following table (see Table 2).

Ontology Changes

The conceptualization modeled in an ontology provides the applications –using the ontology– knowledge about how to use data related to ontology's concepts. However, neither the data, nor the ontology itself is permanent and stable. Ontology changes have direct effect on the way data should be interpreted and data dynamics have to be reflected to ontology. Thus, a better understanding of ontology change management issues is needed.

Change Activities

Ontology change management can be considered as an umbrella of change activities. The distinction between these activities has been, in some cases, confusing. Considering the role of evolution in the ontology lifecycle and adapting the terminology from the database community (Roddick, 1996), it appears that these activities are performed nearly sequentially to produce the ontology lifeline (O'Brien & Abidi, 2006). Four main activities are distinguished in the literature: ontology revision, ontology versioning, ontology adaptation, and ontology evolution.

Table 1. Database schema vs. ontology evolution

Database Schema (DB)	**Ontology**
– Quite frequent change operations during DB system lifecycle. – Managing evolution is not formally handled; DB administrator has to control consistency maintenance. – Manual adaptation of instances.	– Evolution frequency can be substantial especially if change requirements are captured by users. – Ontology engineer guidance for change detection and management. – Automating change resolution and notifying ontology engineer about change effects.
Ontology evolution is more closer to object-oriented DB schema evolution than to relational DB schema as the former schema is semantically richer through inheritance and hierarchy principles.	
– Object-oriented DB schemas reflect structure of data and codes, and take also into account object behavior (methods included in the model). In addition, they integrate physical representation of data (integer, real, etc.).	– Ontologies reflect a domain structure through concepts, relations, and constraints on these primitives.
Several ontology evolution approaches adapted existing approaches in object-oriented DB schemas. However, differences between the two models imply that ontology evolution approaches are a kind of extension rather than an adaptation of existing approaches (Noy & Klein, 2004).	
– Instances (database objects) are not at the same level as classes.	– Terminological level (classes and properties) and assertional level (instances) are not delimited. Classes and instances could be manipulated and used together as in querying ontology. Therefore, change effect on queries should be considered (Klein, 2004).
– The integration of a schema to another is not possible.	– Ontology reuse allows integrating ontologies or part of ontologies. Thus, a change could be an inclusion/exclusion of an ontology in/from another (Stojanovic, 2004).
– Changes are defined on the model itself.	– Besides representing changes according to ontology model, it is essential to specify change semantic through conditions to verify and actions to apply for maintaining consistency (Stojanovic, 2004). Ontology structure brings more inconsistencies and resolutions are more complex.
– Evolution needs the definition of consistency model and the explicit specification of changes to apply.	– As ontology models are richer than object models, defining ontology consistency involves more constraints. In addition, possibilities of applying changes are greater than in object or relational models (e.g. in an ontology, an individual can be added without instantiating a class and a property can be defined without being attached to a class).
– Schema semantic is not sufficiently explicit to apply reasoning mechanisms verifying consistency.	– Ontology semantic is more explicit and allows application of reasoning mechanisms to detect inconsistencies.
Solutions controlling change effects on schemas are based on rules to satisfy to maintain consistency and inference mechanisms based on axioms (Klein, 2004). Ontology change resolutions combine (depending on ontology language) these two complementary solutions.	
– To propagate changes to instances, three solutions are adopted (Stojanovic, 2004): (a) data migration and the immediate adaptation of instances to schema; (b) schema-instance synchronization mechanisms (delayed conversion, propagating change through delayed conversion objects, or versioning if no change propagation and if each object is assigned to the different versions); and (c) combination of the two previous solutions.	– For ontology change propagation, two scenarios are distinguished (Stojanovic, 2004): (a) a centralized scenario based on data migration and (b) a distributed scenario based on delayed conversion. The scenario choice depends on priority degree assigned to two contradictory criteria: global consistency and run-time.
– Change propagation is limited to instances.	– Change propagation is extended to all ontology dependant artifacts i.e. instances, annotations, ontologies, and applications.

Ontology Revision

Ontology revision consists of operations that change the state of an ontology to adapt its representation characteristics to the modeled domain (Klein & Fensel, 2001). It is based on applying logical manipulation to expand the ontology abstraction level like adding/deleting concepts or refactoring properties (Foo, 1995).

Table 2. Knowledge-based system vs. ontology evolution

Knowledge-based System (KBS)	Ontology
– The essential characteristic of KBS is the separation of knowledge representation (rules, propositional logic, predicate logic, etc.) and knowledge manipulation. Knowledge manipulation is handled by inference methods provided by reasoners, generally expressed in procedural codes.	– Domain knowledge of a KBS can be represented by an ontology. Thus, KBS maintenance approaches can be adapted to ontology evolution (Stojanovic, 2004). However, in addition to ontology model richness, integration, and reuse of existing ontologies make ontology evolution more complex, especially in a distributed context.
Maintenance is classified into four groups (Institute of Electrical and Electronics Engineers [IEEE], 1990) (Stojanovic, 2004): adaptive maintenance, perfective maintenance, corrective maintenance, and preventive maintenance.	
– Adaptive maintenance results from KBS environment changes or a better understanding of domain knowledge.	– Adapting ontology to domain changes and new knowledge acquisition.
– Perfective maintenance aims to response to user's requirements.	– Capturing ontology change requirements can be guided by users' requirements (Völker, Vrandecic, & Sure, 2005; Völker, Vrandecic, Sure, & Hotho, 2007a; Völker, Hitzler, & Cimiano, 2007b).
– Corrective maintenance focuses on inappropriate behavior of KBS and aims to resolve errors (syntactic, semantic, or knowledge identification errors).	– Ontology evolution has to be driven in a corrective way to ensure ontology consistency and take into account ontology usage (Stojanovic, 2004).
– Preventive maintenance aims to anticipate and avoid future problems by analyzing the KB structure to reveal possible errors.	– Preventive maintenance is adapted to the refinement of the ontology. Change discovery is driven by ontology structure and ontology usage (Stojanovic, 2004).

Ontology Versioning

Ontology versioning consists in creating and managing different versions of an ontology. It addresses the problem of (partial) incompatibility of new ontology versions with the previous one and thus, with ontology's instances and, applications and dependant ontologies.

Two main cases of version generation can be considered: Different versions of an ontology may result when ontologies are independently developed (Klein & Fensel, 2001). The evolved ontology resulted after applying a change can also be considered as a new conceptualization of the domain and, thus, as a new version. However, in general the decision to consider it so, is taken by the ontology engineer.

Maintaining multiple versions of an ontology requires a means to differentiate between the different versions and to ensure that instances remain valid. Ideally, we should preserve the different versions of an ontology and keep trace of all information about differences and compatibilities between them. This needs methods for version identification and differentiation (based on the same principles of semantic similarity measure in ontology alignment), specification of relationships between versions, ontology update procedures and access mechanisms to the different versions of an ontology (Klein & Noy, 2003).

Ontology Adaptation

Ontology adaptation aims to provide facilities to manage frequent changes in the way concepts are represented in the domain, and to automate the process of producing multiple adapters for versions of interest (O'Brien & Abidi, 2006). It is a local adaptation of an ontology to a specific usage or a sub-set of instances of its environment, without any contradiction with the previous version.

Adaptation can be seen as practice review in which versions are first adopted and then, adapted for a specific application (O'Brien & Abidi, 2006). Adaptation changes are mostly generated by user-driven change discovery approaches (see section on user-driven change discovery).

Ontology Evolution

Ontology evolution consists in managing persistent ontology changes to cope with new requirements, and producing new versions. The modification of an ontology is handled by preserving its consistency, tracking and logging the change to provide mapping between subsequent versions (Stojanovic, Maedche, Motik, & Stojanovic, 2002a), and controlling the use of instances (Stojanovic, Stojanovic, & Handschuh, 2002b).

Change Effects

Changing the behavior of a concept or a relation so that a change requirement comes into effect, one should think about inter-related knowledge and dependant instances; and define mechanisms specifying how knowledge to be change and how to handle consistency. Change effects cannot be considered only by looking at the ontology itself. This also depends on the rationales behind the change and the specificity of the ontology usage.

Change effects were considered in the literature from different perspectives. From an ontology-based data warehouse point of view, two categories of change effects are distinguished (Xuan, Bellatreche, & Pierra, 2006):

- Normal evolution when the ontology evolves without undermining existing knowledge according to the so-called *ontological continuity* principle;
- Revolution when true axioms can be undermined.

According to distributed ontology change management approach (Klein, 2004), change effects depend on what we need to preserve in the ontology:

- Data to maintain instances between the different versions of the ontology;
- The ontology itself to maintain query results i.e. the result of a query q_1 on O_i version is included in the result of the same query q_1 on the version O_{i+1} of the same ontology;
- Consequences to maintain inferred facts i.e. the facts inferred from an axiom a_1 on O_i version are also inferred from the same axiom a_1 on the version O_{i+1} of the same ontology;
- Consistency to ensure that the new version does not contain logical inconsistencies.

With regard to database schema changes, change effects are classified focusing on instances (Noy & Klein, 2004):

- Preserving change: Instances are not lost (e.g. adding a concept or a property);
- Translating changes: Instances can be preserved as knowledge are translated to another form (e.g. when gathering concepts as a union of their super-concepts, sub-concepts, and properties; instances can be preserved.);
- Losing changes: Instances are lost (e.g. deleting a property causes the loss of all its instance values).

Handling change effects involves not only checking ontology consistency but also maintaining it. The maintenance activity consists in generating and proposing/applying a set of additional changes to resolve inconsistencies. Quite often, this is a manual procedure where the expert revises the ontology using an ontology editor (Stojanovic & Motik, 2002; Sure, 2002).

Change Classification

The purpose of change classification is to define taxonomy of changes specifying classes of ontology changes and their properties for a specific ontology language representation. The main classifications defined in literature deal with KAON[1] (Maedche, Stojanovic, Studer, & Volz, 2002; Stojanovic, 2004) and OWL[2] (Klein, 2004) languages.

The ontology of KAON changes classifies KAON changes through three levels of abstraction (Stojanovic, 2004):

- Elementary changes applying modifications to one single ontology entity;
- Composite changes applying modifications to the direct neighborhood of an ontology entity;
- Complex changes applying modifications to an arbitrary subset of ontology entities.

In addition, two change types are considered (Stojanovic, 2004): *Additive changes* adding new entities to an ontology without altering the existing ones, and *subtractive changes* removing some–piece of–entities. Thought of as a complete and minimal set of changes, the change ontology does not include entity modifications. This kind of change is interpreted as a "*rename*" change altering only lexical information about entities or a "*set*" change depending on the entity to modify.

A similar taxonomy for OWL ontology is presented in (Klein, 2004). However, unlike KAON change ontology, OWL change ontology contains *Modify* operations as well as *Set* and *Unset* operations for properties characteristics (e.g., symmetry). Klein (2004) distinguishes basic and complex change operations:

- Basic changes are simple and atomic changes that can be specified by using the structure of the ontology only, and modify only one specific feature of the OWL knowledge model (e.g. add a class, delete a "is-a" relation.);
- Complex changes correspond to composite and rich changes grouping logical sequences of basic changes and incorporating information about their implication on the logical model of the ontology (e.g. raising sub-classes, enlarging the range of an object property to its super-class, merging two classes.). Complexity also deals with change effects. If the effects of a basic change are minor, the cumulative effect of all intermediate changes realizing a complex change can be huge.

Basic changes are exhaustively derived from the underlying ontology language. Complex changes are infinite as new compositions of changes can always be defined.

ONTOLOGY EVOLUTION APPROACHES

Many researches have discussed the characteristics of an ontology evolution process (Klein, 2004; Stojanovic et al., 2002a; Stojanovic et al., 2002b) and several ontology evolution approaches have been proposed in the literature. Some focus on specific change management issues like capturing change requirements (Stojanovic, Stojanovic, Gonzalez, & Studer, 2003a; Cimiano & Völker, 2005; Bloehdorn, Haase, Sure, & Voelker, 2006), change detection and version logging (Klein, Fensel, Kiryakov, & Ognyanov, 2002a; Noy, Kunnatur, Klein, & Musen, 2004; Plessers & De Troyer, 2005; Eder & Wiggisser, 2007), formal change specification (Stojanovic, Stojanovic, & Volz, 2002c; Klein, 2004; Plessers De Troyer, & Casteleyn, 2007), change implementation (Stojanovic, Maedche, Stojanovic, & Studer, 2003b; Stojanovic, 2004; Flouris, 2006), consistency maintenance (Stojanovic, 2004 ; Haase & Stojanovic, 2005; Haase & Völker, 2005; Plessers & De Troyer, 2006), ontology versioning (Klein & Fensel, 2001; Klein et al., 2002a; Klein, 2004), and others propose a more or less global evolution process including change impact analysis and resolution as well as change propagation to dependant artifacts (objects, ontologies and applications referenced by the ontology) (Stojanovic, 2004; Klein, 2004; Bloehdorn et al., 2006). In this section, we present the main approaches and we analyze the functionalities that they support.

Ontology Learning Approach Based on Change Requirement Discovery

An important research area is to explore change sources and capture change requirements. Change requirements could be initiated because of environment dynamics, the evolution of the modeled domain, the users' needs may change, adding new knowledge previously unknown or unavailable, correction and refinement of the conceptualization, or ontology reuse for other applications. In a distributed context, if an ontology changes for any of the preceding reasons, dependant ontologies might also need to be modified to reflect potential changes in terminology or representation (Heflin, Hendler, & Luke, 1999).

It has to be said that detecting changes and capturing change requirements are two distinguished approaches. The former aims to discover ontology changes –already applied– for versioning purposes for example. The latter aims to generate ontology changes from explicit and implicit requirements. Explicit requirements can be defined by ontology engineers to adapt the ontology to new requirements or by users based on ontology usage feedbacks. Changes resulting from explicit requirements are called *top-down* changes (Bloehdorn et al., 2006). Implicit requirements are induced by analyzing the system's behavior and are called *bottom-up* changes (Bloehdorn et al., 2006).

In the approach presented in this section, ontology evolution is considered as dynamics in ontology learning process focusing on change generation (Cimiano, 2007). Learning process is applied as an ontology refinement and, addresses the evolution of data and knowledge. Changes requirements are *data-driven* based on the corpus (term frequency, matches of lexico-syntactic patterns) or *user-driven*. Change traceability (which change, by whom, why, etc.) is also considered to keep a repository of explanations and even references to the segment in a corpus which triggered the change.

Data-Driven Change Discovery

Data-driven change discovery consists in deriving changes from modifications to the knowledge from which the ontology has been constructed (Bloehdorn et al., 2006).

Data-driven change discovery can be organized through four phases (Cimiano & Völker, 2005):

1. Corpus change processing to formulate change request: Many algorithms are implemented (term extraction, taxonomy learning, learning relations etc.), their execution is guided by a change reference store and a change evidence;
2. Change strategy phase studies how corpus changes affect the different types of ontology elements;
3. Change generation and management phase proposes different change possibilities that are stored as a *Model of Possible Ontologies* (POM) used to add or remove object, update relevance and confidence, and to explain each possible change. The phase supports incremental ontology learning for efficiency purposes (to update evidence for ontology elements based on observed corpus changes and generate suggestions and explanations for ontology changes based on new evidence) and includes explicit change management (see section on change management below) ensuring the logical consistency of the learned ontology (Haase & Völker, 2005);
4. Ontology change application by adding or removing ontology elements.

User-Driven Change Discovery

In a user-driven change discovery perspective, ontology learning –applied as refinement– is based on:

- Procedures for formal self-evaluation (Völker et al., 2005);
- Conceptual preciseness based on *Learning Disjointness Axioms* (LeDA) (Völker et al. 2007a) and *Learning Expressive Ontologies* (LExO) (Völker et al., 2007b) in OWL DL language.

Learning Consistent Ontologies

Knowledge extracted from text could be uncertain and potentially contradicting which leads, consequently, to logical inconsistencies in the ontologies learned that the learning process resolve (Haase & Volker, 2008).

Ontology learning generates first, ontologies based on a language independent Model *Learned Ontology Model* (LOM). Knowledge uncertainty is expressed as confidence annotations associated to ontology elements (Haase & Volker, 2008). Then, logical semantics are introduced by transforming the LOM model to a formal ontology expressed in OWL. Transformation algorithm takes into account confidence annotation to avoid generating logical inconsistencies. The purpose is to obtain a consistent ontology capturing the most certain information among different consistent ontologies (Haase & Volker, 2008). Ontology selection is based on evaluation function taking into account the rating annotations.

Consistency maintenance is also based on resolution principles presented in (Haase & Stojanovic, 2005). When confident axioms transformed from LOM to OWL lead to an inconsistent ontology, the inconsistency is localized. Then, the most uncertain axiom (lowest confidence value) is identified and removed to resolve inconsistency.

Multimedia Ontology Evolution Approach BOEMIE[3]

BOEMIE approach (*Bootstrapping Ontology Evolution with Multimedia Information Extraction*) aims to automate the process of knowledge acquisition from multimedia content. Evolving multimedia ontologies are used as background knowledge to guide the extraction of information from multimedia content in networked sources. Besides, reasoning mechanisms are applied on the fused semantic information extracted to populate and enrich the ontologies.

Ontology Population and Enrichment Patterns

A pattern-driven approach was adopted for ontology evolution. Evolution patterns characterize the input of the evolution process –corresponding to the extracted information presented in form of *ABox* (concept and relation instances)– and determine the evolution operation to be performed (Castano et al., 2007; Petasis, 2007): population (adding new instances) or enrichment (extension by new concepts, relations, properties) .

Four evolution patterns are distinguished. Population patterns are used when the interpretation of a multimedia resource (input) can be explained by single (P1) or multiple (P2) ontology concepts (Castano et al., 2007). Enrichment patterns are used when there are no ontology concepts explaining the resource –with (P3) or without (P4) metadata information (Castano et al., 2007). The enrichment is then performed to acquire the missing knowledge. Each pattern is articulated into a set of activities implementing all the required changes as instance matching in population and concept learning in enrichment.

Ontology population aims to identify instances referring to the same real object (or event) based on instances matching and nonstandard clustering techniques (Castano et al., 2007). Ontology enrichment aims to learn new concepts (or relations) by applying clustering techniques and, in some cases (pattern P3), it includes concept enhancement by considering external sources (e.g. external domain ontologies or taxonomies).

Consistency Maintenance

Population and enrichment operations both integrate a consistency validation activity. In population process, consistency maintenance consists in identifying and eliminating redundant information and checking that the instance does not cause any contradicting information using standard reasoning services. In enrichment process, consistency maintenance is limited to inconsistency detection with respect to applied modifications.

Change Versioning

BOEMIE approach also includes a versioning and change audit phase that coordinates the updated ontology to reflect newly inserted knowledge; logs performed changes, and generates an evolved ontology version (Castano, 2006).

Change Management Approach for Distributed Ontology

(Heflin et al., 1999; Klein & Fensel, 2001; Klein et al., 2002a; Klein, Kiryakov, Ognyanov, & Fensel, 2002b; Klein & Noy, 2003; Maedche et al., 2003; Klein, 2004) research focuses on mechanisms and methods required to cope with ontology change in dynamic and distributed environments. They propose a formalism to represent change between ontologies based on a taxonomy of change operations (considering particularly OWL meta-model), and a method for interpreting data from different ontology versions.

Change management is considered within a global framework that includes generating change information, deriving additional change information, and also solving specific problems in specific situations (Klein, 2004). It is assumed that there is not one general procedure for ontology change management, but a need for a variety of methods to consider different types of change information and to support particular tasks. Methods and techniques proposed are (Klein, 2004): comparison algorithms, ontology mapping, reasoning services, human validation, change visualization, effect prediction heuristics, and guidelines for some change-related tasks.

Distributed Change Management

The study brings out three change management characteristics specific to distributed context but related only to the nature of ontology (i.e. independent of distributed settings) (Klein et al., 2002b; Klein, 2004):

- The propagation of a change to other levels of interpretation depends on whether it modifies the specification or the conceptualization of the ontology;
- As distributed ontologies can be used for different tasks, consistency maintenance cannot focus on one specific feature to preserve. Thus, change consequences should be considered for specific ontology use-cases;
- The expressivity and the semantics of the ontology language can sometimes be exploited to solve specific change management problems as for example in matching ontology versions.

An Ontology of Change Operations and a Change Specification Language

The importance of change specification in exchanging information about change between users, tools, or independent processes was also highlighted. Change specification consists of a set of change operations derived from the meta-model of the ontology language and/or specialized and composite operations prescribing the required follow-up steps to transform an old ontology version into a new one (Klein, 2004). It includes conceptual and evolution relations between old and new versions of constructs, meta-data about the change, and change consequences.

An ontology of change operations was proposed as well as a change specification language (RDF-based syntax) based on this ontology. The Ontology of change operations is an ontology change model extending the taxonomy of OWL change operations (see change classification section) to a general change specification language. The change specification language aims to specify transformations of one ontology version into another. In addition, it is proposed to facilitate change reverting or re-execution, change effect analysis, tool interaction based on unambiguous and formal change information, expressiveness in capturing scope of modifications, minimality in change description (i.e. concise format), and description in different granularity levels (Klein, 2004).

A Global Evolution Process for KAON Ontology

A global ontology evolution process is described in (Stojanovic, 2004). The process ensures change semantic specification, consistency maintaining, and change propagation for KAON ontologies. Ontology evolution is defined as the formal interpretation of all change requirements captured from different sources, the application of changes to the ontology, and their propagation to dependent artifacts while preserving consistency (Stojanovic, 2004). Dependent artifacts include objects referenced by the ontology and, dependent ontologies and applications.

A six-phase evolution methodology has been implemented within the KAON ontology management infrastructure, targeted for business-oriented ontology management (Stojanovic, 2004; Oberle, Volz, Motik, & Staab, 2004).

Change Capturing

The starting phase of the process consists in capturing the ontology changes to apply. It is based on explicit requirements and the application of data and usage-driven change discovery methods.

Data-driven changes are derived from the ontology instances by applying techniques like data-mining, Formal Concept Analysis (FCA), and many heuristics (Stojanovic, 2004; Maedche et al., 2003).

Usage-driven change discovery is based on *ontology usage patterns* derived by analyzing the behavior of users and tracking queries in ontology applications to discover the most used part of an ontology and to identify user interests (Stojanovic et al., 2003a).

Change Representation

In this phase, the identified changes are described according to the specification of KAON language. Change can be represented on three granularity levels: elementary change, composite change, and complex change (see section on change classification).

Change Semantics

Evolving an ontology assumes preserving its consistency so that it is still relevant to the modeled domain and to the applications using it (Haase & Völker, 2005). The phase of change semantics aims to evaluate and resolve change effects in a systematic manner by ensuring the consistency of the whole ontology (Stojanovic et al., 2002a). Ontology consistency is defined as following: "*A single ontology* [...] *is defined to be consistent with the respect to its model if, and only if, it preserves the constraints defined for underlying ontology model*" (Stojanovic, 2004, pp.30). The global approach focuses on KAON ontology. In addition, in (Haase & Stojanovic, 2005; Haase, Van Harmelen, Huang, Stuckenschmidt, & Sure,

2005), the authors described the semantics of changes for the consistent evolution of OWL ontologies.

Consistency Maintenance of KAON Ontology

KAON is a constraint-centered language based on closed-world assumption. Its consistency model is defined as a set of constraints describing KAON model invariants (e.g. identity distinction, concept hierarchy, concept closure, concept hierarchy closure, etc.), soft-constraints, and user-defined constraints (Stojanovic et al., 2003b; Stojanovic, 2004). It is mandatory that invariants be satisfied. Soft-constraints could be temporarily invalid to facilitate ontology population, for example. User-defined constraints are rather directives for well-formed ontology construction.

Two type of inconsistency are distinguished: structural and semantic inconsistency. Structural inconsistency relates to a noncompliance with the KAON ontology model constraints. Semantic inconsistency alters the meaning of ontology entities. The method focuses only on structural inconsistency as it enables ontology engineer assistance whereas; semantic inconsistency is heavily dependent on specific semantic information that is not explicitly expressed in standard ontology model (Stojanovic, 2004).

Consistency maintenance is handled in three steps: 1) localizing inconsistency (minimally inconsistent subsets), 2) determining possible resolutions, and 3) choosing and applying the best change.

Consistency Checking. Two approaches are distinguished for consistency checking (Stojanovic, 2004):

- A *posteriori* verification: Only one check is performed for all applied changes;
- A *priori* verification: The check is performed before a change application. For each change, a respective set of preconditions is associated and must be satisfied so that the change can be applied. Besides, it is assumed that the ontology was previously consistent.

A *posteriori* verification is a costly approach, since it is applied to the whole ontology and the resolution needs roll back mechanisms. Besides, it is not possible to explain change impacts and find out which change caused the detected inconsistency after applying changes in batch. To limit the checking to the local range of a change and avoid reverting the ontology into a consistent state, the second approach was adopted (Stojanovic, 2004). A *priori* verification is based on the specification of the necessary preconditions to satisfy to enable the applicability of a change. Besides, it requires the definition of sufficient post-conditions to satisfy after a change application to enable its validation.

Inconsistency Resolution. Two approaches are proposed for automatic inconsistency resolution (Stojanovic et al., 2003b; Stojanovic et al., 2002a; Stojanovic, 2004):

1. ***Procedural approach***: Consistency is maintained by considering the constraints of the consistency model and the definite rules that have to be performed to satisfy them. The approach is organized through two main phases:
   - ***Handling the semantics of change***: The change request is represented as a sequence of changes processed one by one. Preconditions associated to each change are checked and then, inconsistency resolutions are generated. Different resolution possibilities called *evolution strategies* are generated for a change. Each one proposes a set of additional changes resolving the inconsistency in a way that meets particular needs of an ontology engineer. The purpose of generating several strategies is to adapt evolution

policies to different ontology applications. The process is repeated until there is no change that should be handled,

- *Change application*: All the changes are applied to the ontology by considering their respective post-conditions.

2. ***Declarative approach***: Consistency is maintained by considering a comprehensive set of inferred axioms formalizing the evolution. The approach is organized through three phases:
   - ***Request formalization***: The ontology engineer expresses a change request in a declarative manner as a collection of supported ontology changes split into two sets of changes: *must be formed changes* (e.g. remove the concept *a*) and *must not be formed changes* (e.g. do not remove the concept *b*).
   - ***Change resolution***: Only the first set of changes is applied so that inconsistencies caused can be detected. Then, resolution generation is performed by considering the two sets of changes. All possible changes eliminating inconsistencies are generated. The resolution is reduced to a graph searching problem where nodes correspond to evolving ontologies and edges to applied changes. The search is guided by ontology engineer constraints, consistency model rules, and annotations associated to edges and nodes,
   - ***Solutions ordering***: All possible consistent states of the ontology are ranked according to meta-information given by ontology engineer.

Both approaches are able to perform the same set of changes and offer the same possibilities for controlling and customizing the change resolution. Their comparison has to be based on subjective criteria as efficiency of the evolution system (Stojanovic, 2004).

Consistency Maintenance of OWL Ontology

OWL is an axiom-centered language based on open-world assumption. Four different approaches were proposed to handle OWL inconsistencies (Haase et al., 2005): consistent ontology evolution, repairing inconsistencies, reasoning in the presence of inconsistencies, and multi-version reasoning. In (Haase & Stojanovic, 2005), a formal model was proposed for consistent OWL ontology evolution. Consistency is defined as a set of conditions to satisfy and is classified into three levels: structural, logical, and user-defined consistencies.

- Structural consistency refers to ontology language constraints.
- Logical consistency refers to the formal semantic of the ontology and to its satisfiability. It ensures that the ontology is semantically correct and does not present any logical contradiction.
- User-defined consistency refers to ontology usage and application context. Constraints are explicitly defined by users. Two user-defined conditions are considered (Haase & Stojanovic, 2005):
- Generic consistency specifying qualitative criteria for ontology modeling as applying OntoClean meta-properties (Guarino & Welty, 2002);
- Domain dependant consistency specifying particular conditions related to the domain of discourse and depending on user domain expertise.

The definition of OWL semantics is based on a model theory associating OWL syntax to the model of the ontology domain: Satisfying an ontology within an interpretation is constraint by the satisfaction of all its axioms (Haase & Stojanovic, 2005).

Consistency Checking: Consistency is checked only for change operations adding axioms. The rationale behind this choice is based on the monotonic logic of OWL. The purpose is to localize the minimal inconsistent sub-ontology i.e. a minimal set of contradicting axioms (Haase & Stojanovic, 2005).

Inconsistencies Resolution: Rewriting rules was proposed to make axioms compatible with OWL Lite model and resolve structural inconsistencies (Haase & Stojanovic, 2005). For logical inconsistencies, resolution strategies were introduced based on OWL Lite constraints (Haase & Stojanovic, 2005). Each consistency condition is mapped to a resolution function specifying the additional changes to apply. These changes correspond to a set of axioms that have to be removed in order to obtain a logically consistent ontology with 'minimal impact' on the existing ontology. A maximal consistent sub-ontology is obtained (Haase & Stojanovic, 2005). Additional changes are presented to the expert so that he can determine which changes should be generated.

Change Propagation

The propagation phase aims to propagate ontology changes to the possible dependent artifacts in order to preserve the overall consistency. These artifacts can be ontologies reused or extended by the evolved ontology, or distributed applications. Propagation consists in tracking and broadcasting applied changes. Synchronization approaches proposed are described in detail in (Maedche et al., 2003; Stojanovic, 2004).

Change Implementation

This phase consists in the physical implementation of the required change and the derived changes resolving it. First, changes are notified to the ontology engineer to be approved and then, applied. Besides, all performed changes are logged in order to support recovery facilities.

KAON Change logging is based on two specific notions (Stojanovic, 2004): *evolution ontology* and *evolution log*. Evolution ontology defines a model of applicable changes on an ontology, facilitating the management of these changes. Evolution log describes applied-change historic through chronological information sequences about each change. It holds knowledge about the ontology development and maintenance. Modeling change (evolution ontology) and their application historic (evolution log) help in synchronizing the ontology evolution. If a change needs to be cancelled, evolution log based on a formal change model, helps in guiding revoke operations.

Change Validation

This phase consists in the final validation of the applied changes. It ensures the reversibility of the changes if they are finally disapproved by users (may be due to no convincing impacts, divergent points of view in collaborative context, etc.), the rationale explanation of changes, and their usability (Stojanovic, 2004).

At this stage, other problems can be identified, inferring new change requirements in a cyclic change management process.

Ontology Evolution Approach Based on Belief Change Principles

The motivation idea of this work (Flouris, Plexousakis, & Antoniou, 2005; Flouris & Plexousakis, 2006; Flouris, Plexousakis, & Antoniou, 2006; Flouris, 2006) was that the mature field of belief change can provide the necessary formalizations that can be exploited in the ontology evolution research. Belief change deals with the automatic adaptation of a knowledge base to new knowledge, without human participation in the process (Flouris et al., 2006). The starting assumption of this work was that adopting belief change principles and algorithms may reduce knowledge engineer dependence in the ontology evolution process. Indeed, the approach is addressed to applications based on frequently changing ontologies and autonomous applications like software agents i.e. in case it is highly difficult to handle ontology changes by a manual or semi-automatic process.

Belief change theory considered in this study is the AGM theory initiated by the three authors Alchourron, Gärdenfors and Makinson (1985), and applied to belief revision as a method giving minimal properties a revision process should have.

Application Scope

The study shows that AGM theory can be applied only to classical logics like Propositional Logic (PL) and First-Order Logic (FOL) but not (directly) to ontology representation standards like Description Logics DL and OWL (Flouris, 2006). Indeed, belief change techniques are not applicable to DLs under a closed-world assumption. In case of open-world assumption, they are applicable on some DLs but not OWL (Flouris et al., 2005).

Ontology Evolution Operations

Four different ontology evolution operations are distinguished considering belief change literature (Flouris & Plexousakis, 2006): *ontology revision*, *ontology contraction*, *ontology update*, and *ontology erasure*. Ontology revision and contraction occur when the perception of the domain changes i.e. its conceptualization. They are applied regarding a static state of the world. Ontology update and erasure reflect changes in the domain itself.

These four operations do not really match with change operations used in the ontology evolution literature as they are based on different paradigms. Inspired from belief change principles, they have a different viewpoint on how a change should be interpreted and managed. They are *fact-centered* (Flouris, 2006): Each new fact represents a certain need for ontology evolution. By identifying the type of the new fact (static/dynamic world state) and its impact (add/remove knowledge), the type of operation involved can be determined and thus, the system identifies modifications needed and performs them automatically. Standard approaches however, are *modification-centered* (Flouris, 2006): They focus on modifications that should be physically performed in response to a new fact which makes change management process less complicated and gives ontology engineers more control.

Ontology Evolution Algorithm

The ontology evolution process is performed as an evolution algorithm mapping an ontology and a change to an ontology where both the ontology and the change are represented by sets of axioms. The algorithm is close to the axiomatic approach described in (Haase & Stojanovic, 2005). The evolution operations, namely revision, contraction, update, and erasure; are implemented as four evolution functions. In (Flouris, 2006) a first fully automatic contraction algorithm for ontologies based on AGM compliant DLs was introduced. However, the algorithm is based only on syntactic considerations.

Comprehending ontology evolution from a belief revision perspective is actually different from current approaches. The researchers concede that foundations are different (Flouris & Plexousakis, 2006; Flouris, 2006): Belief revision is based on postulation methods, whereas ontology evolution approaches focus on explicit construction involving ontology engineer participation –which cannot be postulated– to cope with technical and practical issues related to change management problem. Applying AGM theory to ontology evolution is presented as a complementary approach to overcome the lack of efficient formalizations of the processes behind ontology evolution in current research.

Consistency Maintenance

Consistency maintenance is also considered in this approach. Two notions are distinguished (Flouris, 2006): consistency and coherence. Consistency is related to the satisfaction of all the ontology DL axioms, whereas coherence deals with satisfaction of predefined constraints or invariants related to efficient ontology design. Ontology coherence is not considered in the evolution algorithm as it deals with design problems. Consistency maintenance is handled according to belief change principles (Flouris, 2006).

Change Detection Approach Using a Version Log

In (Plessers & De Troyer, 2005; Plessers et al., 2006), a change detection approach was proposed within an OWL DL ontology evolution framework. The approach aims to detect changes that were not explicitly requested by an ontology engineer and automatically generate a detailed overview of changes that have occurred based on a set of change definitions. Different overviews of the changes can be provided for the same evolution as each user can have its own set of change definitions.

A *Change Definition Language* CDL was proposed proving several levels of abstraction. Changes are expressed as temporal queries applied to a version log. The version log keeps track of all the different versions of all concepts ever defined in an ontology, while the CDL allows users to define the meaning of changes in a formal way (Plessers et al., 2007).

Evolution Framework Overview

The framework focuses on two kinds of evolution tasks, each one targeting a user role: *Evolution-on-request* for ontology engineers modifying the ontology and *Evolution-in-response* for maintainers of depending artifacts looking for information about the change made. Change detection approach is applied in the two tasks but at different steps.

Evolution-On-Request

This task is organized through five phases: (1) First, the ontology engineer expresses his change request in CDL (see section change definition language below). (2) The consistency maintenance phase deals with inconsistency localization and resolution (see section consistency maintenance below). (3) The change detection phase aims to detect which changes occurred as a consequence of the applied modifications. (4) In the change recovery phase, all unnecessary intermediate changes can be recovered (Plessers, 2006). (5) Finally, in the change implementation phase, the change applied to a local copy of the ontology is implemented in the public version.

Evolution-In-Response

Considering that maintainers of dependant artifacts can have a different viewpoint on the change definitions applied by ontology engineer, this task allows them to approve or not the applied modifications and to decide about their propagation to dependant artifacts. It is organized through three phases: (1) It starts with the change detection phase to obtain evolution log (see section evolution log below). (2) The cost of the evolution phase aims to evaluate the cost of updating dependant artifacts. (3) Finally, if the update is approved, the version consistency phase ensures the consistency of the dependent artifact with the evolved version of the ontology; otherwise, the depending artifact remains consistent with the old version of the ontology.

Change Definition Language

Change Definition Language CDL specifies change definitions in a formal and declarative (differences between past and current versions) way. It is a temporal logic based language defining an ontology change in terms of preconditions and post-conditions. The syntax and the semantics of this language are presented in (Plessers et al., 2007).

Evolution Log Model

Evolution logs aims to express different users' interpretations of an ontology evolution. By applying a temporal query on a version log through a change detection phase, a collection of change definition occurrences –expressed in CDL– is obtained. It corresponds to an evolution log. Version log stores evolution historic of each entity defined in the ontology starting from its creation, over its modifications, until the end of its lifecycle.

To represent version log, a *snapshot* approach was adopted. It captures the different states of an ontology over time and keeps track of the evolution of each individual ontology concept (Plessers et al., 2007).

Consistency Maintenance

In (Plessers & De Troyer, 2006), authors defined an approach and an algorithm localizing axioms causing inconsistencies and proposed a set of rules that ontology engineers can use to resolve inconsistencies. The consistency model refers to OWL DL constraints.

Consistency checking copes with two change scenarios (Plessers & De Troyer, 2006):

- Adding/modifying axioms in the terminological level of OWL DL –the *TBox* (classes and properties): The verification starts with the satisfiability of the *TBox* concepts and then, the consistency of the *ABox* with respect to the modified TBox;
- Adding/modifying axioms in the assertional level of OWL DL –the *ABox* (instances): The verification in this case is applied only to the *ABox*.

Consistency checking is not applied after a delete operation based on the logic monotonicity of OWL DL.

The algorithm proposed to select axioms causing inconsistencies is based on two types of tracking trees (Plessers & De Troyer, 2006): Axiom transformation trees keeping track of axiom transformations that occur in the preprocessing step and, concept dependency trees keeping track of axioms leading to a clash during algorithm execution. Based on the clash information, concept dependency trees, and axiom transformation trees, the algorithm generates a set of axioms causing the inconsistency.

The assumption stated for inconsistency resolution is that inconsistency is a consequence of over-restrictive axioms (contradicting each others) that have to be weakened (Plessers & De Troyer, 2006).

Authors propose a collection of rules guiding ontology engineers in inconsistency resolution. A rule can either call another rule or apply a change to an axiom. Some rules can be applied for weakening axioms (e.g. how to weaken a concept definition or a concept inclusion) others for weakening or strengthening concepts (e.g. disjunction relation, existential quantification) (Plessers & De Troyer, 2006).

After describing existing ontology evolution approaches, we present, in the following section, the main tools supporting ontology evolution functionalities and implementing some of these approaches.

TOOLS SUPPORTING ONTOLOGY EVOLUTION

As described in the previous sections, handling ontology evolution manually is not a trivial task. Ontology engineers can not comprehend all side-effects of the changes, resolve caused inconsistencies, and evaluate change impacts on the ontology. Therefore, appropriate tools providing technical means for supporting ontology evolution are required.

Ontology evolution should be a part of the functionalities of an ontology editor to drive ontology development in an iterative and dynamic process. However, requirements regarding ontology evolution process are not supported by all existing ontology editors. These requirements are mainly (Stojanovic, 2004): functionality, customisation, transparency, reversibility, auditing, refinement, and usability. In (Noy, Chugh, Liu, & Musen, 2006), functional requirements considered particularly for collaborative environment are: change annotation, change history for a concept, change representation from one version to the next, definition of access privileges, querying an old version by using the vocabulary of the new version, and printed summary of changes. Other functional requirements specific to asynchronous collaborative editing, continuous editing, curated (laconic) editing, and non-monitored editing are also described.

Besides some existing ontology editors supporting certain evolution features, current researches on ontology evolution–including some of the approaches described in the previous sections–have proposed more specialized tools whose aim is to guide users to perform the change(s) manually or to perform the change(s) automatically. Some of these tools allow collaborative edits (Duineveld, Stoter, Weiden, Kenepa, & Benjamins, 2000; Noy et al., 2006), others support transactional changes (Haase & Sure, 2004), and others support features related to ontology versioning (Duineveld et al., 2000; Klein et al., 2002a; Noy & Musen, 2002; Noy et al., 2006).

KAON[4] Tool

An ontology evolution system is proposed within the KAON framework (Stojanovic, 2004). Besides automating the evolution process, the KAON tool suite guides ontology engineers in formulating their change request by providing additional information and suggestions for ontology improvement. It includes data-driven change discovery functionality (Maedche et al., 2003), allows users to define some "evolution strategies" that control how changes will be made and helps in adapting the ontology towards needs of end-users that are discovered from the usage of this ontology (Stojanovic et al., 2002a).

Protégé[5] Editor

Protégé is a free, open source ontology editor and knowledge-base framework; developed by the Stanford Medical Informatics group (SMI) at Stanford University. It provides a graphical and interactive ontology-design and knowledge-acquisition environment. Protégé architecture is component-based allowing the enrichment of the editor's functionalities by adding specialized plug-ins (e.g. Protégé-OWL plug-in). Some plug-ins are proposed to support specific evolution features and are presented below.

ONTOLOGY VERSIONING TOOLS

In the change management framework proposed for distributed ontology (see section on change management for distributed ontology), several specialized prototypes were developed (Noy & Klein, 2003; Noy et al., 2004; Klein, 2004):

- OntoView tool implements a change detection procedure for RDF-based ontologies. It uses rules to find specific operations and produces transformation sets between ontology versions (Klein et al., 2002a).

- Two extensions to the PROMPTdiff tool –developed as a plug-in for Protégé finding mappings between frames by means of heuristics (Noy & Musen, 2002)– was proposed in (Klein, 2004). Their role is to produce the evolution relation between the elements of two ontology versions. The user interface allows visualizing some complex changes between ontology versions.

A more comprehensive ontology-evolution system was described in (Noy et al., 2006). The core of the system is the *Change and Annotation Ontology* (CHAO). The instances of CHAO ontology represent changes between two versions of an ontology and user annotations related to these changes. The system is implemented as two related Protégé plug-ins:

- Change Management Plug-in providing access to a list of changes and enabling users to add annotations to individual or grouped changes and to see concept history;
- PROMPT Plug-in providing comparisons of two versions of an ontology, information on users who performed changes and facilitating change acceptance and rejection (Noy & Musen, 2003).

Change Detection and Logging Tools

In (Plessers et al., 2007), two extension prototypes for Protégé ontology editor are presented (for the approach, see section change detection approach using a version log):

- Version Log Generator Plug-in automatically creating a version log by tracking all the changes applied to an ontology. Applied change are caught as events thrown by Protégé, and the version log is updated by setting the end time of the latest concept-versions of the concepts involved in the change, and by creating the appropriate new concept versions representing the new state of the changed concepts;
- Change Detection Plug-in taking as input a set of change definitions and a version log; and providing as output an evolution log by evaluating the given change definitions on the given version log.

Ontology Learning and Data-Driven Change Discovery Tool Text2Onto[6]

Text2Onto an ontology learning system for semi- or fully automatic ontology creating process was presented in (Cimiano & Völker, 2005). Text2Onto was developed as a successor of TextToOnto (Mäedche & Volz, 2001). It supports data-driven change discovery and includes three main components (see section on change requirement discovery):

- GATE[7] (General Architecture for Text Engineering) as an NLP (Natural Language Processing) tool;
- POM (Model of Possible Ontologies) to store the different generated changes proposed and their explanation;
- Change management component to catch change impact on the ontology and choose from the possible changes proposed those who fit the corpus. Change impact management is driven through an incremental learning process.

Text2Onto follows a translation-based approach translating instantiated modeling primitives into OWL (Haase & Volker, 2008).

SYNTHESIS

In this section, we sum up the state-of-the-art presented in the previous sections through a comparative table based on general characteristics, evolution functionalities supported, and specificities of the existing ontology evolution approaches (see Table 3).

FUTURE RESEARCH DIRECTIONS

The purpose of this section is to give an overview on current work in progress in ontology evolution and to discuss open issues and future research directions.

Ontology Debugging and Evolution

Inconsistency resolution cannot be handled without a formal consistency checking, delimiting the detected inconsistencies and giving rational

Table 3. Synthesis of ontology evolution approaches and tools

	Ontology Learning Approach Based on Change Requirement Discovery (Cimiano, 2007) (Cimiano & Völker, 2005)	BOEMIE Approach (Castano et al., 2007) (Petasis, 2007)	Change Management Approach for Distributed Ontology (Klein, 2004) (Maedche et al., 2003) (Noy et al., 2006)	A Global Evolution Process for KAON Ontology (Stojanovic, 2004)	Ontology Evolution Approach based on Belief Change Principles (Flouris, 2006) (Flouris & Plexousakis, 2006) (Flouris et al., 2006)	Change Detection Approach Using a Version Log (Plessers & De Troyer, 2005) (Plessers et al., 2007)
General Characteristics						
Evolution Process	Some phases	Some phases	A process of creating change specification	Global Process	An algorithm that takes a set of axioms as input (ontology + change) and applies them	- Evolution on request - Evolution in response
Ontology Language	OWL DL	OWL	OWL	KAON	Some DL in open-world assumption but not OWL	OWL DL
Tool / Prototype	Text2Onto	Ontology Evolution Toolkit, a component of BOEMIE prototype	- OntoView - PROMPTdiff - Protégé plug-ins	KAON Tool Suite		Two Protégé plug-ins: - Version log generator - Change detection
Change Requirement Identification	- Data-driven - User-driven	By reasoning on multimedia sources and external knowledge sources (existing ontologies or taxonomies, etc.)		- Data-driven - Usage-driven	By new fact identification (fact-centered change operations)	
Change Specification	Learning Ontology Model (LOM)	*ABox* (concept and relation instances)	- Ontology of change operations and version transformation (extended OWL metamodel) - Change specification language (in RDF) - Change and annotation ontology CHAO	- KAON change specification - Evolution ontology	Axioms based on AGM theory	Change definition language CDL (declarative and formal definition based on temporal logic)

continued on following page

Table 3. continued

Functionalities							
Consistency Maintenance	Level	Logical	Structural and logical		- Structural (KAON) - Structural and logical (OWL Lite) (Haase & Stojanovic, 2005)	Logical	Logical
	Checking	Based on confidence annotation	Yes (for enrichment and population)	Compatibility between the different versions	- *A priori* (KAON) - Localization of the minimal inconsistent sub-ontology after axiom adding operations (OWL Lite) (Haase & Stojanovic, 2005)	Axiom satisfaction	Localization algorithm
	Proposition of Resolution	Several solutions ordered by an evaluation function		Derive additional changes	- Resolution strategies proposing axioms to delete (logical level, OWL Lite)		Proposition of rules weakening restrictive axioms
	Automatic Resolution	Deletion of some axioms	If population: Elimination of redundancy and logical inconsistencies	Resolve specific problems (ontology-related tasks)	- Declarative and procedural approaches (KAON) - Rewriting rules (structural level, OWL Lite)	Based on belief change principles	
Change Propagation	Target			- Data sources - Ontologies	- Ontologies - Applications - Instances		- Ontologies - Applications
	Type			- Analysis of compatibilities between ontology versions and data sources and partly translating data - Proposition for ontology synchronization	Different kind of synchronizations with dependant artifacts		Artifact maintainers decide
Change Detection				- Conceptual and evolution relations between versions - Generating change information (RDF ontologies)			Yes
Versioning	Evolved Version		Generation of an evolved version		Saving the evolved version		Saving the evolved version
	Version Comparison			Yes			
	Managing Several Versions			Yes			

continued on following page

Table 3. continued

Evolution Log	Applied Changes	Yes	Yes	Yes (detected changes)	Yes		Saving detected changes: - Version log of each ontology entity - Evolution log of the ontology
	Trace of management Operations	Change explanation traceability			Traceability of all the process (evolution journal)		
Specificities		Evolution integrated to ontology learning	Ontology population and enrichment guided by patterns	A framework of distributed ontology and version synchronization	A global process and a dedicated system	Applying AGM theory as a complementary approach to have a formal formalism and an automatic process	Several interpretations on changes according to the different users

explanations. Existing reasoners are more or less precise in their analysis and do not give sufficient details. Ontology debugging is therefore a promising field to cope with the problem.

In (Moguillansky, Rotstein, & Falappa, 2008), a theoretical approach to handle ontology debugging through a dynamic argumentation framework based on description logics was presented. The purpose of the methodology is to bridge ontology-specific concepts to argumentation notions and to employ argumentation acceptability semantics to restore consistency to ontologies. A Dynamic Argumentation Framework for DLs was detailed.

Other interesting researches focusing on ontology debugging are presented in (Parsia, Sirin, & Kalyanpur, 2005; Wang, Horridge, Rector, Drummond, & Seidenberg, 2005). The purpose is to provide more comprehensible explanations of the inconsistency than standard reasoners' results. Two techniques are distinguished:

- *Black-box* techniques considering the reasoner as a black box and applying inferences to localize inconsistencies;
- *Glass-box* techniques modifying the internal mechanism of the reasoner to explain inconsistencies and complete reasoners' results which really improve consistency maintenance.

A glass-box approach was discussed in (Sirin & Parsia, 2004). It provides information about the contradictions found and axioms causing the inconsistency but does not propose inconsistency resolutions.

Integrating Ontology Evaluation in Ontology Evolution

Evaluation is an important issue for ontology evolution. Ontology evaluation is already employed in capturing change requirements. Based on quality metric assessment, several changes can be proposed for ontology refinement and improvement. Moreover, ontology evaluation can be employed to control the evolution process and even to guide the resolution of change impacts.

In (Dividino & Sonntag, 2008), a controlled evolution of ontologies through semiotic-based evaluation methods was presented. A tool S-OntoEval assessing ontology quality by implementing existing quality metrics was described. It carries out a complete evaluation combining several evaluation metrics categorized into the semiotic levels.

In (Djedidi & Aufaure, 2008; 2010) an ontological knowledge maintenance methodology was proposed. The goal of the methodology is to manage ontology evolution while maintaining consistency and evaluating change impact on ontology quality. A hierarchical model describing and measuring ontology quality through several criteria and metrics was defined. The model is employed to assess the impact of the different inconsistency resolutions proposed for a change, to guide the choice of the most appropriate one regarding ontology quality.

Towards Ontology Evolution Guidelines

The approach described in (Djedidi & Aufaure, 2008), has been enriched by the definition of ontology change management patterns (CMPs).

CMPs are proposed as a solution looking for invariances in change management that repeatedly appear when evolving ontologies. Three categories of patterns are distinguished: *Change Patterns*, *Inconsistency Patterns* and *Alternative Patterns*. The goal of CMP modeling is to offer different levels of abstraction, to establish conceptual links between these three categories of patterns determining the inconsistencies that could be potentially caused by a type of change and the alternatives that may resolve a kind of inconsistency and, thus, to guide an automated change management process (Djedidi & Aufaure, 2009; 2010).

FUTURE ISSUES

By analyzing typical problems in managing ontology changes and studying existing and current research on ontology evolution, the following observations arise:

- An ontology evolution process has to handle the application of a given ontology change by deriving intermediate changes required, capturing change impacts, and ensuring the consistency of the underlying ontology and all dependent artifacts;
- Evolution process should be automated and optimized but also sufficiently flexible to allow the user to easily manage changes and validate or revoke a change application or resolution;
- Evolution process should offer possibilities to catch useful changes for ontology refinement and improvement based on ontology domain, ontology usage and application, and the quality of the ontology itself.

Although the successful realizations that have been performed in the field of ontology evolution, there are still open issues that can be considered more deeply. From our perspective, these issues are based on the following observations:

- Many of the implementations of the described approaches seem to work better for relatively small changes than for complex ones. Complex changes cannot be predefined exhaustively; the question is how to provide, in practice, guidelines for handling their application in a more automated and optimized way? Analyzing ontology evolution use-cases is required to come up with additional specific procedures and guidance for complex change management.
- Distributed and collaborative dimensions of ontology environment have to be

considered more deeply because, in practice, ontologies are still maintained in a centralized way and dependant artifacts are often concerned by their evolution. Research in change propagation, global validation, conflict resolution, and even reasoning on inconsistent ontologies where mutual agreements cannot be reached are required.

- Ontology evolution system should be enriched by a meta-model or a kind of generic layer language-independent to be able to propose generic change management guidelines that can be used for different evolution process. Moreover, this can guide change propagation to dependant ontologies developed in different ontology languages.

CONCLUSION

Ontology evolution is an essential research area for the widespread use of ontologies in industrial and academic applications. In this chapter, we have outlined ontology evolution issues and requirements, and presented a comparative study of ontology, database schema, and knowledge-based system evolution. A state-of-the-art review considering the main existing ontology evolution approaches and tools was presented. Furthermore, a comparative study of this review is given. Before concluding, we have discussed some research in progress on ontology evolution and given some perspectives on future research directions.

REFERENCES

Alchourron, C., Gärdenfors, P., & Makinson, D. (1985). On the logic of theory change: partial meet contraction and revision functions. *Journal of Symbolic Logic*, *50*(2), 510–530. Retrieved from http://www.jstor.org/stable/2274239. doi:10.2307/2274239

Banerjee, J., Kim, W., Kim, H. J., & Korth, H. (1987). Semantics and implementation of schema evolution in object-oriented databases. *ACM SIGMOD Record, 16*(3), 211-322. doi: http://doi.acm.org/10.1145/38714.38748

Bloehdorn, S., Haase, P., Sure, Y., & Voelker, J. (2006). Ontology evolution. In J. Davies, R. Studer, & P. Warren (Eds.), Semantic Web Technologies, Trends and research in Ontology-based Systems (pp. 51-70). New York: John Wiley & Sons Publication. doi: 10.1002/047003033X.ch4

Blundell, B., & Pettifer, S. (2004). Graph visualization to aid ontology evolution in Protégé. In *Proceedings of the 7th International Protégé Conference*, Bethesda, MD.Retrieved from http://protege.stanford.edu/conference/2004/posters/Blundell.pdf

Breche, P., & Wörner, M. (1995). How to remove a class in an object database system. In *Proceedings of the 2nd international conference on applications of databases (ADB-95)*(pp. 235-248). Retrieved from http://citeseerx.ist.psu.edu/viewdoc/download?doi=10.1.1.45.40&rep=rep1&type=url&i=0 doi: 10.1.1.45.40

Castano, S. (2006). Ontology evolution: The BOEMIE approach. In *BOEMIE Workshop, at the conference EKAW 06*. Retrieved from http://www.boemie.org/sites/default/files/pres_castano_boemie.pdf

Castano, S., Espinosa, S., Ferrara, A., Karkaletsis, V., Kaya, A., Melzer, S., et al. (2007). Ontology dynamics with multimedia information: The BOEMIE evolution methodology. In G. Flouris, & M. d'Aquin (Eds.), *Proceedings of the International Workshop on Ontology Dynamics (IWOD-07)* at *ESWC 07 Conference,* (pp. 41-54). Retrieved from http://kmi.open.ac.uk/events/iwod/iwod-proceedings.pdf

Cimiano, P. (2007). *On the relation between ontology learning, engineering, evolution and expressivity.* Invited talk at 7th Meeting on Terminology and Artificial Intelligence TIA 2007, Sophia Antipolis, France. Retrieved from http://www.aifb.uni-karlsruhe.de/WBS/pci/home/Publications/2007/tia07/tia07_slides.pdf

Cimiano, P., & Völker, J. (2005). Text2Onto - a framework for ontology learning and data-driven change Discovery. In A. Montoyo, R. Munoz, & E. Metais (Eds.), Natural Language Processing and Information Systems (LNCS: Vol. 3513, pp. 227-238), Berlin, Germany: Springer. doi: 10.1007/b136569

Coenen, F. & Bench-Capon, T. (1993). Maintenance of knowledge-based systems. *The A.P.I.C. Series,* 40.

Dividino, R., & Sonntag, D. (2008). Controlled ontology evolution through semiotic-based ontology evaluation. In *Proceedings of the 2nd Workshop on Ontology Dynamics, (IWOD-08) at ISWC 08 Conference* (pp. 1-14). Retrieved from http://www.ics.forth.gr/~fgeo/Publications/IWOD-08_Proceedings.pdf

Djedidi, R., & Aufaure, M.-A. (2008). Ontological knowledge maintenance methodology. In I. Lovrek, R. J. Howlett, & Lakhmi C. Jain (Eds.), Knowledge-Based Intelligent Information and Engineering Systems (LNCS: Vol. 5177, Part I, pp. 557-564). Berlin, Germany: Springer. doi: 10.1007/978-3-540-85563-7

Djedidi, R., & Aufaure, M.-A. (2009). Ontology change management. In A. Paschke, H. Weigand, W. Behrendt, K. Tochtermann, T. Pellegrini (Eds.), Semantic Systems (I-Semantics 09), Journal of Universal Computer Science (JUCS), ISBN 978-3-85125-060-2, pp 611-621, Verlag der Technischen Universitt Graz.

Djedidi, R., & Aufaure, M.-A. (2010). Onto-Evo^al an Ontology Evolution Approach Guided by Pattern Modelling and Quality Evaluation. *Proceedings of the the 6th International Symposium on Foundations of Information and Knowledge Systems* (FoIKS 2010), (LNCS: Vol. 5956, pp. 286-305). Berlin, Germany: Springer. doi: 10.1007/978-3-642-11829-6

Duineveld, A. J., Stoter, R., Weiden, M. R., Kenepa, B., & Benjamins, V. R. (2000). WonderTools? A comparative study of ontological engineering tools. *International Journal of Human-Computer Studies, 52*(6), 1111–1133. doi:10.1006/ijhc.1999.0366

Eder, J., & Wiggisser, K. (2007). Change detection in ontologies using DAG comparison. In J.Krogstie, A.L. Opdahl, & G. Sindre, (Eds.), Advanced Information Systems Engineering (LNCS: Vol. 4495, pp. 21-35). Berlin, Germany: Springer. doi: 10.1007/978-3-540-72988-4

Flouris, G. (2006). *On belief change and ontology evolution.* Ph.D. Thesis, University of Crete, Department of Computer Science, Heraklion, Greece.

Flouris, G., & Plexousakis, D. (2006). Bridging ontology evolution and belief change. Advances in Artificial Intelligence (LNCS: Vol. 3955, pp. 486-489). Berlin, Germany: Springer. doi: 10.1007/11752912_51

Flouris, G., Plexousakis, D., & Antoniou, G. (2005). On applying the AGM theory to DLs and OWL. In Y.Gil, E. Motta, V. Benjamins, & M. Musen, (Eds.), The Semantic Web – ISWC 2005 (LNCS: Vol. 3729, pp. 216-231). Berlin, Germany: Springer. doi: 10.1007/11574620

Flouris, G., Plexousakis, D., & Antoniou, G. (2006). *Evolving ontology evolution.* Invited Talk at the 32nd International Conference on Current Trends in Theory and Practice of Computer Science (SOFSEM-06). Retrieved from http://www.sofsem.cz/sofsem06/data/prezentace/23/A/dimitris.pdf

Foo, N. (1995). Ontology revision. In G. Ellis, R. Levinson, W. Rich, & J. F. Sowa (Eds.), *Proceedings of The 3rd International Conference on Conceptual Structures* (pp. 16-31). London: Springer-Verlag.

Franconi, E., Grandi, F., & Mandreoli, F. (2000). A semantic approach for schema evolution and versioning in object-oriented databases. In Computational Logic — CL 2000 (LNCS: Vol. 1861, pp. 1048-1062). Berlin: Springer. doi: 10.1007/3-540-44957-4

Guarino, N., & Welty, C. (2002). Evaluating ontological decisions with OntoClean. In *Communication of the ACM (CACM), 45*(2), 61-65. New York: ACM. Retrieved from http://doi.acm.org/10.1145/503124.503150

Haase, P., & Stojanovic, L. (2005). Consistent Evolution of OWL Ontologies. In A. Gomez-Perez & J. Euzenat (Eds.), The Semantic Web: Research and Applications (LNCS, vol.3532, pp. 182-197). Berlin: Springer. doi: 10.1007/b136731

Haase, P., & Sure, Y. (2004). D3.1.1.b State of the art on ontology evolution. *SEKT Deliverable*. Retrieved from http://www.aifb.uni-karlsruhe.de/WBS/ysu/publications/SEKT-D3.1.1.b.pdf

Haase, P., Van Harmelen, F., Huang, Z., Stuckenschmidt, H., & Sure, Y. (2005). A Framework for handling inconsistency in changing ontologies. In Y. Gil, E. Motta, V. Benjamins, & M. Musen, (Eds.), The Semantic Web – ISWC 2005 (LNCS: Vol. 3729, pp. 353-367). Berlin, Germany: Springer. doi: 10.1007/11574620

Haase, P., & Völker, J. (2008). Ontology learning and reasoning – dealing with uncertainty and inconsistencies. In P. C. G. Costa, C. d'Amato, N. Fanizzi, K. B. Laskey, K. J. Laskey, T. Lukasiewicz et al. (Eds.), Uncertainty Reasoning for the Semantic Web I (LNCS: Vol. 5327, pp. 366-384). Berlin, Germany: Springer. doi: 10.1007/978-3-540-89765-1

Heflin, J., Hendler, J., & Luke, S. (1999). Coping with changing ontologies in a distributed environment. In *Proceedings of the Workshop on Ontology Management of the 16th National Conference on Artificial Intelligence* (AAAI-99), WS-99-13, 74-79. Menlo Park, CA: AAAI Press.

IEEE. (1990). *IEEE Standard Computer Dictionary: A Compilation of IEEE Standard Computer Glossaries*. New York: Author.

Klein, M. (2004). *Change management for distributed ontologies*. Ph.D. Thesis, Dutch Graduate School for Information and Knowledge Systems, Germany.

Klein, M., & Fensel, D. (2001). Ontology versioning on the semantic web. In I. F. Cruz, S. Decker, J. Euzenat, & D. L. McGuinness (Eds.), *Proceedings of the first International Semantic Web Working Symposium (SWWS'01)* (pp. 75-91), Stanford University, CA.

Klein, M., Fensel, D., Kiryakov, A., & Ognyanov, D. (2002a). Ontology versioning and change detection on the web. Knowledge Engineering and Knowledge Management: Ontologies and the Semantic Web (LNCS: Vol. 2473, pp. 197-212). Berlin, Germany: Springer. doi: 10.1007/3-540-45810-7

Klein, M., Kiryakov, A., Ognyanov, D., & Fensel, D. (2002b). Finding and characterizing changes in ontologies. LNCS: Vol. 2503. Conceptual Modeling — ER 2002 (pp. 79–89). Berlin, Germany: Springer. doi: 10.1007/3-540-45816-6

Klein, M., & Noy, N. F. (2003). A component-based framework for ontology evolution. In F. Giunchiglia, A. Gomez-Perez, A. Pease, H. Stuckenschmidt, Y. Sure, & S. Willmott (Eds.). *Proceedings of the IJCAI-2003 Workshop on Ontologies and Distributed Systems* (CEUR Workshop Proceeding series: Vol. 71). Retrieved from http://sunsite.informatik.rwth-aachen.de/Publications/CEUR-WS/Vol-71/Klein.pdf

Mäedche, A., Motik, B., & Stojanovic, L. (2003). Managing multiple and distributed ontologies in the Semantic Web. *The VLDB Journal*, *12*(4), 286–300. .doi:10.1007/s00778-003-0102-4

Mäedche, A., Stojanovic, L., Studer, R., & Volz, R. (2002). Managing multiple ontologies and ontology evolution in OntoLogging. In *Proceedings of the Conference on Intelligent Information Processing* (IIP-2002), Part of the IFIP World Computer Congress WCC2002, (pp. 51-63). Montreal, Canada.

Mäedche, A., & Volz, R. (2001). The ontology extraction and maintenance framework text-to-onto. In F. J. Kurfess, & M. Hilario, (Eds.), *Proceedings of the Workshop on Integrating Data Mining and Knowledge Management,* at The 2001 IEEE International Conference on Data Mining ICDM'01, San Jose, CA. Retrieved from http://users.csc.calpoly.edu/~fkurfess/Events/DM-KM-01/Volz.pdf

Menzis, T. (1999). Knowledge maintenance: The state-of-the-art. *The Knowledge Engineering Review*, *14*(1), 1–46. doi:10.1017/S0269888999134052

Moguillansky, M. O., Rotstein, N. D., & Falappa, M. A. (2008). A theoretical model to handle ontology debugging and change through argumentation. In *Proceedings of the 2nd Workshop on Ontology Dynamics, (IWOD-08) at ISWC 08 Conference* (pp. 29-42). Retrieved from http://www.ics.forth.gr/~fgeo/Publications/IWOD-08_Proceedings.pdf

Noy, N., & Musen, M. (2002). PROMPTDIFF: A fixed-point algorithm for comparing ontology versions. In *Proceedings of the 18th National Conference on Artificial Intelligence and Fourteenth Conference on Innovative Applications of Artificial Intelligence,* Edmonton, Alberta, Canada. Menlo Park, CA: AAAI Press.

Noy, N. F., Chugh, A., Liu, W., & Musen, M. A. (2006). A framework for ontology evolution in collaborative environments. In The Semantic Web - ISWC 2006 (LNCS: Vol. 4273, pp. 544-558). Berlin, Germany: Springer. doi: 10.1007/11926078

Noy, N. F., & Klein, M. (2003). Tracking complex changes during ontology evolution. In *Collected Posters ISWC 2003*, Sanibal Island, FL. Retrieved from http://www.stanford.edu/~natalya/papers/trackingChangesPoster.pdf

Noy, N. F., & Klein, M. (2004). Ontology evolution: Not the same as schema evolution. *Knowledge and Information Systems*, *6*(4), 428–440. doi:10.1007/s10115-003-0137-2

Noy, N. F., Kunnatur, S., Klein, M., & Musen, M. A. (2004). Tracking changes during ontology evolution. The Semantic Web – ISWC 2004 (LNCS: Vol. 3298, 259-273). Berlin, Germany: Springer. doi: 10.1007/b102467

Noy, N. F., & McGuinness, D. (2001). *Ontology development. 101: A guide to creating your first ontology.* Stanford Knowledge Systems Laboratory Technical Report KSL-01-05 and Stanford Medical Informatics Technical Report SMI-2001-0880. Retrieved from http://protege.stanford.edu/publications/ontology_development/ontology101.html

Noy, N. F., & Musen, M. A. (2003). The PROMPT suite: Interactive tools for ontology merging and mapping. *International Journal of Human-Computer Studies*, *59*(6), 983–1024. doi:10.1016/j.ijhcs.2003.08.002

O'Brien, P., & Abidi, S. S. R. (2006). Modeling intelligent ontology evolution using biological evolutionary processes. In *Proceedings of the IEEE International Conference on Engineering of Intelligent Systems ICEIS 06.* Islamabad, Pakistan. doi: 0.1109/ICEIS.2006.1703172

Oberle, D., Volz, R., Motik, B., & Staab, S. (2004). An extensible ontology software environment. In S. Staab, & R. Studer (Eds.), *International Handbooks on Information Systems, Handbook on Ontologies*, (pp. 311-333). Berlin: Springer. Retrieved from http://www.aifb.uni-karlsruhe.de/WBS/dob/pubs/handbook2003a.pdf

Parsia, B., Sirin, E., & Kalyanpur, A. (2005). Debugging OWL ontologies. In *Proceedings of the 14th International World Wide Web Conference (WWW2005)*, 633-640. Retrieved from http://www2005.org/cdrom/docs/p633.pdf

Petasis, B., Karkaletsis, V., & Paliouras, G. (2007). D4.3: Ontology Population and Enrichment: State of the Art. *BOEMIE Deliverable*. Retrieved from http://www.boemie.org/sites/default/files/D%20 4.3.pdf

Plessers, P. (2006). *An approach to web-based ontology evolution*. Ph.D. Thesis, University of Brussels, Belgium.

Plessers, P., & De Troyer, O. (2005). Ontology change detection using a version log. In Y.Gil, E. Motta, V.R. Benjamins, & M. Musen (Eds.), The Semantic Web – ISWC 2005 (LNCS: Vol. 3729, pp. 578-592). Berlin, Germany: Springer-Verlag. doi: 10.1007/11574620

Plessers, P., & De Troyer, O. (2006) Resolving inconsistencies in evolving ontologies. In Y. Sure, & J. Domingue (Eds.), *The Semantic Web: Research and Applications, Proceedings of the 3rd European Semantic Web Conference ESWC 2006* (LNCS: Vol.4011, pp. 200-214). Berlin, Germany: Springer. doi: 10.1007/11762256

Plessers, P., De Troyer, O., & Casteleyn, S. (2007). Understanding ontology evolution: A change detection approach. *Journal of Web Semantics: Science, Services and Agents on the World Wide Web, 5*, 39-49. Retrieved from http://www.sciencedirect.com

Retrieved from http://www.eswc2007.org/pdf/eswc07-voelker2.pdf

Roddick, J. F. (1996). A survey of schema versioning issues for database systems. *Information and Software Technology, 37*(7), 383–393. doi:10.1016/0950-5849(95)91494-K

Sirin, E., & Parsia, B. (2004). Pellet: An OWL DL reasoner. In V. Haaslev & R. Moller (Eds.), *Proceedings of the International Workshop on Description Logics (DL2004)*, (CEUR Workshop Proceedings: Vol. 104), Whistler, BC, Canada. Retrieved from http://sunsite.informatik.rwth-aachen.de/Publications/CEUR-WS/Vol-104/30Sirin-Parsia.pdf

Stojanovic, L. (2004). *Methods and Tools for Ontology Evolution*. Ph.D. Thesis, Karlsruhe University, Germany.

Stojanovic, L., Maedche, A., Motik, B., & Stojanovic, N. (2002a). User-driven ontology evolution management. *Knowledge Engineering and Knowledge Management: Ontologies and the Semantic Web, Proceedings of the 13th International Conference* EKAW2002 (LNCS: Vol. 2473, pp. 285-300). Berlin, Germany: Springer-Verlag. doi: 10.1007/3-540-45810-7

Stojanovic, L., Maedche, M., Stojanovic, N., & Studer, R. (2003b). Ontology evolution as reconfiguration-design problem Solving. In *Proceedings of the 2nd international conference on Knowledge capture K-CAP '03* (pp.162-171). New York: ACM.

Stojanovic, L., & Motik, B. (2002). Ontology evolution within ontology editors. In *Proceedings of the OntoWeb-SIG3 Workshop Evaluation of Ontology-based Tools (EON2002) at the 13th International Conference on Knowledge Engineering and Knowledge Management* (EKAW 2002), (CEUR-WS: Vol. 62, pp. 53-62). Retrieved from http://ftp.informatik.rwth-aachen.de/Publications/CEUR-WS/Vol-62/EON2002_Stojanovic.pdf

Stojanovic, L., Stojanovic, N., Gonzalez, J., & Studer, R. (2003a). OntoManager— A System for the usage-based ontology management. In On The Move to Meaningful Internet Systems 2003: CoopIS, DOA, and ODBASE (LNCS: Vol. 2888, pp 858–875). Berlin, Germany: Springer. 10.1007/b94348

Stojanovic, L., Stojanovic, N., & Handschuh, S. (2002b). Evolution of the metadata in the ontology-based knowledge management systems. In *GI 2002, Proceedings of the 1st German Workshop on Experience Management GWEM 2002* (LNAI: Vol.10 pp. 65-77).

Stojanovic, N., Stojanovic, L., & Volz, R. (2002c). A Reverse engineering approach for migrating data-intensive web sites to the semantic web. In *Proceedings of the Conference on Intelligent Information Processing* (IIP-2002), Part of the IFIP World Computer Congress WCC2002, (pp.141-154), Montreal, Canada.

Sure, Y. (2002). On-to-knowledge – ontology based knowledge management tools and their application. *German Journal Kuenstliche Intelligenz*, 35-37.

Völker, J., Hitzler, P., & Cimiano, P. (2007b). Acquisition of OWL DL axioms from lexical resources. In E. Franconi, M. Kifer, & W. May (Eds.), *The Semantic Web: Research and Applications, Proceedings of the 6th International Semantic Web Conference, ISWC 2007* (LNCS: Vol. 4519, pp. 670-685). Berlin, Germany: Springer. doi: 10.1007/978-3-540-72667-8

Völker, J., Vrandecic, D., & Sure, Y. (2005). Automatic evaluation of ontologies (AEON). In Y. Gil, E. Motta, V. R. Benjamins, & M. A. Musen (Eds.), The Semantic Web – ISWC 2005 (LNCS: Vol. 3729, pp.716-731). Berlin, Germany: Springer. doi: 10.1007/11574620

Völker, J., Vrandecic, D., Sure, Y., & Hotho, A. (2007a). Learning disjointness. In E. Franconi, M. Kifer, & W. May (Eds.), *The Semantic Web: Research and Applications, Proceedings of the 6th International Semantic Web Conference, ISWC 2007* (LNCS: Vol. 4519, pp. 175-189). Berlin, Germany: Springer. doi: 10.1007/978-3-540-72667-8 Retrieved from http://www.eswc2007.org/pdf/eswc07-voelker1.pdf

Wang, H., Horridge, M., Rector, A., Drummond, N., & Seidenberg, J. (2005). Debugging OWL-DL ontologies: A heuristic approach. In Y. Gil, E. Motta, V. R. Benjamins, & M. A. Musen (Eds.), The Semantic Web – ISWC 2005 (LNCS: Vol. 3729, pp. 745-757). Berlin, Germany: Springer. doi: 10.1007/11574620

Xuan, D. N., Bellatreche, L., & Pierra, G. (2006). A versioning management model for ontology-based data warehouses. In *Proceedings of 8th International Conference on Data Warehousing and Knowledge Discovery DaWak'06*. Krakow, Poland. Retrieved from http://www.lisi.ensma.fr/ftp/pub/documents/papers/2006/2006-TICD-WKD-XUAN.pdf

ENDNOTES

1. Karlsruhe ontology: an open-source ontology management infrastructure targeted for business applications: Kaon.semanticweb.org
2. http://www.w3.org/TR/owl-features/
3. http://www.boemie.org/
4. http://kaon.semanticweb.org
5. http://protege.stanford.edu/
6. http://ontoware.org/projects/text2onto/
7. http://gate.ac.uk/

Chapter 9
Large Scale Matching Issues and Advances

Sana Sellami
LIRIS, France

Aicha-Nabila Benharkat
LIRIS, France

Youssef Amghar
LIRIS, France

ABSTRACT

Nowadays, the Information technology domains (Semantic Web, E-business, digital libraries, life science, etc) abound with a large variety of data (e.g. DB schemas, XML schemas, ontologies) and bring up a hard problem: the semantic heterogeneity. Matching techniques are called to overcome this challenge and attempts to align these data. In this chapter, the authors are interested in studying large scale matching approaches. They survey the techniques of large scale matching, when a large number of schemas/ ontologies and attributes are involved. They attempt to cover a variety of techniques for schema matching called Pair-wise and Holistic, as well as a set of useful optimization techniques. They compare the different existing schema/ontology matching tools. One can acknowledge that this domain is on top of effervescence and large scale matching needs many more advances. Then the authors provide conclusions concerning important open issues and potential synergies of the technologies presented.

INTRODUCTION

Recently, we are witnessing an explosive growth of data in the business and scientific area. In fact, there are many databases and information sources available through the web covering different domains: semantic Web, deep Web, e-business, biology, digital libraries, etc. In such domains, the data generated are heterogeneous and voluminous e.g schemas with several thousand elements are common in e-business applications. Currently, the greatest challenge to take up is to perform the integration of such heterogeneous collections of data. Matching techniques are solutions to automatically find correspondences between these data in order to allow their integration in information systems. Matching has found considerable interest in both research and practice. In fact, matching is an opera-

DOI: 10.4018/978-1-61520-859-3.ch009

tion that takes data as input (e.g XML schemas, ontologies, relational database schemas) and returns the semantic similarity values of their elements. However, matching these data at large scale represents a laborious process. The standard approach trying to match the complete input schemas will often lead to shading off performance. Various schema matching systems have been developed to solve the problem semi-automatically. Since schema matching is a semi-automatic task, efficient implementations are required to support interactive user feedback. In this context, scalable matching becomes a problem to be solved.

This chapter describes new research works of large scale schema and ontology matching. In the past years there has been quite an amount of research in the area of matching both for database schemas and more recently for ontologies. Several surveys (Rahm& Bernstein, 2002, Shvaiko & Euzenat, 2005) have been proposed covering many of the existing approaches. The survey proposed by (Rahm& Bernstein, 2002) is devoted to a classification of schema matching approaches and a comparative review of matching systems. The survey exposed by (Shvaiko & Euzenat, 2005) presents, as well, a new classification taking into account some novel schema/ontology matching approaches. A number of approaches and principles have been developed for matching small or medium data (schemas or ontologies). A major challenge that is still largely to be tackled is to scale up semantic matching in two ways: to a large number of data to be aligned or matched and to very large data. While the former is primarily addressed in the database area, the latter has been addressed by researchers in schema and ontology matching. Based on this observation, we provide a survey of work in the large scale area that differs from those proposed by (Rahm& Bernstein, 2002, Shvaiko & Euzenat, 2005). We provide in our study the main features of a large scale matching. We survey, then, the existing matching approaches at large scale called holistic and Pair-wise and we show how these approaches deal with scalability problem. We discuss the several related strategies and topics of optimization techniques, machine learning algorithms, statistical algorithms, etc. We describe the existing schema/ontology matching tools in the literature and compare them. This analysis of state of the art techniques allows us to make some conclusions and observations about the existing matching approaches and systems.

This chapter is organized as follows. Section 2 presents the motivation of large scale matching problem. Section 3 discusses the large scale matching approaches and presents a classification. In section 4, we describe the large scale matching tools. Section 5 reports some future directions and section 6 concludes this chapter.

LARGE SCALE MATCHING PROBLEM

Motivating Example

To motivate the large scale matching problem, let us consider two e-business companies: Company A and B (Figure 1).

These companies try to interoperate with sharing their internal schemas. This comprises a variety of transactions, such as exchanging product information, placing purchase orders, confirming and paying orders, which are carried out by exchanging electronic documents, or messages, between these two business partners. We have then to integrate databases of these two companies. The documents of both companies are described on e-business XML schemas.

The real-life e-business applications often process the XML data that are structured according to some standardized e-business schemas of catalogs and messages, such as OAGIS[1] or XCBL[2]. Such catalogs are developed by various individual, national and public organizations. Table 1 shows some characteristics of these E-business schemas displaying the "amazing scale" about these schemas.

Figure 1. Motivating example

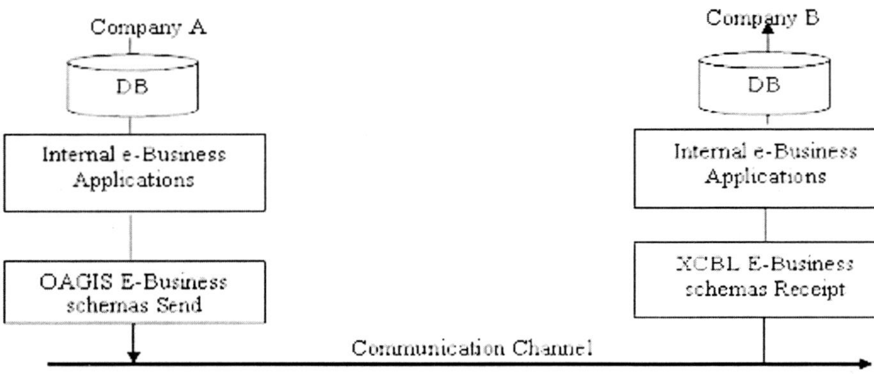

The semantic integration of the considered e-business applications needs some "mutual understanding" of the data exchanged between them. Then one of the most critical steps to integrating heterogeneous e-Business applications using different XML schemas is scalable schema matching.

Scalable Matching

Schema matching is a critical problem for integrating heterogeneous information sources. Large scale matching is characterized by two main features: (1) Data are very large (hundred and thousand components) and (2) large number of data needs to be matched (more than 2). Current matching systems (Rahm& Bernstein, 2002, Shvaiko & Euzenat, 2005) have been performed with simple data holding a small number of components (50-100), whereas in practice, real world data are voluminous (hundred or thousand components). For example, OAEI campaigns (Caracciolo & all, 2008) gave scalability characteristics of the ontology matching technology. The larger tests involve 10.000, 100.000, and 1.000.000 entities per ontology (e.g., UMLS has about 200.000 entities). In consequence, matching algorithms are facing up to more complicated contexts. As a result, many problems can appear, for example: performance decreasing when the matching algorithms deal with large data, their complexity becomes consequently exponential, increasing human effort and poor quality of matching results is observed. In this context, the matching becomes a problem to be solved.

Domains and Applications

Matching issue is a critical problem in many domains, such as heterogeneous data integration, data warehousing, electronic business, semantic web, etc. As the number of these heterogeneous data increase rapidly, large scale matching issues become a great challenge. We discuss in this section the emergent applications of large scale matching

Table 1. Characteristics of e-business schemas

E-business Domain	Number of Schemas	Smallest/largest schema size (number of elements)	Min/Max depth
XCBL schemas	40	22-4000	3-22
OAGIS	100	42-2500	4-19

- **Schema and data integration***:* Data integration is the process of combining data residing at different sources and providing the user with a unified view of these data called global schema. This process emerges in a variety of situations both commercial (when two similar companies need to merge their databases) and scientific (combining research results from different bioinformatics repositories). Data integration appears with increasing frequency as the volume and the need to share existing data explodes. To integrate data from disparate heterogeneous information sources, data-integration systems need to resolve several major issues: (1) heterogeneity (2) scalability, (3) evolution. For example, the real-life e-business applications often process the XML data that are structured according to some standardized e-business schemas of catalogs and messages, such as representation, called the integrated schema, from a set of source schemas that are related to each other. The relationships between the source schemas can be represented via correspondences between schema elements or via some other forms of schema mappings such as constraints or views. The integrated schema can be viewed as a means for dealing with the heterogeneity in the source schemas, by providing a standard representation of the data.
- **E-Business***:* The emerging trends in business-to-business e-commerce require important levels of interoperability. E-business standards provide models for inter organizational data exchange. It is characteristic that these models are not specific for one company, but fulfill the requirements of as much companies as possible. Interoperability is widely regarded as the most critical issue for the future development and growth of e-business. E-business interoperability presents an enormous challenge. To enable systems to exchange messages with each other, application developers need to transform messages from one format to another. This has led to a new motivation for schema matching, namely to support message translation. Like in schema integration, there may be naming and structural conflicts, as schemas often use different names, different data types, and different constraints. However, we have here typically to deal with a much higher number of schemas than in schema integration.
- **Deep Web**: The deep (or invisible) web refers to content that lies hidden behind queryable HTML forms. These are pages that are dynamically created in response to HTML-form submissions, using structured data that lies in backend databases. This content is considered invisible because search-engines rely on hyperlinks to discover new content. The deep web represents a major gap in the coverage of search engines: the deep web is believed to be possibly larger than the current WWW, and typically has very high-quality content. Then the deep Web is qualitatively different from the surface Web. Deep Web sources store their content in searchable databases that only produce results dynamically in response to a direct request. But a direct query is a laborious way to search. On this deep Web, numerous online databases provide data via their query interfaces, instead of static URL links. Schema matching across Web interfaces is essential for mediating queries across deep Web sources.
- **Semantic Web***:* As result of the recent growing of the Semantic Web, a new generation of semantic applications is emerging, focused on exploiting the increasing amount of online semantic data available on the Web. These applications need to handle the high semantic heterogeneity introduced

by the increasing number of available online ontologies, that describe different domains from many different points of view and using different conceptualizations, thus leading to many ambiguity problems. Currently, ontologies applied to the World Wide Web are creating the Semantic Web. Ontologies, which are defined as the formal specification of a vocabulary of concepts and axioms relating them, are seen playing a key role in describing the "semantics" of the data. More and more ontologies are being developed and many of them describe similar domains. The problem of matching ontologies effectively and efficiently is a necessary precondition to integrate information on the Semantic Web.

CLASSIFICATION OF LARGE SCALE MATCHING APPROACHES

In this section, we discuss the solutions of the large scale matching problems described in the previous section. The goal of this study is to show how existing works can deal with scalability problem. The large scale matching problem has been tackled in holistic and pair-wise matching approaches, using fragmentation, clustering and statistical strategies. These approaches represent an effective attempt to resolve large scale matching problem. We describe then these different strategies and we underline the importance of the used techniques to improve scalable matching.

Pair-Wise Matching

Matching has been approached mainly by finding pair-wise attribute correspondences, to construct an integrated schema for two sources. Several pair-wise matching approaches over schemas and ontologies have been developed.

Schema Matching

Being a central process for several research topics like data integration, data transformation, schema evolution, etc, schema matching (figure. 2) has attracted much attention by research community (Bernstein & all, 2004; Lu & Wang, 2005; Smiljanic & all, 2006). We present the main strategies dealing with scalability problem. These strategies represent an effective attempt to resolve large scale matching problem. The used techniques aim at reducing the dimension of the matching problem:

- **Fragment based strategy** (Rahm & all, 2004): This is a divide and conquer approach which decomposes a large matching problem into smaller sub-problems by matching at the level of schema fragments. The fragment can be a schema, or sub-schema that represents parts of a schema which can be separately instantiated, or shared that is identified by a node with multiple parents. Fragmentation is achieved in two matching steps: The first step is the fragments identification of the two schemas that are sufficiently similar and the second step is to match similar fragments. This approach has been implemented in COMA++ (Aumueller & all, 2005) matching tool. The fragment-based approach represents an effective solution to treat large schemas.
- **Extraction of common structures** (Lu & Wang, 2005): The main goal of this approach is to extract a disjoint set of the largest approximate common substructures between two trees. This set of common structures represents the most likely matches between substructures in the two schemas. Identifying these structures aim at improving the efficiency of matching process. However, this approach has not been implemented yet.

Figure 2. Pair-wise schema matching

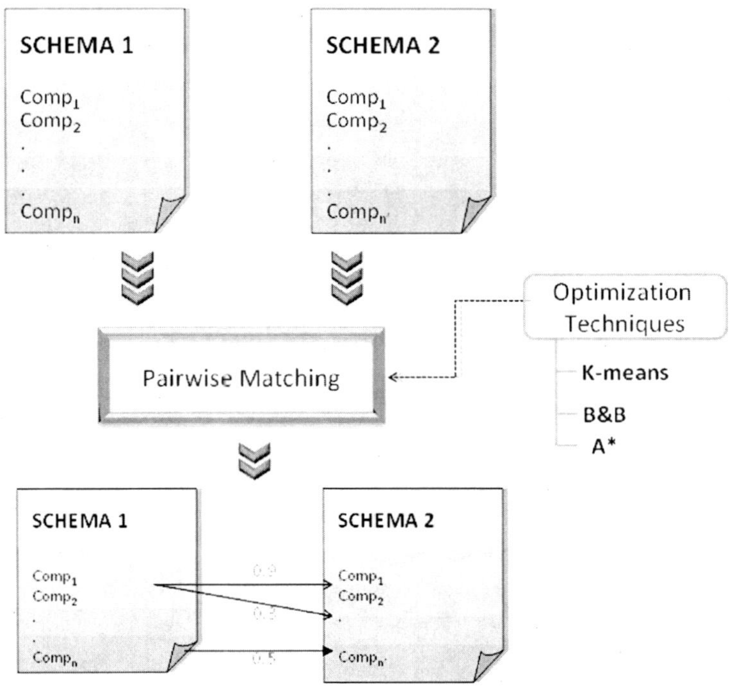

- **Clustered schema matching strategy** (Smiljanic & all, 2006): This is a technique for improving the efficiency of schema matching by means of clustering. In this approach, matching is achieved between a small schema and a schema repository. The clustering is introduced after the generation of matching elements. Clustering is then used to quickly identify regions in the schema repository which are likely to include good matchings for the smaller schema. The clustered schema matching is achieved by the clustering algorithm K-means (Xu & Wunsch, 2005). The authors choose an adaptation of the k-means clustering algorithm. Bellflower system implements this technique. The improved efficiency, however, comes at the cost of the loss of some matching. The loss mostly occurs among the matchings which rank low. However, there is no measure of a cluster's quality that can be used to decide which clusters have better chances to produce good matchings.

Ontology Matching

Ontology matching (figure 3) or aligning is a promising solution to the semantic heterogeneity problem. It finds correspondences between semantically related entities of the ontologies. Matching ontologies enables the interoperability of ontology knowledge and data. In the large scale scenario like a semantic web, the size of these ontologies causes serious problems in managing them. Actually, many approaches (Ehrig & Staab, 2004; Hu & Qu, 2006; Hu &all, 2006; Stuckenschmidt & Klein, 2004; Wang & all, 2006) have been proposed in literature to study the large ontology matching problem. We describe her existing strategies and techniques aiming to improve large scale matching ontologies:

Figure 3. Pair-wise ontology matching

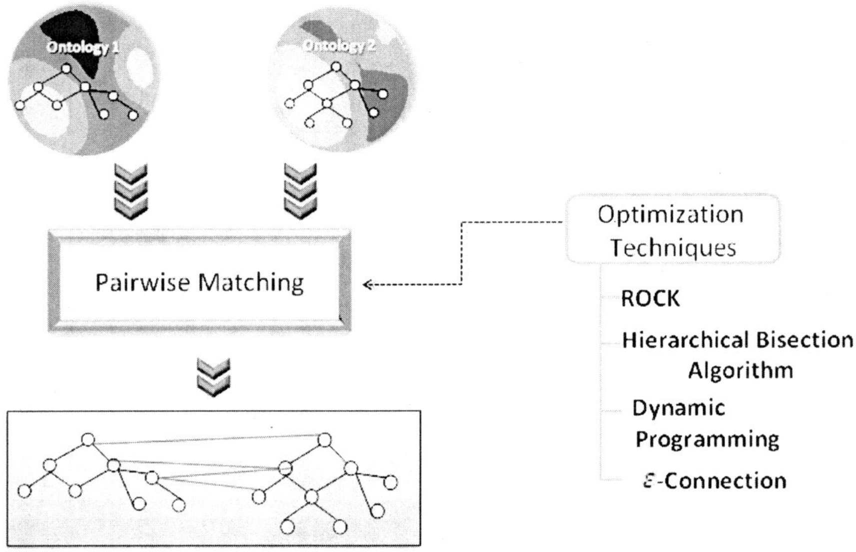

- **Partitioning strategy:** (Hu & Qu, 2006) introduced this strategy as a method for partition-based block matching that is appropriate to large class hierarchies. Large class hierarchies are one of the most common kinds of large-scale ontologies. The two large class hierarchies are partitioned, based on both structural affinities and linguistic similarities, into small blocks respectively. The matching process is then achieved between blocks by combining the two kinds of relatedness found via predefined anchors and virtual documents between them. The partitioning process is realized based on ROCK (Robust Clustering Using Links) algorithm (Xu & Wunsch, 2005). However, this approach is not completely applicable to large ontologies and it partitions two large class hierarchies separately without considering the correspondences between them. In addition, it only assumes matchings between classes, thus it is not a general solution for ontology matching. To cope with large ontologies matching, (Hu & Qu, 2008) then propose a partitioning-based approach to address the block matching problem. This approach considers both linguistic and structural characteristics of domain entities based on virtual documents for the relatedness measure. Partitioning ontologies is achieved by a hierarchical bisection algorithm to provide block mappings.
- **Modularization strategy:** (Wang & all, 2006) propose this approach to deal with large and complex ontologies. The authors propose a Modularization-based Ontology Matching approach (MOM). This is a divide-and-conquer strategy which decomposes a large matching problem into smaller sub-problems by matching at the level of ontology modules. This approach includes sub-steps for large ontology partitioning, finding similar modules, module matching and result combination. This method uses the ε -connection (Grau & all, 2005) to transform the input ontology into an ε -connection with the largest possible number of connected knowledge bases.

Holistic Matching

Traditional schema matching research has been found by pair-wise approach. Recently, holistic schema matching has received much attention due to its efficiency in exploring the contextual information and scalability. Holistic matching (figure.4) matches multiple schemas at the same time to find attribute correspondences among all the schemas at once. These schemas are usually extracted from web query interfaces in the deep Web. The deep Web refers to World Wide Web content not part of the surface Web indexed by search engines. The data sources in the deep Web are structured and accessible only via dynamic queries instead of static URL links.

Several current approaches to holistic schema matching (Chen & all, 2005; He & all, 2006; He & all, 2004; He & all, 2003; He & all, 2005; Madhavan & all, 2005; Pei & all, 2006; Su & all, 2006a; Su & all, 2006b) rely on a large amount of data to discover semantic correspondences between attributes. We describe the most important strategies proposed in the literature and we highlight the used techniques to improve holistic matching process.

- **Statistical strategy:** This approach has been introduced in (He & all, 2004; He & all, 2003) with MGS (for hypothesis modeling, generation, and selection) and a DCM (Dual Correlation Mining) framework. It is based on the observation that co-occurrence patterns across schemas often reveal the complex relationships of attributes. However, these approaches suffer from noisy data. The works suggested in (Chen & all, 2005; He & all, 2006) outperform (He & all, 2004; He & all, 2003) by adding sampling and voting techniques, which are inspired by bagging predictors. Specifically, this approach creates a set of matchers, by randomizing input schema data into many independently down sampled trials, executing the same matcher on each trial and then aggregating their ranked results by taking majority voting.
- **Clustering based approach:** This approach has been presented in (Pei & all, 2006). First, schemas are clustered based on their contextual similarity. Second, attributes of the schemas that are in the same schema cluster are clustered to find attribute correspondences between these schemas.

Figure 4. Holistic matching

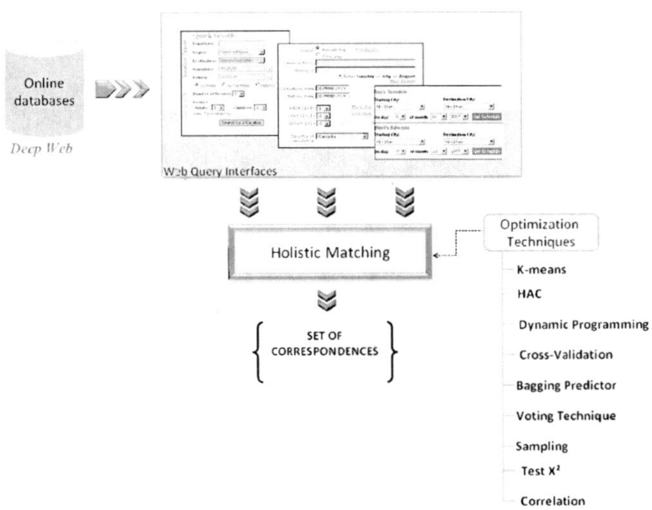

Third, attributes are clustered across different schema clusters using statistical information gleaned from the existing attribute clusters to find attribute correspondences between more schemas. The K-means algorithm has been used in these three clustering tasks and a resampling method (Monti & all, 2003) has been proposed to extract stable attributes from a collection of data.

Optimization Techniques

Most of the proposed matching approaches at a large scale integrate optimization techniques (e.g clustering, sampling, etc). We classify in figure.5 several widely used optimization techniques in four categories: Machine learning techniques, Description logics, heuristic algorithms and statistical algorithms.

- **Machine learning techniques** are supported by unsupervised classification, bagging predictors techniques, Genetic Algorithms, Dynamic Programming. Unsupervised Classification may be hierarchical or partitional. These clustering algorithms are commonly used in several matching approaches to group matching results. Hierarchical algorithms (Xu & Wunsch, 2005) classify data into a hierarchical structure according to the proximity matrix. The results are usually represented by a dendrogram. The hierarchical algorithms that have been used in matching approaches are ROCK (Robust Clustering Using Links) that utilizes the information about links between points when making decisions on the points to be merged into a single cluster. HAC (Hierarchical Agglomerative Clustering), which is Bottom-up hierarchical clustering that treats each document as a singleton cluster at the outset and then successively merge (or agglomerate) pairs of clusters until all clusters have been merged into a single cluster that contains all documents. While

Figure 5. Optimization techniques

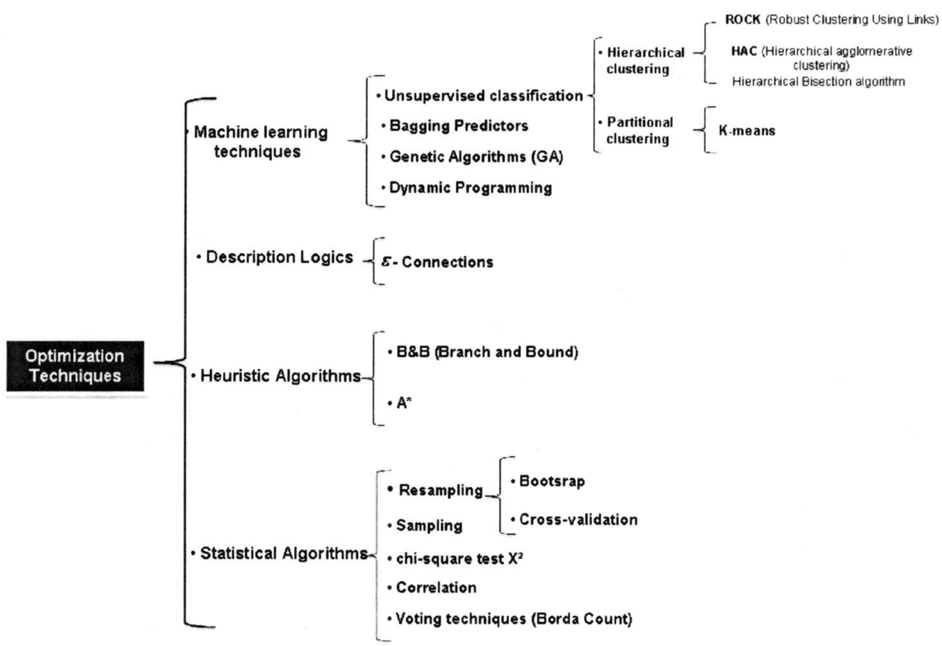

partitional clustering algorithms divide data objects into some pre specified number of clusters without the hierarchical structure. K-means is the most famous clustering algorithm. The main idea of K-means is to assign each point to the cluster whose center (also called centroid) is nearest. Genetic Algorithms (GA) (Berkovsky & all, 2005), on the other hand, are search techniques used in computing to find exact or approximate solutions to optimization and search problems. Genetic algorithms are categorized as global search heuristics. They are a particular class of evolutionary algorithms (also known as evolutionary computation) that use techniques inspired by evolutionary biology such as inheritance, mutation, selection, and crossover (also called recombination).

- **Description logics methods like** ε-Connections (Grau & all, 2005) are defined as a combination of other logical formalisms. They were defined as a way to go beyond the expressivity of each of the component logics, while preserving the decidability of the reasoning services in the combination.
- **Heuristic algorithms such B&B (Branch and bound)** are also used to search the complete space of solutions for a given problem for the best solution. It is the most widely used tool for solving large scale NP-hard combinatorial optimization problems.
- **Statistical Algorithms** are also available to further improve matching quality; examples include resampling (Bootstrap, cross-validation), sampling methods, chi-square X^2, correlation and voting techniques.

Classification of Matching Approaches

In this section, we classify the previous described approaches in (Figure. 6) according to the optimization techniques that have been used in the literature. Figure. 6 can be read from two points of view: In the top down view, we present different input data occurring in both holistic and pair-wise approaches. In the bottom up view, we can base the classification on strategies (Fragmentation, clustering, modularization, partitioning and statistical) related to the optimization techniques. We classify several widely used optimization techniques in four categories: Machine learning techniques, Description logics, heuristic algorithms and statistical algorithms. Our proposed classification is inspired from the one that is projected in (Shvaiko & Euzenat, 2005) by taking into consideration only large scale matching techniques.

REVIEW OF LARGE SCALE MATCHING TOOLS

Several matching tools have been proposed in the literature. We review in this section the matching tools dealing with scalability problem.

Pair-Wise Matching Tools

COMA++ is a composite matching tool that has been developed by (Aumueller & all, 2005). COMA++ supports matching of large real world schemas as well as ontologies using different match strategies (e.g fragmentation). It implements an extensible library of matching algorithms including more than 15 matchers exploiting different kinds of schema (e.g. simple string matchers (Affix, Trigram, EditDistance, etc), reuse oriented matchers, combined matchers) and auxiliary information. It provides a graphical interface that allows the user to interact, match two schemas or ontologies and evaluate match algorithms.

Figure 6. Classification of large scale matching approaches

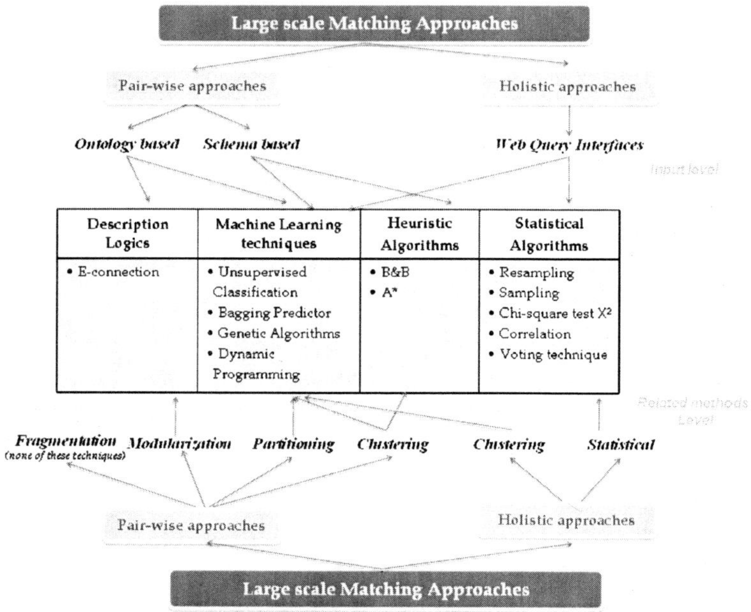

PROTOPLASM (A PROTOtype PLAtform for Schema Matching) (Bernstein &al., 2004) is a matching tool that implements a simplified version of Cupid (Madhavan & all., 2001) and Similarity Flooding algorithms (Melnik & all., 2002) to match the schemas. PROTOPLASM supports numerous operators for computing, aggregating, and filtering similarity matrices. By using a script language, it provides the flexibly for defining and customizing the work flow of the match operators. This tool has been integrated with a prototype version of Microsoft Biztalk Mapper, a visual programming tool for generating XML-to-XML mappings.

Bellflower is a prototype System that implements clustered schema matching approach proposed by (Smiljanic & all., 2006). Bellflower uses a schema repository built by randomly selecting XML schemas available on the Internet. Matching is achieved one personal schema and a repository of schemas with only comparing the element names.

Falcon-Ao (Hu & all, 2008) is an ontology matching system to enable interoperability between Web applications using different but related ontologies expressed in RDFS or OWL. Falcon-Ao implements a partitioning-based approach to divide large ontologies into smaller ones and uses a library of matchers: V-Doc to discover linguistic alignment between entities in ontologies, I-Sub based on string comparison technique and GMO that is an iterative structural matcher which uses RDF Bipartite graphs and computes structural similarities. Falcon-AO provides a graphical user interface (GUI) to make it easily accessible to users.

Malasco (Matching large scale ontologies) (Paulheim, 2008) is an ontology matching system which serves as a framework for reusing existing, non-scalable matching systems on large scale ontologies. It implements a partitioning approach to divide input ontologies into smaller partitions.

QOM (Quick Ontology Matching) (Ehrig & Staab, 2004) is a semi-automatic mapping tool between two ontologies. The ontologies are represented on RDF. QOM avoids the complete pair-wise comparison in favour of top-down strategy. It improves then the quality of mappings and represents a way to trade off between effectiveness and efficiency It shows better quality results than approaches within the same complexity class.

ONTOBUILDER (Roitman et Gal, 2006) is a generic tool for extraction, consolidating, matching and authoring ontologies. OntoBuilder accepts two ontologies as input, candidate ontology and target ontology. It attempts to match each attribute in the target ontology with an attribute in the candidate ontology. OntoBuilder supports an array of matching algorithms and can be used as a framework for developing new schema matchers which can be plugged-in and used via GUI or as an API. OntoBuilder contains also several matching algorithms, that can match concepts (terms) by their data types, constraints on value assignment, and above all, the sequencing of concepts within forms (termed precedence), capturing sequence semantics that reflect business rules.

Holistic Matching Tools

MGS (for hypothesis Modeling, Generation, and Selection) and *DCM* (Dual Correlation Mining) (He & all, 2004; He & all, 2003): The MGS is a holistic framework for global evaluation, building upon the hypothesis of the existence of a hidden schema model that probabilistically generates the schemas we observed. This evaluation estimates all possible "models," where a model expresses all attributes matchings. Nevertheless, this framework does not take into consideration complex mappings. DCM (Dual Correlation Mining) framework has been proposed for local evaluation, based on the observation that co-occurrence patterns across schemas often reveal the complex relationships of attributes. These frameworks are based on the statistical approaches extracted from data mining domain and are implemented in the *MetaQuerier* system.

PSM (Parallel Schema Matching) and *HSM* (Holistic Schema Matching) (Su et al., 2006) are implementations of holistic matching to find matching attributes across a set of Web database schemas of the same domain. HSM integrates several steps: matching score calculation that measures the probability of two attributes being synonym, grouping score calculation that estimates whether two attributes are grouping attributes. PSM form parallel schemas by comparing two schemas and deleting their common attributes. HSM and PSM are purely based on the occurrence patterns of attributes and requires neither domain-knowledge nor user interaction.

WISE-Integrator (He & all, 2005) is an automatic search interface and integration tool. To identify matching attributes between web search interfaces, Wise-integrator applies three levels of schema information: attribute names, field specification and attribute values and patterns. It uses clustering techniques to improve the accuracy of attribute matching. First, all search interfaces in the same domain are considered and attributes are clustered based on exact matches of attributes names/values. Second, further clustering is performed based on approximate matches and meta-information matches. When all potentially matching attributes are clustered together, the global attribute for each group of such attributes is generated.

Matching Tools Comparison

Figure 7 presents a comparison of large scale matching tools. We have compared tools based on different criteria: the input data, the resulted output, the use of auxiliary resources, the implemented strategies and tools availability. We can notice from this table the major differences between pair-wise and holistic tools. First, pair-wise tools (e.g COMA++, Flacon-Ao, QoM, etc) take two large schemas/ontologies as input and produce a matching between elements of these two data that correspond semantically to each other. While, holistic tools (e.g DCM, PSM, Wise-Integrator) are not applied on ontologies and perform matching between many small web query interfaces and finds all matching at once. Then none of these existing tools cope with large and several data

Figure 7. Table of large scale matching tools comparison

Tools	Input	Input Characteristics	Output	Auxiliary Resources	Strategies	Tools availability
COMA++	Relational schemas, XSD, XDR, OWL	2 large schemas/ ontologies	Mappings	Auxiliary thesauri, Alignment reuse repository of structures	Fragmentation Reused oriented approach	Yes
PROTOPLASM	Relational schemas, XML, ODMG	2 large schemas	Mappings	WordNet	None	No
Bellflower	XSD	A small schema And a repository Of schemas	Matchings	None	Clustering	No
Falcon-AO	RDF, OWL	2 large ontologies	Alignment (RDF/XML Formats)	None	Partitioning	Yes
Malasco	RDF	2 large ontologies	Alignment	None	Partitioning	No
QOM	RDF, OWL	2 large ontologies	Alignment	Thesauri	None	No
ONTOBUILDER	RDF	2 large ontologies	Mediated ontology	None	None	Yes
MGS and DCM	Web query interface	Several small Interfaces	Mappings	None	Statistical (Correlation Mining)	No
PSM and HSM	Web query interface	Several small Interfaces	Mappings	None	Statistical (Correlation Mining)	No
Wise Integrator	Web query interface	Several small Interfaces	Integrated Schema	None	Clustering	No

(schemas, ontologies, query interfaces). Second, all tools implementing holistic approaches do not employ any semantic resource for the determination of the correspondences. Whereas, using external resources improves semantic relations between elements or entities. Third, the majority of the described matching tools implement strategies. This aims at improving the performance of large scale matching. Finally, most of these tools are not available as a demo. Some of them are not implemented yet and don't provide a graphical user interface. This unavailability makes difficult their evaluation against a real world schemas or ontologies to determine their performance and the quality of matching results produced.

OPEN DIRECTIONS

Based on the previous described observations, we can outline some open issues that warrant further research:

- **Scalability***:* Most of the proposed matching tools suffer from handling large and several schemas and ontologies. Real problems in specific application contexts require scalable solutions as a first priority. Future schema and ontology matching tools should be able to provide such capability.
- **Combining holistic and pair-wise approaches***:* The combination of holistic and

pair-wise matchers analyzes schemas/elements under different aspects, resulting in a more stable and accurate similarity for heterogeneous schemas and ontologies.

- **Strategies:** Applying the existing strategies and optimization techniques on the large and voluminous schemas or ontologies aims at improving large scale matching. For example, the work proposed by (Sellami & all, 2008) presents an efficient approach based on tree mining techniques to deal with scalability of schema matching.
- **External resources:** We assess that it is essential to employ some auxiliary semantic information to identify finer matching and to deal with the lack of background knowledge in matching tasks. It is also the way to obtain semantic mappings between different input data. For example, the use of domain ontology or the background knowledge represents an efficient solution to semantic ontology matching problems.
- **Evaluation:** Developing metrics and benchmark tools for evaluating scalable matching systems. The quality of matching across different matching systems is defined as the performance of matching algorithms. However, in large scale scenarios, many factors should be evaluated e.g input data, complexity, human interaction, etc.

CONCLUSION

This chapter presented a broad scope of matching at large scale categories and characteristics, and surveyed related work. We have achieved a state of the art study covering existing approaches and strategies in pair-wise and holistic Matching. We have proposed a classification of these approaches based on the input data, proposed strategies and optimization techniques. We have also described the existing large scale schema and ontology matching tools and compared them based on different criteria (Input data, output, auxiliary resources, strategies, tools availability). We have presented different directions for the future works in large ontology/ schema matching.

To resume, scalable matching requires deep domain knowledge: characteristics and representations of data, domain, user's needs, time performance, etc. We hope that this study provides a useful directive for conducting and describing future scalable matching tools and approaches.

REFERENCES

Asai, T., Abe, K., Kawasoe, S., Arimura, H., & Sakamoto, H. (2002). Efficient substructure discovery from large semi-structured data. In *2nd Annual SIAM Symposium on Data Mining (SDM2)*, Arlington, VA, USA.

Aumueller, D., Do, H. H., Massmann, S., & Rahm, E. (2005). Schema and ontology matching with coma++. In Acm sigmod (pp. 906-908).

Berkovsky, S., Eytani, Y., & Gal, A. 2005. Measuring the Relative Performance of Schema Matchers. In *Proceedings of the 2005 IEEE/WIC/ACM International Conference on Web Intelligence,* Compiegne, France, September 19-22, 2005, (WI'05), (pp. 366-371).

Bernstein, P. A., Melnik, S., Petropoulos, M., & Quix, C. (2004). Industrial-Strength Schema Matching. In ACM SIGMOD Record, (pp. 38-43).

Caracciolo, C., Euzenat, J., Hollink, L., Ichise, R., Isaac, A., & Malaisé, V. Meilicke, et al. (2008). Results of the Ontology Alignment Evaluation Initiative 2008. In *Proceedings of the 3rd International Workshop on Ontology Matching (OM-2008) Collocated with the 7th International Semantic Web Conference (ISWC-2008),* Karlsruhe, Germany.

Chen-Chan Chang, K., He, B., & Zhang, Z. (2005). Toward Large Scale Integration: Building a MetaQuerier over Databases on the Web. In *Proceedings of the Second Conference on Innovative Data Systems Research* (CIDR) Asilomar, CA, (pp. 44-55).

Do, H. H., Melnik, S., & Rahm, E. (2002). Comparison of schema Matching Evaluations. In GI-Workshop Web and Databases, (pp 221-237), Erfurt, Germany.

Ehrig, M., & Staab, S. (2004). QOM-Quick Ontology Mapping. In *Proceedings of the Third International Semantic Web Conference (ISWC)* (pp. 683-697), Hiroshima, Japan.

Grau, B. C., Parsia, B., Sirin, E., & Kalyanpur, A. (2005). Automatic Partitioning of OWL Ontologies Using ε-Connections. In *Proceedings of the International Workshop on Description Logics (DL)* Edinburgh, Scotland, UK.

He, B., & Chen-chuan Chang, K.(2006). Automatic Complex Schema Matching Across Web Query Interfaces: A Correlation Mining Approach. In *ACM Transactions on Database Systems* (*TODS*) (pp. 346-395). New York: ACM Press.

He, B., & Chen-Chan Chang, K. (2003). Statistical Schema Matching across Web Query Interfaces. In *Proceedings of the ACM SIGMOD International Conference on Management of Data* (pp. 217-228), San Diego, CA.

He, B., Chen-Chan Chang, K., & Han, J. (2004). Discovering complex matchings across Web Query Interfaces: A Correlation Mining Approach. In *Proceedings of the Tenth ACM SIGKDD International Conference on Knowledge Discovery and Data Mining (KDD),* (pp. 148-157). New York: ACM Press.

He, H., Meng, W., Yu, C., & Wu, Z. (2005). WISE-Integrator: A System for Extracting and Integrating Complex Web Search Interfaces of the Deep Web. In *Proceedings of the 31st International Conference on Very Large Data Bases (VLDB),* (pp. 1314-1317), Trondheim, Norway.

Hu, W., & Qu, Y. (2006). Block Matching for Ontologies. In *Proceedings of the 5th International Semantic Web Conference (ISWC),* (pp. 300-313), Athens, GA.

Hu, W., Zhao, Y., & Qu, Y. (2008). Matching large ontologies: A divide-and-conquer approach. *Journal of Data Knowledge Engineering, 67,* 140–160. doi:10.1016/j.datak.2008.06.003

Lu, J., Wang, S., & Wang, J. (2005). An experiment on the Matching and Reuse of XML Schemas. In *Proceedings of the 5th International Conference on Web engineering (ICWE).* (pp. 273-284), Sydney, Australia.

Madhavan, J., Bernstein, P. A., Doan, A., & Halevy, A. Y. (2005). Corpus-based Schema Matching. In *Proceedings of the 21st International Conference on Data Engineering (ICDE),* (pp. 57-68), Tokyo, Japan.

Madhavan, J., Bernstein, P. A., & Rahm, E. (2001). Generic schema matching with cupid. In *Proceedings of 27th International Conference on Very Large Data Bases*, September 11-14, 2001, Roma, Italy (p. 49-58).

Melnik, S., Garcia-Molina, G., & Rahm, E. (2002). Similarity Flooding: A Versatile Graph Matching Algorithm and Its Application to Schema Matching. In *Proceedings of the 18th International Conference on Data Engineering,* San Jose, CA, (pp. 117-128).

Paulheim, H. (2008). On Applying Matching Tools to Large-Scale Ontologies. In *Proceedings of the 3rd International Workshop on Ontology Matching (OM-2008) Collocated with the 7th International Semantic Web Conference (ISWC-2008),* Karlsruhe, Germany.

Pei, J., Hong, J., & Bell, D. A. (2006). A Novel Clustering-based Approach to Schema Matching. In *Proceedings of the 4th International Conference on Advances in Information Systems (ADVIS),* (pp. 60-69), Izmir, Turkey.

Rahm, E., & Bernstein, P. A. (2001). *A survey of approaches to automatic schema matching*. The International Journal on Very Large Data Bases.

Rahm, E., Do, H. H., & Maßmann, S. (2004). Matching Large XML Schemas . In *SIGMOD Record* (pp. 26–31). New York: ACM Press.

Roitman, H., & Gal, A. (2006). Fully automatic extraction and consolidation of ontologies from web sources using sequence semantics . In *EDBT workshops* (pp. 573–576). Ontobuilder.

Sellami, S., Benharkat, A.-N., & Amghar, Y. (2008) From simple to large scale Matching: A hybrid approach. In *23rd International Symposium on Computer and Information Sciences (ISCIS)*, Istanbul, Turkey, (pp. 1-4).

Shvaiko, P., & Euzenat, J. (2005). A Survey of Schema-based Matching approaches. *Journal on Data Semantics IV, 3730*, 146–171. doi:10.1007/11603412_5

Smiljanic, M., Keulen, M., & Jonker, W. (2006). Using Element Clustering to Increase the Efficiency of XML Schema Matching. In *Proceedings of the 22nd International Conference on Data Engineering Workshops (ICDE Workshops).*

Stuckenschmidt, H., & Klein, M. (2004). Structure-based Partitioning of large concept hierarchies. In *Proceedings of the 3rd International Semantic Web Conference (ISWC),* (pp. 289-303), Hiroshima, Japan.

Su, W., Wang, J., & Lochovsky, F. (2006a). Holistic Schema Matching for Web Query Interface. In *Proceedings of the 10th International Conference on Extending Database Technology (EDBT),* (pp. 77-94), Munich, Germany.

Su, W., Wang, J., & Lochovsky, F. (2006b). Holistic Query Interface Matching using Parallel Schema Matching. In *Proceedings of the 22nd International Conference on Data Engineering (ICDE),* Atlanta, GA.

Wang, Z., Wang, Y., Zhang, S., Shen, G., & Du, T. (2006). Matching Large Scale Ontology Effectively. In *Proceedings of the First Asian Semantic Web Conference (ASWC),* (pp. 99-106), Beijing, China.

Xu, R., & Wunsch, D. (2005). Survey of Clustering Algorithms. *IEEE Transactions on Neural Networks*, 645–678. doi:10.1109/TNN.2005.845141

KEY TERMS AND DEFINITIONS

Matching: It is an operation that takes data as input (e.g XML schemas, ontologies, relational database schemas) and returns the semantic similarity values of their elements.

Large Scale: represents a context characterized by various (more than 2) and voluminous data hundred and thousand components).

Schemas: are the data used in the matching process. The schemas can be: XML schemas, relational schemas, OWL, RDF, etc…

Ontologies: is a formal representation of a set of concepts within a domain and the relationships between those concepts.

Optimization Techniques: Represent the used techniques in different large scale matching approaches to improve the quality of matching.

ENDNOTES

[1] www.oagi.org

[2] www.xcbl.org

Chapter 10
From Temporal Databases to Ontology Versioning: An Approach for Ontology Evolution

Najla Sassi
MIRACL Laboratory, Tunisia

Zouhaier Brahmia
MIRACL Laboratory, Tunisia

Wassim Jaziri
MIRACL Laboratory, Tunisia

Rafik Bouaziz
MIRACL Laboratory, Tunisia

ABSTRACT

The problem of versioning is present in several application areas, such as temporal databases, real-time computing and ontologies. This problem is generally defined as managing changes in a timely manner without loss of existing data. However, ontology versioning is more complicated than versioning in database because of the usage and content of ontology which incorporates semantic aspects. Consequently, ontology data models are much richer than those of database schemas. In this chapter, the authors are interested in developing an ontology versioning system to express, apply and implement changes on the ontology.

INTRODUCTION

In computer science, several application areas are especially concerned with versioning such as software development, CAD/CAM applications, temporal databases and ontologies.

Temporal databases support time-varying information and maintain the history of the modelled data (Jensen et al., 1998) (Özsoyoğlu et al., 1995) (Tansel et al., 1993). Versions of temporal data are kept along one or two time dimensions: valid time and transaction time (Jensen et al., 1998). The valid time concerns the time of the modelled real world and denotes the time a fact was, or will be true, whereas the transaction time is the one of the system and concerns the one during which the fact was or is current in the database as a stored data.

DOI: 10.4018/978-1-61520-859-3.ch010

The interest on schema versioning in temporal databases arose as a logical extension of the work formerly done on data.

Ontologies are defined as an explicit representation based on the identification of conceptual entities (concepts or relationships) and their semantics. Since ontology has to be continually changed for many reasons, it is interesting to manage its evolution and to take into account the different versions of this ontology.

Ontologies are like database schema and schema versioning in temporal databases can be useful in order to propose an approach for ontology versioning. In fact, we can benefit from principles and tools we previously defined for schema versioning in such databases, in order to ensure an efficient management of versions in ontological databases.

We are interested in developing an ontology versioning system to express, apply and implement changes on the ontology. The adopted ontology versioning approach is based on three steps: evolution changes, ontology coherence and versioning management. Our goal is to assist users in expressing evolution requirements, observing their consequences on the ontology and comparing ontology versions.

This paper is structured as follows. Sections 2, 3 and 4 present the versioning in software development, databases and ontology. In section 5, we propose an approach of ontology evolution based on three steps: evolution changes, ontology coherence and versioning management. Section 6 is dedicated to the process of ontology versioning storage. Finally, we conclude in section 7.

VERSIONING IN SOFTWARE DEVELOPMENT AND CAD/CAM APPLICATIONS

Software systems are rarely stable following initial implementations. They have complex structures which are likely to continually undergo changes during their lifetime. Software versioning is the process of assigning unique versions to unique states of computer software. These versions correspond to new developments in the software; sometimes they are also called revisions. A software version can be identified by a sequence of numbers and/or letters (for example, "Oracle 10g"), date (for example, "Wine 20040505"), year (for example, "Windows 2000"), or alphanumeric code (for example, "Adobe Dreamweaver CS4").

CAD/CAM, CIM, CASE tools and other engineering applications first put forward the requirement of managing multiple design versions (Kim et al., 1990). The primitive concepts of versioning were introduced to support different users concurrently working on parallel, or even merging, versions of the same piece of data (objects) (Katz, 1984) (Katz, 1990). These non-conventional applications not only often demand the support of many database states, but also design alternatives. To fulfil such requirement, works like (Kim et al., 1989) (Talens et al., 1993) have focused on the question of version support. A version describes an object in a period of time or from a certain point of view. Although some design alternatives are stored as versions, not all the history about data modification is recorded. The full history is just accessible if a temporal model is used. The majority of the above proposals have extended their data model using either temporal concepts or just version control mechanisms.

Temporal databases support time-varying information and maintain the history of the modelled data. Versions of temporal data are kept along one or two time dimensions: valid time and transaction time (Jensen et al., 1998). Versioning can be temporal or non-temporal. In a temporal versioning context, any version has a temporal connotation (a temporal interval or a temporal point); versions are necessarily successive and there are no overlapping temporal connotations of two versions. Temporal versioning is usually used in temporal databases for data and schema versioning. In a non-temporal versioning environment, versions coexist with any temporal pertinence; versions can be considered as parallel or alternative versions. Non-temporal versioning is usually used in software development and in advanced application domains like CAD/CAM and CIM.

TEMPORAL VERSIONING IN DATABASES

The database schema and raw data are evolving entities which require adequate support for past, present and even future versions. Temporal databases supporting schema versioning were developed with the aim of satisfying this requirement. In the following subsections, we will present respectively data versioning and schema versioning in such databases.

Data Versioning

Temporal databases allow the maintenance of data histories through the support of time semantics at system level. In such databases, data are versioned along one or two time dimensions: valid time and transaction time. In (Tansel et al., 1993), three categories of temporal databases are identified:

- **Transaction time databases:** these databases, also called rollback, use the transaction time of data as timestamp.
- **Valid time databases:** these databases use the valid time of data as timestamp.
- **Bi-temporal databases:** these databases use both transaction and valid time as timestamps.

In (De Castro et al., 1997), another category of temporal database is identified: the multi-temporal. In a multi-temporal database, entities (relations, classes, etc.) of different temporal formats coexist.

Several temporal data models were proposed as extensions of the relational data model. Clifford et al. (Clifford et al., 1995) classified them into two main categories: temporally ungrouped and temporally grouped. In temporally ungrouped models, the temporal representation is realized at extensional level, by means of timestamps added to data values as additional attributes to represent their temporal pertinence. Table 1 shows a temporally ungrouped valid time history of employees. In temporally grouped models, the temporal dimension is implicit in the structure of data representation: attributes are represented as histories considered as a whole and without the introduction of distinguished attributes. Attribute histories can be regarded as functions which map time into attribute domains. Table 2 shows a temporally grouped valid time history of employees.

The second representation is said to have more expressive power and to be more natural since it is history-oriented (Clifford et al., 1995).

Schema Versioning

A schema describes the structure of the data that are stored in a database. Whereas careful and accurate the initial design may have been, a database schema is likely to undergo changes and revisions after implementation. In order to avoid the loss of data after schema changes, many database

Table 1. Temporally ungrouped valid time history of employees

Name	Title	Dept	Salary	Validity start time	Validity end time
Abdullah	Engineer	RD	1000	2004-01-01	2006-12-31
Abdullah	Sr Engineer	RD	1200	2007-01-01	2008-12-31
Abdullah	Sr Engineer	RD	1300	2009-01-01	Now

Table 2. Temporally grouped valid time history of employees

<table>
<tr><th>Name</th><th>Title</th><th>Dept</th><th>Salary</th></tr>
<tr><td rowspan="3">2004-01-01
Abdullah
Now</td><td>2004-01-01
Engineer
2006-12-31</td><td rowspan="3">2004-01-01
RD
Now</td><td>2004-01-01
1000
2006-12-31</td></tr>
<tr><td rowspan="2">2007-01-01
Sr Engineer
Now</td><td>2007-01-01
1200
2008-12-31</td></tr>
<tr><td>2009-01-01
1300
Now</td></tr>
</table>

management systems support schema evolution, which provides (partial) automatic recovery of the extant data by adapting them to the new schema. However, if only the updated schema is retained, all the applications compiled with the past schema may cease to work. In order to let applications work on multiple schemata, schema evolution is not sufficient and the maintenance of more than one schema is required. This leads to the notion of schema versioning and schema version which is the persistent outcome of the application of schema modifications.

A database management system supports schema versioning if it allows (1) the modification of the database schema without loss of existing data, and (2) the accessing of all data through user-definable version interfaces (Jensen et al., 1998).

Schema versioning is a powerful technique not only to ensure reuse of data and continued support of legacy applications after schema changes, but also to add a new degree of freedom to database designers, application developers and final users. In fact, different schema versions actually allow one to represent, in full relief, different points of view over the modelled application reality.

Most proposals deal with schema versions as natural extensions of studies on temporal data modelling and temporal databases. The same considerations on temporal dimensions applied for data can be applied at schema level; schema versioning can then be done along one or two time dimensions. Thus, three temporal schema versioning mechanisms were defined in (De Castro et al., 1997): transaction-time schema versioning, valid-time schema versioning and bi-temporal schema versioning. Another temporal schema versioning mechanism is identified in (Brahmia et al., 2008): the application-time schema versioning.

In the following, we briefly describe the main differences between the fourth temporal schema versioning mechanisms.

Transaction-time schema versioning: it implies the creation of a new version of the schema each time a schema change is issued by a transaction. The old schema and its corresponding data are archived and a new current version is produced together with its data. Afterwards, the old schema version can still be accessed but not modified; only the current one may undergo modifications. This kind of schema versioning allows on-time schema changes, that is to say schema changes that are effective when applied. Due to the semantics of transaction-time, the management of time is completely transparent to the user: schema changes are effected in the usual way, without any reference to time; however, support of past schema versions is granted by the system so that the user can always rollback the full database to a past state of its life.

Valid-time schema versioning: it is necessary only when retroactive or proactive schema modifications have to be supported. With this type of schema versioning, all the existing versions of the schema can be accessed and modified. The designer can choose any of the stored schema versions for further modification and assign any validity to the resulting new schema version. Such validity can also be in the past or the future in order to perform retro- and pro-active schema modifications, respectively. For each valid-time instant, a single version of the schema can exist. For this reason, the (portions of) existing schema versions overlapped by the validity of the new schema version are overwritten.

Bi-temporal schema versioning: it allows the maintenance along transaction time of all the valid-time schema versions created by successive schema changes. In addition to transaction-time schema versioning, the bi-temporal schema versioning supports retro- and pro-active schema updates. With respect to valid-time schema versioning, the complete history of schema changes is maintained as no schema version is ever discarded (overlapped portions are "archived" rather than deleted). In a system where full auditing/traceability of the maintenance process is required, only bi-temporal schema versioning allows verifying whether a schema version was created by a retro- or pro-active schema modification.

Application-time schema versioning: it maintains schema versions along a new time dimension, the application time, which is defined only for schema version management, in an environment of enterprise information system. The application time denotes the time a schema version is applied and used by applications to manipulate and to query underlying data. Each schema version has an application start time (i.e. time from which this version is being applied by the database administrator as the current version) and possibly an application end time (i.e. time from which this version is being considered by the database administrator as a past version). These two times are always assigned by the database administrator. When applying a new schema version, its application start time can be the current time or another time which is anterior to current time, and the application end time of the previous schema version is necessarily equal to the instant which precedes at once the application start time of the new version.

Schema versions are successive and their application intervals cannot be overlapped. Application time is different from transaction time and valid time for the following reasons:

1. It cannot be used for data versioning; it is appropriate for schema versioning.
2. It does not allow defining a past schema version (i.e. a schema version whose application end time is anterior to current time) or a future schema version (i.e. a schema version whose application start time is superior to current time), since in an environment of enterprise information system, schema versions are always applied from the time they are defined.
3. It allows defining at a time t a schema version whose application start time is equal to a time t' which is anterior to t, when this new version must actually be applied from t' and not from t. For example, suppose that the current date is 2009-03-10, the current schema version of the EMPLOYEE relation is V3_EMPLOYEE, and a schema change must be done on this relation in the same date, but the database administrator is absent from 2009-03-08 to 2009-03-12. When he/she comes back, he/she will define a new schema version for the EMPLOYEE relation, V4_EMPLOYEE, with an application start time equal to 2009-03-10, and will assign the application end time of V3_EMPLOYEE to 2009-03-09. Then, he/she will ask the system to redo, under V4_EMPLOYEE and during the period [2009-03-10, 2009-03-12], all the treatments which had been realized under V3_EMPLOYEE and during the same period. To do this, the system must be able to keep the history of program executions (i.e. the system supports transaction versioning).

Briefly, the application time can be considered as "the transaction-time with more flexibility" and "the valid-time with constraints".

In information systems, ontologies are large and complex structures holding information. They are like database schema and need to change every time the modelled real world has changed. Thus, schema versioning in databases can be useful in order to propose an approach for ontology versioning. In fact, we can benefit from principles and tools that we defined for schema versioning in temporal databases, in order to ensure an efficient management of ontology versions in ontological databases.

ONTOLOGY VERSIONING

The concept of ontology has become a key technology to represent an agreement of a vocabulary and relationships between words in the vocabulary about a given topic. Ontologies can be used to represent information systems and to take into account the semantics for the automatic handling of information. In our work, we use the ontology as explicit representation of the domain in terms of concepts, relationships and properties by providing a semantic dimension.

Using ontology for modeling information systems require taking into account its evolution. There are several reasons for the evolution of ontology. The evolution of a domain can lead to an incompatibility of the initial ontology in relation to new users' requirements. Users' requirements may evolve in response to the emergence of new users' profile or in response to new knowledge.

In the literature, several approaches have been taken into account to address ontology evolution: the ontology evolution approach and the versioning ontology approach. The evolution of the ontology is defined in (Stojanovic et al., 2002) as the ability to update the existing ontology following the emergence of new needs and maintain its consistency and coherence. It maintains a unique ontology (the last one). This approach has a disadvantage because we can not retrieve old versions of ontology (only the last created).

The second approach provides access to data through different versions of ontology. To control the various versions, it is also important to monitor the relationship between them. This approach allows accessing all versions of ontology rather than only the last one. However, establishing links between versions is a complex task and requires an investment. These links must respect the order of versions and the changes that have occurred (Klein, 2002) (Noy et al., 2004).

In (Flouris et al., 2006), the authors distinguish the management of versions and changes in the ontology. They consider the evolution as the process of updating ontology while maintaining its validity. The version management is the process of managing several versions of the ontology while maintaining interoperability between them and allowing access to each version as required by the access element (data, service, application or other ontology).

Objectives of Versioning Support

Klein et al. (Klein et al., 2001) define ontology versioning as *the ability to handle changes in ontologies by creating and managing different variants of it*. In fact, ontology versioning implies the preservation of the old and new versions of ontology by giving transparent access to these various versions. It is thus necessary to identify the relationships between versions and their ontological entities (concepts, relationships and properties). By using these relationships, it becomes easy to identify the modifications between various versions.

The general goal of a versioning approach is to provide interoperability. Interoperability can be defined as the ability to use different versions of ontologies and possible data sets together in a meaningful way (not conflicting with the intended meaning of the ontology).

The most often mentioned objective of versioning is the ability to access instance data via an ontology version which can be different from the one used to describe it.

Requirements of Versioning Support

Ontology versioning typically involves the storage of several ontology versions and takes into account identification issues (i.e., how to identify the different versions), the relationship between different versions (i.e., a tree of versions resulting from the various ontology modifications) as well as compatibility information (i.e., information regarding the compatibility of any pair of ontology versions).

We identify the following requirements for a process of ontology versioning:

Identification of versions: Identification of ontology versions is very important in order to distinguish them.

Version relationship: between the definitions of concepts and properties in the original version of the ontology and those in the new version.

Links*:* between ontology versions that specify how the knowledge in the different versions is related. These links can be used to re-interpret data and knowledge during the evolution process.

Coherence: represent the conservation of semantic compatibility between ontology versions.

Optimization: of the ontology versioning process in order to remove the irrelevant versions

Approaches for Ontology Versioning

Klein and Noy (Klein, 2002) (Noy et al., 2004) use the term *versioning* to describe their approach of ontology change. They define ontology versioning as *the ability to manage ontology changes and their effects by creating and maintaining different variants of the ontology.*

Adequate methods and tools must be used to distinguish and identify the versions, in combination with procedures of ontology change and a mechanism of interpretation of their effects on the ontology (Klein, 2002).

According to these same authors, a methodology of versioning must provide a mechanism to clear up the interpretation of the concepts for the users of the various versions of ontology, and must make explicit the effects of changes on the various tasks. Their methodology provides methods to distinguish and recognize versions, procedures for updates and changes in ontologies, and an interpretation mechanism for effects of change. The authors compared ontology evolution with database schema evolution. The framework they proposed contains a set of operators, on the form of ontology, useful for modifying another evolving ontology. Klein (Klein et al., 2001) also proposes a change specification language based on the ontology of change operations.

A versioning methodology gives access to the data and concepts modeled via various versions of ontology. To control these versions, it is necessary to control the links of derivation between them. The derivation links allow defining and checking the compatibility between versions and following the transformations of data from a version to another one (Haase et al., 2004).

A methodology of ontology versioning must comprise a model of analysis of the relationships between two versions in order to (Noy et al., 2004b):

- Clarify what has changed in the ontology;
- Specify the semantic relationships between the elements of ontology whose definition has changed (e.g. entities equivalent or conceptually different);
- Describe the changes by a set of metadata specifying the date, the author and the goal of each change;

- Describe the context in which the changes are considered valid.

This mechanism of ontology versioning must give transparent access to various versions of ontology. A link must be ensured between the ontological elements in various versions of ontology. This link, called specification of change in (Klein et al., 2001), aims to explicit the relationship between the ontological elements in various versions.

Klein (Klein, 2002) cites additional opportunities to access and represent the changes between two versions of ontologies: (1) perform a structural diff, (2) represent the differences in the form of conceptual changes and (3) represent the differences in the form of a set of transformations. The first method is to perform a structural diff as described in the work of Noy and Musen (Noy et al., 2004b). According to these authors, this approach can not be applied to compare two versions of ontology because it is possible that two ontologies are identical in terms of their conceptualization, but differ completely in terms of their representation.

The second method is to represent the differences between ontologies in the form of conceptual changes. Such changes specify conceptual relationships between ontological structures in the old and the new version. For example, a conceptual change may declare that a concept A was a sub-concept B in the old version before being placed in the new version. The third method is to represent the differences between ontologies in the form of a container processing operation changes that specify the transition from an old version to the new version of the ontology.

AN APPROACH FOR ONTOLOGY EVOLUTION

In an evolution context, different versions of ontology should coexist. Thus, the evolution of ontology is a part of the ontology versioning process, treating only the determination of the next version. We propose in this work a hybrid approach based on the ontology evolution and versioning to manage the evolution of ontology. This approach is composed of three phases to allow monitoring the evolution of ontology by creating a new version better adapted to the required changes (figure 1).

Phase 1: ***Evolution changes***. This phase consists in identifying the changes and in representing them.

- ***Change identification:*** identify the needs of evolution and the compatible changes to apply on the existing ontology. These modifications are expressed informally by different ontology actors (user, expert and ontology designer).
- ***Change representation:*** in order to resolve changes, they should be identified and represented clearly and in a suitable format. Changes must be

Figure 1. The process of ontology versioning

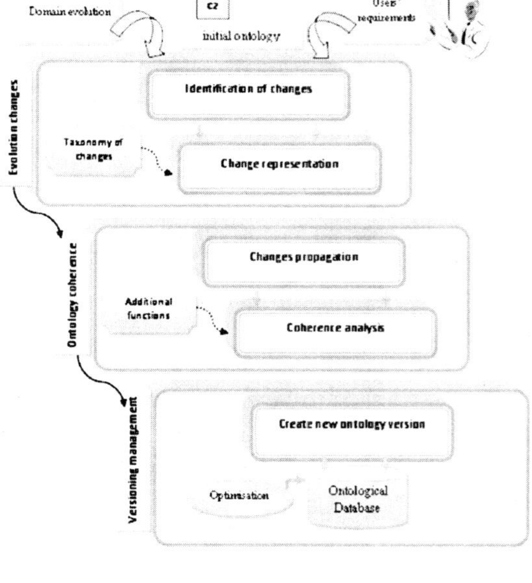

formally expressed through types of changes (Sassi et al., 2008).

Phase 2: ***Ontology coherence***. This phase is composed of change propagation and coherence analysis.

- ***Change propagation***: each type of change can generate additional changes on the other parts of a given ontology. These changes are called derived changes. During this step, it is necessary to determine the direct and indirect types of changes to be applied.
- ***Coherence analysis***: consist in verifying the effects of changes on the ontology coherence. In this step, we provide a proactive approach to anticipate possible inconsistencies and correct them by means of corrective operations.

Phase 3: ***Versioning management***. In this step, a new ontology version is created. At this level, we focus on the relevance to preserve the old version of ontology in addition to the new version or to remove it. This choice is conditioned by the types of implemented changes (subtractive or not subtractive changes). In the case of a subtractive evolution, the old version of the ontology will be stored and added to the *ontological database*.

Evolution Changes

The ontology is composed of concepts, properties, conceptual relationships (simple aggregation, composition, inheritance) and semantic relationships (synonymy, homonymy, equivalence ...) between the ontology concepts. Each ontological entity can be update by simple changes such as: add and delete. Each change need is represented using a specific operator expressing its type of change. Several types of changes express the needs of ontology evolution. The analysis of the consequences of changes allows us to expect the effects of each type on the ontology structure.

In our work, we identify two types of changes: elementary and composite changes (Sassi et al., 2008). *An elementary change* is an ontology change that modifies (adds or removes) only *one* entity of the ontology model. The elementary changes are not sufficient to express the types of evolution requirements. Often, the need of changes may be expressed at a higher level of abstraction. A composite change expresses a sequence of several elementary changes (figure 2). Composite changes specify coarse-grained changes. They are more powerful since an ontology engineer does not need to go through every step of the sequence of elementary changes to achieve the desired effect. For example, for an ontology engineer, it may be more useful to know that a concept was moved from one place in the hierarchy to another than to know that it was detached from one concept and attached to another concept. Therefore, composite changes make the ontology evolution much easier, faster and more efficient since they correspond to the one "conceptual" operation that someone wants to apply without understanding the details (i.e. a set of elementary changes) that an evolution system has to perform.

Ontology Coherence

The preservation of the coherence of ontology requires the preservation of the integrity of the model and the constraints of ontology by preventing the effect of each type of change on the ontology. We defined some coherence rules that evolution changes must respect.

Definition: *a type of change of ontology preserves the coherence of ontology if it preserves all the rules of ontology coherence.*

In an evolution process, the application of types of changes should have as consequence an ontology which is in conformity with the whole coherence rules. In fact, each type of change can generate inconsistencies on the ontology. We

Figure 2. Taxonomy of types of changes

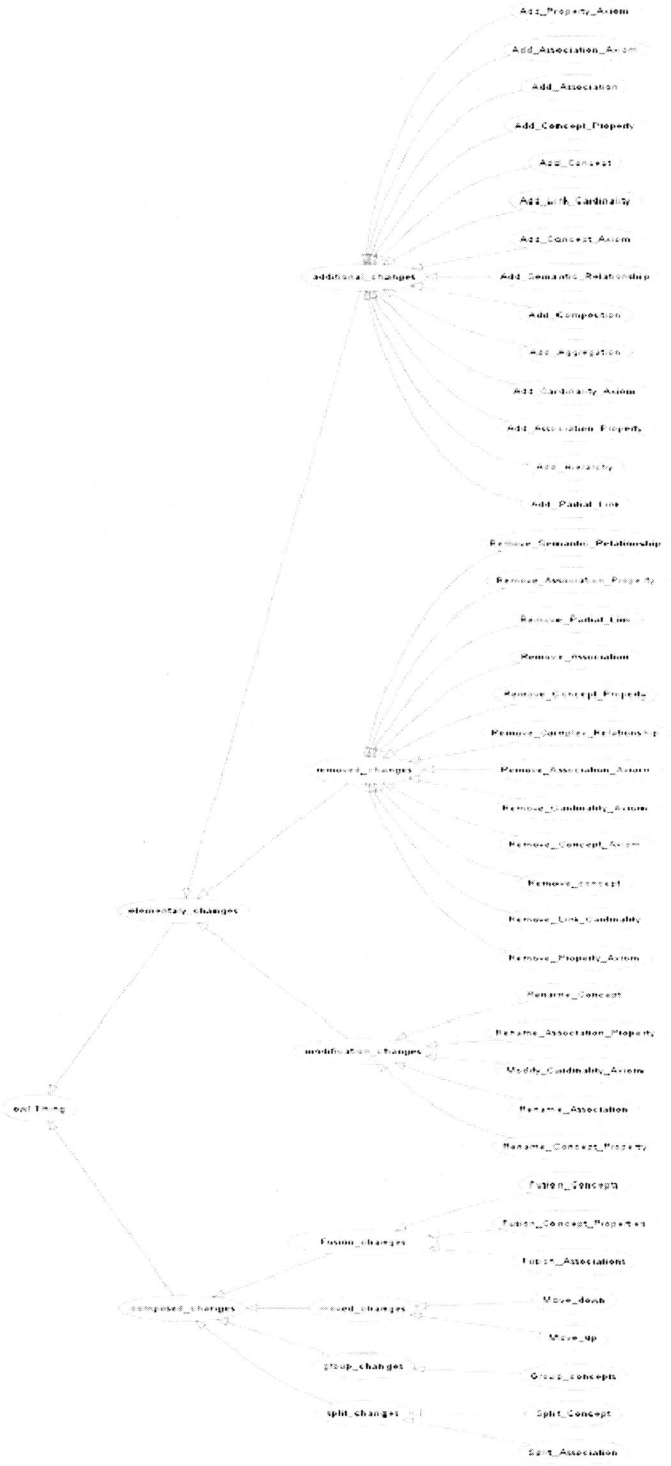

defined corrective operations, able to correct these inconsistencies and to bring ontology in a coherent state.

- Examples of coherence rules:
- Define for each domain the key concepts which should not be removed from the ontology.
- Ontology should not have isolated concepts.
- A concept should not be empty (must comprise at least a property).
- Ontology should not contain semantically contradictory information.
- The semantics of information should not be reversed between ontology versions.
- Ontology should not contain redundancies of data.
- Ontology should preserve the transitivity of hierarchical links.

To identify the adequate corrective operations related to each type of change, it is necessary to determine the types of change likely to generate consistencies and to identify these inconsistencies. If several possibilities exist, i.e., different correct operations can be applied by users with different effects. In order to assist users in the choice of the correct operations to apply, the various operations and their impact on the quality of ontology will be presented to them. Thus, to ensure coherence of the ontology after evolution, we defined coherent changes kit. A coherent change kit is composed of changes and corrective operations that allow to correct the potential inconsistencies caused by the considered change. For every evolution requirement, the corresponding coherent change kit is applied rather than only the type of change. More details about the ontology coherence process are available in (Sassi et al., 2009).

VERSIONING MANAGEMENT

After identifying and representing different types of changes, they will be implemented to create a new version V_{N+1}. This version must be identified in a unique way. The identification is to assign a URI to the ontology and also to each version of ontology. Thus, we must use an agreement for the construction of URIs versions of ontology. We will use the convention as date of creation of the ontology versions for the identification of each one.

Links between Versions

The trace of changes carried out must be kept to indicate how new version V_{N+1} was obtained from the first version V_N and to control the history of ontology evolution. Also, for each created version, we must know the version source of evolution and the following version. This will allow the passage between ontology versions (figure 3).

Figure 3 shows how the various versions are connected by indicating for each one the previous and the following version.

Figure 3. Links between versions

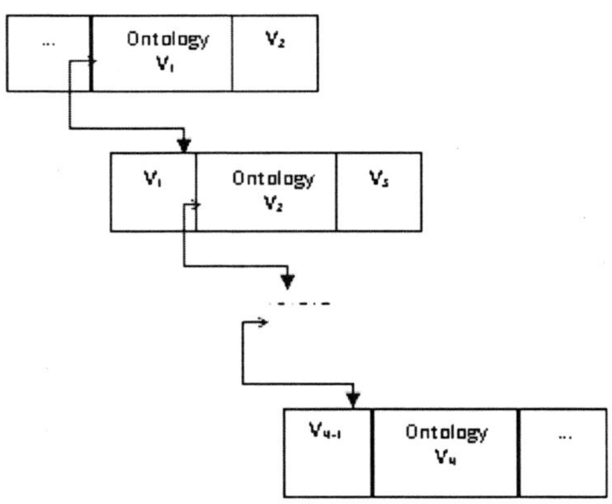

Optimization of the Ontology Versioning Process

We proposed an approach to take into account changes in the ontology based on the creation of a new ontology to take into account the evolution requirements. The new ontology can be seen as an evolved version of the original ontology and is characterized by duration of time (until the application of a new change). The first version of ontology represents the original one not yet updated. A link must be introduced between versions. The latest version is called current version, all other versions are historical versions.

However, the preservation of a large number of versions (especially in highly evolving environments), quickly becomes a handicap because of the space cost and the quantity of information handled. For this reason, we conserve the most relevant versions. The relevance reflects the importance of information and semantics in a version. When the number of versions becomes very important, a mechanism must be associated to reorganize the various versions while deleting the irrelevant one. Deleting a version of the ontology is expressed by removing all ontological entities that are not propagated to other versions of ontology. We fixed a maximum number of versions to store. When this number is reached, we delete the least relevant version. In case of equal value of relevance between two or more versions, we eliminate the oldest one. We also retain the original created version of the ontology. The list of versions kept at a given time represents a list whose first element is the original version and the last is the current version. The other elements are the most relevant versions ordered chronologically.

The evaluation of the relevance of the versions of ontology depends on two criteria: the number of uses of each version and the importance of information contained in each one. The second criterion consists in supporting the conservation of the richest (semantically and structurally) versions. Each version will be compared with the others to check if the information which it contains already exists in the other versions. In such case, its suppression will not involve a loss of information, i.e., all the ontological entities created at a given step of the evolution could be found and recovered starting from the other versions.

Ontology Versions Storage

The most often mentioned objective of versioning is the ability to access instance data via an ontology version which can be different from the one used to describe it. This mechanism of ontology versioning must give transparent access to various versions of ontology. A link must be ensured between the ontological elements (class) in various versions of ontology. This relation is called specification of change in (Klein et al., 2001) and its role is to submit the relationship between the ontological elements of the various versions explicitly. By using this relation, one can identify the changes of any element which plays a part between the various versions.

Dynamic schema evolution in databases is defined as managing schema changes in a timely manner without loss of existing data. Particular problems addressed are cascading changes (changes required to other parts of the schema as a result of a change), ensuring consistency of the schema, and propagation of the changes to the corresponding database.

Although there are significant differences between schema evolution and ontology evolution, many of the methods and technologies developed for schema evolution can be applied or adapted to ontology evolution. Our research in the ontology evolution can benefit from the many research works in database systems. Thus, we aim at exploiting the techniques of data-bases to create versions of ontologies and to develop an Ontological Management System that allows representing, saving, evolving and accessing to ontologies.

However, the usage and content of ontology are more complicated than database schemas:

ontologies incorporate semantic aspects and consequently the data models of ontology are much richer than those of database schemas. These aspects must be taken into account in our future work.

To store ontology versions, we based on the VERSIOS tool (Brahmia et al., 2008). In fact, we express and apply changes on the ontology at the level of the ontology conceptual schema. The analysis of the coherence of ontology versions is also made at the conceptual level regarding the ontology meta-model. However, to store ontology versions, we translate the conceptual schema of ontology to a relational schema in order to store data in conformity with the relational databases rules. The semantic relationships are taken into account as additional relations and/or foreign keys, according to the translation rules from conceptual to relational schemas (Roques et al., 2003).

Architecture of VERSIOS

VERSIOS is a prototype for the management of versions of schemas in a multi-temporal relational environment.This prototype was implemented under the DBMS Oracle 9i by using the language Java (JDK 1.4), which allows its portability. The interaction between the programs, on the one hand, and the meta-database and the databases, on the other hand, is done through the API JDBC.

VERSIOS is structured in three layers (figure 4):

- ¶*Presentation Layer*: it gathers the interfaces intended to the designers and the administrator of Databases.
- *Trade* **Layer:** it contains the modules ensuring on the one hand the management (creation, modification and suppression) of the provisional versions prepared by the database designer and on the other hand the management of versions (application of a version, declaration of the end of a version, etc.) officialized by the database administrator.
- ***Data storage Layer***: it ensures the establishment and the management of meta-base that stores the versions prepared by the database designer, the meta-bases of officialized versions (current and exceeded) and the spaces of data storage of databases, managed in accordance with the official meta-base.

Figure 4. Architecture of versios

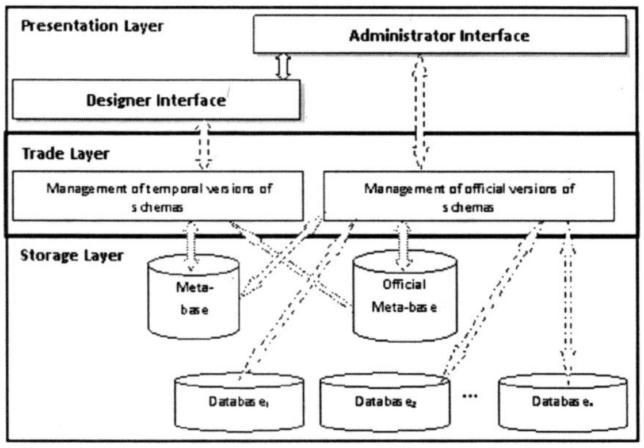

VERSIOS is intended for two categories of users: the designers and the administrator of the database. The designer could (i) prepare, in its environment of work, a version of the schema of a temporal relation (new or already existing), (ii) declare the applicability of a new version or the removal of a version in application, (iii) modify or remove a prepared version. The administrator could (i) apply a version of the schema of a temporal relation prepared by a designer, (ii) finish the application of a version of the schema of a temporal relation.

In addition, VERSIOS offers common functionalities, such as the consultation of the prepared versions and the versions put in application (current and exceeded), and the development of technical dashboards relating to the situation and the evolution of maintained databases (figures 5 and 6).

Figure 5. Definition of a new schema version of relation

Figure 7 shows a class diagram which represents the conceptual meta-model for ontology management. To store the versions of ontology, we should translate this conceptual meta-model in a meta-model which can be managed by an existing DBMS (e.g. Oracle, O2, eXist). The new meta-model can be relational, object-oriented, relational-object or XML.

Figure 6. Application of a new version of relation schema

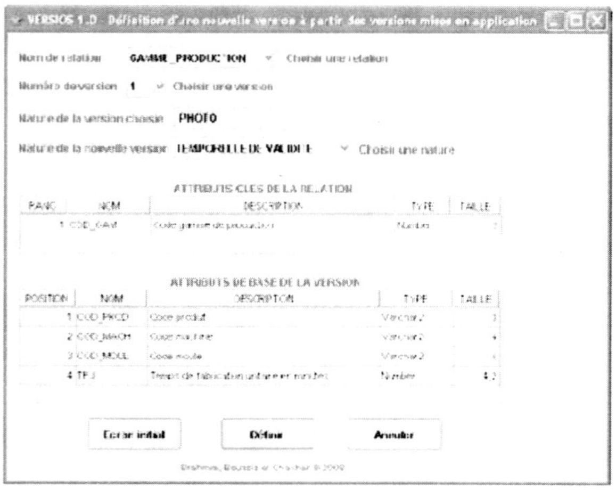

The translation to the relational environment leads to the following meta-relations (Roques et al., 2003):

ONTOLOGY (ONT_ID, C_O, RAS, RA, RH, RS, R, A, AR, AC_O, KEY_C, KEY_SEM, L, #AXIO_ID, #CONC_ID, #PROP_ID, #RS_ID)

CONCEPT (CONC_ID, NAME_C, P_C, #PAR_LIN_ID)

AXIOM (AXIO_ID, EXPRERSSION, AXI_TYPE)

AXIOM_RELATIONSHIP (#AXIO_RS_ID, R1, R2)

AXIOM_CONCEPT (#AXIO_CONC_ID, C1_AC, C2_AC)

AXIOM_CARDINALITY (#AXIO_CARD_ID, C_AC)

PROPERTY (PROP_ID, NAME_P, PROP_TYPE)

PROPERTY_C (#PROP_C_ID, AP, #AXIO_PROP_ID)

PROPERTY_R (#PROP_R_ID, #ASS_RS_ID)

RELATIONSHIP (RS_ID, NAME_RS, C1_RS, C2_RS, RS_TYPE)

ASSOCIATION_RELATIONSHIP (#ASS_RS_ID, P_AR, AC_AR, CL, #AXIO_CARD_ID, #PAR_LIN_ID)

PARTIAL_LINK (PAR_LIN_ID, CONCEPT_PL, REL, AC_PL, #AXIO_CARD_ID)

CONCEPT-AXIOM_CONCEPT (#CONC_ID, #AXIO_CONC_ID)

CONCEPT-RELATIONSHIP (#CONC_ID, #RS_ID)

ASSOCIATION_RELATIONSHIP-AXIOM_RELATIONSHIP (#ASS_RS_ID, #AXIO_RS_ID)

The translation to the XML environment leads to an XML schema expressed in a chosen XML schema language (e.g. XML Schema, DTD, XDR, DSD, RELAX). We choose XML Schema since it is a powerful XML schema language, supports the major features available in other languages, and is backed by the W3C. The translation of the conceptual meta-model presented above into this language leads to the following XML schema:

```
<?xml version="1.0" encoding="UTF-8"
standalone="no"?>
<xsd:schema xmlns:xsd="http://www.w3.
```

Figure 7. The conceptual meta-model for ontology management

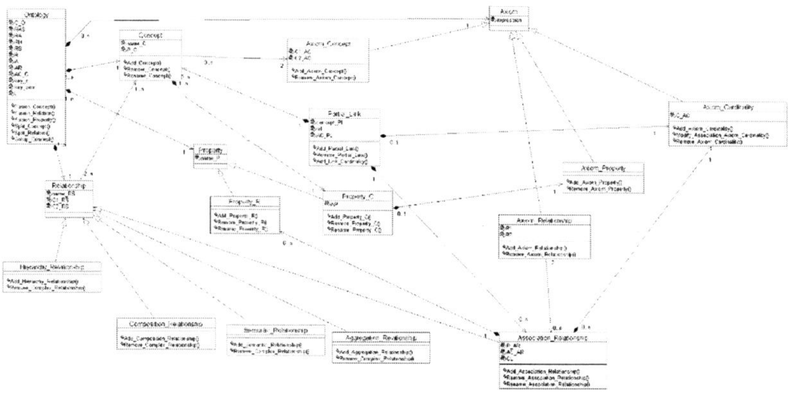

```
org/2001/XMLSchema" attributeFormDefault=
"unqualified"
elementFormDefault="qualified">
  <xsd:include schemaLocation="EcorePrimi
tiveTypes.xsd"/>
  <!-- ==================================
================ -->
  <!-- =====  Complex Type Definitions
-->
  <!-- ==================================
================ -->
  <!-- ~~~~~~~~~~~~~~~~~~~~~~~~~~~~~~~~~~
~~~~~~~~~~~~~~~~ -->
  <!--  Aggregation_Relationship
<<Class>>  -->
  <!-- ~~~~~~~~~~~~~~~~~~~~~~~~~~~~~~~~~~
~~~~~~~~~~~~~~~~ -->
  <xsd:complexType name="Aggregation_Re-
lationship">
    <xsd:complexContent>
      <xsd:extension
base="Relationship"/>
    </xsd:complexContent>
  </xsd:complexType>
  <!-- ~~~~~~~~~~~~~~~~~~~~~~~~~~~~~~~~~~
~~~~~~~~~~~~~~~~ -->
  <!--  Association_Relationship
<<Class>>  -->
  <!-- ~~~~~~~~~~~~~~~~~~~~~~~~~~~~~~~~~~
~~~~~~~~~~~~~~~~ -->
  <xsd:complexType name="Association_Re-
lationship">
    <xsd:complexContent>
      <xsd:extension base="Relationship">
        <xsd:sequence>

<xsd:element minOccurs="0" name="P_AR"
type="String"/>

<xsd:element minOccurs="0" name="AC_AR"
type="String"/>

<xsd:element minOccurs="0" name="CL"
type="String"/>

<xsd:element maxOccurs="unbounded" minOc-
curs="0" name="Property_R"
type="Property_R"/>

<xsd:element maxOccurs="unbounded" minOc-
curs="0" name="Partial_Link"
type="Partial_Link"/>

<xsd:element name="Axiom_Cardinality"
type="Axiom_Cardinality"/>
        </xsd:sequence>
      </xsd:extension>
    </xsd:complexContent>
```

```
  </xsd:complexType>
  <!-- ~~~~~~~~~~~~~~~~~~~~~~~~~~~~~~~~~~
~~~~~~~~~~~~~~~~ -->
  <!-- Axiom <<Class>> -->
  <!-- ~~~~~~~~~~~~~~~~~~~~~~~~~~~~~~~~~~
~~~~~~~~~~~~~~~~ -->
  <xsd:complexType name="Axiom">
    <xsd:sequence>
      <xsd:element minOccurs="0"
name="expression" type="String"/>
    </xsd:sequence>
  </xsd:complexType>
  <!-- ~~~~~~~~~~~~~~~~~~~~~~~~~~~~~~~~~~
~~~~~~~~~~~~~~~~ -->
  <!-- Axiom_Cardinality <<Class>> -->
  <!-- ~~~~~~~~~~~~~~~~~~~~~~~~~~~~~~~~~~
~~~~~~~~~~~~~~~~ -->
  <xsd:complexType name="Axiom_Cardinal-
ity">
    <xsd:complexContent>
      <xsd:extension base="Axiom">
        <xsd:sequence>

<xsd:element minOccurs="0" name="C_AC"
type="String"/>
        </xsd:sequence>
      </xsd:extension>
    </xsd:complexContent>
  </xsd:complexType>
  <!-- ~~~~~~~~~~~~~~~~~~~~~~~~~~~~~~~~~~
~~~~~~~~~~~~~~~~ -->
  <!-- Axiom_Concept <<Class>> -->
  <!-- ~~~~~~~~~~~~~~~~~~~~~~~~~~~~~~~~~~
~~~~~~~~~~~~~~~~ -->
  <xsd:complexType name="Axiom_Concept">
    <xsd:complexContent>
      <xsd:extension base="Axiom">
        <xsd:sequence>

<xsd:element minOccurs="0" name="C1_AC"
type="String"/>

<xsd:element minOccurs="0" name="C2_AC"
type="String"/>
        </xsd:sequence>
      </xsd:extension>
    </xsd:complexContent>
  </xsd:complexType>
  <!-- ~~~~~~~~~~~~~~~~~~~~~~~~~~~~~~~~~~
~~~~~~~~~~~~~~~~ -->
  <!-- Axiom_Property <<Class>> -->
  <!-- ~~~~~~~~~~~~~~~~~~~~~~~~~~~~~~~~~~
~~~~~~~~~~~~~~~~ -->
  <xsd:complexType name="Axiom_Property">
    <xsd:complexContent>
      <xsd:extension base="Axiom">
        <xsd:sequence>

<xsd:element maxOccurs="unbounded" minOc-
curs="0" name="Property_C"
type="Property_C"/>
        </xsd:sequence>
      </xsd:extension>
    </xsd:complexContent>
  </xsd:complexType>
  <!-- ~~~~~~~~~~~~~~~~~~~~~~~~~~~~~~~~~~
~~~~~~~~~~~~~~~~ -->
  <!-- Axiom_Relationship <<Class>>
-->
  <!-- ~~~~~~~~~~~~~~~~~~~~~~~~~~~~~~~~~~
~~~~~~~~~~~~~~~~ -->
  <xsd:complexType name="Axiom_Relation-
ship">
    <xsd:complexContent>
      <xsd:extension base="Axiom">
        <xsd:sequence>

<xsd:element minOccurs="0" name="R1"
type="String"/>

<xsd:element minOccurs="0" name="R2"
type="String"/>

<xsd:element maxOccurs="unbounded" minOc-
curs="0" name="Association_Relationship"
type="Association_Relationship"/>
        </xsd:sequence>
      </xsd:extension>
```

```
    </xsd:complexContent>
  </xsd:complexType>
  <!-- ~~~~~~~~~~~~~~~~~~~~~~~~~~~~~~~~~~
~~~~~~~~~~~~~~~~ -->
  <!--  Composition_Relationship
<<Class>>  -->
  <!-- ~~~~~~~~~~~~~~~~~~~~~~~~~~~~~~~~~~
~~~~~~~~~~~~~~~~ -->
  <xsd:complexType name="Composition_Re-
lationship">
    <xsd:complexContent>
      <xsd:extension
base="Relationship"/>
    </xsd:complexContent>
  </xsd:complexType>
  <!-- ~~~~~~~~~~~~~~~~~~~~~~~~~~~~~~~~~~
~~~~~~~~~~~~~~~~ -->
  <!--  Concept  <<Class>>  -->
  <!-- ~~~~~~~~~~~~~~~~~~~~~~~~~~~~~~~~~~
~~~~~~~~~~~~~~~~ -->
  <xsd:complexType name="Concept">
    <xsd:sequence>

<xsd:element minOccurs="0" name="name_C"
type="String"/>

<xsd:element minOccurs="0" name="P_C"
type="String"/>
      <xsd:element maxOccurs="unbounded"
minOccurs="0" name="Ontology"
type="Ontology"/>
      <xsd:element name="Property_C"
type="Property_C"/>
      <xsd:element maxOccurs="2" minOc-
curs="2"
name="Axiom_Concept" type="Axiom_Con-
cept"/>
    </xsd:sequence>
  </xsd:complexType>
  <!-- ~~~~~~~~~~~~~~~~~~~~~~~~~~~~~~~~~~
~~~~~~~~~~~~~~~~ -->
  <!--  Hierarchy_Relationship  <<Class>>
-->
  <!-- ~~~~~~~~~~~~~~~~~~~~~~~~~~~~~~~~~~
~~~~~~~~~~~~~~~~ -->
  <xsd:complexType name="Hierarchy_Rela-
tionship">
    <xsd:complexContent>
      <xsd:extension
base="Relationship"/>
    </xsd:complexContent>
  </xsd:complexType>
  <!-- ~~~~~~~~~~~~~~~~~~~~~~~~~~~~~~~~~~
~~~~~~~~~~~~~~~~ -->
  <!--  Ontology  <<Class>>  -->
  <!-- ~~~~~~~~~~~~~~~~~~~~~~~~~~~~~~~~~~
~~~~~~~~~~~~~~~~ -->
  <xsd:complexType name="Ontology">
    <xsd:sequence>

<xsd:element minOccurs="0" name="C_O"
type="String"/>

<xsd:element minOccurs="0" name="RAS"
type="String"/>

<xsd:element minOccurs="0" name="RA"
type="String"/>

<xsd:element minOccurs="0" name="RH"
type="String"/>

<xsd:element minOccurs="0" name="RS"
type="String"/>
      <xsd:element minOccurs="0" name="R"
type="String"/>
      <xsd:element minOccurs="0" name="A"
type="String"/>

<xsd:element minOccurs="0" name="AR"
type="String"/>

<xsd:element minOccurs="0" name="AC_O"
type="String"/>

<xsd:element minOccurs="0" name="key_c"
type="String"/>
```

```
<xsd:element minOccurs="0" name="key_sem"
type="String"/>
      <xsd:element minOccurs="0" name="L"
type="String"/>
      <xsd:element maxOccurs="unbounded"
minOccurs="0" name="Concept"
type="Concept"/>
      <xsd:element name="Axiom"
type="Axiom"/>
      <xsd:element name="Property"
type="Property"/>
      <xsd:element name="Relationship"
type="Relationship"/>
    </xsd:sequence>
  </xsd:complexType>
  <!-- ~~~~~~~~~~~~~~~~~~~~~~~~~~~~~~~~~~
~~~~~~~~~~~~~~~~ -->
  <!--  Partial_Link  <<Class>>  -->
  <!-- ~~~~~~~~~~~~~~~~~~~~~~~~~~~~~~~~~~
~~~~~~~~~~~~~~~~ -->
  <xsd:complexType name="Partial_Link">
    <xsd:sequence>

<xsd:element minOccurs="0" name="concept_
PL" type="String"/>

<xsd:element minOccurs="0" name="rel"
type="String"/>

<xsd:element minOccurs="0" name="AC_PL"
type="String"/>
      <xsd:element maxOccurs="unbounded"
minOccurs="0" name="Concept"
type="Concept"/>
      <xsd:element maxOccurs="unbounded"
minOccurs="0" name="Association_Relation-
ship" type="Association_Relationship"/>

<xsd:element name="Axiom_Cardinality"
type="Axiom_Cardinality"/>
    </xsd:sequence>
  </xsd:complexType>
  <!-- ~~~~~~~~~~~~~~~~~~~~~~~~~~~~~~~~~~
~~~~~~~~~~~~~~~~ -->
```

```
  <!--  Property  <<Class>>  -->
  <!-- ~~~~~~~~~~~~~~~~~~~~~~~~~~~~~~~~~~
~~~~~~~~~~~~~~~~ -->
  <xsd:complexType name="Property">
    <xsd:sequence>

<xsd:element minOccurs="0" name="name_P"
type="String"/>
    </xsd:sequence>
  </xsd:complexType>
  <!-- ~~~~~~~~~~~~~~~~~~~~~~~~~~~~~~~~~~
~~~~~~~~~~~~~~~~ -->
  <!--  Property_C  <<Class>>  -->
  <!-- ~~~~~~~~~~~~~~~~~~~~~~~~~~~~~~~~~~
~~~~~~~~~~~~~~~~ -->
  <xsd:complexType name="Property_C">
    <xsd:complexContent>
      <xsd:extension base="Property">
        <xsd:sequence>

<xsd:element minOccurs="0" name="AP"
type="String"/>

<xsd:element maxOccurs="unbounded" minOc-
curs="0" name="Axiom_Property"
type="Axiom_Property"/>
        </xsd:sequence>
      </xsd:extension>
    </xsd:complexContent>
  </xsd:complexType>
  <!-- ~~~~~~~~~~~~~~~~~~~~~~~~~~~~~~~~~~
~~~~~~~~~~~~~~~~ -->
  <!--  Property_R  <<Class>>  -->
  <!-- ~~~~~~~~~~~~~~~~~~~~~~~~~~~~~~~~~~
~~~~~~~~~~~~~~~~ -->
  <xsd:complexType name="Property_R">
    <xsd:complexContent>
      <xsd:extension base="Property"/>
    </xsd:complexContent>
  </xsd:complexType>
  <!-- ~~~~~~~~~~~~~~~~~~~~~~~~~~~~~~~~~~
~~~~~~~~~~~~~~~~ -->
  <!--  Relationship  <<Class>>  -->
  <!-- ~~~~~~~~~~~~~~~~~~~~~~~~~~~~~~~~~~
```

```
~~~~~~~~~~~~~~~~ -->
  <xsd:complexType name="Relationship">
    <xsd:sequence>

<xsd:element minOccurs="0" name="name_RS"
type="String"/>

<xsd:element minOccurs="0" name="C1_RS"
type="String"/>

<xsd:element minOccurs="0" name="C2_RS"
type="String"/>
      <xsd:element maxOccurs="unbounded"
name="Concept" type="Concept"/>
    </xsd:sequence>
  </xsd:complexType>
  <!-- ~~~~~~~~~~~~~~~~~~~~~~~~~~~~~~~~~~~
~~~~~~~~~~~~~~~~ -->
  <!--  Semantic_Relationship  <<Class>>
-->
  <!-- ~~~~~~~~~~~~~~~~~~~~~~~~~~~~~~~~~~~
~~~~~~~~~~~~~~~~ -->
  <xsd:complexType name="Semantic_Rela-
tionship">
    <xsd:complexContent>
      <xsd:extension
base="Relationship"/>
    </xsd:complexContent>
  </xsd:complexType>
</xsd:schema>
```

CONCLUSION

This paper deals with the problem of ontology evolution and versioning. We presented the ontology evolution process based on three phases: evolution changes, ontology coherence and versioning management. The proposed approach allows monitoring the evolution of ontology by creating a new version better adapted to the changes involved. The new version is an update of the original ontology and is characterized by a period of validity that ends when new evolution changes appear. To create a new ontology version, we defined elementary and composed changes to express the different possibilities of evolution requirements. The inconsistencies that can be generated on the ontology after evolution are identified for each type of changes. Some corrective operations are defined and combined with the types of changes to form the coherent changes kits.

Two levels are used in the ontology management approach: the evolution changes are expressed and applied at the conceptual level as well as the analysis of the ontology coherence. However, the storage of ontology versions is made at the logic level based on the relational schema. To store and query the ontology versions, we based on the VERSIOS tool VERSIOS which is developed to manage versions of schemas in a multi-temporal relational environment.¶

REFERENCES

Böhlen, M., Clifford, J., Elmasri, R. A., Gadia, S. K., Grandi, F., Hayes, P., et al. (1998), The Consensus Glossary of Temporal Database Concepts (C. S. Jensen, C. E. Dyreson, Eds.). In O. Etzion, S. Jajodia, S. Sripada, (Eds.), Temporal Databases—Research and Practice, (LNCS vol. 1399, pp. 367–405). Berlin: Springer-Verlag.

Brahmia, Z., & Bouaziz, R. (2008, May 14-16). Schema Versioning in Multi-Temporal XML Databases. In *Proceedings of the 7th IEEE/ACIS International Conference on Computer and Information Science (IEEE/ACIS ICIS 2008)*, Oregon, USA, (pp. 158-164).

Clifford, J., Croker, A., Grandi, F., & Tuzhilin, A. (1995, September 17-18). On Temporal Grouping. In *Proceedings of the International Workshop on Temporal Databases*, Zürich, Switzerland, (pp. 194–213).

De Castro, C., Grandi, F., & Scalas, M. R. (1997). Schema versioning for multitemporal relational databases. *Information Systems, 22*(5), 249–290. doi:10.1016/S0306-4379(97)00017-3

Flouris, G., Plexousakis, D., & Antoniou, G. (2006). Evolving Ontology Evolution. In *Proceedings of the 32nd International Conference on Current Trends in Theory and Practice of Computer Science (SOFSEM 06)*, (pp. 14-29), Czech Republic.

Haase, P., & Sure, Y. (2004). *State of the Art on Ontology Evolution*. Retrieved from http://www.aifb.uni-karlsruhe.de/WBS/ysupublications/SEKT-D3.1.1.b.pdf

Katz, R. H. (1990). Toward a Unified Framework for Version Modeling in Engineering Databases. *ACM Computing Surveys, 22*(4), 375–408. doi:10.1145/98163.98172

Katz, R. H., & Lehman, T. J. (1984). Database Support for Versions and Alternative of Large Design Files. *IEEE Transactions on Software Engineering, 10*(2), 191–200. doi:10.1109/TSE.1984.5010222

Kim, W., Banerjee, J., Chou, H.-T., & Garza, J. F. (1990). Object-Oriented Database Support for CAD. *Computer Aided Design, 22*(8), 469–479. doi:10.1016/0010-4485(90)90063-I

Kim, W., Bertino, E., & Garza, J. F. (1989). Composite objects revisited. In *Proceedings of the 1989 ACM SIGMOD International Conference on Management of Data*, Portland, Oregon, 31 May – 2 June, pp. 337–347.

Klein, M. (2002). *Versioning of distributed ontologies*. EU/IST Project WonderWeb.

Klein, M., & Fensel, D. (2001). Ontology versioning for the Semantic Web. *International Semantic Web Working Symposium*, USA.

Noy, N., & Klein, M. (2004). Ontology evolution: Not the same as schema evolution. *Knowledge and Information Systems, 6*(4), 428–440. doi:10.1007/s10115-003-0137-2

Noy, N. F., & Musen, M. A. (2004). Ontology versioning in an ontology management framework. *IEEE Intelligent Systems, 19*(4), 6–13. doi:10.1109/MIS.2004.33

Özsoyoğlu, G., & Snodgrass, R. T. (1995). Temporal and real-time databases: A survey. *IEEE Transactions on Knowledge and Data Engineering, 7*(4), 513–532. doi:10.1109/69.404027

Sassi, N., & Jaziri, W. (2009). Vérification de la cohérence de l'ontologie par analyse des effets de bords. *Proceedings of the Second International Conference on Web and Information Technology (ICWIT 2009)*, Kerkennah, Tunisia, Sassi, N., Jaziri, W. & Gargouri, F. (2008). Formalisation of evolution Changes to update domain ontologies. In *Proceedings of International Arab Conference on Information Technology (ACIT'2008)*, Hammamet, Tunisia.

Stojanovic, L., Stojanovic, N., & Handschuh, S. (2002). Evolution of the Metadata in the Ontology based Knowledge Management Systems. In *Proceedings of the 1st German Workshop on Experience Management*, (pp. 65 – 77), Germany.

Talens, G., Oussalah, C., & Colinas, M. F. (1993, August 24-27). Versions of simple and composite object. In *Proceedings of the 19th International Conference on Very Large Data Bases (VLDB 1993)*, Dublin, Ireland, (pp. 62–72).

Tansel, A. U., Clifford, J., Gadia, S. K., Jajodia, S., Segev, A., & Snodgrass, R. T. (Eds.). (1993). Temporal Databases: Theory, Design and Implementation. Redwood City, CA: Benjamin/Cummings Publishing Company. Roques, P., & Vallée, F. (2003). UML en action (2nd Ed.). Paris: Eyrolles edition.

Section 4

Ontology Applications and Experiences

Chapter 11
Ontology Learning from Thesauri:
An Experience in the Urban Domain

Javier Nogueras-Iso
Universidad de Zaragoza, Spain

Javier Lacasta
Universidad de Zaragoza, Spain

Jacques Teller
Université de Liège, Belgium

Gilles Falquet
Université de Genève, Switzerland

Jacques Guyot
Université de Genève, Switzerland

ABSTRACT

Ontology learning is the term used to encompass methods and techniques employed for the (semi-)automatic processing of knowledge resources that facilitate the acquisition of knowledge during ontology construction. This chapter focuses on ontology learning techniques using thesauri as input sources. Thesauri are one of the most promising sources for the creation of domain ontologies thanks to the richness of term definitions, the existence of a priori relationships between terms, and the consensus provided by their extensive use in the library context. Apart from reviewing the state of the art, this chapter shows how ontology learning techniques can be applied in the urban domain for the development of domain ontologies.

INTRODUCTION

The activity of knowledge acquisition constitutes one of the most important steps at the beginning of the ontology development process. This activity is essential in all the different methodologies for ontology design as a previous step to the conceptualization and formalization phases. As its name indicates, this activity is devoted to gather all avail-

DOI: 10.4018/978-1-61520-859-3.ch011

able knowledge resources describing the domain of the ontology and to identify the most important terms in the domain (Gandon, 2002).

To alleviate the work of knowledge acquisition there is an emerging interest in the study of methods and techniques for the (semi-)automatic processing of knowledge resources. The main aim of this automatic processing, known as ontology learning (Gómez-Pérez, Fernández-López & Corcho, 2003; Antoniou & van Harmelen, 2004), is to apply the most appropriate methods to transform unstructured (e.g., text corpora), semi-structured (e.g., folksonomies, HTML pages) and structured data sources (e.g., databases, thesauri) into conceptual structures (Gómez-Pérez and Manzano-Macho, 2003). The methods of ontology learning are usually connected with the activity of ontology population which also relies on (semi-)automatic methods to transform unstructured, semi-structured and structured data sources into instance data (i.e., instances of ontology concepts).

Among all the knowledge resources to be used as an input for ontology learning, thesauri, hierarchical classification standards and such taxonomies are likely the most promising sources for the creation of domain ontologies at reasonable costs (Hepp & de Bruijn, 2007). A thesaurus defines a set of terms describing the vocabulary of a controlled indexing language, formally organized so that the a priori relationships between concepts (e.g., synonymous terms, broader terms, or narrower terms) are made explicit. Additionally, the applicability of thesauri for search and retrieval in digital libraries has promoted the creation and diffusion of well-established thesauri in many different domains. Therefore, thesauri reflect some degree of community consensus and contain, readily available, a wealth of category definitions plus a hierarchy.

During the last years and even within the context of digital libraries and information retrieval, there is a general consensus about promoting the use of more elaborated ontologies. Ontologies with formal *is-a* hierarchies, frame definitions or even general logical constraints can improve the performance of retrieval systems. As (Fisher, 1998) remarks, the advantage of doing this transformation work between models is that combining formal ontologies with concept-oriented lexical databases can cover a spectrum of functionality which in principle includes all the traditional services of a classical thesaurus, and can offer more. (Soergel et al, 2004) remark that we need to change the use of thesauri into other more formal representations when at least one of the following requirements is needed:

- Improved user interaction with thesauri on both the conceptual and the term level for improved query formulation and subject browsing, and for more user learning about the domain.
- Intelligent behind-the-scenes support for query expansion, both concept expansion and synonym expansion, within one language and across languages.
- Intelligent support for human indexers and automated indexing/categorization systems.
- Support for artificial intelligence and semantic Web applications.

This chapter will be devoted to review the state of the art in ontology learning from thesauri, and to show how these techniques can be applied to practical examples in the urban domain. This domain is a quite interesting one since, although quite technical, it usually involves a number of different disciplines, from architecture to law or transport engineering. Accordingly, it is not obvious to delineate a priori the set of concepts to be captured in an urban ontology which is well reflected by the intricacy and looseness of existing urban thesauri. Furthermore, urban systems have evolved quite rapidly over the last decades due to growing environmental concerns and metropolisation processes. Urban knowledge is hence a very "active" material, which constitutes a key challenge for existing thesauri and further justifies their transformation into formal ontologies.

This chapter will present two use cases showing the experience of transforming two urban thesauri employed in two closely-related bibliographic databases, called URBAMET and URBISOC. On the one hand, URBAMET is a bibliographic database developed by the French Centre for Urban Documentation for indexing bibliographic notes. The first version of the thesaurus for this bibliographic database was released in 1969 and it contained 2,300 terms. Nowadays, it contains around 4,200 terms and has been used for indexing some 230,000 technical documents related to urban development. On the other hand, URBISOC was developed by the Spanish National Research Council for the indexing of scientific and technical journals on Geography, Town Planning, Urbanism and Architecture. The thesaurus created for this database contains around 3,600 different concepts labelled in Spanish.

The rest of this chapter is organized as follows. The following section analyzes the existent methods for ontology learning classified according to the type of source data. Then, there is a section to describe two experiences of transforming sources to ontologies in the urban domain. Finally, this chapter ends with some conclusions and ideas for future work.

STATE OF THE ART IN ONTOLOGY LEARNING FROM THESAURI

Ontologies are usually classified according to the amount and type of structure of their conceptualization. Following this criteria, (Sowa, 1996) distinguishes two main families: terminological ontologies (also called lexical ontologies) and axiomatized ontologies (also called formal ontologies).

- **Terminological/Lexical ontology:** An ontology whose concepts and relations are not fully specified by axioms and definitions that determine the necessary and sufficient conditions of their use. The concepts may be partially specified by relations such as *subtype/supertype* or *part/whole*, which determine the relative positions of the concepts with respect to one another, but do not completely define them.
- **Axiomatized/Formal ontology:** A terminological ontology whose concepts and relations have associated axioms and definitions that are stated in logic or in some computer-oriented language that can be automatically translated to logic. There is no restriction on the complexity of the logic that may be used to state the axioms and definitions.

The distinction between terminological and formal ontologies is one of degree rather than kind. According to (Lassila, 2001), the difference between the different ontology models is the degree of semantics they provide. Therefore, it is possible to transform simpler models into more formal ones by adding the semantic they lack. Thesauri are terminological ontologies that, as indicated in the introduction, are interesting to be formalized in order to increase the functionality of the systems using them.

Among the works related to the transformation of thesauri into ontologies, we must first cite a set of works that transform thesauri from its native format into Semantic Web languages such as RDF (Resource Description Framework), OWL (Web Ontology Language) or SKOS (Simple Knowledge Organization System). The most basic one is RDF (Manola & Miller, 2004), an XML based language used to make statements about resources in the form of subject-predicate-object expressions. The subject denotes a resource, the object can be a property or another resource, and the predicate expresses the relationship between the subject and the object. SKOS and OWL are both based on RDF but they differ in their scope. On the one hand, SKOS (Lacasta et al, 2007) has been designed to represent terminological ontologies such as thesauri, classification schemes, subject heading lists, taxonomies, and other types of controlled vocabulary. Therefore, it limits the set of types for subjects, predicates, and objects that can be used to maintain a simple representation. On the other hand, OWL (Bechhofer et al, 2004) has been thought for formal models and provides the constructs required to express concepts in description logic (Baader et al, 2003).

The output of these first set of works cannot be directly categorized as a formal ontology because the relations between concepts are still ambiguous, but at least it is a step forward. We move from the term-based approach recommended in ISO standards, in which terms are related directly to one another, into a concept-based approach. In the concept-based approach concepts are interrelated, while a term is only related to the concept for which it stands; i.e. a lexicalization of a concept. Some examples of these works are (van Assem et al., 2004) and (van Assem et al., 2006), which describe the methods applied for transforming MeSH (Medical Subject Headings) and GTAA (Common Thesaurus for Audiovisual Archives) and IPSV (Integrated Public Sector Vocabulary) thesauri into RDF/OWL and SKOS. (Golbeck et al., 2003) describe the conversion of the NCI (National Cancer Institute) thesaurus into OWL format. Every thesaurus concept is translated into an OWL class. This thesaurus holds specific roles or relations (not usual BT/NT, USE/UF, RT) that are translated into specific RDF properties. (Wielinga et al., 2001) describes the transformation of the Art and Architecture Thesaurus (AAT) into an ontology expressed in RDFS. The full AAT hierarchy was converted into a hierarchy of concepts, where each concept has a unique iden-tifier and slots corresponding with the main term and its synonyms.

A second set of works are more ambitious and try to transform the ambiguous BT/NT relations of thesauri into more formal relations such as *is-a* or *part-of* hierarchies. The ISO 2788 guidelines for monolingual thesauri contain a differentiation of the hierarchical relation into generic, partitive and instance relations. However, because the main purpose of thesauri was to facilitate document retrieval, the standards allow this differentiation to be neglected or blurred. But in contrast to thesauri, ontologies are designed for a wider scope of knowledge representation and need all these logical differentiations in relationships (Fisher, 1998). As stated in (van Assem et al., 2006) a major difference between thesauri and ontologies is that the latter feature logical *is-a* hierarchies, while in thesauri the hierarchical relation can represent anything from *is-a* to *part-of*. (Fisher, 1998) identifies several cases where this no differentiation of the BT/NT relation may be a source of fallacies or problems when transforming a thesaurus into an ontology. In particular this work focuses on the problems of identifying subsumption and instance relations behind the ambiguous BT/NT.

For instance, (Hepp & de Bruijn, 2007) describes an algorithm called GenTax to derive an RDF-S or OWL ontology from most hierarchical classifications available in the SKOS exchange format. This algorithm, implemented in a tool called SKOS2Gentax, derives OWL classes from the instances of SKOS concepts and their broader and narrower relations. The algorithm assumes that SKOS concepts can be used in different contexts with varying semantics of the concepts and their relations. The algorithm has two main steps. Firstly, it creates two ontology classes per SKOS concept: one for the context of the original hierarchy, and a related second class (subclass of the first one) for the narrower meaning of the concept in a particular context. Secondly, GenTax inserts *subClassOf* relations between the classes in the original hierarchy context. However, since SKOS broader and narrower relations are translated by default to an *is-a* taxonomy, the output of the algorithm requires many corrections.

Other works use natural language processing to refine the hierarchical relation of thesauri.

For example, (Clark et al., 2000) describes the experience of transforming a technical thesaurus (Boeing's technical thesaurus) into an initial ontology. In particular, this work introduces algorithms for enhancing the thesaurus connectivity by computing extra subsumption and association relations. An important characteristic of technical thesauri is that many concept names are compound (multi-word) terms. They implemented a graph enhancement algorithm for this task, which automatically inferred these missing links using word-spotting/natural language processing technology. Additionally, they also used natural language processing to refine the RT relation into finer semantic categories.

Another remarkable work with the aim of automating the refinement of relations is the one of (Soergel et al., 2004; Kawtrakul et al., 2005). It introduces a semi-automatic approach for detecting problematic relations, especially BT/NT and USE/UF relations, and suggesting more appropriate ones. Upon the experience obtained with the transformation of AGROVOC into an ontology, their approach is mainly based on the identification of patterns and the establishment of rules that can be automatically applied. The method is based on three main ideas. Firstly, they try to find expert-defined rules. Assuming that concepts are associated with categories (e.g., geographic term, taxonomic term for animals ...), experts may define rules that can be generally applied to transform BT/NT relations of concepts under the same category into *is-a* or *part-of* hierarchies. Secondly, they propose noun phrase analysis to detect *is-a* hierarchies. If two terms in a BT/NT relation share the same headword, this relation can be transformed into *is-a*. Alternatively, if two terms are in the same hierarchy of hypernyms in Wordnet, their relation is also transformed into *is-a*. Thirdly, in the case of RT relations, which are usually under-specified relations, refinement rules, acquired from experts and machine learning, are applied. If we identify a particular case of conversion of an RT relation between two terms, we may derive a general rule for the hypernyms of these two particular terms and apply it again to all their hyponyms related through RT.

APPLICABILITY OF ONTOLOGYLEARNING IN THE URBAN DOMAIN

The URBAMET Use Case

This subsection presents a methodology for the analysis of an urban thesaurus that should lead to the incremental development of a shared urban ontology.

URBAMET (Centre de documentation de l'urbanisme, 2009) is a bibliographic database created and maintained by the French Centre for Urban Documentation. The corpus currently includes 280'000 documents and is fed with an additional 8'000 documents each year. Originally designed in 1969 with 2'300 terms, the URBAMET thesaurus currently includes 4'200 terms in French, English and Spanish, which are used to index the document corpus. It is a hierarchy of terms with 24 main themes (top level categories). Figure 1 shows the main themes and an excerpt of the hierarchy of sub-domains in the field of transportation.

It can be observed on the figure that the terms in URBAMET denote either concepts or sub-domains. For instance, the term *"utility vehicle"* may denote a concept that has an intension (the properties of a utility vehicle) and an extension (the set of all utility vehicles). Conversely, the term *"road and traffic"* can hardly denote a concept: it is difficult to figure out what is an instance of *"road and traffic"*. Moreover *"road and traffic"* cannot be considered as a specialization of its parent term *"land transport"*. Hence, the URBAMET thesaurus, at least on the first levels, is mostly a hierarchy of sub-domains. As a consequence, it does not provide an immediate starting point or structural backbone for the construction of an urban ontology.

Figure 1. Main themes (domains) of the urbamet thesaurus

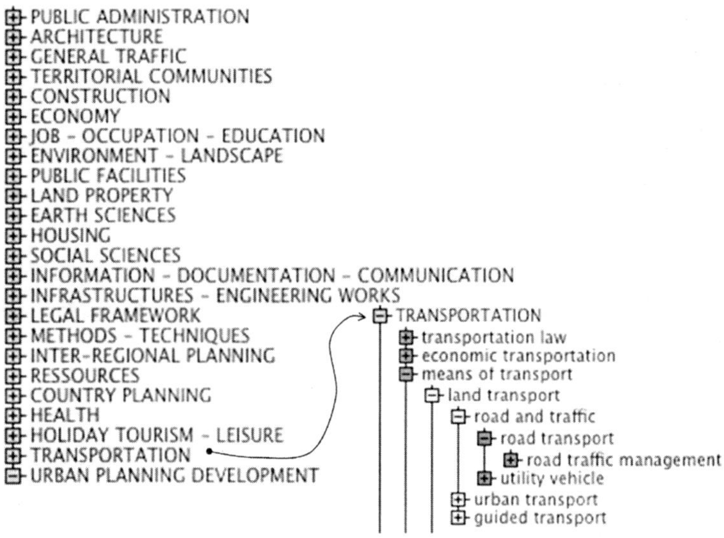

Since the thesaurus cannot be directly used to build an ontology, the proposed methodology relies on the existing thesaurus and the indexed document corpus. The document classification induced by the thesaurus is analyzed with an automated document classifier. This tool operates on document contents. Initially a training corpus is used to teach the classifier the class concepts. Then the tool can start classifying other documents. The analysis is performed in the following steps:

1. Extracting the corpus.
2. Building up the training catalogue.
3. Training and validating a classifier.
4. Generating and analyzing the confusion matrix (list of mistakes made by the classifier).
5. Generating the Top-50 terms (list of the most classifying terms).

For the creation of the training corpus, around 10,000 abstracts, together with their manual assigned themes, have been extracted from URBAMET. This means about 70 indexed words per document and a final vocabulary of about 18,000 words (stems). Then, the classifier built a neural network by reading the training files and applying the Winnow learning technique. Figure 2 depicts an excerpt of the neural network classifier employed to analyze the correspondence between the set of terms in the abstracts of the URBAMET database and the main themes (domains) assigned to the documents. The neural network contains weighted arcs from a word or pair of words to a domain. The weight of term *i* for domain *j* represents how strongly *i* draws a document to *j*. A neural network was trained with 80% of the corpus, using the remaining 20% for testing purposes. The generated classifier found the main domain of each tested document with probability: 59% for the first proposed domain; 16% for 2nd choice; 7% for 3rd choice. Hence, this classifier had a probability of 82% to find the correct class in the first 3 proposals (a random choice would give a 23% probability). We did not carry out extensive evaluation of the classifier because the goal is not to produce the best classifier or to determine its generalization accuracy but to generate information that will help the ontology developer. Indeed, the analysis we present below could be repeated with other classifiers generated with other training sets.

Figure 2. Neural network classifier for the urbamet thesaurus

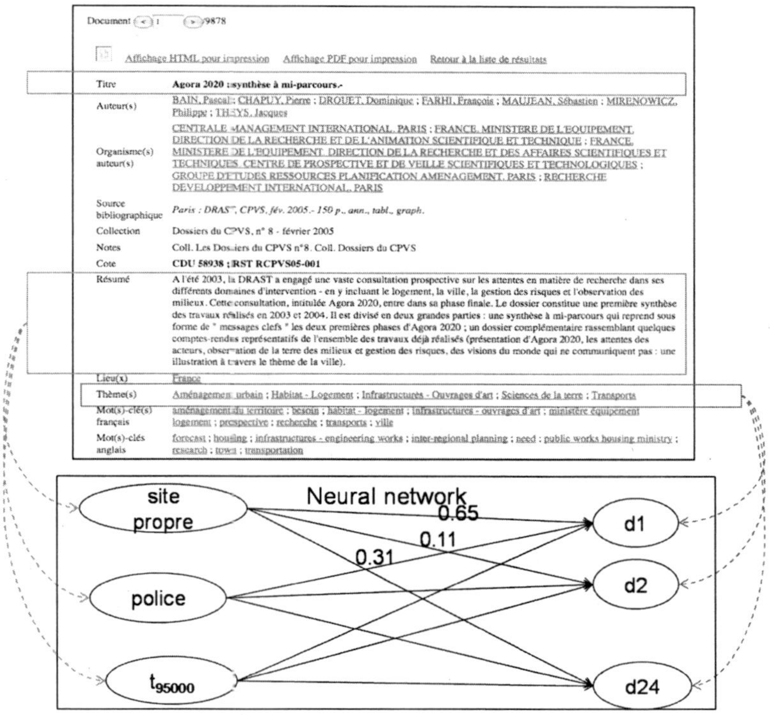

In general, it can be stated that the classifier is effective: the URBAMET classification corresponds to the text contents. However, to detect possible problems and restructure the domains, the methodology proposes an analysis based on the creation of confusion matrices. The objective is to find domains which are poorly classified. Tables 1 and 2 show two excerpts of the complete in-out 24x24 matrix. Each cell M_{ij} represents the percentage of document in domain *i* classified in *j*. Ideally M_{ii} should be 100%. On the one hand, this confusion matrix indicates not clearly separated domains. For instance, see the confusion between *Traffic* and *Transportation* in table 1. Probably, it would be a good idea to merge the domains and create new subdomains. On the other hand, this matrix also shows orthogonal or transverse domains. For instance, *Legal framework* and *Methods* are orthogonal to the other domains (see table 2). Documents are rarely only about *Law* or *Methods*, they usually present legal aspects of *Urbanism*, *Transportation*, etc.

Neural networks can be criticized because of the lack of explanation on the fact that the classifier chose a particular class (as compared to rule-based engines which can explain their reasoning).

Table 1. Excerpt of the confusion matrix. results for "traffic" and "transportation"

In \ out	*Transportation*	*Traffic*	*Tourism*	...
Transport	45%	24%	3%	
Circulation	10%	40%	1%	
Tourism	1%	1%	49%	
...				

Table 2. Excerpt of the confusion matrix. results for "legal framework" and "methods"

In \ out	*Legal*	*Methods*	*Urbanism*	*Infra...*
Legal	*8%*	*3%*	*5%*	*3%*
Methods	*2%*	*4%*	*4%*	*13%*
Urbanism	*17%*	*14%*	*24%*	*4%*
Infrastructure	*2%*	*11%*	*1%*	*22%*

However, it is always interesting to analyze, for each class, the list of the most heavily weighted terms. This list is a selection of the "champion terms" of the domain. They are good candidates to designate concepts of the domain.

Comparing this list with the thesaurus terms provided interesting in-sights into each domain. For instance, among the 50 most classifying terms of the Environment domain, 34 were not found in the thesaurus. Among them one can find terms such as *ecology*, *biodiversity*, *garden*, *swamp*, *guidelines*, *species*, *forestry*. In addition to providing good candidate concepts, this list indicates that this domain has evolved and that this evolution is not reflected in the current thesaurus.

This study shows that even if a thesaurus is far from having a clear ontological structure, it can be exploited to create an ontology if it is considered together with the document corpus it indexes. In fact, such thesauri are the superposition of a domain/sub-domain hierarchy and some ontological elements. Some terms, like means of transport designate at the same time a domain (everything that is related to means of transport) and a concept (a system for transporting people or objects). The confusion between domains and concepts and between different URBAMET domains may be related to the incremental development of the thesaurus, which currently reflects diverging rationales and methodologies. A methodology for extracting an ontology from such a thesaurus must literally *extract* the ontological elements, as was done by selecting the most classifying terms of each domain.

The URBISOC Use Case

This subsection presents the work done to transform an urban thesaurus into a more formalized model. The urban thesaurus employed as use case has been the one developed by the Spanish National Research Council to facilitate the classification of a bibliographic database called URBISOC, which is specialized in scientific and technical journals on Geography, Town Planning, Urbanism and Architecture (Alvaro-Bermejo, 1988). This thesaurus, called URBISOC from now on, contains around 3,600 different concepts labelled in Spanish. It is very close to URBAMET in its scope and use though it has been developed in a much shorter period and its design has been trusted to a reduced number of domain experts in charge of safeguarding its consistency.

Apart from the formalization objective, there were two additional goals for the transformation of this thesaurus. On the one hand, we wanted to convert it into a multilingual resource. On the other hand, we wanted to enrich it with more concepts. It must be taken into account that urbanism can be considered an intersection of different domain areas such as economics, politics culture or civil engineering.

The transformation methodology proposed is based on the merging of source thesauri containing concepts from cross-domain areas. The method takes as input a set of different thesauri and obtains as a result a more consistent and formalized ontology. Figure 3 remarks the 5 main steps involved in this process, showing the inputs and the produced results:

Figure 3. Workflow for the generation of an urban domain ontology

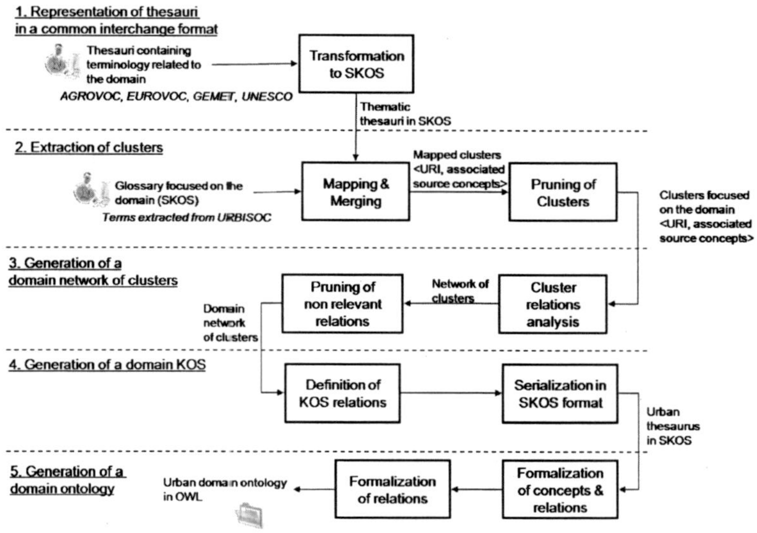

1. Representation of input thesauri in a common format. This task is devoted to the transformation of the input thesauri into SKOS. The thesauri used as input for the method are: GEMET (the GEneral Multilingual Environmental Thesaurus of the European Environment Agency), AGROVOC (the FAO Agricultural Vocabulary), EUROVOC (the European Vocabulary of the European Communities) and the UNESCO thesaurus. They provide a shared conceptualization in the areas of economics, politics, culture and environment.
2. Extraction of clusters. This is the main step and consists in the detection of intersections between concepts in the different input thesauri, through the analysis of their lexical similarities and making profit of their multilingual support. Each set of mapped concepts is grouped into a cluster, which is the name given to a concept in the output ontology. A cluster represents a group of equivalent concepts and is identified with one of the URIs of the original concepts. But previous to this and because top terms of input thesauri are usually very generic, we must identify core concepts specific to the knowledge area in the cross-domain thesauri. Thus, a reduced set of terms in the knowledge area, which is extracted from URBISOC (the *urban planning* term and the recursive chain of *related* and *narrower* terms), is added as another input in the merging process to focus on the domain. Additionally, not all the clusters obtained in the mapping process are useful; many clusters contain terms not related to the desired domain. Therefore, only the clusters that contain a concept from the selected list of terms and those with at least one concept directly related (through *broader*, *narrower* and *related* relations) to another one in a cluster of the first case are kept. The rest are considered not relevant and they are pruned from the system.
3. Generation of a domain network of clusters. This step consists in connecting the clusters previously extracted. The relations between the concepts assigned to the different clusters are converted into relations between the clusters that contain them. The relations between clusters are labelled with: the types of relations, which are derived from the

original types of relations between concepts; and a weight that represents the number of occurrences for each original relation type between the concepts of the inter-related clusters. Besides, it must be noted that the output network may be still too complex and/or contain spurious clusters. Therefore, a process to prune the less relevant relations has been created. This process receives as input the complete network of concepts and a weight threshold to determine if a relation is maintained. All the relations with a weight below the threshold are pruned. After the pruning, all the clusters that do not have at least one relation with another one are also eliminated.

4. Generation of a new thematic thesaurus. The generation of the thesaurus consists in taking the clusters of the network and organizing them into a hierarchical model. The clusters are transformed into concepts of the new thesaurus; one of the labels of the original concepts within the cluster is selected as preferred label. With respect to the thesaurus structure, each relation is marked with the type that has more occurrences. Additionally, those concepts that do not have *broader* relations are marked as top terms. Finally, the generated structure is reviewed to verify that the BT/NT relations structure does not contains cycles. If any cycle is found, it has to be removed. To do so, starting from the top term of the branch that contains the cycle, all the BT/NT relations are reviewed until a concept previously analyzed is found (there is a cycle). Then, the BT/NT relation that has been used to access the problematic concept (concept considered to be the cause of the cycle) is replaced by a *related* relation.
5. Formalization of the thematic thesaurus. In order to transform the obtained thesaurus into a formal model the following tasks have been performed: transformation of each thesaurus concept into a class, identification of relations with higher semantics (*is-a*), and serialization into OWL format. The transformation of the thesaurus concepts into OWL classes requires the transformation of their identifiers, and the registration of their preferred and alternative labels as *rdfs:label* properties. With respect to the relations, to determine which narrower relations can be transformed into *is-a* relations, the following heuristic has been used: "a narrower relation is transformed into an *is-a* relation if the related concepts contain the same headword (substantive) in at least one of their labels (preferred or alternatives) in any of the available languages". The relations that are not transformed are left as they were and must be manually converted.

Table 3 shows the results obtained from the formalization process. Each row represents a possible output ontology according to the weight threshold used for pruning non relevant relations in the third step of the process. For each output ontology, table 1 informs about: the number of

Table 3. Features of the output ontologies

Weight threshold	Nr concepts	Nr *RT*	Nr *BT/NT*	Nr *is-a*	% *is-a*
1	4698	13992	4297	890	20%
2	2568	4150	1480	698	47%
3	1514	1266	857	455	53%
4	1082	566	552	318	57%
5	681	302	317	195	61%

concepts, the number of RT relations, the number of original BT/NT relations obtained in step 4, and the percentage of *is-a* relations derived from these BT/NT relations in step 5.

The output ontologies cannot be considered a final work, but they are a helpful resource for ontologists and experts in the domain. On the one hand, the ontology with weight thresshold 1 can be used to explain the relation between *urban planning* (seed concept in URBISOC to focus on the domain, see step 2) and other concepts that, at first sight, might seem far related. For instance, Figure 4 shows three possible paths to explicate the connection between the *recycling* concept and the *urban planning* concept. On the other hand, as the thresshold increases, the derived ontologies help to discard spurious concepts, which are only considered in some of the original input thesauri. However, it must be taken into account that the increase of weight threshold implies as well a decrease of contributions from other domains. On the positive side, it can be observed that an increasing threshold leads to a higher ratio of BT/NT relations that can be clearly identified as is-a relations.

CONCLUSION

Although to build high-quality ontologies, some kind of manual proc-essing is always required, there are ontology learning methods that can alleviate the task of ontology construction. This chapter has been devoted to present different ontology learning methods that make profit of existent thesauri for building ontologies. In general, we must say that there are not industrial applications for ontology construction. Quite the opposite, depending on the application domain and the availability of sources, ontologists must choose the best ontology learning method in each case.

Additionally, this chapter has shown two different use cases in the context of the urban domain. On the one hand, the URBAMET use case has demonstrated the use of automated classification, with a neural network, for: evaluating the quality of the thesaurus hierarchy (in terms of concept overlap and confusion), finding parts that must be re-structured; and identifying new emerging terms that correspond to new concepts already present in the documents but not yet introduced in the thesaurus. On the other hand, the URBISOC use case has presented a method that takes as input a set of different thesauri and obtains, as a result of a merging and pruning process, a

Figure 4. Three possible paths to understand the connection between recycling and urban planning

Path 1:	Path 2:	Path 3:
recycling	**recycling**	**recycling**
BT	*RT*	*RT*
waste treatment	pollution control	pulp industry
RT	*BT*	*RT*
sewers	environmental policy	industry
RT	*NT*	*RT*
public utility	land use planning	industrial areas
BT	*NT*	*RT*
urban planning	**urban planning**	**urban planning**

more consistent and formalized ontology with multilingual support.

The two proposed approaches complement each other. The first one focuses on identifying the areas of a thesaurus that have to be improved, while the second one focuses on extending a model with concepts and relations of the desired area of interest. These approaches can be used to improve the quality of thesaurus models in the same way as the works described in the state of art section, but they have some relevant differences. In contrast to (van Assem et al., 2004; Golbeck et al., 2003; Wielinga et al., 2001) the proposals described in this paper do not focus on the translation of the models to a more formal representation model, but in the enrichment of the models with additional concepts and relations. In this sense they are more similar to the proposals of (Soergel et al, 2004) and (Kawtrakul et al., 2005) because external knowledge sources are also used to improve the model. The main difference is that their proposals use expert-defined rules to improve the relations and our approaches use knowledge extracted from data collections and models of other domains. Finally, with respect to the methodology used to improve the relationships in our second approach, we must remark the similarity with the work of (Clark et al., 2000), where natural language processing techniques are used to improve the relations. The difference falls in the fact that while they focus on identifying additional relations, our proposal focuses on improving the characterization of the type of the existent relations.

It has not yet been possible to apply the same methodology to both thesauri because required data was not available in the same way in the two use cases. However, applying a neural network classifier approach to both URBAMET and URBISOC document corpus would provide very interesting information about possible convergences and divergences between the content of these bibliographic databases and the ontology resulting from their analysis.

Comparing the concepts obtained in both use cases would further constitute a promising avenue for building multi-lingual ontologies in the urban domain. Multi-lingual issues are especially challenging in this domain. Urban conceptualisations are traditionally based on a mix between science and practice, which makes it hard to find perfect matches for high-level concepts like planning documents or types of interventions between different regions.

Finally ontologies resulting from automatic classification exercises may be submitted to a review amongst a sample of domain experts as well as by persons in charge of the indexation of new documents. This would certainly provide additional information about the relevance of designed methodologies for end-users as well as about possible divergences between domain experts and thesauri designers/managers.

ACKNOWLEDGMENT

This work has been supported by the COST UCE C21 Action (Urban Ontologies for an improved communication in Urban Civil Engineering projects) of the European Science Foundation. The work of J.Nogueras and J. Lacasta has been partially supported by the Spanish Government through the projects España Virtual" (ref. CENIT 2008-1030), TIN2007-65341; and by the Aragon Government through the project PI075/08.

REFERENCES

Alvaro-Bermejo, C. (1988). *Elaboración del Tesauro de Urbanismo URBISOC. Una Cooperación Multilateral*. Salamanca: Encuentro Hispano-Luso de Información Científica y Técnica II.

Antoniou, G., & van Harmelen, F. (2004). Ontology engineering . In *A Semantic Web Primer* (pp. 205–222). Cambridge, MA: Massachusetts Institute of Technology Press.

Baader, F., Calvanese, D., McGuinness, D. L., Nardi, D., & Patel-Schneider, P. F. (Eds.). (2003). *Description Logic Handbook: theory, Implementation, and Applications*. New York: Cambridge University Press.

Bechhofer, S., van Harmelen, F., Hendler, J., Horrocks, I., McGuinness, D. L., Patel-Schneider, P. F., & Stein, L. A. (2004). *OWL Web Ontology Language Reference, W3C, W3C Recommendation 10 February 2004*. Retrieved July 6, 2009, from W3C Web site http://www.w3.org/TR/2004/RECowl-ref-20040210/

Centre de documentation de l'urbanisme. (2009). *URBAMET home page*. Retrieved July 6, 2009, from http://www.urbamet.com/

Clark, P., Thompson, J., Holmback, H., & Duncan, L. (2000). Exploiting a thesaurus-based semantic net for knowledge-based search. In R.S. Engelmore, & H. Hirsh (eds.) *Proceedings of the Twelfth Conference on Innovative Application of Artificial Intelligence* (pp. 988–995). Menlo Park, CA: Association for the Advancement of Artificial Intelligence Press.

Fisher, D. H. (1998). From thesauri towards ontologies? In W.M. el Hadi, J. Maniez & S.A. Pollitt (Eds.), *Structures and relations in knowledge organization: proceedings of the fifth International ISKO Conference* (pp. 18–30). Wurzberg, Germany: Ergon Verlag.

Gandon, F. (2002). *Distributed Artificial Intelligence and Knowledge Management: ontologies and multi-agent systems for a corporate semantic web*. Doctoral dissertation, Institut National de Recherche en Informatique et Automatique (INRIA) and University of Nice Sophia Antipolis, France.

Golbeck, J., Fragoso, G., Hartel, F., Hendler, J., Parsia, B., & Oberthaler, J. (2003). The National Cancer Institute's thesaurus and ontology. *Journal of Web Semantics*, *1*(1), 1–5. doi:10.1016/j.websem.2003.07.007

Gómez-Pérez, A., Fernández-López, M., & Corcho, O. (2003). Methodologies and Methods for Building Ontologies . In *Ontological Engineering* (pp. 107–198). London: Springer-Verlag.

Gómez-Pérez, A., & Manzano-Macho, D. (Eds.). (2003). *Deliverable 1.5: A survey of ontology learning methods and techniques*. Retrieved July 6, 2009, from the OntoWeb Consortium Web site http://www.sti-innsbruck.at/fileadmin/ documents/deliverables/Ontoweb/D1.5.pdf

Hepp, M., & de Bruijn, J. (2007). Gentax: A generic methodology for deriving OWL and RDF-S ontologies from hierarchical classifications, thesauri, and inconsistent taxonomies. In *Proceedings of the 4th European Semantic Web Conference,* (LNCS, Vol. 4519, pp. 129–144). Berlin: Springer-Verlag.

Kawtrakul, A., Imsombut, A., Thunkijjanukit, A., Soergel, D., Liang, A., Sini, M., et al. (2005). *Automatic Term Relationship Cleaning and Refinement for AGROVOC, Workshop on The Sixth Agricultural Ontology Service*. Vila Real, Portugal.

Lacasta, J., Nogueras-Iso, J., López-Pellicer, F. J., Muro-Medrano, P. R., & Zarazaga-Soria, F. J. (2007). ThManager: An Open Source Tool for creating and visualizing SKOS. *Information Technology and Libraries*, *26*(4), 40–53.

Lassila, O., & MacGuinness, D. (2001). *The Role of Frame-Based Representations on the Semantic.* Web Knowledge Systems Laboratory Report. Retrieved July 6, 2009 from http://www.ksl.stanford.edu/KSL_Abstracts/KSL-01-02.html

Manola, F., & Miller, E. (Eds.). (2004). *RDF Primer, W3C, W3C Recommendation 10 February 2004*. Retrieved July 6, 2009, from W3C Web site http://www.w3.org/TR/2004/REC-rdf-primer-20040210/

Soergel, D., Lauser, B., Liang, A., Fisseha, F., Keizer, J., & Katz, S. (2004). Reengineering Thesauri for New Applications: the AGROVOC Example. *Journal of Digital Information*, *4*(4), 1–19.

Sowa, J. F. (1996) Ontologies for Knowledge Sharing. *Terminology and Knowledge Engineering Congress (TKE '96).*

Studer, R., Benjamins, V. R., & Fensel, D. (1998). Knowledge engineering: principles and methods. *Data & Knowledge Engineering*, *25*(1-2), 161–197. doi:10.1016/S0169-023X(97)00056-6

van Assem, M., Malaisé, V., Miles, A., & Schreiber, G. (2006). A Method to Convert Thesauri to SKOS. In *Proceedings of the 3rd European Semantic Web Conference*, (LNCS, Vol. 4011, pp. 95-109).

van Assem, M., Menken, M. R., Schreiber, G., Wielemaker, J., & Wielinga, B. (2004). A method for converting thesauri to RDF/OWL. In S.A. McIlraith, D. Plexousakis, & F. van Harmelen, (eds), *Proceedings of the Third International Semantic Web Conference*, (LNCS, Vol. 3298, pp. 17-31).

Wielinga, B. J., Schreiber, A. T., Wielemaker, J., & Sandberg, J. A. C. (2001). From Thesaurus to Ontology. In *Proceedings of the 1st international conference on Knowledge capture* (pp. 194–201). New York: Association for Computing Machinery Press.

Chapter 12
Applications of Ontologies and Text Mining in the Biomedical Domain

A. Jimeno-Yepes
European Bioinformatic Institute, UK

R. Berlanga-Llavori
Universitat Jaume I, Spain

D. Rebholz-Schuchmann
European Bioinformatic Institute, UK

ABSTRACT

Ontologies represent domain knowledge that improves user interaction and interoperability between applications. In addition, ontologies deliver precious input to text mining techniques in the biomedical domain, which might improve the performance in different text mining tasks. This chapter will explore on the mutual benefits for ontologies and text mining techniques. Ontology development is a time consuming task. Most efforts are spent in the acquisition of terms that represent concepts in real life. This process can use the existing scientific literature and the World Wide Web. The identification of concept labels, i.e. terms, from these sources using text mining solutions improves ontology development since the literature resources make reference to existing terms and concepts. Furthermore, automatic text processing techniques profit from ontological resources in different tasks, for example in the disambiguation of terms and the enrichment of terminological resources for the text mining solution. One of the most important text mining tasks that exploits ontological resources consists of the mapping of concepts to terms in textual sources (e.g. named entity recognition, semantic indexing) and the expansion of queries in information retrieval.

DOI: 10.4018/978-1-61520-859-3.ch012

INTRODUCTION

The development of ontologies is a time-consuming task. The formal representation of domain knowledge is one important step and the acquisition and confirmation of terms representing relevant concepts is another one. If textual and semantic resources such as the scientific literature can be exploited for the design and development of ontologies, work efficiency can be improved.

Large document collections like the World Wide Web or the biomedical scientific literature (Medline) are readily available, however neither one is currently available in a semantically structured representation. The extraction of information from both sources requires text mining solutions and could be beneficial for the ontology development since such literature resources contain a significant portion of the domain knowledge.

Text mining does not only contribute to the generation of ontological resources, but also text mining profits from the use of ontological resources. For example, the use of an ontology can support the disambiguation of terms that represent different concepts and furthermore, the use of synonyms linked to individual concepts can enlarge the coverage of a text mining solution. The most basic text mining tasks that integrate ontologies into text mining are the mapping of concept labels to terms in textual sources (e.g. named entity recognition) and the expansion of query terms in information retrieval. As shown later, the combination of ontologies with text mining solutions leads to benefits in different IT approaches and their combined exploitation is developing into a dedicated research topic.

In this chapter, we analyze the symbiosis between text mining and the ontological field, namely: which text mining techniques are profitable for ontological development and, how ontologies can enhance the coverage and precision of existing text mining solutions. In this context, we introduce and discuss the different techniques from both fields as well as their interaction. Accordingly, the chapter is divided into two main parts: one part is dedicated to explore the relevance of ontologies for text mining, and the other one to describe the contributions of text mining solutions to the ontology development lifecycle.

The contents of this chapter are specially aimed to bioinformatics and computer science researchers, showing them the state-of-the-art and the new opportunities that are arising from the combination of text mining and ontology-based technology.

USE OF ONTOLOGIES IN TEXT MINING

Text mining is the processing and analysis of data stored in textual representation. Text mining extracts facts from text to fill databases or to improve exploitation of document content through better retrieval or navigation in the document. Text mining consists of two main sub-tasks: information retrieval (IR) and information extraction (IE). IR techniques aim at recovering relevant documents from a large textual repository in order to satisfy the user's information need expressed by his retrieval query. Information extraction solutions glean facts from a set of documents.

In text mining systems, IR and IE are usually interlinked (e.g. Figure 1). IR is used to retrieve relevant documents or parts of the document (e.g., paragraphs or sentences) to be possibly further processed by IE methods. The other way around, IE may feed identified results into an IR system to produce better results. For example, the IR system can generate an enriched index based on the results from the IE system to allow better performance. In the following sections, we present in more detail all involved text mining components and demonstrate different usages and exploitations of the ontological resources to this end.

Figure 1. Information retrieval and information extraction interaction

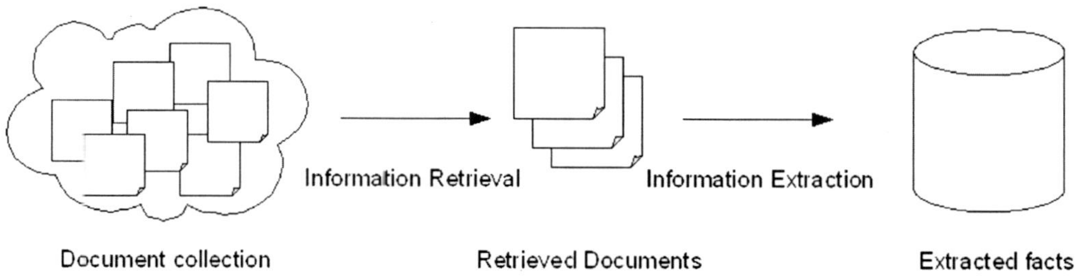

ONTOLOGIES AND INFORMATION RETRIEVAL

The main task of an Information Retrieval (IR) system is the recovery of documents from a collection of documents to respond to the user's information need with the most relevant set of document available. Figure 2 shows the schema of a typical setup of an IR system. The input to the system is a collection of documents and a query. The output is a selection of documents (ranked or not) that matches the relevance criterion for retrieval. Relevance feedback, i.e. user feedback on the relevance of the retrieved documents, might improve the retrieval performance.

Any IR system pre-processes the documents and thus prepares the retrieval for high performance concerning time. All documents are tokenized and the tokens are normalized. Usually the normalization transforms the characters into lower case representations and the word form is stemmed using publicly available stemmers (e.g. Porter, Levin or Krovetz). Occasionally the normalization decreases the retrieval performance, in particular if the morphology of a term is discriminative, i.e. the upper case and the lower case version of a term have different meanings. Therefore, special care is required to adopt the selected normalization to the required needs. Often a list of stop words is used to filter out tokens that are either not discriminative such as general English terms or decrease the discriminative performance of the retrieval index. Such words have low information content and thus add unnecessary noise to the index (e.g. prepositions). Standard stop word lists are available for many languages.

Once the documents are pre-processed, they are represented with the selection of words that have been identified in the text (called bag of

Figure 2. Ir system workflow

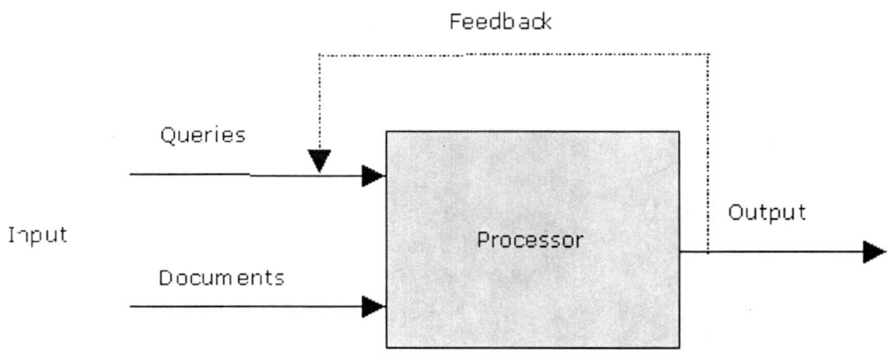

words (BOW)). The outcome of the analysis is a collection of terms from the documents and a long list (called the index or document index) which for each document has a link from the terms in the list to the document, i.e. upon querying the index any term can be resolved to the selection of documents that make reference to the term.

Any query fetched to the retrieval engine is decomposed in a similar way. Thus, the query is also expressed as a BOW, and it is resolved against the index generating a list of documents complying with it. Figure 3 shows the tokenization and normalization of two documents that serve as a sample:

D1: Inhibition of apoptosis by Heliothis virescens ascovirus.

D2: Role of apoptosis in biology and pathology.

In principle, the document retrieval solution uses the tokens from the query to search through all document representations, i.e. through all BOW representations of the documents. This process is very time-consuming since the process has to query across all BOWs of the whole document collection for each query. To improve the speed, the BOWs are processed and an *inverted index* is built. An inverted index is a structure that links all terms to documents being relevant to the term. Efficient data structures for the retrieval of terms exist such as the TRIE structures that locate terms in the index efficiently. The inverted index presented in Figure 3 demonstrates how the query containing the term *apoptosis* is resolved to the two relevant documents D1 and D2.

The bag of words approach is often the most effective approach for document retrieval but there are conditions under which a solution based on an ontological resource could produce even better retrieval performance.

It is worth mentioning that some search engines go beyond this simple representation of documents and terms and adopt some well-known IR model. Basically, an IR model maps queries and documents to a known representation space where it is possible to rank documents according to queries. For example, the vector space model (VSM) represents the queries and documents as multidimensional vectors, where each dimension represents a possible token of the collection. Thus, the rank of a document d w.r.t. a query q is derived from the distance between them (e.g. cosine). Alternatively, Language Models are aimed

Figure 3. Inverted index

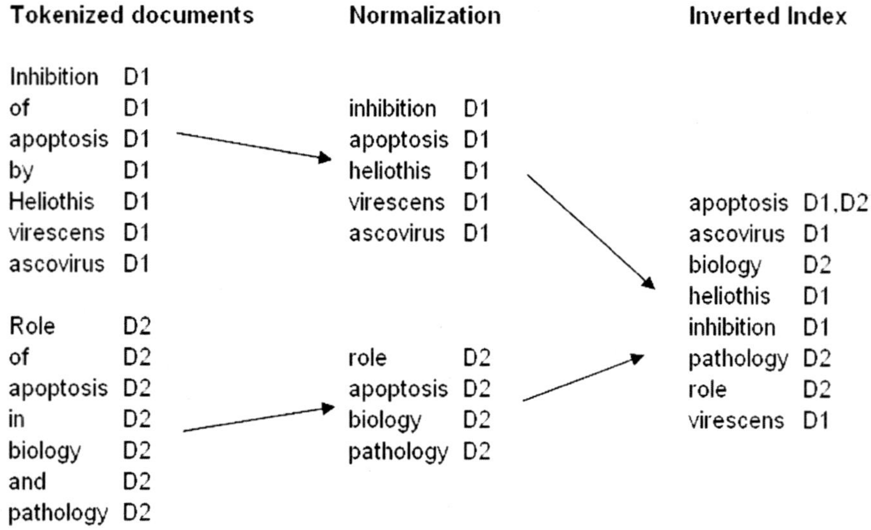

at estimating the similarity of the (unigram) word models of both d and q.

In the next sections, we review the use of ontologies in information retrieval to support query reformulation, semantic indexing and improved navigation of the search results.

QUERY REFORMULATION

In Query Reformulation the query is transformed into a new representation. Query Reformulation refers to the different operations that are applied to the original user query in order to improve its performance. A user facing an information retrieval system has to consider how to transform an information need into a query representation in the system's query language in such a way that it is effective in terms of retrieval performance.

QUERY EXPANSION

Query expansion (QE) has been studied from the very beginning of information retrieval research. QE uses thesauri and statistical techniques for the benefit of improved IR. QE techniques can be distinguished concerning the original source of the expansion terms (Efthimiadis, 1996), i.e. whether they have been gathered from a document collection or from other knowledge sources.

Collection Dependent QE

In relevance feedback approaches, the user proposes a query and after retrieval, the user marks relevant documents. Then all tokens are selected from all relevant documents D_r and from all non-relevant documents D_{nr}. Then both sets of tokens are integrated into an approach that generates an expanded query using the method from Rocchio or Ide (Buckley et al., 1994, 1995). According to Rocchio the expanded query is defined as follows:

$$\vec{q_m} = \alpha \vec{q_0} + \beta \frac{1}{|D_r|} \sum_{\vec{d_j} \in D_r} \vec{d_j} - \gamma \frac{1}{|D_{nr}|} \sum_{\vec{d_j} \in Dn_r} \vec{d_j}$$

where $\vec{q}_0$ represents the vector of the original query that is transformed with the input from the token vectors of relevant documents D_r and irrelevant documents D_{nr}. The parameters (α, β and γ) are used to adjust the relevance of the original query, the vector of the relevant documents and the vector of the non-relevant documents respectively.

Gathering feedback from a user is costly and is very often not available. To overcome this problem, pseudo-relevance feedback (or blind feedback) has been introduced. It regards the first *k* documents from the retrieved set as relevant before applying relevance feedback. Even though this approach has shown to be effective, it sometimes can decrease the retrieval performance. (Mitra et al., 1998) suggests filtering the first k retrieved documents before applying pseudo-relevance feedback. (Lavrenko and Croft, 2001) have proposed a similar relevance feedback method for language models.

Other solutions use statistical evidence from document collections and are called global strategy techniques. For example, the co-occurrence of terms can be an indicator of relevance of terms (van Rijsbergen, 1979). In this approach, one important parameter is the window size that determines the selection and number of words that have to be considered for the co-occurrence estimates. (Qui and Frei, 1993) propose a statistical thesaurus that enables selection of words for query expansion.

Unfortunately, co-occurrence does not identify associations between semantically related words that do not appear in the same document such as *astronaut* and *cosmonaut*. Latent semantic indexing (LSI) (Deerwester et al., 1990) solves this problem, i.e. it separates the textual features categories that are separated by distinguishable "dimensions". However, the dimensions gathered from LSI are by definition meaningful concerning the mathematical approach but are difficult to understand semantically. Regarding Latent Semantic Analysis (LSA), which adopts a probabilistic model, the problem stems from the definition of the underlying word distribution (Roelleke, 2003). However, the main drawback of both, LSI and LSA, is that they tend to be computationally expensive and induce time-consuming updates to the index.

Furthermore, co-occurrences can be estimated given the retrieved documents (local strategy) which are examined at query time to determine words for expansion. Clustering techniques are used to find clusters from which the words will be selected. As it is applied at retrieval time, the fetching and processing of the documents make it difficult to adapt to on-line systems. A combination of global and local analysis is proposed in (Xu and Croft, 1996).

Knowledge Source Dependent

As mentioned before, knowledge sources have been widely exploited for information retrieval tasks. (Voorhees, 1994) used WordNet and documents from a TREC collection to achieve manual query expansion. The selection of the terms is based on the extended vector space model by (Fox, 1980) which is a convenient solution to combine multiple sources. She found that query expansion improves the performance in the case of short queries. However, she also measured that, in principle, larger queries can better specify the information need, but leading to a decrease in the retrieval performance. This is mainly due to query drift, i.e. added terms contribute to the overall ambiguity induced by every single term and thus increase the likelihood for the inclusion of documents that do not properly fit to the query. In short, she noticed that each query term contributes its specific peculiarities linked to its inherent ambiguity and thus influences the performance of the IR problem. (Bodner and Song, 1996) have shown an improvement in IR performance if manual query expansion is applied using general and domain-specific knowledge sources.

(Nie et al., 2002) integrated the use of logical operators into a vector space representation and exploited the content of WordNet in this approach. They added the expansion terms into the original query similar to the approach proposed by Voorhees but then applied fuzzy logic when resolving the query. Finally, they could not demonstrate that the addition of the expansion terms into the original query improves retrieval performance because the added terms introduce too much emphasis in the query and the retrieval. This lead to the final result that the Boolean "OR" operator has to be used in combination with the added terms to improve the performance because the added terms have to act as alternatives to the original query terms. Only this approach does not put significant bias on the user needs due to the expansion terms.

For the biomedical domain, (Aronson and Rindflesch, 1997) and (Liu and Chu, 2005) explored on retrieval improvements for documents that refer to genes and proteins. They made use of UMLS and the content from the LocusLink database to achieve their goal. (Chu et al., 2002) have improved their query expansion through the integration of terms from UMLS and from the OSHUMED document collection into the queries. They have proposed to fit user queries to template specific queries in which the expansion terms are selected according to their relation to the original query terms. Such relations between terms have been identified either in UMLS or in the document collection.

QUERY REFINEMENT

In the case of large document collections, like Medline or the Web, a query may return a large numbers of documents, in particular if the query contains polysemous terms. This induces the interest in increasing the precision of the IR system to avoid an *information overload* generated by an unspecific initial user query. If an unspecific query allows different interpretations, then it results to a large number of possible resolutions of the query and therefore retrieves a large number of documents. One particular case is the resolution of acronyms to several long-forms, i.e. theirs different expansions. For instance, if there is an interest in documents from Medline with reference to *APC*, the IR system will produce documents that report on the gene *adenomatous polyposis coli (APC)* or on the biological structure *anaphase-promoting complex (APC)*. In both cases, the user query does not clearly specify which result is the expected one.

Query refinement (QR) has the goal to optimize the query in such a way that the retrieved set is specific to one of the many subsets associated to the unspecific query. In query refinement, the query is modified in order to filter out irrelevant documents and thus to increase the precision. Different techniques exist that make assumptions on the appropriate interpretation of the user query taking into consideration content of the documents in the underlying collection.

Term polysemy is not the only reason for a strong increase of false positive results upon query resolution. Another reason is the condition that a document collection contains a large number of documents that either report on very general topics or on very specific ones. Therefore, the query terms do not match precisely to the main concepts contained in the documents and thus the query produces irrelevant results.

In all these cases, QR solutions make suggestions for improvements to the query. One suggestion is to select a subpart of the terms in the query or to specify better the meaning of the given terms in the query. Unfortunately, these solutions do not always provide satisfactory results, since they again produce side effects. Proposed techniques used in IR solutions include the one from Scatter/Gather (Hearst et al., 1995) where the retrieved documents are clustered and an informative label is attached to each cluster. The user then selects a label and the associated cluster as his query refinement and receives the documents in the cluster as the retrieval result. (Pratt et al., 1999, 2000) have investigated into research on the classification of the user queries and on the expected clustering results according to the different categories. For the biomedical domain, the IR solution called SOPHIA represents a similar solution (Patterson et al., 2005).

(Gauch and Smith, 1993) developed an interactive expert system for QR but could not prove significant improvements. Other systems that rely on the integration of ontological resources, e.g. OntoRefiner (Safar et al., 2004), post-process the retrieved documents to display them in an alignment along a given lattice (e.g. Galois Lattice).

In several studies, the researchers rely on building a statistical user model that exploits the information provided by the user during the use of the system instead of filtering the terms from the document collection (called the document model). (Amati and Bruza, 1999) analyzed the changes from the original query to the refined query upon user input and identified belief changes in the sets of query terms that they gathered from the query log. This resulted in the generation of user profiles instead of standard information filtering, where information need remains stable and the documents are gathered from a continuous document stream in contrast to a fixed document collection.

SEMANTIC INDEXING

A number of phenomena like term ambiguity are difficult to cope with during the retrieval of documents, even upon integration of query reformulation. We have to remember that the use of bag of word approaches excludes any exploitation of the structure of the text, for example the word order. However, in many cases the co-occurrence and order of two words can help to better disambiguate the correct meaning of the words. The overall goal would be for example, to represent an ambiguous word with all of its senses. This leads into the ambition to represent concepts in the index instead of simple terms.

The semantic index that refers to concepts from an ontological resource in addition to lexical tokens (e.g. words and phrases) should enable disambiguation and normalization of terms to concepts. The ontological and textual sources are combined during the indexing process to improve the specificity of the tokens and to normalize terms to the same concept if the identified terms are synonymous.

Information extraction components like named entity recognition (NER) are used to identify the concepts in the documents; which are integrated into the semantic index. The query has to be mapped in a similar way to a conceptual representation, i.e. the concept identifier. This last step can be problematic if the query context is not sufficient to find the appropriate concept identifier or if several alternatives have to be resolved.

Several examples of semantic indexing have been proposed, i.e. (Basiz et al., 2004) for general document collections, (Ambroziak and Woods, 1998) for technical documents, (Mayfield, 2003) to integrate inference in the retrieval process and (Rebholz et al., 2007) for biomedical scientific literature.

ORGANIZATION OF SEARCH RESULTS

The large amount of documents returned by a retrieval system can be organized using a categorization scheme according to available taxonomic resources. The categorization scheme enables improved navigation of the search results based on the underlying taxonomic resources (e.g., ontologies), for example, the user can address directly and explore the documents attributed to the different subtopics and can match them to his information need. One example of an application that post-processes the retrieval results based on ontological and terminological resources is EBIMed (Rebholz et al., 2007). This solution analyzes the complete set of retrieved documents and identifies the associations between the concepts contained in the documents (co-occurrence). All associations are then delivered in a table and, for every association it is possible to recover the documents that support the evidence. In another solution, GoPubMed (Doms and Schroeder, 2005), retrieved documents are categorized into sets labeled with concepts that best represent each set. All labels come from existing taxonomic resources such as MeSH and the Gene Ontology (GO). This categorization enables the user to navigate to his/her preferred topics referring to a set of retrieved documents and to explore the used taxonomy overall to identify other subtopics of interest.

Both applications put a focus to the use of domain knowledge represented by taxonomic resources that are well established in the scientific community and both move one step further in contrast to applications where the organization of the documents depends solely on the retrieved document set without use of external resources.

DISCUSSION

The presented techniques use ontologies and terminological resources in different ways. Query reformulation is intended to provide and improved representation of the information need but might be sensible to query drift due to term ambiguity. Semantic indexing provides an improved representation of the documents, but this technique may ignore important parts of the document by transforming the document into the conceptual space. Finally, the organization of the search results provide a better organization of the retrieved documents but this organization may not suit the user either because of the complex task of categorizing documents or because the categories are too shallow or too much specific.

ONTOLOGIES AND INFORMATION EXTRACTION

Information extraction (IE) is the engineering science leading to solutions that gather facts from unstructured textual sources (e.g. documents). IE processes text to filter out terms, preferably named entities, to identify characteristics of named entities and to extract relations between named entities, which form a subpart of events. The complexity of fact representations in the text can be high, and advanced information extraction solutions make use of a large number of semantic and computational resources to enable precise identification of facts.

IE is different to IR in the sense that IE extracts facts that then are used as individual entities, whereas IR does not deliver content from the documents but the whole document as such. The following example sentence demonstrates an IE task. The information extraction need is expressed as a template that has to be filled with content from the document, i.e. the template slots have to be filled with pieces of text from the document by the IE system. Several steps in the analysis have to be performed (cf. Figure 4) to produce a structured output expressed by the template.

The development of an IE system either focuses on solutions proposed by research in either knowledge engineering (KE) or machine learning (ML) approaches. Combined solutions also exist (Rosenfeld et al., 2004) but tend to have a strong focus in one of the two trends. In the KE approach, the IE patterns (e.g., syntactical or grammar rules) are developed by a knowledge engineer, usually a domain expert. The quality of the produced rules depends on the ability of the engineer to render the language features in the rule-based system.

Machine learning approaches use available training data to infer extraction patterns. In other words, the domain expert annotates the representations for the information need in the documents and the ML expert trains a classification technique to reproduce the feature sets at the appropriate quality level. The annotation of the features of interest in the documents is the main challenge of this approach. A sufficient amount of training data is required to produce a classifier that identifies the complexity of language patterns with the intended quality. Compared to the knowledge engineering approach we only require training data, oftentimes cheaper to collect.

The KE approach needs to rebuild the set of extraction patterns if the interest moves to a different domain. On the other hand, the domain knowledge of the knowledge engineer may produce extraction patterns that provide better performance than ML ones. The patterns are built based on experience and are easier to maintain for the same task since the language rules express explicitly the logic in the IE approach. Therefore, the patterns from the KE approach form a contrast to the patterns produced by the ML approach, i.e. in the latter case the patterns are only implicitly encoded into the final solution and thus cannot be filtered out by an approach of reverse engineering.

Figure 4. Information extraction example

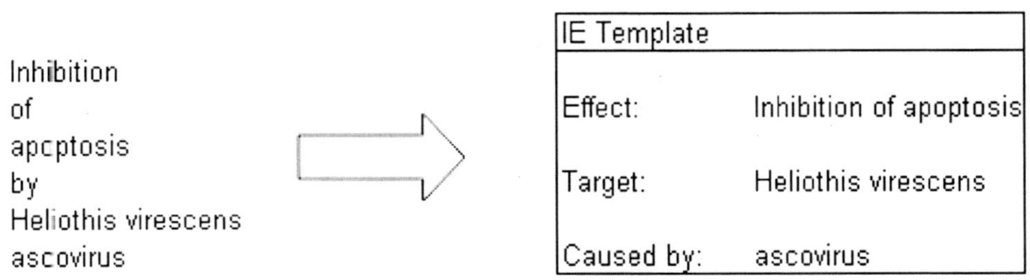

Figure 5. Information extraction components

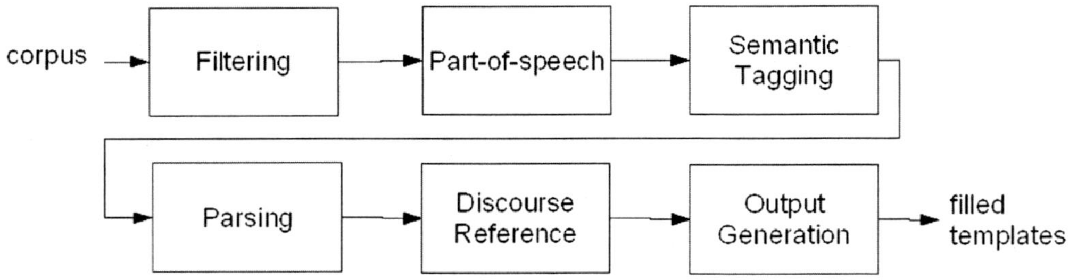

Several IE solutions have been proposed as either general-purpose solutions or as specialized solutions. The solutions with a general purpose integrate different available IE components, but in general, these solutions are not interoperable amongst each other. A well-established system is GATE[1] that integrates many natural language processing and IE components. The UIMA framework[2] enables the integration of a number of IE components.

In a different proposal, the harmonization of documents has been suggested. The IeXML initiative (Rebholz et al., 2006) targets the harmonization of annotations in the processed documents to integrate different tools purely on the standardization of the document format. Thus, documents are enhanced with additional contents. In this initiative, tools only need to comply with the standard document representation and could work in a pipeline mode leading to an easy interchange of the components.

IE solutions are difficult to compare and to assess through comparison since each one deals with a different extraction need, reflected in the available data sets (Bunescu et al., 2004). In the biomedical domain, several data sets are now freely available for selected and standardized IE tasks like the Biocreative I and II corpus, the GENIA corpus, the BioInfer dataset, the AIMed corpus and Prodiser. All these datasets cover only part of the information needs in the biomedical domain and furthermore, their focus is mainly limited to the identification of protein and gene names (PGN), the functional annotation of proteins and the interaction between proteins (protein-protein interaction, PPI).

Even though the requirements for information extraction need depend on the application, IE systems are usually composed of the same components (Cowie and Lehnert, 1996) that can be combined in a pipeline of modules as shown in Figure 5. This pipeline approach modularizes the IE systems allowing the interchange of several components. Typical components in an IE system are shown in Figure 5.

Filtering

Filtering is a basic task in information extraction. In the filtering step, the document is subdivided into parts that are relevant from the remainder. Only the relevant document parts proceed to further processing. Traditional IR systems or text categorizers may be used to do this selection.

Part-Of-Speech

This processing step assigns the part-of-speech (POS) to the words in the filtered text. This additional information about the syntactical role of a token allows, in later stages, to perform more advanced text processing tasks such as word sense disambiguation or parsing. Table 1 shows the example sentence tokenized (tag *w*). For each token, we find the POS in the Penn Treebank annotation (attribute *c*).

Table 1. Part-of-speech example

```
<w c='NN'>Inhibition</w>
<w c='IN'>of</w>
<w c='NN'>apoptosis</w>
<w c='IN'>by</w>
<w c='NP'>Heliothis</w>
<w c='NP'>virescens</w>
<w c='NN>ascovirus</w>
```

Semantic Tagging

In the semantic tagging step, phrases in the text are annotated with concept ids. In the first step, relevant phrasal units are identified. This requires shallow parsing or deep parsing techniques driven by the POS annotations (see Table 2). Semantic annotation follows the named entity recognition step (see Table 3) and attributes the concept id to the named entity.

Parsing

The ultimate goal is the identification of the dependencies between the constituents in the text, in particular between the named entities. The parser delivers the relations between the different units identified by the components of the previous analysis at the sentence level. Several parsers are available trained usually on standard corpora like the Penn tree bank. Similar solutions exist in the biomedical domain like the Enju parser that uses the GENIA corpus (http://www-tsujii.is.s.u-tokyo.ac.jp/enju/). In Figure 6, we find the example sentence processed by Enju.

DISCOURSE ANALYSIS

Some facts, for example relations between named entities, are not expressed in a single sentence but span across several sentences. The discourse analysis component brings together the output from different parsed sentences. Apart from gathering the different outputs, it recognizes related entities and unifies expressions that refer to each other. This approach requires techniques such as co-reference resolution that identifies phrases denoting the same entity expressed across a number of sentences.

OUTPUT GENERATION

The overall IE task is the extraction of facts from text. The task is completed if the pieces of information composing the fact can be filled into a template where the different slots in the template represent the same components of the related entities. The outcome of the previous components is prepared to fill the specified template.

Table 2. Shallow parser example

```
<c c='NP'>Inhibition</c>
of
<c c='NP'>apotheosis</c>
by
<c c='NP'>Heliothis virescens</c>
ascovirus
```

Figure 6. Enju parse example

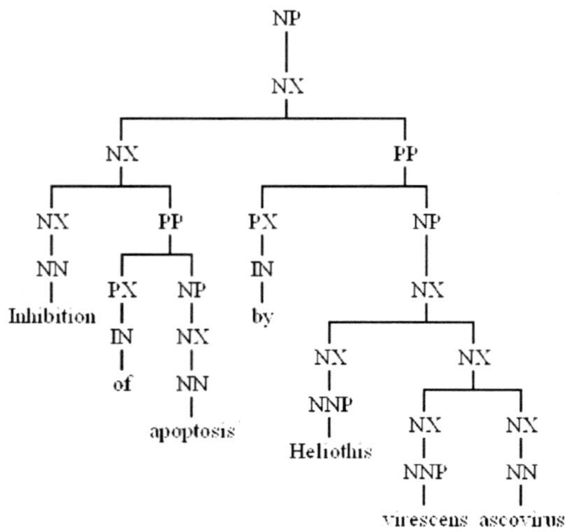

The Use of Ontologies in Information Extraction

Information extraction deals with the extraction of facts (e.g. entities, events, relations) from textual sources (see above). Information extraction components deliver information relevant to information retrieval. Combined with ontological resources, IE solutions enable better identification of concepts in the documents.

In the identification of named entities, which is an IE task, we can distinguish the identification of an entity mention from the normalization of the named entity. In the first case, the IE system delivers the boundaries of the named entities. This is the classic named entity recognition task. In the latter case, the named entity has to be mapped to a concept identifier. This requires that the named entity mention is identified, that any ambiguity is resolved, that all synonymous expressions of the named entity are considered and that finally the right concept identifier is attributed to the named entity.

The use of ontologies and thesauri is mandatory to achieve named entity normalization and semantic indexing. Both resources provide the concept identifiers and the IE task has to achieve the appropriate mapping to the semantic resource. Once the mapping has been achieved, the IR solution will deliver better coverage at higher precision.

The normalization of named entities from the text to concepts in the ontology, i.e. the mapping of the potential surface form of a concept label in text to the concept identifier in the ontological resource, has to tackle the following two basic problems. The first one is that a given concept is represented using different surface forms in the text, for example different representations of a given term (morpho-syntactic variation) or simply different terms for the same concept (synonymy). Both cases require different surface forms to be linked to the same concept identifier. The other basic problem arises from the situation that the same term may denote more than one concept (polysemy). In this case, the context of the term is further analyzed to resolve any ambiguity of the term in the text in order to derive the appropriate concept.

Different techniques can be applied to normalize given surface forms to a concept label. Natural language processing techniques can be applied (Jacquemin, 2001) to include the missing entries. It may happen that the terminological or ontological resources in the biomedical domain lack the required terminology for the mapping since not all terms for the concepts have been made available (Beisswanger et al., 2008). This gap in the resources is solved by adding the missing terms using term extraction tools, which process large amounts of text to identify statistically overrepresented terms (Spasic et al., 2008).

Resolving the ambiguity of polysemous terms requires special solutions. A large lexical resource can directly contribute to the resolution of ambiguous cases (Pezik et al., 2008). In other cases, it is necessary to process the contextual information provided by the documents. Recently, special approaches take the topology of the ontology into consideration to disambiguate terms (Spasic et al., 2005). This example reinforces the need of

Table 3. Entity annotation example

```
Inhibition of
<e id='GO:0006915' onto='biological_process'>apoptosis</e>
By
<e id='species:7102'>Heliothis virescens</e>
<e id='species:43680'>ascovirus</e>
```

having a lexicon in combination with an ontology (Jimeno et al., 2008).

Contextual information of a term is commonly used to perform disambiguation (Gaudan et al., 2005). In this document, we focus on the mutual benefits between ontologies and text mining for the sake of term disambiguation. Several disambiguation algorithms exist that exploit the ontology topology and the context of the co-occurring terms to estimate the conceptual distance between the associated terms (Agirre and Rigau, 1996). The contextual information is compared with a model of the concept based on its terminology and relations as expressed in the ontology.

The extraction of relations between entities is usually based on a set of rules applied on annotated text (e.g. based on part-of-speech and named entities). This set of rules can be improved by applying inference using domain knowledge producing simpler rules with similar or higher efficiency and easier to maintain.

TEXT MINING AND THE ONTOLOGY LIFECYCLE

This section describes the different developments steps in the ontology lifecycle from the first analysis of requirements to the final refinement of the ontology following the schema presented by (Maedche and Staab, 2001). This lifecycle representation appeals to us since its can be used to show the integration of text mining and to demonstrate the benefits of text mining in the lifecycle. This lifecycle covers all steps from the design of the ontology, the reuse of existing resources up to the refinement of the ontology.

In the development step, concerned with the import and reuse of resources, the first objective is to define the structure of the ontology based on the input from domain experts. Furthermore, the content of existing data sources such as databases and other existing ontologies has to be kept in mind. Due to the increasing amount of data in public knowledge sources, there is a strong interest to gather the content from all complementary sources. Unfortunately, a solution that enables automatic import of content from complementary resources is not yet available.

Further input for the completion of the ontological resource is delivered from the literature through text mining. This processing step profits from automatic term collection for ontology and lexicon completion. It has to be kept in mind that this task is constrained by the fact that content (e.g., facts) from the scientific literature is, in many cases, not factual but hypothetical (Tsujii, 2005) and requires confirmation in the future. For example, experimental results give first evidence on drug-gene or gene-diseases relations but require further confirmation through future experiments.

Ontology Lifecycle

This section covers the steps in the ontology lifecycle from the first analysis to the final refinement of the ontology following the schema presented in (Maedche and Staab, 2001). Other proposals for the ontology lifecycle have been proposed but we favor this approach since it is better suited to integrate text mining solutions for ontology refinement.

Import and Reuse

The first objective of this step is the definition of the layout of the ontology based on the input from domain experts and based on the input (e.g., the structure) from existing data sources like databases or other ontologies. The increasing number of knowledge sources in the biomedical domain requires integration to enable their automatic exploitation. Automatic means for the integration and exploitation are still not readily available.

The combination of different sources may be trivial if an explicit link between two related concepts in different resources exists, similar to a foreign key in relational databases. Unfortunately, such links are sparse and semi-automatic solutions have been proposed to solve this problem. For example, techniques based on semantic distances have been described in the literature (Guarino, 1997). (Stoilos et al., 2005) have tested several techniques in combination with text mining solutions. These solutions have then been applied to the biomedical domain (Gangemi et al., 1998) and to the anatomical domain knowledge (Zhang and Bodenreider, 2004). Special care is still needed when relying on these links since is not always clear that the ontologies share the same conceptualization.

Another step is the integration of the ontologies, i.e. their combined representation and use. Two major techniques have been proposed by (Noy and Musen, 1999): ontology alignment and ontology merging. In ontology alignment, both ontologies keep their original structure and content and the integration step only introduces links between related concepts in the two ontologies. This technique is used whenever the ontologies cover complementary domains. (Thompson et al., 2005) created the Mao system to align nucleic acid with protein sequences. (Rosse et al., 2005) used the system OBR to integrate domain ontologies in anatomy, physiology and pathology and (Smith and Rosse, 2004) used such solutions for the ontology called "Foundational Model of Anatomy" (FMA). In ontology merging, the content and the structure of both ontologies is merged into a single representation. This technique is applied whenever final objective is the provision of a consistent and unified ontology.

Ontology Learning

As stated in the introduction, only a small portion of information is contained in structured representations. The major portion of information is delivered in an unstructured form of which a large portion is textual data. These unstructured sources still contain precious information that can be exploited by data mining solutions to create a new ontology or to enlarge an existing ontology. Several techniques have been proposed that enable the extraction of specific information pieces that can well be integrated in an ontology.

Term and Concept Extraction

Several techniques are available to identify new terms or concept labels from text that can be integrated into ontology learning solutions. (Jacquemin, 2001) used a term extractor in combination with a component to decompose the term structure (identification of head dependency) to generate a taxonomy of terms that, in addition, are linked by meronymy. (Navigli and Velardi, 2002) used the system OntoLearn to generate a specialized version of WordNet for a defined domain. Furthermore, statistical techniques have been proposed for the same objectives (Maedche and Staab, 2001).

Extracted terms might be synonyms of existing concepts. The different techniques to identify synonyms are either, based on the inner structure of the term (Hole and Srinivasan, 2000; Wilbur and Kim, 2001) or based on the context of the term (Dagan et al., 1993; Li and Abe, 1998; Lin,

1998). (Pearson, 1998) identified several patterns like *known as, called as* useful for synonym identification. (Yu et al., 2003) introduce new semi-supervised methods on the basis of large corpora and few examples, using bootstrapping based on the SNOWBALL system to learn extraction patterns indicating synonymous relations. Acronyms are another source of synonyms. Unsupervised methods (Schwartz and Hearst, 2003) exist that might be useful to deal with new acronyms. The low performance of these systems shows that there is still a lot to investigate on this subject. In addition, these methods do not make a clear distinction between synonyms and hyperonyms.

Hierarchical Clustering

Hierarchical clustering can be applied to generate groups of terms based on their contextual information. An ontology engineer or domain expert can verify the resulting structures. For example, (Blaschke and Valencia, 2002) have applied hierarchical clustering to group gene-products (proteins) from the literature. The result was a collection of disjoint trees that were merged by a knowledge expert. (Faure and Nédellec, 1998) used syntactical verb and preposition language patterns in the ASSIUM system to add terms to a structural representation. This did not only generate a taxonomical structure to build an ontology but in addition, the different verbs were also put into relation to concepts if both tend to co-occur. (Caraballo, 1999) uses co-occurrence of appositives and conjunctions within the document collection in order to build a tree in a bottom-up fashion.

Term composition, like head-modifier, can give additional input that helps to add a taxonomical structure to the term repository. For instance, *colon cancer* is more specific than *cancer*, i.e. it is the sub-specification of the cancer to the organ type labeled the *colon*. (Navigli and Velardi, 2002) and (Missikoff et al., 2003) extended WordNet by including domain specific information about the tourism domain based on term composition of multi-word terms extracted from domain specific documents.

The context of the terms can be used to identify similar and related terms. The terms within a fixed length window surrounding a term can be used to gear a *word sense* model. (Hearst and Pedersen, 1996) collected from the context of terms additional tokens. These tokens were integrated into a vector space model that was thereafter reduced to its main components based on principal component analysis (PCA). This approach led to the identification of hierarchical relations at 58% precision. A similar solution from (Widdows, 2003) added the POS information to discriminate better the words defining the context.

Other solutions to generate content for ontologies are based on language patterns. However, the lack of training data is limiting these solutions in their effectiveness. This lack in training data has inspired several researchers to overcome the problem by proposing methods that enable bootstrapping of ontological resources. (Riloff and Jones, 1997) and (Roark and Charniak, 1998) have built dictionaries from extracted terms and have automatically assigned the terms to general categories. Unfortunately, their method does not scale to more refined categories. (Hearst et al., 1992, 1998) initiated a solution with an initial set of reliable extraction patterns that then have been combined with bootstrapping methods to identify additional extraction patterns. Their objective was the extraction of hyponym and hypernym relations without using a dictionary. They achieved to demonstrate high precision with their system at the expense of very low recall.

Methods relying on the contextual information achieve good recall but suffer from low precision. Furthermore, it remains to be still a difficult problem to differentiate relations such as synonymy, part-of and hierarchical relations. These distinctions are relevant if a domain expert revises the produced tree in the case of using a hierarchical clustering approach (Blaschke and Valencia, 2002).

Table 4. Hearst patterns

<table>
<tr><td>NP0 such as {NP1, NP2 …,(and|or)} NPn</td></tr>
<tr><td>such NP as {NP,}* {(or|and)} NP</td></tr>
<tr><td>NP {,NP}*{,} or other NP</td></tr>
<tr><td>NP{,NP}*{,} and other NP</td></tr>
<tr><td>NP{,} including {NP,}*{or|and} NP</td></tr>
<tr><td>NP{,}especially {NP,}* {or|and} NP</td></tr>
</table>

Term composition, like the extraction of head-modifier, is relevant to the classification of several semantic types like diseases where term composition can be used to identify the ancestors in the hierarchy (e.g. colon cancer *is a* cancer). Normalization techniques applied in the previous section contribute to improve the coverage of this technique.

The Hearst patterns (see table 4) have higher precision in contrast to the statistical methods presented above.

Ontology Pruning

The steps of ontology development presented above collect information from the literature. The generated ontology representation possibly requires pruning the content of the ontology to gather the information that fits best to the domain representation. We have to keep in mind that a good solution keeps a balance between the completeness of the ontological resource and its scarcity. A complete solution is favored but induces problems in managing the whole content and increases the complexity in all processing steps. A scarce solution can be better maintained but it limits the expressiveness of the ontology. (Khan and Luo, 2002) have tested an approach that pruned WordNet to reduce the information overhead when looking for information. It is based on a self-organizing tree algorithm called SOTA.

Ontology Refinement

Ontology refinement has the objective to transform an existing ontological resource to perform better in a task that requires a more specialized ontological resource. Different techniques based on IE have been proposed in the scientific literature to achieve fine-tuning of an ontology. Ontology refinement is one of the main ontology lifecycle steps. Techniques that process textual data like IE play an essential role. We can distinguish automatic from semi-automatic techniques. In the latter case, the ontologist is still involved in the ontology refinement process.

Semi-Automatic Approaches

This type provides to the ontologist evidences of knowledge found in the data sources that do not exist in the ontology and that appear to be relevant. In these approaches, the ontologist is supported by the retrieved information from relevant corpora.

Available solutions are focused to produce extensive evidence (more recall) since they are more concerned about not loosing relevant information instead of reducing the noise level. Typically, these techniques identify statistically overrepresented terms with indication of domain relevance in contrast to frequently occurring common terms. A drawback of these techniques is the large number of useless associations produced, which have to be revised manually. All methods provide means to refine more efficiently an existing ontology.

Proposed methods exploit term co-occurrence to identify novel information supported by statistical means. (Faatz and Steinmetz, 2002) suggest a system where the result is the identification of a link between existing terms given a certain quality or likelihood. This association is then analyzed, even though the relation between the terms is not defined semantically. (Maedche and Staab, 2001) selected association rules that specify relations between concepts based on the analysis of a set of documents describing hotels. The association study is based on the work of (Srikant and Agrawal, 1997) where the confidence and the support of the associations are used to choose the related candidates. (Köhler et al., 2006) have compared GO to other ontologies to identify conflicting concepts (e.g. circular definitions) and new synonyms that are then presented to the ontologists.

Automatic Approaches

Automatic methods replace the work of the ontologist by a process that decides which terms from the processed corpora are relevant enough for integration and ontology refinement. These automatic methods rely either on heuristics, similar to ontological quality measures (Hahn and Schnattinger, 1998), or on IE from unstructured sources. The main assumption supporting these approaches is that the documents refer to the same conceptualization as the ontological resource.

The evaluation of the results from such techniques is performed either by the domain expert or by comparing the results against an existing ontology (i.e. the gold standard). These techniques propose ways to refine the taxonomy of the ontology but do not offer mechanisms to refine the relations between the concepts.

(Navigli and Velardi, 2002) worked on the adaptation of WordNet in the domain of tourism. Their idea is the extraction of new terms from text and then placing them in the taxonomy or identifying taxonomic relations between existing concepts. The system uses heuristics that analyze the term composition and that processes the head dependency of the terms found in the document. The evaluation is done using a gold standard ontology previously built by ontologists helped by semi-automatic methods. In the biomedical domain, (Lee et al., 2006) propose an automated method to refine the Gene Ontology. The idea is to find rules based on GO terms variations for automatic expansion that is validated with the literature.

(Agirre et al., 2000) refined WordNet to improve their results in word sense disambiguation (WSD). The method consists of adding new terms to the ontology identifying the different senses of the terms and then builds the sense signature using clustering techniques. WSD is used to improve the performance of the overall system. On the other side, the refinement is not driven directly to improve the WSD directly.

Other techniques combine different approaches that extract the similar information from the same corpus (Alfonseca and Manandhar, 2002), increasing the confidence in the extracted information that is then integrated into the ontology. These techniques focus on the refinement of the taxonomic structure and their evaluation is usually performed using a gold standard ontology or measuring the performance in a given task (e.g. WSD) but there is no specific optimization approach to refine it for the given task. Recently, (Jimeno et al., 2009) have developed a method to refine a biomedical ontology to improve information retrieval.

ONOTOLOGY EVALUATION

Several criteria exist to evaluate the quality of ontologies (Brank et al., 2005). Despite of logically evaluating the coherence of the ontology (e.g. incoherence in the ontology axioms), it is interesting to consider techniques that assess the ontology learning process; considering as well information that is not too general or too specific.

A general way of evaluating a refined ontology is to compare it to a reference ontology or gold standard. This applies only to a limited number of cases that support ontology learning with text mining means. The reason is that the evaluation of text mining support to ontology generation requires evaluating the extracted and integrated information against a gold standard, i.e. a complete and curated ontology, which is usually difficult to obtain. Other evaluation techniques use manual evaluation of the added information to the ontology, this process is tedious to complete and could be biased by the information added to the ontology.

Other techniques exist to compare an ontology with a refined version based on task-based evaluation (Porzel and Malaka, 2004). We will compare the refined ontology to the original ontology given their performance on a given task, assuming that it is an improved ontology if it has a better performance on this task. This technique has been extensively used to evaluate and improve an ontology for information retrieval (Jimeno et al., 2009).

SUMMARY

We have presented the possible interactions between ontologies and text mining and the benefits that we might obtain by its combination.

Text mining profits from ontologies in that it can increase the coverage of existing algorithms given the information encoded in the ontology. This includes expanding the original user query in information retrieval and a larger set of terms for the identification of entities in information extraction. In addition, the ontology encodes and organizes domain knowledge in a way that it can be

Furthermore, ontology development is time consuming while a lot of information is available in unstructured sources. We have presented several approaches that allow us to profit from unstructured sources using information extraction. These approaches might provide easier ways to reuse existing information and to improve ontology development.

REFERENCES

Agirre, E., Ansa, O., Hovy, E., & Martinez, D. (2000). Enriching very large ontologies using the www. In *Proceedings of the Ontology Learning Workshop, ECAI*, Berlin, Germany

Agirre, E., & Rigau, G. (1996). Word sense disambiguation using conceptual density. In *Proceedings of the 16th conference on Computational linguistics*, August, (pp. 05–09).

Alfonseca, E., & Manandhar, S. (2002). Improving an ontology refinement method with hyponymy patterns . In *Language Resources and Evaluation*. Las Palmas: LREC.

Alfonseca, E., & Manandhar, S. (2002). An unsupervised method for general named entity recognition and automated concept discovery. In *Proceedings of the 1st International Conference on General WordNet.*

Amati, G., & Bruza, P. (1999). A logical approach to query reformulation motivated from belief change. In *Workshop on logical and uncertainty models for Information Systems,* University College London (UCL), London.

Aronson, A. R., & Rindflesch, T. C. (1997). *Query expansion using the UMLS Metathesaurus*. Amia.

Baziz, M., & Boughanem, M. Aussenac, & Gilles, N. (2004). The Use of Ontology for Semantic Representation of Documents. In proceedings of Semantic Web and Information Retrieval Workshop, SIGIR, (pp. 38-45).

Beisswanger, E., Poprat, M., & Hahn, U. (2008). Lexical Properties of OBO Ontology Class Names and Synonyms. In *3rd International Symposium on Semantic Mining in Biomedicine.*

Blaschke, C., & Valencia, A. (2002). Automatic Ontology Construction from the Literature. In Genome Informatics Series, (pp. 201–213).

Bodner, R. C., & Song, F. (1996). Knowledge-based approaches to query expansion in information retrieval. In *Canadian Conference on AI*, (pp. 146–158).

Brank, J., Grobelnik, M., & Mladeni, D. (2005). A survey of ontology evaluation techniques. In *Proceedings of SIKDD.*

Buckley, C., & Salton, G. (1995). Optimization of relevance feedback weights. In *Proceedings of the 18th annual international ACM SIGIR conference on Research and development in information retrieval*, (pp. 351–357). New York: ACM Press.

Buckley, C., Salton, G., Allan, J., & Singhal, A. (1994). Automatic query expansion using SMART: TREC 3. In *Text REtrieval Conference.*

Bunescu, R., Ge, R., Kate, R., Marcotte, E., Mooney, R., Ramani, A. & Wong, Y. (2004). Comparative experiments on learning information extractors for proteins and their interactions in Comparative Experiments on Learning Information Extractors for Proteins and their Interactions. *Journal of Artificial Intelligence in Medicine.*

Caraballo, S. A. (1999). Automatic construction of a hypernym-labeled noun hierarchy from text. In *Proceedings of the 37th annual meeting of the Association for Computational Linguistics on Computational Linguistics*, (pp. 120–126). Morristown, NJ: Association for Computational Linguistics.

Chu, W. W., Liu, Z., & Mao, W. (2002). Textual document indexing and retrieval via knowledge sources and data mining . In *Communication of the Institute of Information and Computing Machinery* (*Vol. 5*, p. 2). Taiwan: CIICM.

Cowie, J., & Lehnert, W. (1996). Information extraction. *Communications of the ACM, 39*(1), 80–91. doi:10.1145/234173.234209

Dagan, I., Marcus, S., & Markovitch, S. (1993). Contextual word similarity and estimation from sparse data. In *Proceedings of the 31st conference on Association for Computational Linguistics,* (pp. 164–171). Morristown, NJ: Association for Computational Linguistics

Deerwester, S., Dumais, S. T., Furnas, G. W., Landauer, T. K., & Harshman, R. (1990). Indexing by latent semantic analysis. *Journal of the American Society for Information Science American Society for Information Science, 41*(6), 391–407. doi:10.1002/(SICI)1097-4571(199009)41:6<391::AID-ASI1>3.0.CO;2-9

Doms, A., & Schroeder, M. (2005). GoPubMed: exploring PubMed with the gene ontology. *Nucleic Acids Research,*33.

Efthimiadis, E. (1996). Query expansion . In Williams, M. E. (Ed.), *Annual Review of Information Systems and Technologies* (*Vol. 31*, pp. 121–187). ARIST.

Faatz, A., & Steinmetz, R. (2002). Ontology enrichment with texts from the www. In Semantic Web Mining, WS02.

Faure, D., & Nédellec, C. (1998). A corpus-based conceptual clustering method for verb frames and ontology acquisition. In LREC workshop on adapting lexical and corpus resources to sublanguages and applications, (pp. 707–728).

Faure, D., & Nédellec, C. (1998). ASIUM: learning subcategorization frames and restrictions of selection. In Y. Kodrato, (ed.), *10th European Conference on Machine Learning (ECML 98), Workshop on Text Mining.*

Fox, E. A. (1980). Lexical relations: Enhancing effectiveness of information retrieval systems. *SIGIR Forum, 15*(3), 5–36. doi:10.1145/1095403.1095404

Gangemi, A., Pisanelli, D. M., & Steve, G. (1998). Ontology Integration: Experiences with Medical Terminologies. In *Formal Ontology in Information Systems: Proceedings of the First International Conference (FOIS'98),* June 6-8, Trento, Italy. Amsterdam: IOS Press, Inc.

Gauch, S., & Smith, J. B. (1993). An expert system for automatic query reformation. *Journal of the American Society for Information Science American Society for Information Science, 44*(3), 124–136. doi:10.1002/(SICI)1097-4571(199304)44:3<124::AID-ASI2>3.0.CO;2-C

Gaudan, S., Kirsch, H., & Rebholz-Schuhmann, D. (2005). Resolving abbreviations to their senses in Medline. *Bioinformatics (Oxford, England), 21*(18), 3658–3664. doi:10.1093/bioinformatics/bti586

Guarino, N. (1997). *Semantic matching: Formal ontological distinctions for information organization, extraction, and integration* (pp. 139–170). SCIE.

Hahn, U., & Schnattinger, K. (1998). Towards text knowledge engineering. In *Proceedings of the fifteenth national/tenth conference on Artificial intelligence/Innovative applications of artificial intelligence*, (pp. 524–531). Menlo Park, CA: American Association for Artificial Intelligence.

Hearst, M. A. (1992). *Automatic acquisition of hyponyms from large text corpora.* Technical Report S2K-92-09.

Hearst, M. A. (1998). Automated discovery of wordnet relations . In *WordNet: An Electronic Lexical Database and Some of its Applications.* Cambridge, MA: MIT Press.

Hearst, M. A., Karger, D. R., & Pedersen, J. O. (1995). Scatter/gather as a tool for the navigation of retrieval results . In *Working Notes AAAI Fall Symp.* AI Applications in Knowledge Navigation.

Hearst, M. A., & Pedersen, J. O. (1996). *Reexamining the cluster hypothesis: Scatter/gather on retrieval results* (pp. 76–84). SIGIR.

Hole, W. T., & Srinivasan, S. (2002). *Discovering missed synonymy in a large concept-oriented metathesaurus.* AMIA.

Jacquemin, C. (2001). *Spotting and discovering terms through natural language processing.* Cambridge, MA: The MIT Press.

Jimeno-Yepes, A., Berlanga-Llavori, R., & Rebholz-Schuhmann, D. (In press). Ontology refinement for improved information retrieval. In Information Processing & Management: Special Issue on Semantic Annotations in Information Retrieval.

Jimeno Yepes, A., Jimenez-Ruiz, E., Berlanga-Llavori, R., & Rebholz-Schuhmann, D. (2008). Use of shared lexical resources for efficient ontological engineering. In Semantic Web Applications and Tools for Life Sciences.

Khan, L. R., & Luo, F. (2002). Ontology construction for information selection. In *ICTAI '02: Proceedings of the 14th IEEE International Conference on Tools with Artificial Intelligence (ICTAI'02),* (pp. 122). Washington, DC: IEEE Computer Society.

Köhler, J., Munn, K., Ruegg, A., Skusa, A., & Smith, B. (2006). Quality control for terms and definitions in ontologies and taxonomies. *BMC Bioinformatics, 7*, 212. doi:10.1186/1471-2105-7-212

Lavrenko, V., & Croft, W. B. (2001). Relevance based language models. In *Proceedings of the 24th annual international ACM SIGIR conference on Research and development in information retrieval.*

Lee, J. B., Kim, J. J., & Park, J. C. (2006). Automatic extension of gene ontology with flexible identification of candidate terms. *Bioinformatics (Oxford, England)*, *22*(6), 665–670. doi:10.1093/bioinformatics/btl010

Li, H. & Abe, N. (1998). *Word clustering and disambiguation based on cooccurrence data.* CoRR, cmp-lg/9807004.

Lin, D. (1998). Automatic retrieval and clustering of similar words. In *Proceedings of the 17th international conference on Computational linguistics*, (pp. 768–774). Association for Computational Linguistics.

Liu, Z., & Chu, W. W. (2005). Knowledge-based query expansion to support scenario-specific retrieval of medical free text. In *SAC '05: Proceedings of the 2005 ACM symposium on Applied computing*, (pp. 1076–1083). New York: ACM Press.

Maedche, A., & Staab, S. (2001). Ontology learning for the semantic web. *IEEE Intelligent Systems*, *16*(2), 72–79. doi:10.1109/5254.920602

Mayfield, J. (2003). Information retrieval on the semantic web: Integrating inference and retrieval. In *Proceedings of the SIGIR Workshop on the Semantic Web.*

Missikoff, M., Velardi, P., & Fabriani, P. (2003). Text mining techniques to automatically enrich a domain ontology. []. Amsterdam: Kluwer Academic Publishers.]. *Applied Intelligence*, *18*, 323–340. doi:10.1023/A:1023254205945

Mitra, M., Singhal, A., & Buckley, C. (1998). Improving automatic query expansion. In Research and Development in Information Retrieval, (pp. 206–214).

Navigli, R., & Velardi, P. (2002). Automatic adaptation of wordnet to domains. In *Proceedings of 3rd International Conference on Language Resources and Evaluation.*

Nie, J. Y., & Jin, F. (2002). Integrating logical operators in query expansion in vector space model. In *Workshop on Mathematical/Formal Methods in Information Retrieval, 25thACM-SIGIR*, Tampere, Finland, (Vol. 8).

Noy, N. F., & Musen, M. A. (1999). An algorithm for merging and aligning ontologies: automation and tool support. In *the Proceedings of the Workshop on Ontology Management at Sixteenth National Conference on Artificial Intelligence (AAAI)*, Orlando, FL.

Patterson, D., Rooney, N., Dobrynin, V. & Galushka, M. (2005). Sophia: A novel approach for textual case-based reasoning. *IJCAI*, 15–20.

Pearson, J. (1998). *Terms in Context. Studies in Corpus Linguistics, 1.* Philadelphia: John Benjamins.

Pezik, P., Jimeno-Yepes, A., Lee, V., & Rebholz-Schuhmann, D. (2008). *Static dictionary features for term polysemy identification. Building and evaluating resources for biomedical text mining.* LREC Workshop.

Porzel, R., & Malaka, R. (2004). A Task-based Approach for Ontology Evaluation. In *ECAI Workshop on Ontology Learning and Population*, Valencia, Spain.

Pratt, W., Hearst, M. A., & Fagan, L. M. (1999). *A knowledge-based approach to organizing retrieved documents* (pp. 80–85). AAAI/IAAI.

Pratt, W., & Wasserman, H. (2000). QueryCat: Automatic Categorization of MEDLINE Queries. *Journal of the American Medical Informatics Association, 7*, 655–659.

Qiu, Y., & Frei, H. P. (1993). Concept-based query expansion. In *Proceedings of SIGIR-93, 16th ACM International Conference on Research and Development in Information Retrieval*, (pp. 160–169), Pittsburgh.

Rebholz-Schuhmann, D., Kirsch, H., Arregui, M., Gaudan, S., Riethoven, M., & Stoehr, P. (2007). EBIMed – Text crunching to gather facts for proteins from Medline. *Bioinformatics (Oxford, England), 23*(2), e237–e244. doi:10.1093/bioinformatics/btl302

Rebholz-Schuhmann, D., Kirsch, H., & Nenadic, G. (2006). *IeXML: towards an annotation framework for biomedical semantic types enabling interoperability of text processing modules. SIG BioLink.* ISMB.

Riloff, E., & Jones, R. (1999). Learning Dictionaries for Information Extraction by Multi-level strapping, Boot-. In *Proceedings of the Sixteenth National Conference on Artificial Intelligence*, (pp. 1044–1049). Menlo Park, CA: The AAAI Press/MIT Press.

Riloff, E., & Shepherd, J. (1997). A corpus-based approach for building semantic lexicons. In C. Cardie & R. Weischedel, (Eds.), *Proceedings of the Second Conference on Empirical Methods in Natural Language Processing*, (pp. 117–124). Somerset, NJ: Association for Computational Linguistics.

Roark, B., & Charniak, E. (1998). *Noun-phrase co-occurence statistics for semiautomatic semantic lexicon construction* (pp. 1110–1116). COLING-ACL.

Roelleke, T. (2003). A frequency-based and a poisson-based definition of the probability of being informative. In *Proceedings of the 26th annual international ACM SIGIR conference on research and development in information retrieval*, (pp. 227–234). New York: ACM Press.

Rosenfeld, B., Feldman, R., Fresko, M., Schler, J., & Aumann, Y. (2004). *Teg: a hybrid approach to information extraction* (pp. 589–596). CIKM.

Rosse, C., Kumar, A., Mejino, J. L. V., Jr., Cook, D. L., Detwiler, L. T., & Smith, B. (2005). A strategy for improving and integrating biomedical ontologies. In proceedings of AMIA, Washington, DC.

Safar, B., Kefi, H., & Reynaud, C. (2004). OntoRefiner, a user query refinement interface usable for Semantic Web Portals . In *Applications of Semantic Web technologies to web communities.* Workshop ECAI.

Schwartz, A. S., & Hearst, M. A. (2003). A simple algorithm for identifying abbreviation definitions in biomedical text. *Pacific Symposium on Biocomputing, 451*, 62.

Smith, B., & Rosse, C. (2004). The role of foundational relations in the alignment of biomedical ontologies. *Medinfo, 11*(Pt 1), 444–448.

Spasic, I., Ananiadou, S., McNaught, J., & Kumar, A. (2005). Text mining and ontologies in biomedicine: Making sense of raw text. *Briefings in Bioinformatics, 6*(3), 239–251. doi:10.1093/bib/6.3.239

Srikant, R., & Agrawal, R. (1997). Mining generalized association rules. *Future Generation Computer Systems, 13*(2–3), 161–180. doi:10.1016/S0167-739X(97)00019-8

Stoilos, G., Stamou, G., & Kollias, S. (2005). A string metric for ontology alignment. In *4th International Semantic Web Conference (ISWC),* Galway.

Thompson, J. D., Holbrook, S. R., Katoh, K., Koehl, P., Moras, D., Westhof, E., & Poch, O. (2005). Mao: a multiple alignment ontology for nucleic acid and protein sequences. *Nucleic Acids Research*, *33*, 4164. doi:10.1093/nar/gki735

Tsujii, J. I., & Ananiadou, S. (2005). Thesaurus or logical ontology, which do we need for mining text? *Language Resources and Evaluation*, *39*(1), 77–90. doi:10.1007/s10579-005-2697-0

van Rijsbergen, C. J. (1979). *Information retrieval* (2nd ed.). London: Butterworths.

Voorhees, E. M. (1994). Query expansion using lexical-semantic relations. In *Proceedings of the 17th annual international ACM SIGIR conference on Research and development in information retrieval*, (pp. 61–69). New York: Springer-Verlag, Inc.

Widdows, D. (2003). *Unsupervised methods for developing taxonomies by combining syntactic and statistical information*. HLT-NAACL.

Wilbur, W. J., & Kim, W. (2001). Flexible Phrase Based Query Handling Algorithms. In . *Proceedings of the ASIST Annual Meeting*, *38*, 438–449.

Xu, J., & Croft, W. B. (1996). Query expansion using local and global document analysis. In *Proceedings of the Nineteenth Annual International ACM SIGIR Conference on Research and Development in Information Retrieval*, (pp. 4–11).

Yu, A. C. (2006). Methods in biomedical ontology. *Journal of Biomedical Informatics*, *39*(3), 252–266. doi:10.1016/j.jbi.2005.11.006

Zhang, S., & Bodenreider, O. (2004). Investigating implicit knowledge in ontologies with application to the anatomical domain. In *Pacific Symposium on Biocomputing*, (pp. 250–261).

ENDNOTES

1 http://gate.ac.uk

2 http://www.research.ibm.com/UIMA

Chapter 13
Ontology Based Multimedia Indexing

Mihaela Brut
Institut de Recherche en Informatique de Toulouse, France

Florence Sedes
Institut de Recherche en Informatique de Toulouse, France

ABSTRACT

The chapter goal is to provide responses to the following question: how the ontologies could be used in order to index and manage the multimedia collections? Alongside with reviewing the main standard formats, vocabularies and ontology categories developed especially for multimedia content description, the chapter emphasis the existing techniques for acquiring ontology-based indexing. Since a fully automatic such technique is not possible yet, the chapter also proposes a solution for indexing a multimedia collection by combining technologies from both Semantic Web and multimedia indexation domains. The solution considers the management of multimedia metadata based on two correlated dictionaries: a metadata dictionary centralizes the multimedia metadata obtained through an automatic indexation process, while the visual concepts dictionary identifies the list of visual objects contained in multimedia documents and considered in the ontology-based annotation process. This approach facilitates as well the multimedia retrieval process.

INTRODUCTION

The multimedia indexing and management is a very important issue in the actual context where various domains such as news gathering, TV, banks of resources for commercial or consumer applications, collaborative work, video surveillance are flooded by a huge amount of multimedia sources.

DOI: 10.4018/978-1-61520-859-3.ch013

The traditional multimedia indexation techniques are focused on the effective multimedia content processing, being mainly in charge with low-level multimedia features analysis. They could capture some information about the content description (such as shapes or faces recognition), but not in terms of high-level concepts (such as ontology or vocabulary concepts). The chapter is focused on possible solutions for the problem of bridging the "semantic gap" between low-level multimedia features and high-level concepts describing the multimedia content, in order to reach a semantically enhanced multimedia indexing.

From the representational point of view, the main goal of transforming multimedia materials into machine-compatible content is ensured both by the Semantic Web activity of the Web Consortium[1], and by the ISO's efforts in the direction of complex media content modeling, in particular the Multimedia Content Description Interface (MPEG-7)[2]. The two directions are syntactically and semantically different, and some solutions to unify them were proposed. Moreover, a set of tools were developed to enable the automatic extraction of the visual features in multimedia materials (as MPEG proposes) and the manual association of them with ontology concepts (as a semantic web approach requires). Efforts were made also towards a multimedia annotation interoperability framework[3] and towards a core ontology for multimedia framework[4], aiming a uniform use of the multimedia ontologies, according to their intended use and context.

In order to capture and express a high-level semantics for multimedia objects, some domain specific vocabularies were established for describing multimedia content, as well as some specialized multimedia ontologies. A quick overview is included in Section 2. Because of the binary character of the multimedia content, the main issue concerning the automatic ontology-based multimedia indexing still remains an unsolved problem. However, some steps were accomplished, and our chapter emphasizes the main such techniques: exploiting the textual descriptions accompanying the multimedia content, using the free tagging community effort, applying multiple automatic indexing techniques (in cascade), combining human and machine efforts.

Focused on the last two technique types, the present chapter proposes a solution for indexing a multimedia collection by combining technologies from both semantic Web and multimedia indexation domains. The solution considers the management of multimedia metadata based on two correlated dictionaries: a metadata dictionary centralizes the multimedia metadata obtained through an automatic indexation process, while the visual concepts dictionary identifies the list of visual objects contained in multimedia documents and considered in the ontology-based annotation process. This architecture facilitates as well the multimedia retrieval process.

The second section presents the state of the art in the domain of developing standard formats, vocabularies and ontologies for being adopted to specify and organize the multimedia metadata, as well as the interoperability problems raised by the heterogeneity of vocabularies. the third section provides an overview of the most important ontology-driven frameworks for multimedia semantic annotation, while the fourth section exposes the semantic multimedia indexation proposed solution. In final, the conclusions and further work directions are presented.

VOCABULARIES FOR MULTIMEDIA ANNOTATION

In the recent years, multiple standards were developed for storing each type of multimedia content. Among these, the most popular are:

- **Video:** MPEG-1, MPEG-2, MPEG-4, QuickTime, Sony DV, AVI, ASF, RealMedia, etc.
- **Audio:** Raw PCM, WAV, MPEG-1, MP3, GSM, G.723, ADPCM, etc;
- **Image:** JPEG, TIFF, BMP, GIF, etc.;
- **Multimedia presentations:** SMIL, MHEG, SVG.

Alongside with the binary information that constitutes the content itself, some basically metadata (creation date, file size, title, author etc.) are also stored in these formats.

In addition, a lot of multimedia features could be expressed through specialized metadata vocabularies, which could be classified in the following three types:

1. **Descriptive or Content metadata:** provide information about the objects captured in the multimedia content (the object's name, title, materials, dates, physical description, etc.). These metadata provide valuable semantic information for the multimedia search and retrieval operations.
2. **Technical metadata:** provide information about the multimedia document itself (not about objects depicted in the document). For example, technical metadata for images could include information about color histogram, file format, resolution, technical processes used in image capture or manipulation, etc.
3. **Administrative metadata** includes information related to the management of multimedia documents (such as rights management, creation date, author).

We provide further a brief overview of the most common metadata vocabularies:

For images:

- *Exchangeable Image File Format* (Exif)[5] includes metadata related to the image data structure (e.g., height, width, orientation), capturing information (e.g., rotation, exposure time, flash), recording offset (e.g., image data location, bytes per compressed strip), image data characteristics (e.g., transfer function, color space transformation), as well as general tags (e.g., image title, copyright holder, manufacturer);
- *VRA* Core set of metadata elements defined by Visual Resource Association and used to describe works of visual art as well as the images that represent them (http://www.vraweb.org/projects/vracore4/); in order o acquire uniform descriptions, two complementary dictionaries were established: Art & Architecture Thesaurus (AAT) and ULAN - Union List of Artist Names;
- *NISO Z39.87* includes metadata elements for raster digital images: basic image parameters, image creation, imaging performance assessment, history, etc. (http://www.niso.org/)
- *DIG 35* includes metadata regarding image creation, content description in terms of "who, what, when and where", or intellectual property rights; (http://xml.coverpages.org/FU-Berlin-DIG35-v10-Sept00.pdf);
- *PhotoRDF* is intended to provide description support for personal photo collections through three different metadata schemas: a Dublin Core, a technical schema and a content schema (http://www.w3.org/TR/2002/NOTE-photo-rdf-20020419).

For audio content:

- *ID3* metadata are embedded with an MP3 audio file format and provides multiple information about a song: title, artist, album, genre, as well as involved people list, lyrics, band, ownership, recording dates (http://www.id3.org/Developer_Information);
- *MusicBrainz* is a RDF-S based vocabulary, including a core set with basic music related metadata such as artist, album, track, etc., as well as two other sets for the extended music related metadata such as contributors, roles, lyrics, etc. (http://musicbrainz.org/MM/);
- *MusicXML* metadata provide support for universally translating and encoding the common Western musical notation from the 17th century onwards (http://www.recordare.com/xml.html).

For audio-visual content:

- *MPEG-7* (Multimedia Content Description Interface) represents one of the biggest

ISO effort in the direction of complex media content modeling (including support for spatio-temporal descriptions), aiming to be an overall for describing any multimedia content. It includs a large set of annotations on different semantic levels (the set of MPEG-7 XML Schemas define 1182 elements, 417 attributes and 377 complex types). MPEG-7 standard is organized into Descriptors (Ds), Description Schemes (DSs) and the relationships between them. Descriptors are used to represent specific features of the content, generally low-level features such as visual (e.g. texture, camera motion) or audio (e.g. melody), while description schemes refer to more abstract description entities (usually a set of related descriptors): Standard No. ISO/IEC 15938:2001 - http://www.iso.org/iso/en/prods-services/popstds/mpeg.html.

- *AAF* (Advanced Authoring Format) provides support for describing the authoring information: timeline-based compositions (i.e. motion picture montages), event-related information (e.g. time-based user annotations and remarks) or specific authoring instructions (http://www.aafassociation.org/html/techinfo/);
- *MXF* (Material Exchange Format) supports the interchange of material for the content creation industries. MXF is a wrapper/container format intended to encapsulate video, sound, pictures, etc. and allows applications to know the duration of the file, what essence codecs are required, what timeline complexity is involved and other key points to allow interchange.

As well, a set of domain-specific standard vocabularies were developed by various communities and initiatives:

- *LSCOM (Large-Scale Concept Ontology for Multimedia)*[6] organizes more than 800 visual concepts for which extraction algorithms are known to exist;
- *CIDOC/CRM (ICOM-CIDOC* Data Model Working Group, 1995) for museum objects,
- *FGDC (Federal Geographic Data Committee, 2003)* for geospatial data,
- *NewsML* (International Press Telecommunications Council, 2003) for news objects
- *IEEE LOM* (IEEE Learning Technology Standards Committee, 2002) for educational resources;
- *INDECS* for rights management;
- *TV*-Anytime for TV digital broadcasting;
- *MPEG-21* to enable transparent and augmented use of multimedia resources contained in digital items across a wide range of networks and devices;
- *EBU P/Meta* for the broadcast industry.

Despite the metadata richness supported by the above exposed standards and vocabularies, the meaning produced by these metadata still needs for a semantic enhancement in order to become explicit and transparent for automatic processing. For example, <dc:title> generic metadata from DCMI vocabulary only indicates the content as being the title of the current resource, but nothing about the meaning of this title. The solution consists in correlating this low-level metadata with certain ontology constructs in order to locate its semantics (Lee et al., 2005).

Some *specialized multimedia ontologies* were also developed in order to capture and express the high-level semantics for multimedia objects: aceMedia Visual Descriptor Ontology, mindswap Image Region Ontology, MSO - Multimedia Structure Ontology, VDO - Visual Descriptor Ontology, AIM@SHAPE ontology for representing, modeling and processing knowledge which derives from digital shapes, Music Information ontology, Semantic User Preference Ontology developed to be used in conjunction the MPEG-

7 MDS Ontology and with domain ontologies, in order to interoperate with MPEG-7 and allow domain knowledge utilization, CIDOC CRM core ontology for all multimedia objects, especially concerning cultural heritage items and events .

An effort could be noticed towards common, shareable harmonized multimedia ontology. A set of core features and specifications was defined, which is a "minimum set" for multimedia ontology to be used as a basis for all multimedia related applications[7]. An important conclusion concerns the separation into a Visual Descriptors Ontology, a Multimedia Structures Ontology, and a Spatiotemporal Ontology inside the common ontology framework.

The Interoperability and Semantic Description Problems

Two major problems concern the adoption of multimedia metadata standards inside real applications: the interoperability problem between the existing XML-based multimedia vocabularies, and their capacity to provide semantic enhanced description of the multimedia content.

The interoperability problem is a major problem in managing the multimedia metadata, which concerns many aspects:

- Metadata created and enhanced by different tools and systems follows different standards and representations; the existing tools and standards are not necessarily interoperable (Tzouvaras et al., 2007);
- Yet, none of the multimedia metadata standards used in practice for images, audio or video content can fully describe all categories of multimedia features corresponding to that multimedia content type. A complex description requires metadata from multiple vocabularies. For example, Dublin Core and EXIF do not allow to specify and annotate regions within images, and FotoNotes, or JPEG-2000 metadata image vocabularies should be used for this purpose, or even MPEG-7 (which is suitable for all multimedia types);
- The semantics of the information encoded in the XML are only specified within each standard's framework, using that standard's structure and terminology (Stamou et al., 2006): it is hard to re-use MPEG-7 metadata in environments that aren't based on MPEG-7 or to integrate non-MPEG metadata in an MPEG-7 application.
- The different standards include synonyms – metadata with different syntactic forms but referring the same semantic information. The W3C Media annotation working group developed a set of mappings between various multimedia metadata formats, and a core vocabulary, named MAWG, which represents the pivot for these synonymy relations (http://dev.w3.org/2008/video /mediaann/mediaont-1.0/mapping_table_common.htm).

So, in order to use together multiple standards for describing a multimedia document, at least two important issues should be addressed: to define a general structure for incorporating the multiple particular structures while keeping each particular semantic, as well as to define a solution for the synonymy problem.

(Chan, 2005) presents five models adopted by the multimedia application developers in order to achieve interoperability among different metadata schemas:

- *Uniform standard*: to use the same schema for describing all multimedia features;
- *Application profiling/adaptation/modification*: an existing schema is used as the basis for description, while the other schema are adapted or modified in order to match it;
- *Derivation*: an existing complex schema may be used as the "source" or "model" from which new and simpler individual

schemas may be derived. For example, both the MODS (Metadata Object Description Schema) and MARC Lite (Machine-Readable Cataloging) are derived from the MARC21 standard.

- *Crosswalk/mapping*: a mapping is accomplished for elements, semantics, and syntax from one metadata scheme to those of another. Some examples are: MARC21 to Dublin Core, MARC to UNIMARC, VRA to Dublin Core, ONIX (ONline Information eXchange) to MARCXML, FGDC to MARC, EAD (Encoded Archival Description) to ISAD(G) (General International Standard Archival Description), ETD-MS to MARCXML, as well as common mappings between Dublin Core/MARC/GILS or between ADL/FGDC/MARC/GILS, MARC/LOM/DC;
- *Switching schema*: an existing schema is used as the switching mechanism among multiple schemas. As example, Dublin Core is used as the switching schema by Picture Australia project as well as by Open Archive Initiative (OAI).
- *Lingua franca*: multiple existing metadata schemas are treated as satellites of a superstructure (lingua franca) which consists in common elements or most widely used by individual metadata schemas. An example of lingua franca superstructure is the ROADS template (Resource Organisation And Discovery in Subject-based services), which uses a set of broad, generic attributes.
- *Metadata framework/container*: metadata framework is used as a shell or container within which elements from multiple metadata schemas can be accommodated. Two prominent examples are:
- Resource Description Framework (RDF) – a data model that provides a mechanism for integrating multiple metadata schemes. As example, for the audio content semantic description, the metadata belonging to ID3 and OGG Vorbis vocabularies are mapped to RDF (Tzouvaras et al., 2007) by using also the DCMI and FOAF vocabularies as well as the music ontology (http://musicontology.com/). Multiple examples could be noticed in Section 2.2.
- Metadata Encoding and Transmission Standard (METS) – provides a framework for combining several internal metadata structures with external schemas.

As could be noticed, all the exposed techniques remain at the XML-based level of multimedia metadata, except the last technique, which consider the possibility of multimedia description based on RDF. Considered as the basis of the semantic Web, RDF is a semantically enriched language, having a formal, machine-processable semantics. As illustrated further, instead of RDF, many current approaches adopt OWL (Web Ontology Language), the basic ontology language representation in the Semantic Web community, with more accurate and enriched facilities for describing the resource semantics. This is a major step towards the integration of semantic Web technologies into multimedia indexation and processing domain.

RDF-Based Solutions for Solving the Interoperability and the Semantic Description Problems

The syntactic and semantic differences between XML-based and RDF-based multimedia description were widely identified and discussed (Nack et al., 2005):

- *Syntactically*: because one RDF assertion could be XML serialized in many ways, it is hard to process RDF using generic XML tools, and reverse;
- *Semantically*, the semantic Web approach makes use of different layers that define semantic structures and ontologies as third

parties, while an XML-based specification such as MPEG is a monolithic specification, including a large number of schemata.

In addition, particular XML-based multimedia metadata formats face the following problems (Arndt et al., 2007):

- Are not open to Web standards: do not allow to integrate the existing vocabularies for multimedia or to point out to domain-specific ontologies in order to semantically describe multimedia content;
- Could encounter interoperability problems, as the case of MPEG-7 is: several descriptors, semantically equivalent and representing the same information while using different syntax can coexist (Arndt et al., 2007).

Despite these differences and inconveniences, multiple solutions to unify the XML-based and RDF-based representations were proposed in order to gain multimedia metadata interoperability and to make multimedia content available for the semantic Web applications.

A first type of such solution considered the abandon of XML-based initial format of the multimedia metadata standards in the favor of their RDF-based equivalent or even their OWL-based equivalent. Thus, multiple XML-based multimedia vocabularies were translated into RDF or OWL - see especially (Hausenblas et al., 2007):

- The MPEG-7 gained the most numerous translations to RDF: (Hunter, 2001) or to OWL: (Tsinaraki et al., 2004), (Garcia & Celma, 2005), COMM (Arndt et.al., 2007). etc. The differences between them concern the coverage of the entire MPEG7 specification, as well as the maintenance of the initial MPEG7 structure.
- EXIF (Exchangeable Image File Format) specification encountered two notable conversions to RDF. The Kanzaki EXIF RDF Schema (http://www.kanzaki.com/test/exif2rdf) provides an encoding of the basic EXIF metadata tags (especially the section 4.6 of EXIF specification) in RDFS (RDF Schema). It additionally provides an EXIF conversion service, EXIF-to-RDF, which extracts EXIF metadata from images and automatically maps into the RDF encoding. As well, the Norm Walsh EXIF RDF Schema (http://nwalsh.com/java/jpegrdf/) provides another encoding of the basic EXIF metadata tags in RDFS. It additionally provides JPEGRDF, which is a Java application providing an API to read and manipulate EXIF metadata stored in JPEG images. Currently, JPEGRDF can extract, query, and augment the EXIF/RDF data stored in the file headers.
- The XBRL (Extensible Business Reporting Language - http://www.xbrl.org/) acquired multiple representations as OWL taxonomies: (Declerck & Krieger, 2006), (Lara et a., 2006), (Knublauch et al., 2004);
- VRA vocabularies knows two RDF/OWL representation: one accomplished by Mark van Assem (http://www.w3.org/2001/sw/BestPractices/MM/vra-conversion.html) and the other performed in the framework of SIMILE project (http://simile.mit.edu/2003/10/ontologies/vraCore3);
- For DIG35 specification, an RDF/OWL ontology was developed at IBBT Multimedia Lab (University of Ghent);

It could be noticed that the exposed approaches are focused mainly on the translation to RDF and to OWL of one particular XML-based multimedia metadata standard, while a general approach which to adopt and integrate multiple such RDF-based representations of different standards was not yet proposed.

Moreover, the majority of these approaches are focused only on a metadata mapping to RDF/

OWL, not also to a structural mapping, ignoring the advantages of an XML-based structural layer (Stamou et al., 2006).

Ontology-Based Solutions for Solving the Interoperability Problem

In order to define a solution for adopting a semantic description of multimedia resources while keeping also their XML-based metadata, (Stamou et al., 2006) consider three different levels of abstraction: subsymbolic, symbolic, and logical multimedia annotations could be accomplished (see Figure 1):

- The subsymbolic abstraction level is constituted by the established binary formats for storing the video, image, audio and text data, incorporating general metadata such as creation date, size, used tool;
- The symbolic abstraction level includes metadata that provides information about the internal structure or other specific features of the media stream in XML formats: Dublin Core, MPEG-7, MPEG-21, Visual Resource Association, International Press Telecommunications Council, and so on. The interoperability is the big problem of these standards.
- The logical abstraction level provides semantics for the above level, actually defining mappings between the structured information sources and the domain's formal knowledge representation (for example, by using OWL – Web Ontology Language).

As could be noticed, the metadata interoperability problem is solved by making explicit the implicit knowledge represented by the multimedia document description; as well, the multimedia semantics is enriched through its correlation with domain knowledge (encoded into ontologies) and through available reasoning operations that enable to acquire new knowledge about multimedia document.

Existing Tools for Ontology-Based Multimedia Annotation

The explicit multimedia ontology-based annotation through a visual interface is the simplest but hence the most expensive method to acquire

Figure 1. The different levels of multimedia information and the type of annotation provided for each level, according

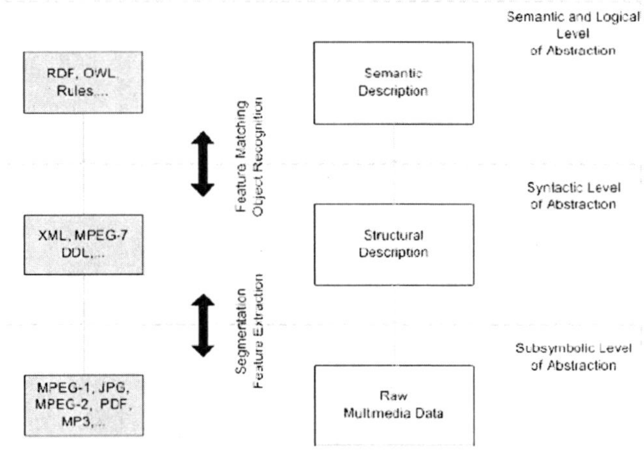

semantically enhanced metadata. Some specialized tools were developed in order to support this type of manual annotation.

Protégé allows a user to load OWL ontologies, annotate data, and save annotation markup. Protégé provides only simple multimedia support through the Media Slot Widget, which allows general description of multimedia files like metadata entries, but not also description of multimedia document spatio-temporal fragments.

PhotoStuff allows annotating images and contents of specific regions in images according to several OWL ontologies of any domain (http://www.mindswap.org/2003/PhotoStuff/). Also designed for images, *AKtive Media* is an ontology based annotation system. *ImageSpace* provides support DAML+OIL language, and integrate image ontology creation, image annotation and display into a single framework.

The *IBM MPEG-7 Annotation Tool* provides support for annotating video sequences with MPEG-7 metadata based on the shots of the video. The annotations do not use ontologies but an editable lexicon from which a user can choose keywords to annotate shots. Except the automatic video decomposition in shots, the facilities for video time alignment and time segmentation are not provided. Annotations are saved based on MPEG-7 XML Schema.

Advene (Annotate Digital Video, Exchange on the NEt) provides tools to edit and visualize the hypervideos generated from both the annotation and the audiovisual do-cument.

ELAN (EUDICO Linguistic Annotator) provides support for linguistic annotation (analysis of language, sign language, and gesture) of multimedia recordings, inclu-ding support for time segmentation and multiple annotation layers, but not the support of ontology. OntoELAN (Chebotko et al., 2005) extends ELAN with an ontology-based annotation approach: OWL linguistic ontologies could be used in annotations, while the ontological tiers should be linked to general multimedia ontology classes. With this role, GOLD (General Ontology for Linguistic Description) ontology is adopted (Farrar & Langendoen, 2003).

ONTOLOGY-DRIVEN FRAMEWORKS FOR MULTIMEDIA SEMANTIC ANNOTATION

A set of requirements were defined for a framework that enables a standardized way to create, store, and maintain semantic annotated multimedia content (Hentschel et al., 2007):

- To provide support for storing information about the multimedia content segments together with their spatial and hierarchical relations;
- To provide means for storing low and high level features as well as semantic annotations for each multimedia segment or document itself;
- All annotation types have to be stored in a data structure that can be easily modified, is well documented and possibly already supported by existing software tools.

Ontologies for Modeling the Multimedia Semantic Annotations

Some solutions were proposed for defining ontology in order to provide semantic expression for the multimedia structural metadata expressed in XML vocabularies. Such solutions include *COMM* ontology (http://comm.semanticweb.org/), *ABC* (Lagoze & Hunter, 2001) or *aceMedia* ontology (http://www.acemedia.org/), all of them being developed on the top of MPEG7 particular. These ontologies aim to provide support for describing all the multimedia features (no matter of their initially XML-based expression), but concretely no approach illustrates the integration and interoperability of multimedia vocabularies. We present bellow the most important such solutions.

The *aceMedia Ontology Framework* (Bloehdorn et. al, 2005) define an integrated multimedia annotation framework based on a core ontology (DOLCE), two multimedia MPEG-7 based ontologies (VDO - Visual Descriptor Ontology - and MSO - Multimedia Structure Ontology), as well as domain ontologies such as PCS (Personal Content Management) and CCM (Commercial Content Management) Ontologies.

DELOS II Network of Excellence (Tsinaraki et al., 2004) defined an MPEG-7 upper ontology, which was extended with Semantic User Preference Description ontology and harmonized with MPEG 21 DIA Ontology, as well as with *SUMO* and *DOLCE* core ontologies in order to acquire an integrated annotation framework. *GraphOnto* was adopted as visual ontology-based annotation tool for multimedia content.

The goal of *COMM (Common Ontology for Multimedia)* ontology is to describe the semantics of multimedia content in terms of current semantic Web languages (Arndt et al., 2007). The approach is based on MPEG-7 (which provides support for complexly describing multimedia content) and provides a formal semantics for MPEG-7 by developing the COMM ontology:

- Uses DOLCE (Descriptive Ontology for Linguistic and Cognitive Engineering) as a modeling basis, especially two its main design patterns:
- Descriptions & Situations (D&S): to formalize contextual knowledge;
- Ontology of Information Objects (OIO): implements a semiotics model of communication theory.

Some patterns are defined:

- **Decomposition:** exploits MPEG-7 descriptors for spatial, temporal, spatiotemporal and media source decompositions of multimedia content into segments; for example, LocalizationDescriptor includes an ontological representation of the MPEG-7 RegionLocatorType for localizing regions in an image.
- **Annotation:** exploits the MPEG-7 very large collection of descriptors that can be used to annotate a segment. The annotations are associated to a particular media content region (or to the entire media document):
- **Content Annotation:** for annotating the features of a multimedia document, which means for expressing its associated metadata (media-specific metadata). E.g. DominantColorAnnotation expresses the connection between a MPEG-7 DominantColorType with a segment. The annotations are obtained by applying algorithms (see Algorithm pattern).
- **Media Annotation Pattern:** for describing the physical instances of multimedia content (general metadata): e.g. MediaFormatType (FileSize="462848", FileFormat="JPEG");
- **Semantic Annotation Pattern (semantic metadata):** allow the connection of multimedia descriptions with domain descriptions provided by independent domain-specific ontologies
- **Digital Data** pattern is used to formalize most of the complex MPEG-7 low-level descriptors. E.g., the MPEG-7 RegionLocatorType which mainly includes two elements – Box and Polygon – is formalized through the RegionLocatorDescriptor concept and through its BoundingBox and StructuredDataParameters subconcepts. Also, DigitalData entities express Descriptions such as Regions described by Parametres.
- Algorithm pattern defines:
- **Methods:** for the manual (or semiautomatic) annotations;
- **Agorithms:** for automatically computed features (e.g. dominant colors)

Every Algorithm defines at least one InputRole and one OutputRole which both have to be played by DigitalData.

So that, the COMM ontology exploits and extends the structure of the MPEG-7 specifications in order to provide support for organizing the multimedia metadata; also, the COMM ontology provides support for expressing all the multimedia features covered by the MPEG-7 specification, which forms a really huge set. The advantage of its formal semantics consists in enabling to express these features, independently of the XML-based vocabulary through which the features were expressed initially. In other words, COMM provides support to express all the XML-based multimedia metadata having synonyms in MPEG-7 specification. However, the set of the remaining multimedia metadata is not covered.

In order to respect the COMM approach, a concrete implementation should adopt the MPEG-7 based structure for organizing multimedia metadata and should match all the used multimedia features to the features included in this structure. It must be noticed that the synonymy between MPEG-7 originate features and those specified by other XML-based multimedia metadata vocabularies remains an unsolved problem, to be decided by the developer himself.

The Problem of Integrating Indexation Algorithms into Annotation Frameworks

It must be also noticed the *algorithm pattern* considered by the COMM ontology. Its role consists in modeling the indexation algorithms applied to multimedia content in order to produce metadata. There is a very large palette of such algorithms, characterized by a great heterogeneity concerning their input and output data, preconditions and effects, implementation details, hosting platforms or architectures. These algorithms are usually integrated into different tools (e.g. for multimedia content production, management, storage, or even indexation – as the case of search engines). In order to make these algorithms interoperable, the COMM ontology provides support for uniformly describing these algorithms, especially in terms of their input data (the processed multimedia content) and their output data (the produced multimedia metadata).

The distinction made by the COMM ontology between algorithms and methods was dictated by the existing "semantic gap" between low-level multimedia features and high-level concepts describing the multimedia content: while algorithms could be executed to automatically extract the low-level multimedia features, in order to acquire a high-level semantic annotation, the methods mentioned in COMM should combine the human intervention with multiple complex techniques such as content analysis, knowledge databases, machine learning, or semantic Web. An ordinary method for multimedia semantic description is constituted by the manual annotation accomplished with the support of specialized tools such those mentioned above.

The algorithms for automatic indexation are specific for each media type. Algorithms for image indexation are based on feature extraction, clustering/segmentation, object descriptor extraction, object recognition (Chen et al., 2000). Audio analysis is accomplished in some main directions (Foote, 1999): segmentation (splitting an audio signal into intervals for determining the sound semantics or composition), classification (categorizing audio segments according to predefined semantic classes, such as speech, music, silences, background noise) and retrieval by content (using similarity measures in order to retrieve audio items that syntactically or semantically match the query). Video indexing techniques enable to extract metadata such as the content type, shot-boundaries, audio keywords, textual keywords or caption text keywords (Donderler et al., 2005). There is also some research, e.g., (Viola & Jones, 2001), in determining the class of video scene objects (e.g., human, vehicle, type of vehicle, animal) and detecting activities (e.g., carrying a bag, raising arms in the air, the manner of running or walking).

More usually, a complex semantic multimedia indexation involves the application of multiple algorithms into a given order and under certain conditions. The description of each algorithm input and output data (as facilitated by COMM) is very important for this purpose; however a modality of defining the order and conditions for algorithms' combination is required. A solution is provided by (Haidar et al., 2005), where a set of rules for proper algorithms sequential combination is defined. Another solution (Brut et al., 2009) separates the description of the input/output data of the indexation algorithm from the semantic description of its functionality and from the orchestration rules for algorithm combination, expressing this type of information in terms of WSMO (Web Service Modeling Ontology) – (Roman et al., 2005).

Ontology-Driven Multimedia Annotation and Management Concrete Approaches

As could be noticed, there is a lot of support for expressing and organizing the multimedia metadata, despite the lack of a generally accepted integrated framework. We are focusing now on some concrete approaches where techniques for obtaining such metadata are also considered, in the framework of concrete multimedia systems.

In (Song et al., 2006), a video content annotation architecture built on PhotoStuff (see section 2.3) is used to link MPEG-7 visual descriptors (obtained through automatic multimedia processing) to high-level, domain-specific concepts. The obtained multimedia semantic metadata is further used in order to improve the browsing and searching capabilities.

(Marques & Barman 2003) used machine-learning techniques to semantically annotate images with descriptors defined through ontological constructs.

In (Hunter & Little, 2005) is presented a system were the multimedia content is annotated through three ontologies: the developed otology on the top of MPEG-7 (Hunter, 2001), and two domain-specific ontologies. In order to enable the semantic interoperability, the three ontologies are merged with the support of ABC top-level ontology (Lagoze & Hunter, 2001). Alongside with the manual annotation, domain-specific inferencing rules are defined by domain-experts through an intuitive user-friendly interface in order to automatically produce supplementary semantic metadata.

In (Garcia & Celma, 2005), the MPEG-7 OWL ontology[8] is used as upper-level multimedia ontology where three different music ontologies have been linked in order to annotate the multimedia content. System architecture is proposed that facilitates multimedia metadata integration and retrieval, while the semantic metadata themselves are considered available to be fed up into system.

In SAFIRE project (Hentschel et al., 2007), MPEG-7 structure is used as basis for organizing multimedia features. Alongside with automatically extracted features, the semantic annotations are accomplished manually, using WordNet ontology in order to acquire disambiguate annotations. The synonyms are also exploited for increasing the efficiency of the further query process.

In (Yildirim & Yazici, 2007) ontology is used in order to define the video database model. Such ontology must be previously developed for a certain modeled domain, containing definitions of objects, events and concepts in terms of attributes and components. The system applies in a first phase a set of automatic multimedia processing techniques in order to segment the video into regions, and to extract features for each region (color, shape, color distribution etc.). If some regions have similar properties for a period of time (consecutive keyframes), the possible occurrence of an object could be inferred. By using similarity functions, objects identified from regions are assigned to their actual names by using information gained from the training set developed by experts according the considered ontology. The ontology-based data model enables the system to support ontology-based queries, able to specify objects, events, spatio-temporal clauses, trajectory clauses, as well as low-level features of objects.

In METIS project (King et al., 2007), the multimedia content is organized into a database, characterized by customizable media types, metadata attributes, and associations, which constitutes a highly expressive and flexible model for media description and classification. The multimedia ontology-based annotations could be also defined, due to the developed plug-in for the open-source Protégé ontology editor.

The project aceMedia adopts manual ontology-based multimedia annotations, with the support of M-OntoMat-Annotizer. As well, the project developed a multimedia analysis system for automatically annotate the multimedia content based on the developed aceMedia Visual Descriptor Ontology. The system includes methods that automatically segment images, video sequences and key frames into a set of atom-regions while visual descriptors and spatial relations are extracted for each region (Bloehdorn et al., 2005). A distance measure between these descriptors and the ones of the prototype instances included in the domain ontology is estimated using a neural network approach for distance weighting. Finally, a genetic algorithm decides the labeling of the atom regions with a set of hypotheses, where each hypothesis represents a concept from the above mentioned domain ontology. This approach is generic and applicable to any domain as long as specific domain ontologies are designed and made available.

A PROPOSAL FOR MULTIMEDIA DISTRIBUTED INDEXING AND ANNOTATION

We consider the case of a distributed Web environment where there is necessary to manage the following collections:

- A multimedia collection, containing images, video and audio documents, in different formats;
- A metadata collection, where some metadata fragments with different structures and formats are available for each multimedia document;
- A collection of indexing algorithms which could be applied to multimedia content in order to produce new specific metadata;
- A set of domain ontologies that could be used in order to semantically enhance the multimedia content description.

In order to respond to the user queries for retrieving specific multimedia content, the problem of harmonizing the metadata stored in the second collection raises. As well, for detecting the algorithms suited for a new indexation according to the user query requirements, a uniform handling of these algorithms should be possible. Supplementary, when intending to use some domain ontologies in order to acquire a semantically enhanced annotation, the obtained metadata should be handled uniformly and should be harmonized with the metadata collection.

Harmonizing the Multimedia Metadata

A user query usually refers to a set of requested multimedia features. In more complex situations, some relations could de demanded between certain such features. As presented in the second section, each multimedia vocabulary organizes a subset of the multimedia features into a particular structure. Moreover, some multimedia features receive different names in some distinct vocabularies, and a set of mappings was established between various multimedia metadata formats in the framework of W3C media annotation working group, considering the MAWG core vocabulary as pivot[9]. This synonymy table will be considered in our proposal as an instrument in order to harmonize the multimedia metadata collection, as well as the metadata specified by the user queries:

- The multimedia metadata from the specific collection are "serialized" according to the MAWG vocabulary, receiving the corresponding synonym from the MAWG vocabulary as value for a new attribute, namely "id";

For example, a MPEG7 fragment that defines the duration of a video file will receive a new "id" attribute indicating the "ma:duration" as corresponding synonym:

```
<MPEG7> <Video id="Movie01.avi">
   <MediaTime>
        <MediaDuration id="ma:duration">
</MediaDuration>
   </MediaTime>
</Video> </MPEG7>
```

- As well, the user queries will be processed in order to detect the desired multimedia features. A textual analysis will be accomplished based on the list of descriptive properties considered by the core set of MAWG vocabulary[10].

For example, a query such as "Find me the videos about Paris taken by Paul and having a duration of about 10 minutes" should detect the following features: ma:title="Paris", ma:keywords="Paris", ma:duration="10min", ma:creator="Paul".

Our approach considers the MAWG vocabulary as a dictionary where the meaning of metadata from different vocabularies could be retrieved. Ideally, these metadata should be included inside a uniform structure. Such structure is provided by the COMM ontology, where the Annotation pattern provides support for expressing the multimedia features considered by the MPEG7 specification. The combination between COMM structure and MAWG vocabulary (which covers the problem of synonymy between multiple metadata formats) could be a solution for addressing the problem of interoperability between multimedia metadata formats for harmonizing them. The issue that should be addressed in the future concerns the alignment between COMM ontology and each particular multimedia metadata format: the features expressed by such format could be easily identified in order to be transferred into COMM structure, but however their initial structure is lost, despite it could include some useful information.

Harmonizing the Indexation Algorithms

The available algorithms for multimedia documents indexing are characterized by a great heterogeneity concerning their input and output data, as well as implementation details, hosting platforms or architectures. An indexation algorithm is applied to a multimedia document and produces some

Figure 2. Proposal to harmonize the multimedia metadata

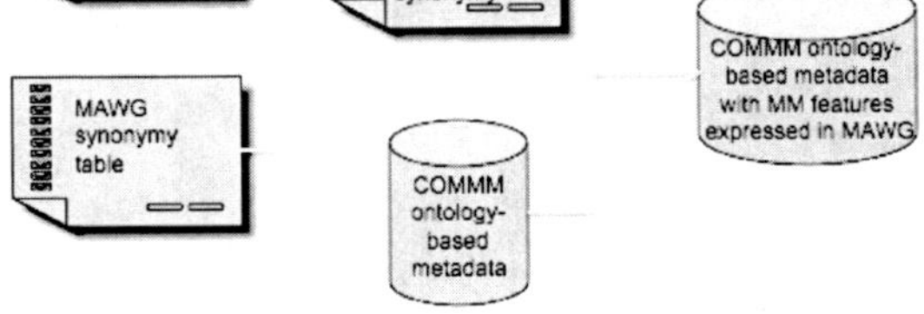

information about certain multimedia features. In order to uniformly handling these algorithms, a comprehensive description should be associated to each of them, concerning:

- the input and output data;
- the preconditions required for the algorithm execution, e.g. the words recognition algorithm could be applied if the English Language detection was previously performed;
- the postconditions, describing the state of the metadata collection that is guaranteed to be reached after the successful execution of the algorithm;
- the effects over the multimedia items, such as color enhancement.

In (Brut et al., 2009) we provided a solution for such indexation algorithm description, based on the WSMO framework (*Web Service Modeling Ontology*) – (Roman et al., 2005). While keeping the initial implementation of each algorithm, it should be embedded into a generic interface (e.g. a Web service) where:

- The input data consists in metadata which identify the input multimedia file (e.g. "ma:identifier", "ma:title", "ma:location");
- The output data consists in metadata structures where each feature is accompanied by an identifier for its MAWG synonym.

This solution enables also to determine the combination of multiple algorithms that could produces complex multimedia metadata required in order to respond to a particular user query. For example, a query which demands for the "Tim Berners Lee's speeches about the Semantic Web" should necessitate the multimedia indexation by using multiple algorithms into a certain order, e.g. first the algorithm for detecting the human speech, then the voice of Tim Berners Lee, then the English Language, and finally the algorithm for recognizing the particular English words. This sequence could be acquired due to the WSMO-based description of the algorithms (mentioned through WSMO *Capability* class).

The metadata produced as result of simple or multiple indexations are stored inside the metadata collection. At the storage time, these metadata are also indexed according the metadata dictionary, considering the synonymy table (see Figure 3).

Harmonizing the Ontology-Based Semantic Annotations

The typical example of semantically enhanced metadata produced by the automatically executed multimedia indexation algorithms concerns the detection of some particular objects or persons, as well as some particular audio information. Such possible metadata associated to multimedia objects having visual characteristics was gathered and organized into the Large-Scale Concept Ontology for Multimedia (LSCOM) in order to facilitate the multimedia representation and classification (Naphade et al., 2006).

On the other hand, the semi-automatically produced ontology-based semantic annotations provide, in fact, enhanced descriptions of the

Figure 3. Proposal to harmonize algorithms-based multimedia indexation

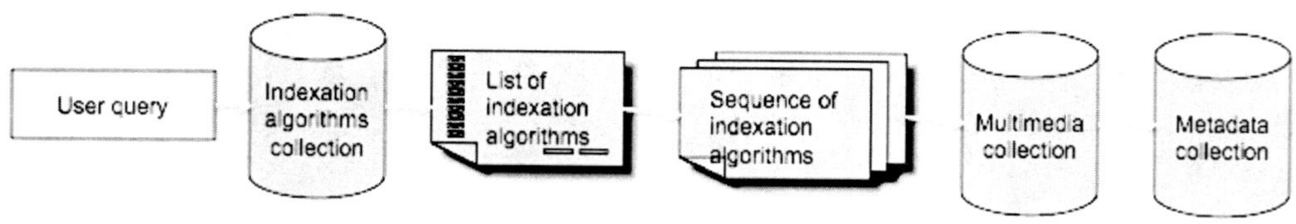

same visual objects, which could be identified as instances for LSCOM ontology. For example, about a castle detected by indexation algorithms and identified through LSCOM ontology, a semantic enhanced annotation could make use of a geographic ontology in order to provide supplementary information about the name and the role of that particular castle, the city where it is located, the possibilities to be visited, etc.

Departing from these considerations, our approach for semantic multimedia annotation consists in:

- To use the LSCOM ontology as a dictionary of visual objects comprised into multimedia documents;
- To include the identifier (with its value from LSCOM ontology) of a particular visual object inside the automatically acquired annotations;

For example, since in LSCOM the castle object has the identifier "042" (and the definition "Exterior shots of a castle (building with turrets)"), the corresponding MPEG7 description should include:

```
<MPEG7> <MultimediaContent id="042">
       Castle - Exterior shots of a cas-
tle (building with turrets)
                    </MultimediaCon-
tent></MPEG7>
```

- As well, since the visual objects were previously identified with the support of indexation algorithms, the enhanced ontology-based annotations will refer these objects, while keeping their LSCOM-based identifier. A semantic annotation tool (e.g. M-OntoMat-Annotizer) could be used for manually acquiring these annotations. Further, some classification methods or reasoning techniques could be developed in order to semi-automate this process. For example, if the location of the castle could be retrieved among the existing metadata, then the mentioned supplementary information could be inferred from a corresponding geographical ontology.

This approach adopts the LSCOM-based visual objects dictionary as a pipe between the automatic and manual multimedia indexation: the results of indexation algorithms are correlated with visual concepts, and these concepts will be further detailed based on supplementary domain ontologies.

The user query could be analyzed in this case from two perspectives, in order to detect the requested multimedia features centralized inside the MAWG dictionary (as discussed in the previous section) as well as to detect the requested visual objects, centralized inside the LSCOM ontology.

CONCLUSION

While the domain of ontology-based annotation of textual documents is still incipient in acquiring a solution based on automatic tasks, the domain of ontology-based annotating multimedia documents is much more complex and a solution which to avoid the human intervention is far from achieved. The big challenge of this domain consists in integrating the manual ontology-based annotation with the benefits acquired by the solid experienced multimedia indexing domain.

The chapter presented an overview of the accomplishments gained in developing standards, vocabularies and ontologies for describing multimedia metadata, as well as in indexing multimedia collections in order to develop metadata collections. A solution for integrating the exposed metadata structures and indexing techniques was proposed in Section 4. While preserving the original metadata structures, the solution proposes

Figure 4. Visual objects-based indexation process

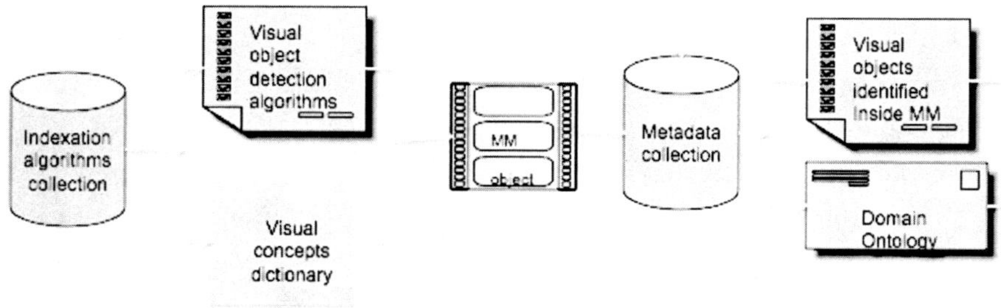

the metadata management with the support of a metadata dictionary. As well, in order to harmonize the different ontology-based annotation, the exposed solution considers a dictionary of visual objects. In addition, the correlation between the two dictionaries facilitates the management of the multimedia indexation process and the user query analysis.

We will further focus our research on implementing the proposed approach and testing it on concrete multimedia collections.

ACKNOWLEDGMENT

This research work is supported by the FP7 PIEF-GA-2009-235229 SOMIR project.

REFERENCES

Arndt, R., Troncy, R., Staab, S., Hardman, L., & Vacura, M. (2007). COMM: Designing a well-founded multimedia ontology for the web. In ISWC/ASWC Proceedings, (LNCS 4825, pp. 30-43). Berlin: Springer.

Bloehdorn, S. Petridis, K. Saathoff, C. Simou, N. Tzouvaras, V. Avrithis, et al. (2005). Semantic Annotation of Images and Videos for Multimedia Analysis. In *Proc. 2nd European Semantic Web Conference, ESWC 2005*, Heraklion, Greece.

Brut, M., Sedes, F., & Manzat, A.-M. (2009). A Web Services Orchestration Solution for Semantic Multimedia Indexing and Retrieval. In *Proceedings of CISIS 2009*. Washington, DC: IEEE Computer Society Press.

Chan, L. M. (2005). Metadata Interoperability: A Study of Methodology. In *Proceedings of the 3rd China-US Library Conference.*

Chebotko, A., Deng, Y., Lu, S., Fotouhi, F., & Aristar, A. (2005, Jan-March). An Ontology-Based Multimedia Annotator for the Semantic Web of Language Engineering. *International Journal on Semantic Web and Information Systems, 1*(1), 50–67.

Chen, S.-C., Ghafoor, A., & Kashyap, R. L. (2000). *Semantic Models for Multimedia Database Searching and Browsing*. Norwell, MA: Kluwer Academic Publishers.

Declerck, T., & Krieger, H.-U. (2006). Translating XBRL Into Description Logic. An Approach Using Protege, Sesame & OWL. In W. Abramowicz, & H. Mayr, (Eds.), *Proceedings of the 9th International Conference on Business Information System.*

Donderler, M. E., Saykol, E., Arslan, U., Ulusoy, O., & Gudukbay, U. (2005). BilVideo: Design and implementation of a video database management system. *Multimedia Tools and Applications, 27*(1), 79–104. doi:10.1007/s11042-005-2715-7

Farrar, S., & Langendoen, D. T. (2003). A linguistic ontology for the Semantic Web. *GLOT International*, *7*(3), 97–100.

Foote, J. (1999). An overview of audio information retrieval. *Multimedia Systems*, *7*(1), 2–10. doi:10.1007/s005300050106

García, R., & Celma, O. (2005) Semantic integration and retrieval of multimedia metadata. In *Proceedings of the Fifth International Workshop on Knowledge Markup and Semantic Annotation at the Fourth International Semantic Web Conference,* Galway, Ireland.

Haidar, B., Joly, P., & Haidar, S. A Graph-based Approach to Automatically Chain Distributed Multimedia Indexing Services. In *Proceedings of 9th Int. Conf. on Internet and Multimedia Systems and Applications*, Switzerland. Boston: ACTA Press.

Hausenblas, M., Troncy, R., Halaschek-Wiener, C., Bürger, T., Celma, O., Boll, S., & Mannens, E. (2007). Multimedia Semantics on the Web: Vocabularies. *W3C technical reports,* Boston. Retrieved from http://www.w3.org/2005/Incubator/mmsem/wiki/Vocabularies

Hentschel, C., Nurnberger, A., Schmitt, I., & Stober, S. (2007). SAFIRE: Towards Standardized Semantic Rich Image Annotation. In S. Marchand-Maillet et al. (Eds.), *Proceedings of AMR 2006*, (LNCS 4398, pp. 12–27). Berlin: Springer.

Hunter, J. (2001). Adding multimedia to the semantic web – building an MPEG-7 Ontology. *International Semantic Web Working Symposium (SWWS)*, Stanford.

Hunter, J., & Little, S. (2005). A framework to enable the semantic inferencing and querying of multimedia content. *Int. J. Web Engineering and Technology*, *2*(2/3), 264–286. doi:10.1504/IJWET.2005.008487

King, R., Popitsch, N., & Westermann, U. (2007). METIS: a flexible foundation for the unified management of multimedia assets. *Multimedia Tools and Applications*, *33*, 325–349. doi:10.1007/s11042-007-0101-3

Knublauch, H., Musen, M. A., & Rector, A. L. (2004). Editing description logic ontologies with the Protege OWL plugin. In *Proceedings of the International Workshop on Description Logics (DL)*.

Lagoze, C., & Hunter, J. (2001). The ABC ontology and model (v3.0). *Journal of Digital Information*, *2*, Retrieved from http://jodi.ecs.soton.ac.uk/Articles/v02/i02/Lagoze/.

Lara, R., Cantador, I., & Castells, P. (2006). BRL Taxonomies and OWL Ontologies for Investment Funds R . In *ER* (pp. 271–280). Workshops.

Lee, J. K., Oh, J.-H., & Hwang, S. (2005). Strg-index: Spatio-temporal region graph indexing for large video databases. In *Proceedings of SIGMOD Conference*, (pp. 718–729).

Marques, O., & Barman, N. (2003). Semi-automatic semantic annotation of images using machine learning techniques. *International Semantic Web Conference*, Florida, October 2003.

Nack, F., Van Ossenbruggen, J., & Hardman, L. (2005, January - March). That Obscure Object of Desire: Multimedia Metadata on the Web (Part II). *IEEE MultiMedia*, *12*(1), 54–63. doi:10.1109/MMUL.2005.12

Naphade, M., Smith, J. R., Tesic, J., Chang, S., Hsu, W., Kennedy, L., et al. (2006). A Large-Scale Concept Ontology for Multimedia. *IEEE MultiMedia, 13* (3), 86-91. Retrieved from DOI=http://dx.doi.org/10.1109/MMUL.2006.63

Roman, D., Keller, U., Lausen, H., de Bruijna, J., Lara, R., Stollberg, M., et al. (2005). Web Service Modeling Ontology. Applied Ontology, 1, 77–106. Amsterdam: IOS Press.

Song, D., Cho, M., Choi, C., Shin, J., Park, J., & Kim, P. (2006). Knowledge Representation for Video Assisted by Domain-Specific Ontology. In A. Hoffmann et al. (Eds.), *Proceedings of PKAW 2006*, (LNAI 4303, pp. 144 – 155). Berlin: Springer-Verlag.

Stamou, G., van Ossenbruggen, J., Pan, J. Z., & Schreiber, G. (2006, January-March). Multimedia Annotations on the Semantic Web. In IEEE Multimedia.

Tsinaraki, C. Polydoros, P. & Christodoulakis, S. (2004, July 21-23). Interoperability support for Ontology-based Video Retrieval Applications. In *Proc. of 3rd International Conference on Image and Video Retrieval (CIVR 2004)*, Dublin, Ireland.

Tzouvaras, V., Troncy, R., & Pan, J. (2007). *Multimedia Annotation Interoperability Framework*. W3C Technical reports, Boston. Retrieved from http://www.w3.org/2005/Incubator /mmsem/XGR-interoperability/

Viola, P. & Jones, M. (2001). Robust real-time object detection. *International Journal of Computer Vision*.

Yildirim, Y., & Yazici, A. (2007). Ontology-Supported Video Modeling and Retrieval. In S. Marchand-Maillet et al, (Eds.), *Proceedings of AMR 2006*, (LNCS 4398, pp. 28–41). Berlin: Springer.

ENDNOTES

[1] http://www.w3.org/2001/sw/

[2] http://www.iso.org/iso/prods-services/popstds/mpeg.html

[3] http://www.w3.org/2005/Incubator/mmsem/XGR-interoperability/

[4] http://comm.semanticweb.org/

[5] Technical Standardization Committee on AV & IT Storage Systems and Equipment, Exchangeable image file format for digital still cameras: Exif Version 2.2, Japan Electronics and Information Technology Industries Association: http://www.digicamsoft.com/exif22 /exif22/html/exif22_1.htm

[6] http://www.lscom.org/

[7] Eleftherohorinou, H., Zervaki, V., Gounaris, A., Papastathis, V., Kompatsiaris, Y., Hobson, P., Towards a Common Multimedia Ontology Framework: http://www.acemedia.org/aceMedia/reference/multimedia_ontology/index.html

[8] http://rhizomik.net/ontologies/mpeg7ontos/

[9] http://www.w3.org/2008/WebVideo/Annotations/drafts/ontology10/WD/mapping_table.html

[10] http://www.w3.org/2008/WebVideo/Annotations/drafts/ontology10/WD/summary.html

Chapter 14
Semantic Enrichment of Web Service Architecture

Aicha Boubekeur
University of Tiaret, Algeria

Mimoun Malki
University of Sidi Bel-Abbes, Algeria

Abdellah Chouarfia
University of Oran, Algeria

Mostefa Belarbi
University of Tiaret, Algeria

ABSTRACT

The SOA: Service Oriented Architecture is a paradigm which allows the unification in the approaches of integration of the information systems. This data integration of shared semantic description of handled by the services. This integration of the data is more flexible considering the limited number of the concepts used by the services. Therefore, architecture is suggested in order to reduce domain ontologies development and integration complexity. It allows also finding and automatic invocation of the services. Ontologies are integrated without doing major changes in operating mode of web services like HTTP, SOAP. This chapter presents an architecture which is a step towards its automation through the semantic web services without redefining the information system completely.

INTRODUCTION

Exchanged information on the Web becomes very important. Information research is therefore vague and slow. XML common language enriched by the metadata was insufficient. In order to satisfy the enormous number of users, it will be necessary to introduce the data semantics concept. This latter allows defining the semantic Web project also called "Web 2.0". Web 2.0 uses the Artificial Intelligence (AI) tools especially Knowledge Engineering (KE) where ontology formalism was used.

Moreover, some research works added functional level to the Web. On this level, Web services satisfy interoperability. Therefore, an explicit semantic annotation (Benslimane, Amar Bensaber, & Malki, 2004) is necessary to improve interoperation. In this

DOI: 10.4018/973-1-61520-859-3.ch014

case ontologies will be used. Semantic integration of the applications is a very active research field (Izza, Vincent, & Burlat, 2006; Harriche, Hatchabi, & Maiga, 2008). It concerns integration with the data and treatments. Semantic integration by the data requires the design of voluminous and heavy ontology which take into account enormous concept number. Rigorous analysis can establish that service oriented computing represents federative element for the whole approaches. Indeed, service oriented paradigm allows unification during integration of information systems approaches. It includes the integration of shared semantic description of the data handled by the services. This integration of the data (Keita, Catherine & Robert, 2006) is more flexible considering the limited number of the concepts used by the services. This leads us to focus on the semantic Web services, which constitute one of the most effective means to implement systems semantic integration.

In this chapter architecture is suggested for semantic enrichment of SOA. It consists of reduce complexity in the development and the integration of domain ontologies It also ensure finding an automatic invocation of the services, by integrating ontologies without doing major changes in the operating mode of the web services (HTTP, SOAP, WSDL, UDDI). The purposes of the present research work consist of defining interoperable system using web services semantic without using traditional web services and redefining the information system completely.

Web services add a new level of functionality on top of current web, transforming the Web from a distributed source of information to a distributed source of functionality. Current web service technologies based on UDDI[1], WSDL[2] and SOAP[3] provide very limited service automation support. In this context, Web services are enriched using semantic information; automatic location, composition, invocation and interoperation, bring the new concept of Semantic Web Services (Fensel, & Bussler, 2002). The use of web services and semantic technologies to enhance these services must be evaluated as it reflects to what extent; a given portal exposes its functionality as services accessible over the Web.

This chapter is organised as follows: Section 2 presents the state of the art of the various approaches aiming to satisfy the integration of the information systems. Comparative study of these approaches is performed. Section 3 discusses SWS (Semantic Web Services) technologies. Sections 4 and 5 present our architecture where appears the specification and the semantic description of the VTA[4] case. Sections 6, 7, 8 and 9 describe the four top-level elements for describing several aspects of SWS. These are: ontologies, goals, web services, and mediators. Section 10 discusses open directions. Finally, the last section establishes comparison between our work and existing work and makes suggestions for future works.

STATE OF ART

Application enterprise integration can be performed on several layers level, mainly on data and treatments[5] level. Based ontologies approaches (Keita, Catherine, & Robert, 2006) represent a framework on which we can validate semantic integration.

Approaches of integration are characterized after selecting layer by using three main relevant activities (Izza, Vincent, & Burlat, 2006) as follows:

- **Description:** represents an activity which semantically describes system information aspects.
- **Discovery:** represents an activity which allows to find information on which functionalities can be provided by the system.
- **Composition:** represents an activity which allows to compare and to combine information, functionalities and processes.

Data Semantic Integration

Data applications semantic integration consists of defining one or more ontologies describing field semantic. These ontologies are used during conciliation and unification of semantic which resolve semantic conflicts related to exchanged data and/or handled by applications. Such approaches are often called by the databases community: context mediation approaches (Bussler, 2004). Classification and comparison of ontological entities allow to resolve conflicts.

Description

Description main purposes consist of liking ontologies to information system data. Therefore, the mainly used generic approaches are (Wache, Scholz, Stieghahn, & Konig-Ries, 1999):

- **Structural similarity:** consists of create mapping between data base schema and the ontology. We can give the approach used in SMIS systems as an example, (Arens, Chee, Hsu, & Knoblock, 1993) and TSIMMIS (Chawathe, Garcia-Molina, Hammer, Ireland, Papakonstantinou, Ullman, & Widom, 1994);
- **Terms definition:** it consists of using ontology and therefore defining terms which belong in the data base. For example, BUSTER system is based on this kind of approach;
- **Structural enrichment:** is considered as the combination of earlier approaches such as it allows enriching structural elements of the data base by using terms definitions. For example, OBSERVER system (Mena, Kashyap, Sheth, & Illarramendi, 1996) and KRAFT (KRAFT, 2004) are based on this kind of semantic description approach.
- **Meta - annotation:** consists of adding semantic annotations to data bases. For example, the OntoBroker system is based on the using of semantic annotations.

Another important aspect of architecture description of chosen ontology consists of considering mono-ontological architecture or multi-ontological architecture or also hybrid architecture.

Discovery

Information discovered consists of finding link between information contained in several systems. The main principle is based on the using of similarity measures which can exist between ontology terms. There exist several approaches and algorithms which can performed in order to do comparison and similarity measure. It is important to underline that request languages can be established by client applications. Requests formulation is done by using defined ontology which plays the canonical model role. Languages such as RDQL (RDF Query Language) or OQL (OWL Query Language) based on RDF and OWL respectively are often used to formulate requests (Izza, Vincent, & Burlat, 2006; Harriche, Hatchabi, & Maiga, 2008).

Composition

Composition mechanisms are realized once information similarities were found. It allows to combine these information in order to make more complexes elements. This activity is often called by data bases community data integration.

Semantic Integration by the Treatments

Semantic web technology has represented important interests for researchers in the domain of distributed information systems, artificial intelligence, and so on. Therefore, semantic web technology is considered as central component of software constructions. As we know that the web services are lacking the semantic description, the semantic web researchers have proposed to enrich web services with a semantic description of their functionality and make easy their discovery and integration. This technology viewed as combination of web services with semantic web technology, is referred by semantic web services (SWS).

SWS are, therefore, an extension of web service with explicit semantic representation. SWS will support the automatic discovery, composition, and execution of web services. Hence, it has the potentiality to alter the way knowledge and business services are provided and used on the web.

Since the existing of these technologies for web services only describe in syntactical level, it is difficult for service requestors and service providers to interpret or represent non obvious statements such as the semantic of inputs and outputs or applicable constraints.

Semantic description of web services can make possible for automatic discovery, composition and execution across heterogeneous users and domains.

So, SWS is a web service which is semantically rich and defined through service ontology, performing automatic discovery, execution, composition and interpretation.

Description

Services description focuses mainly on the capacities furnished by the service, the quality of service as well as service process representation. Two basic forms of service capacity description are generally implemented. The first approach uses an explicit description based on tasks ontology, where each task of the service is described by an ontology concept, and where each service capacity is modelled by tasks set. The second approach use implicit description based on transformation states, i.e. it is based on the formalization of inputs, outputs, pre-conditions and post conditions.

The explicit approach is simple but it can lead to great ontologies. The second approach is less intuitive, but it is more compact. Example representation based on implicit approach technology is ontology OWL-S or formalism WSMO.

Discovery

Discovered purposes consist to be able to reach dynamically and automatically services for a given need. For example, search for a service allowing to carry out particular type of analysis on data sample.

Discovered services process generally includes generally the following steps (Cardoso, & Sheth, 2004):

- Define target service (request) by using implicit or explicit description of the desired service;
- Bind request elements (input, output, etc) to ontologies;
- Identify possible services objects which satisfy the target service;
- Compute distances between each service object and target service.

Several approaches suggested in the literature try to improve the process of discovered by enriching service description. (METEOR-S, 2005; Paolucci, Kawamura, Payne, & Sycara, 2002).

Composition

Services and process composition have the aim of combining services to create more complex services. Among contributions on this subject, we can find (Rao, 2004; Casati, Illnicki, & Jin, 2000) which cover assembling services problem in the context of the B2B.

Discussion

Semantic integration of the applications is a very active research field. It can as well relate to, as we have just seen it, integration with the data as that by the treatments. Semantic integration by the data requires the design of heavy ontology because number of concepts which it must be taken into account on data level is enormous. Relevant analysis made possible to deduce that orientation services can constitute a federator element for all approaches. Indeed, the orientation service is a paradigm which allows the unification in the approaches of integration of the information systems. It includes the integration of the data of shared semantic description handled by the services. This "implicit" integration of the data more flexible considering the limited number of the concepts is used by the services. This thus, led us to focus us mainly on the semantic services Web, which in our case constitutes one of the most effective means to implement the semantic integration of the systems. For more precision, we completely devote the next section to the semantic services study.

SWS TECHNOLOGIES

There are many approaches or technologies have been driving the development of SWS frameworks, like, OWL Service Ontology (OWL-SC, 2004), WSMF (Gruber, 1993; WSMF, 2005), Web Service Modeling Ontology (WSMO, 2005), METEOR-S (METEOR-S, 2005; Fensel, & Bussler, 2002), First-order Logic for SWS (FLOWS), IRS-III, and so on (McIlraith, Son, & Zeng, 2001; Kellert, & Toumani, 2003; Cardoso, & Sheth, 2004).

OWL-S and WSMO are the most important initiatives to describe SWS. This section presents OWL-S, WSMF, WSMO and METEOR-S.

WSMO

WSMO (Web Services Modelling Ontology) (Bussler, 2004) resulted from design WSMF principles dedicated to semantic services realization. It focused on the users and offers close user language. Therefore WSMO is an extension and a refinement of WSMF. It is a formal language and an ontology which describes various aspects of the Web semantic services. The specification of WSMO includes the definition of three various ontologies: WSMO Lite (basic ontology), Standard WSMO (intermediate ontology) and WSMO Full (which contains all the concepts defined in WSMO).

The architecture of WSMO is based on the use of mediators. One distinguishes two classes from mediators: refiners and bridges. The refiners are used to generate new components as refinements of those which exist. One distinguishes the refiners from goals (ggMediator) and the refiners from ontologies (ooMediator). Refiners of goals (goal refiners) which are used to refine a goal source with a target aim and refiners of ontologies (ontology refiners) which is used to resolve problems at the time of ontologies importing. The bridges, second class of mediators, inter-operate given two components. One is distinguished from the bridges between services and goals (wgMediator) and from the bridges between services (wwMediator). The bridges between services and goals (service-to-goal bridges) resolve the problems between services and goals, while the bridges between services (service-to-service bridges) carry out the link between two Web services. In approach WSMO, one distinguishes two main components which are WSML (Web service Modeling Language) and WSMX (Web Service Environment Execution). WSML provides a formal language for the description of the elements defined in architecture WSMO. WSML is based on various logical languages, which are descriptive logic, first order logic and the logic of programming. This language is structured in five subsets: WSML-Core, WSML-DL, WSML-Light, WSML-Rule and WSML-Full. Each one of these subsets provides increasing expressivity level. WSMX is an execution environment which offers a certain number of modules usable in run-time making it possible to deal with the discovery, the selection, the mediation and the invocation of the semantic services.

The most important characteristic of WSMO is compatibility with the standards web services like WSDL (web services description language). Although it uses its own language, the concept of "Grounding" define mapping with WSDL entries (WSDL-S, 2005) which imply obvious integration with an existing SOA system.

WSMF

WSMF (Web Service Modelling Language) is a generic framework for the realization of the semantic services. It offers much more methodological framework than an operational infrastructure.

OWL-S

OWL-S (Ontology Web Language for Services) (formerly DAML-S) is the first well researched web services ontology, and has numerous users from industry and academe. It provides the mechanism to describe the services offered by the service providers and the services needed by the services requesters (OWL-SC, 2004).

OWL-S represents services generic ontology which mainly describe the web services within general framework. It is also compatible with the standards web services via "ServiceGrounding" which define mappings between OWL-S and WSDL. It lacks from execution environment which makes the mediation of the services to be implemented.

Three kinds of information/knowledge about the web services are provided:

- What kind of service does this web service provide? Provides by service profile. It is basically the advertisement about the capabilities of the services
- How this web service works. Provides by service model and
- How to use or access this web service. Provides by service grounding.

METEOR-S

METEOR-S (Managing End to End Operations) focuses on workflow management techniques SWS (McIlraith, Son, & Zeng, 2001). This project involves creation and application of a broad variety ontologies related to data, function, non-functional/QoS and execution semantics to support the complete web process lifecycle encompassing annotation, discovery, composition, optimization and execution.

METEOR-S consists of an approach based on the extension of standards Web services such as WSDL and UDDI (Universal Description, Discovery, and Integration Protocol). The semantic principle of enrichment is based mainly on the WSDL-S files resulting from WSDL files semantic annotation. Its represents an approach closer to implementation level than conceptual level. Although this approach is based on standards, it is closer to the level of implementation than of the conceptual level.

Discussion

OWL-S provides generic ontology of services; it allows to describe capacities and properties of Web Services. WSMF represents generic framework, which supports the concept of semantic Service Web. WSMO consists of refining WSMF; it also makes it possible to describe the capacities and the properties of the Services Web. METEOR-S is based on the extension of web services standards like WSDL and UDDI (WSDL-S, 2005).

We can say that OWL-S is a generic approach, based on OWL standard (Keita, Catherine, & Robert, 2006). WSMF constitutes much moreover methodological framework. WSMO is an approach which focuses on users and based on the language (Web Service Modelling Language). WSML (WSMO, 2005) combine several logics. METEORS-S is a pragmatic approach which misses abstraction. None of these approaches deals completely with semantic process. It suffers more or less from the lack of methodologies, the lack of flexible architectures, which can effectively describe semantics related to the applications, and also from the lack of mechanisms of effective description and mediation. Based on all elements presented earlier, WSMO approach is that which has the most advantages and specifically abstraction. It leads us to focus on this approach (WSMO) and take as basis for the work undertaken within the framework of the semantic SOA system. In the next section, architecture is presented, which is the solution that we propose to satisfy semantic web services. Thus, the following aspects will be evaluated:

Functionality: The different functionalities available on a SW portal – e.g. content search, content publication, etc. – can be made accessible by using web service technologies. The functionality degree presentation via web services determines to what extent a portal can be used not only through the user interface, but also in programming context. Thus, a comparison, between the portal functionalities accessible through its user interface and the ones accessible through web services, has been developped.

Semantic Web Services: Enriching web services with semantic information allows automated location, composition, invocation, and interoperation of services. Therefore not only the portal functionality exposed through web services must be considered, but also to what extent these services include automated support.

ARCHITECTURE PRESENTATION

The architecture that we propose in this chapter operates on semantic descriptions based on conceptual model Web Services Modelling Ontology (WSMO). These descriptions are defined using the web service modelling language (WSML). Architecture design is oriented using certain principles. These principles reflect the fundamental requirements characterized by distributed environment, service oriented computing and focusing on user, according to the indications of the Figure 1.

SOA represents particular approach for analysis, design, and implementation. It introduces specific principles which control communication, architecture and logic processing aspects. That includes the re-use of the service, the weak coupling and the abstraction. The services can be divided from two points of view. In first, we distinguish two types of services according to functionalities, which are provided in architecture, namely the trade services and the middleware services. The trade services are provided by several suppliers of services. These services give access to the

Figure 1. The proposed architecture

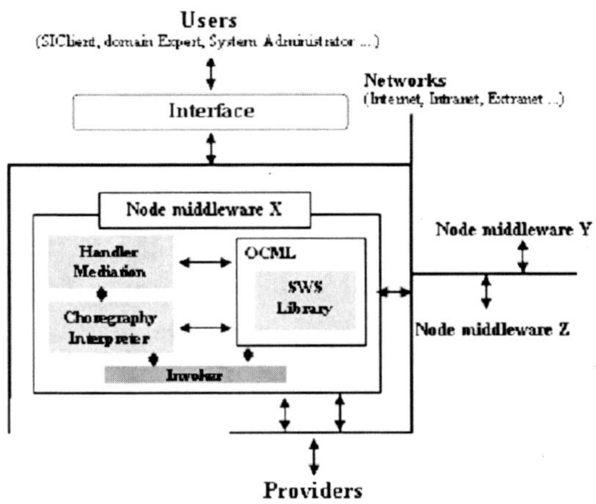

information systems of their suppliers. In addition, the middleware services are the principal orchestrator of integration and the interoperation of the services trades.

Also, we distinguish two types of services from the point of view of their abstraction in architecture, namely the web services, and the services. The web service is a general service which can take several forms when it is instantiated (for example flight reservation) whereas the service is a real authority of the web service which is consumed by a user and provides a concrete value to the user (for example flight reservation from Algiers to Oran).

In general, the semantic is viewed as a rich and formal description of information and models of behaviours allowing the automation of certain tasks based on logical reasoning. The semantic combined with the principle of the service oriented computing, it allows definition of rich and formal models of service. Ontologies allow automation of the total or partial tasks such as the discovery of the service, mediation, invocation, etc.

Distributing principle consists of integration process of integration of several data-processing entities collaborating for the execution of a computational task in a transparent and coherent way like a single and centralized system. Architecture middleware distributing principle is also applied to the services trades which allow the execution of a process through the composition of several services trades distributed on the network.

The principle "oriented user" puts the user at the center of architecture. It refers to the concepts like personalization of the services trades, to make easy the using and distributing of the service, in order to achieve with effectiveness user purposes. The architecture establishes the infrastructure which by means of its functionalities deals to the principle of the orientation towards the user in the supply of the service.

The IRS III[6] (Internet Reasoning Service) is an infrastructure for publishing, locating, executing and composing semantic web services, organized according to the WSMO conceptual model.

IRS-III is framework and platform for semantic web services.

Middleware (IRS III), the core of its architecture, is composed by:

- **Semantic web services (SWS) Library:** In the center of waiter IRS-III, we have a bookshop SWS where semantic descriptions of the web services are stored and established using OCML[7]. Domain ontologies are also available in the Bookshop.
- **Choreography Interpreter:** This component interprets the grounding and the transitions in the description of choreography when they are invoked by the component mediation handler.
- **Mediation Handler:** The activities of IRS-III are the selection or the invocation. Each one carried out by specific mediation component using in the component mediation handler. These activities can imply service mediation execution. It makes easy, mediation between various services interfaces implied in the process in progress, during the execution. It is used during the processes of discovery to evaluate the possibility of conversation between the service and its user. During this analysis, it discovers the possible conflicts to resolve for example unnecessary message withdrawal.
- **Invoker:** The component invoker of the server communicates with supplier platform, by sending customer input and receiving customer result output.

Our suggestion consists of a solution to semantic integration problems which add new abstract layer into web services classical architecture. This new architecture is opened and extensible and allows adding new components. The suggested architecture performs enrichment to the description of the services and the data, which they use through ontologies integration.

SAMPLE APPLICATION

Our scenario outlines general structure for the VTA which can be extended to more complex scenarios where the customer can be viewed as web service. Therefore, network of compounds services is created which offers tourism complex services. For example, a VTA can provide flight reservation services for an integration of the air lines which represents another reservation services VTA for a chain of hotels on a worldwide scale, and also provides reservation services of rented cars by combining the services of car hiring several agencies. Then, the last uses these services to provide an end-user service for voyage reservation in the world.

Actor, Role and Goal

There are four actors in the use case. We define how can participate and their particular interactions (roles) see Figure 2:

The customer: the end-user requesting a service provided by the TVA.

- **Goal:** request automated resolution by a convivial voyage service.
- **Role:** end-user and TVA react each other for service using.

Suppliers of Service: commercial companies of voyage.

- **Goal:** to sell the service for customers, to maximize the profit.
- **Role:** to provide the service of voyage as a Web Service (provides also the semantic description necessary of the Web Service), possibility of having a contract of usage/d' allowance with the TVA.

Administrator: administrator of the system.

- **Goal:** to make well function of the system.
- **Role:** to manage and configure the system.

TVA: the intermediate between Customer and Suppliers of the voyage services. It allow voyage services to the customers by integrating separated services provided by service suppliers.

- **Goal:** consists *of* providing service of quality to the end-user, by integrating the existing services of voyage in new services and maximizing the profit.
- **Role:** interaction with the customer via user interface (the customers can be the

Figure 2. Use case diagram

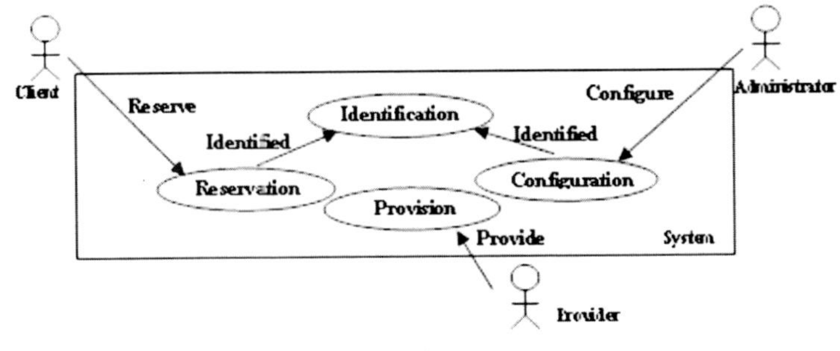

human ones or web services), the use and SLA[8] of the web services are offered by service suppliers, by holding all there functionalities to manage these services (mechanisms for the discovery, the composition, the execution, etc).

Scenario of Use

We identify the scenarios of the following use in Figure 3:

Figure 3. Sequence diagram

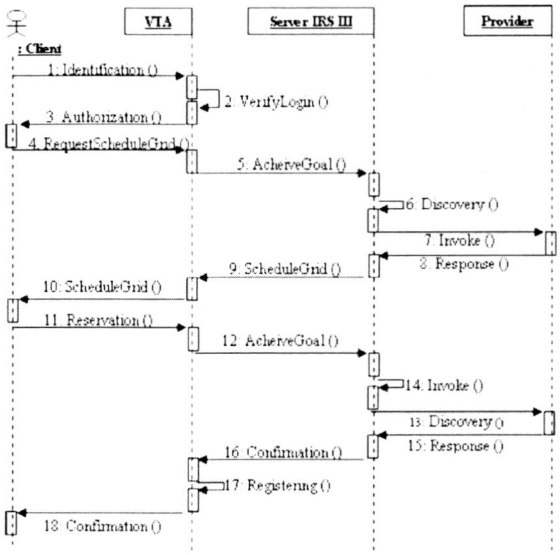

1. TVA reacts with service suppliers on the basis of contract, the use and the allowance of the web service:
   ◦ **The present actors:** TVA and service suppliers.
   ◦ **Activities:** negotiation of service contract.
   ◦ **Technological requirements:** information on the requirements of the contract is defined in the system, for example using web service police.
   ◦ **Possible extensions:** contract negotiation can be supported by automated mechanisms.
2. Ask of user TVA to find voyage service offers, TVA detects and calls upon the suitable web service and turns over the results to the customer.
   ◦ **The present actors:** Customer, TVA, voyage service suppliers.
   ◦ **Activities:**
      ▪ The customer sends his request to TVA.
      ▪ TVA discovers, calls upon and carries out the Web corresponding Service.
   ◦ **Technological requirements:**
      ▪ TVA has to predefine the functionalities which can be required by a customer.
      ▪ Suppliers Web Service of voyage service must be described in such a manner that dynamic discovery semantically was supported.
      ▪ TVA must provide mechanisms for service automated discovery.
   ◦ **Possibly extensions:** the customer specifies his request in natural language which will be translated in readable form readable and manageable automatically by TVA service.
3. The customer selects concrete offer and asks for reservations for this offer (to react reciprocally with the TVA), TVA detects and calls upon the Web Service. The results are posted at the customer (left computational).
   ◦ **The present actors:** Customer, TVA, voyage service suppliers,
   ◦ **Activities:**
      ▪ The customer selects an offer according to the result of scenario 2.
      ▪ TVA discovers and calls upon the Web Service to satisfy user request.

- **Technological requirements:** information on contract requirements is defined in the system, for example a police force of Web Service use.
    - Web Service must be described in such manner that it supports both dynamic discovery and semantic execution.
    - TVA must take account mechanisms for service automated discovery.
    - TVA must provide contingent interaction interface with the user during the service execution.
- **Possibly extensions:** advanced mechanisms for automated execution of the Web Services (services compositions).

4. TVA reacts reciprocally with customer and service supplier for non computational parts (for example real tickets distribution).
    - **The present actors:** Customer, TVA, service Suppliers,
    - **Activities:** customer notification, accountancy, good distribution (out of computational system), etc.
    - **Technological requirements:** mechanisms for notification and accountancy.
    - **Possible extensions:** the Web Service can be used for: the notification of customer - TVA - Service interaction of the supplier on the accountancy and delivery order.

Basic Architecture of the VTA

In the use case, the VTA is the central point of interaction between the Customer and the other Web services. It is obvious for technological requirements to start from descriptions of the scenario of the use. These web services offered by service provider must be furnished sufficient descriptive information to support web services. VTA must provide also all mechanisms handling semantic web services.

The TVA basic architecture viewed as central entity for semantic web services execution is shown in Figure 4. The directed semantic web services main VTA functionalities by considering semantic web services technology requirements are:

- To provide a user interface for the interaction with the customer (for a human user or a machine).
- To discover the suitable web services for a user request.
- To call upon the external services.
- To provide semantic execution environment of semantic web service.
- To deal with heterogeneous resource taking account the suitable mediation.
- To provide interfaces for the co-operation with services providers.

The VTA is a directed application B2C SWS which provides service to end-user who follows a customer-server model. In order to support a coherent operation of the VTA and to ensure web services descriptions compatibility, generic platform for the semantic web services is necessary. That is provided by our architecture through WSMO. Technology semantic web service can

Figure 4. Architecture of the vta

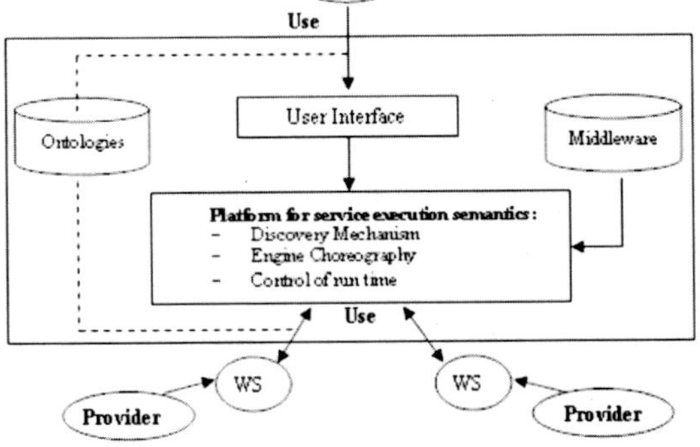

Figure 5. Stage of design of the ontology of services

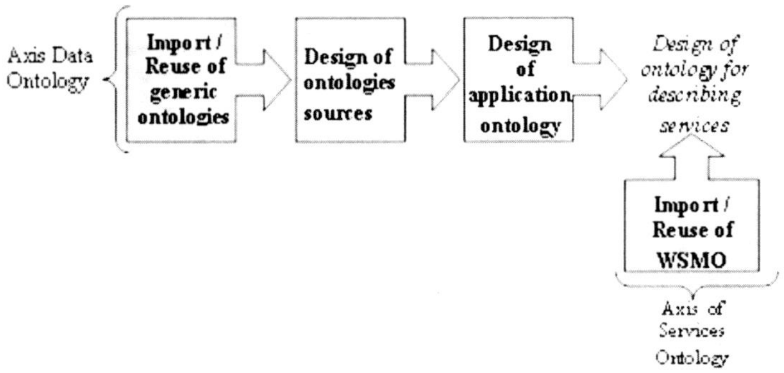

be applied to make easy the use of the system (to find the VTA and to achieve its goal) and also for the VTA which takes advantages from semantic descriptions of the services offered by its associates (providers of services).

The VTA integrates the web services of four different airline companies. In short, it makes it possible for a customer to carry out flight reservations. The VTA discovers and calls upon the web services offered by the airline companies. Contract called "umbrella" can be established between service suppliers and the VTA for a regularized use and an allowance of the web service.

SERVICES DESCRIPTION ONOTOLOGY DESIGN

The Current solutions analysis of semantic integration, and in particular of approaches based on the semantic services, presents certain number of limits. They are related to mainly architectures of current ontologies. It presents lack of methodology defining semantic ontologies and services and finally not powerful mechanisms of description and mediation of services.

Development problem is resolved by using a modular approach of development. Complete ontology through several modules integration can be made by this approach. This integration establishes relations between the various concepts. This modular approach fixes some constraints on the developer of ontology in particular; the fact of referring to a single ontology in their design: consensual ontology (Benslimane, Amar Bensaber, & Malki, 2004).

Such design approach requires finding ontologies which mainly formalize generic concepts of the field. We call them level 1 ontologies. We import ontologies sources in ontology lately created which we call "ontology of application". The latter contains the whole of the concepts making it possible to define the data used in the system. It allows flexibility of adding or withdrawal of a service, so it ensures architecture flexibility. Ontologies sources and the ontology of application constitute ontologies of level 2. Level 3 is composed of a single ontology importing the ontology of application and the generic ontology of service description[9]. The Figure 5 gives the stages of development and the Figure 6 gives the general outline of modelling in WSMO of our use case resulting from such a conceptualization.

A widespread definition of the ontology is given by (Harriche, Hatchabi, & Maiga, 2008) where an ontology is defined like "explicit and formal specification of a shared conceptualization". In WSMO, ontologies provide a comprehensible semantics for information system employed by all the actors implied in the use of the Web services.

Figure 6. WSMO modelling of use case

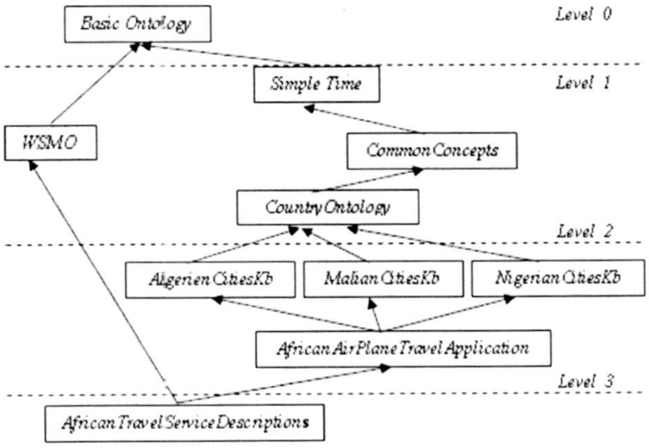

So it allows the exchange and the interoperation of the components. The basic types of an ontology are the concepts, the relations, the functions, the authorities, and the axioms.

We define five ontologies in knowing AlgerianCitiesKb, NigerianCitiesKb, MalianCitiesKb, AfricanAirplaneTravelApplication, AfricanTravelServicesDescriptions. They are connected to three already exiting ontologies commonConcepts countryOntology and simpleTime.

SEMANTIC DESCRIPTION OF THE WEB SERVICES

The description of a Web services in WSMO includes five components: non-functional properties, imported ontologies, mediators used, capacity and interface. The most important components are capability and interface.

Web service capability: a Web service is defined by only one capability. It is defined by specifying the following elements:

- Non-functional properties
- Imported ontologies
- Mediators used
- **Precondition:** This that the Web service imposes on the customer
- **Post condition:** This that the Web service offers to the customer
- **Assumption:** represents conditions to be satisfied before service execution
- **Effect:** represents changes in the environment after service execution.

Web service Interfaces: service interface provides more information on how the functionalities of the service are realized. It contains:

- A model of communication which allows the service functionalities design, named ***choreography***.
- A description of how the whole of the functionalities can be consumed through a co-operation with other services, named ***orchestration***

Thus the interface is defined by specifying the following elements:

- Non-functional properties
- Imported ontologies
- Mediators used
- Choreography
- Orchestration

We present in this chapter these two components for services implemented in our application:

Service Offering the Flight Times (GetPlaneTime)

Each company offers to its customers the schedules of flight through a service named getPlaneTime:

- getPlaneTime capability:
  - preconditon:
    - Origin city
    - Destination city
    - Date of travel
  - postcondition:
    - Schedule
  - Assumption:
    - An example of assumption: In the case of the service African Air Lines the awaited cities must be natives of Niger, Malians or Algerian and belong to different countries.
  - getPlaneTime interface
  - Choreography:
    - In _url of the service[10]
    - Out _url of the Service[11]

Reservation Services Flight

Each company offers its customers the ability to book a flight through a service named bookPlane:

- bookPlane capability:
  - preconditon:
    - Name of the passenger
    - Origin city
    - Destination city
    - Date of travel
  - postcondition:
    - Possible confirmation
  - Assumption:
    - An example of assumption: In the case of the service Air Algeria the awaited cities must be Algerian
  - bookPlane interface:
  - Choreography:
    - In _url of the service
    - Out _url of the service

GOALS

A goal specifies the objectives which a customer wants to have by consulting a service, i.e. the functionalities he expects on service. Goals and Web services represented as different entities provides decoupling between the application and the web service. The customer, human or machine, defines the goal to resolve and in counter part the web service discovery detects suitable services Web to resolve the goal automatically. This kind of problems, in which the applicant formulates independently its purposes without knowing the service Web resolving, it is known as "goal-driven approach", derived from the AI.

A goal in WSMO is described by the following elements: Non-functional properties, imported ontologies, mediators used, capability and interface.

The goals indicate what user wants as a result of web service. To model the goal, WSMO describes the data which the user wants to obtain from the service (post condition) as well as the data that it will provide (precondition). It can also possibly define the obtained state (effects) after the service execution. It is also possible to describe generic goals in application specific context. We define two generic goals one for voyages schedule consultation and another for flight reservation.

The goal for consultation travel time:

- getTimeGoal:
  - Precondition
    - Origin city

        - Destination city
        - Date
    - postcondition
        - Schedule

The goal for booking a flight:

- bookPlaneGoal:
    - Precondition
        - Name of the passenger of the string type
        - Origin city
        - Destination city
        - Date
        - Hour
    - postcondition
        - The result is a notification type string.

MEDIATORS

The mediators presence allows to link between heterogeneous resources. Mediators resolve incompatibilities that arise at the following levels:

Level data: performs mediation between the different terminologies used, specifically addressing the problem of integration of ontologies.

At level process: mediator is used between communication heterogeneous models. This kind of heterogeneity appears during the communication between different web services or between web services and goal.

Process mediators provides the functionalities for an analysis during the execution of two given models, and to resolve the possibly incompatibilities, in, for example, generating virtual messages, by grouping several messages in one message, by changing their order or by removing some of the messages to make easy communication between the two receptions.

Generally WSMO defines mediators by means of non functional properties, imported ontologies, source component, target component and mediation service such that the component source and target component may be a mediator, a web service, an ontology or a goal. The current version of the specification WSMO distinguishes between four types of mediators: OOMediators, GGMediators, WGMediators and WWMediators.

WSMO can either import ontology elements or other WSMO. WSMO framework includes mediator notion in order to resolve conflicts and interoperability issues possibly arising. More specifically, WSMO defines:

- **ooMediator:** which links two ontologies, resolving possible mismatch issues between them;
- **ggMediator:** which links two goals in the context of refinement of the source goal into the target goal;
- **wwMediator:** which links two services and
- **wgMediator:** which links services with goals, meaning that a service fulfills the goal that is linked to.

WgMediators are mediators who bind web services to the goals. A wgMediator binds a goal to a web service via choreography, it means that the web service achieves (completely or partially) the goal on which it is dependent. A wgMediator can be defined in the AVV binding the goal "reservation of a ticket for a voyage from Algiers to Tiaret" (source) with the service "bookPlane" (target). Only the source (the goal) is specified at the time of the design, when with the target, it found and is charged after discovered suitable service during the execution. The property service of mediation is used in the case or the goal and the web service use different concepts in Figure 7.

Like mediators defined in our appplication, getTimeMediator and bookPlaneMediator having respectively for getTimeGoal source and bookPlaneGoal. The web services targets are given in runtime after the process of discovery.

Figure 7. Defining a wg-mediator

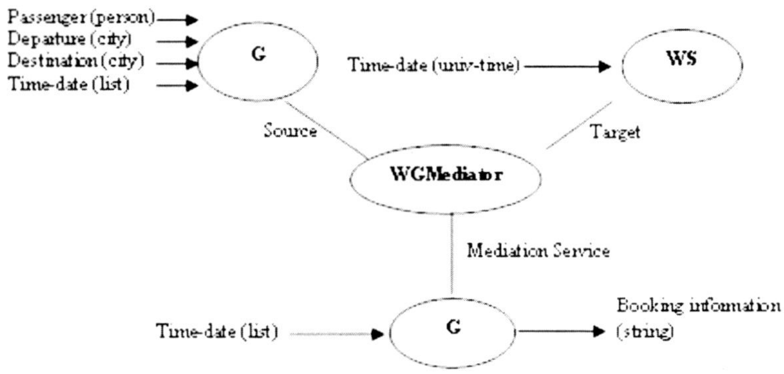

- The wgMediator getTimeMediator
- Source: getTimeGoal.
- The wgMediator bookPlaneMediator
- Source: bookPlaneGoal

OPEN DIRECTIONS

The suggested architecture does not take into account service composition. The integration semantic complexity of the services requires an incremental approach during architecture design. We limited ourselves to problems related to discovered and invocation of the services. Various ideas are then to be considered:

- Discovery semantic service improvement: This improvement must make the discovery service dealt with more complex requests. It includes numerical values constraints which allow us to have better targeting during certain services research.
- Developing problems which add complexity level to architecture such as the composition of the semantic services and the user interfaces improvement through request mechanism improvement.

CONCLUSION

We presented an implemented flexible architecture dedicated to services semantic integration. This architecture was based on WSMO approach. This latter approach is, among the various approaches proposing to resolve the services semantic integration and which is strong and relevant because it proposes conceptualization whose implementation is possible with the current means.

The suggested architecture is based on principles which are: semantic services oriented paradigm (through an explicit formalization), distributed and centered user. The service oriented paradigm ensures architecture flexibility and an easy integration of several SOA systems. The semantic aspect is due to the integration of ontology in the system. The distribution ensures the upgradeability of architecture. The principles of centred user make possible the end-user through the integration of user interface.

Each one of these principles is fundamental and allows applying solution to real use case. It represents first contribution of the suggested architecture. Secondly, we have proposed domains ontologies development approach. We chose the modular development, which allows ontologies integration at design-time, thus it reduces to the maximum the number of mapping necessary during the execution (run-time).

Figure 8. Comparative table of the research

	Our architecture	*PoliMi-CeFriel*[1]	*DERI at & DERI IE*[2]	*FSUJena*[3]
Approach	WSMO	WSMO	WSMO	-
Ontology	Multi-ontological (modular developpement) (OCML Language)	-	Multi-ontological (WSML Language)	Multi-ontological (DSD Language)
Middleware oriented service	Yes	Yes	Yes	Yes
User oriented	Yes	No	No	No
Distributed	Yes	No	Yes	No
Middleware	IRSIII	SWE-ET	WSMX	DIANE

[1] http://www.cefriel.it/
[2] http://www.deri.at/
[3] http://hnsp.inf-bb.uni-jena.de/DIANE

The instan:iation of our approach is related to the use case satisfaction suggested by the W3C. It is about a virtual travel agency dealing with all the mechanisms necessary to voyage process. It offers semantics necessary so that the user can secure a route while satisfying the unit of the collateral constraints (reservation of ticket, hotel, car…). This case illustrates strongly possible application of the SWS. Our application integrates the services s:ggested by three travel agencies. It satisfies a first level of the use case.

The SWS represent a basic active research field. That means that important research works are at the conceptual step. We listed some research works which are compared with our architecture in Figure 8.

REFERENCES

Arens, Y., Chee, C. Y., Hsu, C. N., & Knoblock, C. A. (1993). Retrieving and integrating data from multiple information sources. *International Journal of Intelligent and Cooperative Information Systems*, *2*(2), 127–158. do:10.1142/S0218215793000071

Benslimane, S. M., Amar Bensaber, D., & Malki, M. (2004, June). Vers une médiation de contexte à base d'ontologies multi-représentation . In *Colloquy on Optimization and Information Systems* (pp. 101–112). Tizi Ouzou, Algeria: COSI.

Bussler, C. (2004). Semantic Web Services. In *The International Conference on Web engineering (ICWE), ICWE'04*, Munich, Germany.

Cardoso, J., & Sheth, A. (2004). *Semantic Web services and web processes composition*. New York: Springer-Verlag, Inc.

Casati, F., Ilnicki, S., & Jin, L. (2000). *Adaptive and dynamic Advanced Information Systems Engineering (CAiSE), Stockholm, Sweden*. Berlin: Springer Verlag.

Chawathe, S., Garcia-Molina, H., Hammer, J., Ireland, K., Papakonstantinou, Y., Ullman, J., & Widom, J. (1994). The Tsimmis project: Integration of heterogeneous information sources. In *Conference of the Information Processing Society*, Japan, (pp. 7-18).

Fensel, D., & Bussler, C. (2002). The Web Service Modeling Framework WSMF. [New York: Elsevier Sciences, B. V.]. *Electronic Commerce Research and Applications*, *1*(2), 113–137. doi:10.1016/S1567-4223(02)00015-7

Gruber, T. (1993). A Translation Approach to Portable Ontology Specifications. *Knowledge Acquisition*, *5*(2), 199–220. doi:10.1006/knac.1993.1008

Harriche, K., Hatchabi, M., & Maiga, A. (2008). *Vers une approche d'integration semantique via les web services. Research report.* Ibn Khaldoun University of Tiaret - Algeria.

Izza, S., Vincent, L. & Burlat, P. (2006). Dealing with Semantic Application Integration With in Large and Dynamic Enterprises. *International Journal of Cooperative Information systems*, *15*(4), 507-533.

Keita, A. K., Catherine, R., & Robert, L. (2006). Un outil d'aide à la construction d'ontologies préconsensuelles: Towntology project . In *Acts of Congress computer science Organizations and Information Systems and Decision XXIVème* (pp. 911–926). Hammamet, Tunisia: INFORSID.

Kellert, P., & Toumani, F. (2003). *Les web services sémantiques*. Research Report *LIMOS/RR* 03-*15*. Computer science Laboratory, modeling and optimization systems, Clermont-Ferrand, France.

KRAFT. (2004). *Knowledge Re-use and Functional Transformation*. Retrieved from http://www.cs.cf.ac.uk/kraft/

McIlraith, S., Son, T., & Zeng, H. (2001). Semantic Web Services. *IEEE Intelligent Systems*, *16*(2), 46–53. doi:10.1109/5254.920599

Mena, E., Kashyap, V., Sheth, A., & Illarramendi, A. (1996). OBSERVER: An approach for query processing in global information systems based on interoperability across pre-existing ontologies. In *Proceedings 1st International Conference on Cooperative Information Systems (IFCIS),* Brussels.

METEOR-S. (2005). *Deliverables of METEOR-S project.* Retrieved from http://lsdis.cs.uga.edu/Projects/METEOR-S/

OWL-SC. (2004). OWL-S: Semantic Markup for Web Services. *OWL-S Coalition*. Retrieved from http://www.daml.org/services/

Paolucci, M., Kawamura, T., Payne, T. R., & Sycara, K. (2002). Semantic Matching of Web Services Capabilities. In *Proceedings the First International Semantic Web Conference (ISWC),* Sardinia, Italy.

Rao, J., & Su, X. (2004). A survey of automated Web service composition methods. In *Proceedings of the First International Workshop on Semantic Web Services and Web Process Composition,* (SWSWPC'2004, LNCS), San Diego, CA. Berlin: Springer-Verlag.

Wache, H., Scholz, T., Stieghahn, H., & Konig-Ries, B. (1999). An integration method for the specification of rule–oriented mediators. In [Kyoto, Japan.]. *Proceedings of DANTE, 99,* 109–112.

WSDL-S. (2005). *Web Service Semantics - WSDL-S, Version 1.0.* Recovered W3C Member. Retrieved from http://www.w3.org/Submission/WSDL-S/

WSMF. (2005). *Semantic Web Services Framework (SWSF) Overview.* Recovered W3C Member Submission. Retrieved from http://www.w3.org/Submission/SWSF/

WSMO. (2005). *Web Service Modeling Ontology Project Deliverables D1v1.0*. WSMO Project, CMS-WG Final Draft. Retrieved from http://cms-wg.sti2.org/TR/d1/v1.0/

KEY TERMS AND DEFINITIONS

Semantic Web Services (SWS): Enrichment of web services with semantic information. SWS allows automated location, composition, invocation, and interoperation of services.

Ontology: provides common vocabulary to define concepts and relationships between concepts.

Mediator: is a component, which allows us to resolve mismatching types of used heterogeneous components.

WSMO: The Web Service Modeling Ontology describes all relevant aspects related to services which are accessible through Web service interface. The goal is the automation of the tasks such as discovery, selection, composition, mediation, execution, monitoring, etc.), which are involved in both intra- and inter-enterprise integration of Web services.

WSDL: describes interface for using Web Service: interface, (input and output operations), access with binding protocol, endpoint (location of service)

SOA: Structured Oriented Architecture represents a new threshold of connectivity and computing power.

Integration: means anything ranging from integration, merges, use, mapping, extending, approximation, unified views and more; sometimes interchanging the words as if all are synonyms, although these concepts are used as homonyms as well.

ENDNOTES

1 http://www.uddi.org.

2 http://www.w3.org/TR/wsdl.html .

3 http://www.w3.org/TR/SOAP/

4 Virtual travel Agency: It is about a case of use proposed by the W3C.

5 Some authors also speak about integration on the level of the processes. From our point of view this level east includes on the level treatment.

6 Language OCML is a combination of the frames and logic of first order and makes it possible to represent the concepts, the taxonomy of the concepts, the relations, the functions, the axioms and the authorities.

7 Supported by the AKT (Advanced Knowledge Technologies) under a research collaboration with several European universities including the University of Aberdeen, Edinburgh etc…

8 Service Agreements Level.

9 WSMO in our case.

10 One specifies here the exact URL of the port which allows the consumption of the Web service by specifying the method and the entry concerned.

11 One specifies here the exact URL of the port which allows the consumption of the Web service by specifying the method and the output concerned.

Compilation of References

Abel, M.-H., & Leblanc, A. (2008). An operationnalization of the connections between e-learning and knowledge management: the memorae approach. In *Proceedings of the 6th IEEE International Conferences on Human System Learning*, Toulouse, France (pp. 93--99).

Abel, M.-H., Lenne, D., & Leblanc, A. (2007). Organizational learning at university. In *Second European Conference on Technology Enhanced Learning, EC-TEL2007*, (pp. 408—413).

Abels, S., Haak, L., & Hahn, A. (2005). Identification of Common Methods Used for Ontology Integration Tasks. In A. Hahn, S. Abels & L. Haak (Ed.), *Proceedings of the 1st International ACM Workshop on Interoperability of Heterogeneous Information Systems* (pp. 75-78). New York: ACM.

Agirre, E., & Rigau, G. (1996). Word sense disambiguation using conceptual density. In *Proceedings of the 16th conference on Computational linguistics*, August, (pp. 05–09).

Agirre, E., Ansa, O., Hovy, E., & Martinez, D. (2000). Enriching very large ontologies using the WWW. In *Proceedings of the Workshop on Ontology Construction of the European Conference of AI (ECAI-00).*

Agrawal, R., & Srikant, R. (1994). Fast algorithms for mining association rules. In *20th Int. Conf. Very Large Data Bases (VLDB).*

Akkiraju, R., Farrell, J., Miller, J., Nagarajan, M., Schrnidt, M.-T., Sheth, A. & Verma K. (2005). *Web service semantics – wsdl-s.* W3C Member Submission, November 2005

Alavi, M., & Leidner, D. E. (1999). Knowledge Management Systems: Issues, Challenges, and Benefits. *Communications of the Association for Information Systems, 1*(7), 1–37.

Alchourron, C., Gärdenfors, P., & Makinson, D. (1985). On the logic of theory change: partial meet contraction and revision functions. *Journal of Symbolic Logic, 50*(2), 510–530. Retrieved from http://www.jstor.org/stable/2274239. doi:10.2307/2274239

Aleman-Meza, B., Halaschek-Wiener, C., Arpinar, I. B., & Sheth, A. P. (2008). Context-aware semantic association ranking. In *Proceedings of SWDB'03, the first International Workshop on Semantic Web and Databases, Co-located with VLDB 2003*, Humboldt-Universit¨at, Berlin, Germany, (pp. 33-50).

Aleman-Meza, B., Nagarajan, M., Ramakrishnan, C., Ding, L., Kolari, P., Sheth, A. P., et al. (2006). Social networks: Semantic analytics on social networks: Experiences in addressing the problem of conflict of interest detection. In *Proceedings of the 15th international conference on World Wide Web, WWW 06*, Edinburgh, Scotland, (pp. 407-416). New York: ACM.

Alfonseca, E., & Manandhar, S. (2002). An unsupervised method for general named entity recognition and automated concept discovery. In *Proceedings of the 1st International Conference on General WordNet*, Mysore, India.

Alfonseca, E., & Manandhar, S. (2002). Improving an ontology refinement method with hyponymy patterns. In *Language Resources and Evaluation.* Las Palmas: LREC.

Alvaro-Bermejo, C. (1988). *Elaboración del Tesauro de Urbanismo URBISOC. Una Cooperación Multilateral.* Salamanca: Encuentro Hispano-Luso de Información Científica y Técnica II.

Amati, G., & Bruza, P. (1999). A logical approach to query reformulation motivated from belief change. In *Workshop on logical and uncertainty models for Information Systems,* University College London (UCL), London.

Antoniou, G., & Harmelen, F. (Eds.). (2004). *A Semantic Web Primer.* Cambridge, MA: The MIT Press.

Antoniou, G., & van Harmelen, F. (2004). Ontology engineering. In *A Semantic Web Primer* (pp. 205–222). Cambridge, MA: Massachusetts Institute of Technology Press.

Anyanwu, K., & Sheth, A. P. (2002). The rho Operator: Discovering and Ranking Associations on the Semantic Web. *SIGMOD Record, 31*(4), 42–47. doi:10.1145/637411.637418

Anyanwu, K., Maduko, A., & Sheth, A. P. (2005). Semrank: Ranking complex relationship search results on the semantic web. In *Proceedings of International World Wide Web Conference*, Chiba, Japan, (Vol.14, pp. 117–127). New York: ACM.

Arens, Y., Chee, C. Y., Hsu, C. N., & Knoblock, C. A. (1993). Retrieving and integrating data from multiple information sources. *International Journal of Intelligent and Cooperative Information Systems, 2*(2), 127–158. doi:10.1142/S0218215793000071

Arndt, R., Troncy, R., Staab, S., Hardman, L., & Vacura, M. (2007). COMM: Designing a well-founded multimedia ontology for the web. In ISWC/ASWC Proceedings, (LNCS 4825, pp. 30-43). Berlin: Springer.

Aronson, A. R., & Rindflesch, T. C. (1997). *Query expansion using the UMLS Metathesaurus.* Amia.

Arpirez-Vega, J. C., Gomez-Perez, A., Lozano-Tello, A., & Sofia Pinto, H. (1998). (ONTO)2Agent: an ontology-based WWW broker to select ontologies. In *Proceedings of the workshop on Applications of Ontologies and Problem-Solving Methods*, (pp. 16–24).

Artala, A., Franconi, E., Guarino, N., & Pazzi, L. (1996). Part-Whole Relations in Object- Centered Systems: an Overview. *Data & Knowledge Engineering, 20*(3), 347–383. doi:10.1016/S0169-023X(96)00013-4

Asai, T., Abe, K., Kawasoe, S., Arimura, H., & Sakamoto, H. (2002). Efficient substructure discovery from large semi-structured data. In *2nd Annual SIAM Symposium on Data Mining (SDM2)*, Arlington, VA, USA.

Assadi, H., & Bourigault, D. (2000). Analyses syntaxique et statistique pour la construction d'ontologies à partir de textes, Ingénierie des connaissances, évolution récentes et nouveaux défis, chapter 15, Eyrolles, collection technique et scientifique des Télécommunications, (pp. 243-255).

Aumueller, D., Do, H. H., Massmann, S., & Rahm, E. (2005). Schema and ontology matching with coma++. In Acm sigmod (pp. 906-908).

Aussenac, N. (1989). *Conception d'une méthodologie et d'un outil d'acquisition des connaissances expertes.* Ph.D Thesis, University P. Sabatier, Toulouse, France.

Aussenac-Gilles, N., Biébow, B., & Szulman, S. (2000). Corpus Analysis For Conceptual Modelling, Workshop on Ontologies and Text, Knowledge Engineering and Knowledge Management: Methods, Models and Tools. In *12th International Conference EKAW'2000*, Juan-les-pins, France. Berlin: Springer- Verlag.

Aussenac-Gilles, N., Biébow, B., & Szulman, S. (2000). Revisiting Ontology Design: A Methodology Based on Corpus Analysis (LNCS, pp.172–188). Berlin: Springer-Verlag.

Ayadi, N., & Lacroix, Z. (2008, November). Biomap: a deductive approach for resource discovery. In *UWAS'2008 - The Tenth International Conference on Information Integration and Web-based Applications Services*, 24-26 November 2008, Linz, Austria, pages 477-482

Baader, F., Calvanese, D., McGuinness, D. L., Nardi, D., & Patel Schneider, P. F. (Eds.). (2007). *The Description Logic Handbook: Theory, Implementation, and Applications.* Cambridge, UK: Cambridge University Press.

Baader, F., Calvanese, D., McGuinness, D., Nardi, D., & Patel-Schneider, P. (2003). *The Description Logic Handbook: Theory, Implementation and Applications*. Cambridge, UK: Cambridge University Press.

Bachimont, B. (2000). Engagement sémantique et engagement ontologique: conception et réalisation d'ontologies en ingénierie des connaissances. In J. Charlet, M. Zacklad, G. Kassel, D. Bourigault (Ed.), Ingénierie des connaissances: évolutions récentes et nouveaux défis, Eyrolles, (pp. 305-323).

Bachimont, B. (2001), Modélisation linguistique et modélisation logique des ontologies: l'apport de l'ontologie formelle. In *Proceedings of IC 2001, Plate-forme AFIA*, (pp. 349-368), Grenoble, France.

Bachimont, B., Isaac, A., & Troncy, R. (2002). Semantic commitment for designing ontologies: a proposal. In A. Gomez-Perez & V.R. Benjamins (Eds.), EKAW 2002, (LNAI 2473, pp. 114–121).

Bailey, C., & Clarke, M. (2001). Managing knowledge for personal and organisational benefit. *Journal of Knowledge Management*, *5*(1), 58–67. doi:10.1108/13673270110384400

Banerjee, J., Kim, W., Kim, H. J., & Korth, H. (1987). Semantics and implementation of schema evolution in object-oriented databases. *ACM SIGMOD Record, 16*(3), 211-322. doi: http://doi.acm.org/10.1145/38714.38748

Baral, C. (2003). *Knowledge Representation, Reasoning and Declarative Problem Solving*. Cambridge, UK: Cambridge University Press. doi:10.1017/CBO9780511543357

Baralis, E., & Psaila, G. (1997). Designing templates for mining association rules. *Journal of Intelligent Information Systems*, (9): 7–32. doi:10.1023/A:1008637019359

Batini, C., Lenzerini, M., & Navathe, S. B. (1986). A comparative analysis of methodologies for database schema integration. *ACM Computing Surveys*, *15*, 323–363. doi:10.1145/27633.27634

Baziz, M., & Boughanem, M. Aussenac, & Gilles, N. (2004). The Use of Ontology for Semantic Representation of Documents. In proceedings of Semantic Web and Information Retrieval Workshop, SIGIR, (pp. 38-45).

Bechhofer, S., van Harmelen, F., Hendler, J., Horrocks, I., McGuinness, D. L., Patel-Schneider, P. F., & Stein, L. A. (2004). *OWL Web Ontology Language Reference, W3C, W3C Recommendation 10 February 2004*. Retrieved July 6, 2009, from W3C Web site http://www.w3.org/TR/2004/RECowl-ref-20040210/

Beisswanger, E., Poprat, M., & Hahn, U. (2008). Lexical Properties of OBO Ontology Class Names and Synonyms. In *3rd International Symposium on Semantic Mining in Biomedicine.*

Benslimane, D., Arara, A., Yetongnon, K., Gargouri, F., & Ben Abdallah, H. (2003). Two approaches for ontologies building: From-scratch and From existing data sources. In *the 2003 International Conference on Information Systems and Engineering (ISE 2003)*, Montreal, Canada.

Benslimane, S. M., Amar Bensaber, D., & Malki, M. (2004, June). Vers une médiation de contexte à base d'ontologies multi-représentation. In *Colloquy on Optimization and Information Systems* (pp. 101–112). Tizi Ouzou, Algeria: COSI.

Bergamaschi, S., Castano, S., Vimercati, S., & Vincini, M. (1998). *An Intelligent Approach to Information Integration, Formal Ontology in Information System.* Amsterdam: IOS Press.

Berkovsky, S., Eytani, Y., & Gal, A. 2005. Measuring the Relative Performance of Schema Matchers. In *Proceedings of the 2005 IEEE/WIC/ACM International Conference on Web Intelligence,* Compiegne, France, September 19-22, 2005, (WI'05), (pp. 366-371).

Bernaras, A., Laresgoiti, I., & Corera, J. (1996). Building and reusing ontologies for electrical network applications. In *Proceedings of the European Conference on Artificial Intelligence (ECAI_96),* Budapest, Hungary, (pp. 298–302).

Berners-Lee, T. (1998). *What the Semantic Web can represent.* Retrieved from http://www. w3. org/DesignIssues/RDFnot.html

Berners-Lee, T., Hendler, J., & Lassila, O. (2001). *The Semantic Web: A new form of Web content that is meaningful to computers will unleash a revolution of new possibilities*. Scientific American.

Berners-Lee, T., Hendler, J., & Lassila, O. (2001). The semantic web. *Scientific American*, (5): 35–43.

Bernstein, P. A., Melnik, S., Petropoulos, M., & Quix, C. (2004). Industrial-Strength Schema Matching. In ACM SIGMOD Record, (pp. 38-43).

Biébow, B., & Szulman, S. (1999). TERMINAE: a linguistic-based tool for the building of a domain ontology. In *Proceedings of the 11th European Workshop on Knowledge Acquisition, Modelling and Management*, (pp. 49-66), Germany.

Blaschke, C., & Valencia, A. (2002). Automatic Ontology Construction from the Literature. In Genome Informatics Series, (pp. 201–213).

Blazquez, M., Fernandez, M., Garcia-Pinar, J. M., & Goméz-Pérez, A. (1998). Building ontologies at the knowledge level using the ontology design environment. In *Proceedings of the Eleventh Workshop on Knowledge Acquisition, Modeling and Management*, Banff, Canada, Borgida A., (1995). Description Logics in Data Management. *IEEE Transactions, Knowledge and Data Engineering, 7*(5), 671-682.

Bloehdorn, S. Petridis, K. Saathoff, C. Simou, N. Tzouvaras, V. Avrithis, et al. (2005). Semantic Annotation of Images and Videos for Multimedia Analysis. In *Proc. 2nd European Semantic Web Conference, ESWC 2005*, Heraklion, Greece.

Bloehdorn, S., Haase, P., Sure, Y., & Voelker, J. (2006). Ontology evolution. In J. Davies, R. Studer, & P. Warren (Eds.), Semantic Web Technologies, Trends and research in Ontology-based Systems (pp. 51-70). New York: John Wiley & Sons Publication. doi: 10.1002/047003033X.ch4

Blundell, B., & Pettifer, S. (2004). Graph visualization to aid ontology evolution in Protégé. In *Proceedings of the 7th International Protégé Conference*, Bethesda, MD.Retrieved from http://protege.stanford.edu/conference/2004/posters/Blundell.pdf

Bodner, R. C., & Song, F. (1996). Knowledge-based approaches to query expansion in information retrieval. In *Canadian Conference on AI*, (pp. 146–158).

Böhlen, M., Clifford, J., Elmasri, R. A., Gadia, S. K., Grandi, F., Hayes, P., et al. (1998), The Consensus Glossary of Temporal Database Concepts (C. S. Jensen, C. E. Dyreson, Eds.). In O. Etzion, S. Jajodia, S. Sripada, (Eds.), Temporal Databases—Research and Practice, (LNCS vol. 1399, pp. 367–405). Berlin: Springer-Verlag.

Borst, W. N. (1997). *Construction of Engineering Ontologies for Knowledge Sharing and Reuse*. Enschede, Netherlands: University of Tweenty, Centre for Telematica and Information Technology.

Brahmia, Z., & Bouaziz, R. (2008). Schema Versioning in Multi-Temporal XML Databases. In *Proceedings of the 7th IEE/ACIS International Conference on Computer and Information Science (IEEE/ACIS ICIS 2008)*, (pp. 158-164), Oregon, OR.

Brank, J., Grobelnik, M., & Mladeni, D. (2005). A survey of ontology evaluation techniques. In *Proceedings of SIKDD.*

Breche, P., & Wörner, M. (1995). How to remove a class in an object database system. In *Proceedings of the 2nd international conference on applications of databases (ADB-95)*(pp. 235-248). Retrieved from http://citeseerx.ist.psu.edu/viewdoc/download?doi=10.1.1.45.40&rep=rep1&type=url&i=0 doi: 10.1.1.45.40

Breuker, J., & Van de Velde, W. (1994). *CommonKADS Library for Expertise Modeling: reusable problem solving components*. Amsterdam, Tokyo: IOS-Press/Ohmsha.

Brewka, G., Dix, J., & Konolidge, K. (1997). *Nonmonotonic Reasoning: an Overview. Center for the Study of Language and Information, Lecture Notes Number 73.* CA: Stanford University.

Brézillon, P. (1999). Context in problem solving: A survey. *The Knowledge Engineering Review, 14*(1), 1–34. doi:10.1017/S0269888999141018

Brézillon, P., & Pomerol, J.-C. (2001). *Some comments about knowledge and context.* Research Report 2001-022, LIP6, University of Paris VI, Paris, France, Brézillon, P. (2002). *Modeling and using context: Past, present and future.* Research Report, LIP6 2002/010, University of Paris 6, France, Châabane, S. & Jaziri, W. (2009). Méta-modélisation du processus de construction d'ontologies géographiques: Application au domaine routier. In *Proceedings of ICWIT '2009*, Kerkennah, Tunisia, (pp. 653-661).

Brut, M., Sedes, F., & Manzat, A.-M. (2009). A Web Services Orchestration Solution for Semantic Multimedia Indexing and Retrieval. In *Proceedings of CISIS 2009.* Washington, DC: IEEE Computer Society Press.

Buckley, C., & Salton, G. (1995). Optimization of relevance feedback weights. In *Proceedings of the 18th annual international ACM SIGIR conference on Research and development in information retrieval*, (pp. 351–357). New York: ACM Press.

Buckley, C., Salton, G., Allan, J., & Singhal, A. (1994). Automatic query expansion using SMART: TREC 3. In *Text REtrieval Conference.*

Bunescu, R., Ge, R., Kate, R., Marcotte, E., Mooney, R., Ramani, A. & Wong, Y. (2004). Comparative experiments on learning information extractors for proteins and their interactions in Comparative Experiments on Learning Information Extractors for Proteins and their Interactions. *Journal of Artificial Intelligence in Medicine.*

Bussler, C. (2004). Semantic Web Services. In *The International Conference on Web engineering (ICWE), ICWE'04*, Munich, Germany.

Caraballo, S. A. (1999). Automatic construction of a hypernym-labeled noun hierarchy from text. In *Proceedings of the 37th annual meeting of the Association for Computational Linguistics on Computational Linguistics*, (pp. 120–126). Morristown, NJ: Association for Computational Linguistics.

Caracciolo, C., Euzenat, J., Hollink, L., Ichise, R., Isaac, A., & Malaisé, V. Meilicke, et al. (2008). Results of the Ontology Alignment Evaluation Initiative 2008. In *Proceedings of the 3rd International Workshop on Ontology Matching (OM-2008) Collocated with the 7th International Semantic Web Conference (ISWC-2008)*, Karlsruhe, Germany.

Cardoso, J., & Sheth, A. (2004). *Semantic Web services and web processes composition.* New York: Springer-Verlag, Inc.

Caroll, J.-M. (1997). Scenario-Based Design. In Helander, M., Landauer, T. K., & Prabhu, P. (Eds.), *Handbook of Human-Computer Interaction* (2nd ed.). Amsterdam: Elsevier Science B.V.

Casati, F., Ilnicki, S., & Jin, L. (2000). *Adaptive and dynamic Advanced Information Systems Engineering (CAiSE), Stockholm, Sweden.* Berlin: Springer Verlag.

Castano, S. (2006). Ontology evolution: The BOEMIE approach. In *BOEMIE Workshop, at the conference EKAW 06.* Retrieved from http://www.boemie.org/sites/default/files/pres_castano_boemie.pdf

Castano, S., Espinosa, S., Ferrara, A., Karkaletsis, V., Kaya, A., Melzer, S., et al. (2007). Ontology dynamics with multimedia information: The BOEMIE evolution methodology. In G. Flouris, & M. d'Aquin (Eds.), *Proceedings of the International Workshop on Ontology Dynamics (IWOD-07)* at *ESWC 07 Conference,* (pp. 41-54). Retrieved from http://kmi.open.ac.uk/events/iwod/iwod-proceedings.pdf

Centre de documentation de l'urbanisme. (2009). *URBAMET home page.* Retrieved July 6, 2009, from http://www.urbamet.com/

Chan, L. M. (2005). Metadata Interoperability: A Study of Methodology. In *Proceedings of the 3rd China-US Library Conference.*

Chandrasekaran, B., Josephson, J. R., & Benjamins, R. V. (1999). What are ontologies, and why do we need them? *IEEE Intelligent Systems and Their Applications, 14*(1), 20–26.

Chawathe, S., Garcia-Molina, H., Hammer, J., Ireland, K., Papakonstantinou, Y., Ullman, J., & Widom, J. (1994). The Tsimmis project: Integration of heterogeneous information sources. In *Conference of the Information Processing Society*, Japan, (pp. 7-18).

Chebotko, A., Deng, Y., Lu, S., Fotouhi, F., & Aristar, A. (2005, Jan-March). An Ontology-Based Multimedia Annotator for the Semantic Web of Language Engineering. *International Journal on Semantic Web and Information Systems*, *1*(1), 50–67.

Chen, S.-C., Ghafoor, A., & Kashyap, R. L. (2000). *Semantic Models for Multimedia Database Searching and Browsing*. Norwell, MA: Kluwer Academic Publishers.

Chen-Chan Chang, K., He, B., & Zhang, Z. (2005). Toward Large Scale Integration: Building a MetaQuerier over Databases on the Web. In *Proceedings of the Second Conference on Innovative Data Systems Research* (CIDR) Asilomar, CA, (pp. 44-55).

Chikofsky, E. J., & Cross, J. H. (1990). Reverse engineering and design recovery: a taxonomy. *IEEE Software*, 13–17. doi:10.1109/52.43044

Chira, O. (2003). Ontologies. *IDIMS Report*.

Chu, W. W., Liu, Z., & Mao, W. (2002). Textual document indexing and retrieval via knowledge sources and data mining. In *Communication of the Institute of Information and Computing Machinery* (*Vol. 5*, p. 2). Taiwan: CIICM.

Cimiano, P. (2007). *On the relation between ontology learning, engineering, evolution and expressivity*. Invited talk at 7th Meeting on Terminology and Artificial Intelligence TIA 2007, Sophia Antipolis, France. Retrieved from http://www.aifb.uni-karlsruhe.de/WBS/pci/home/Publications/2007/tia07/tia07_slides.pdf

Cimiano, P., & Völker, J. (2005). Text2Onto - a framework for ontology learning and data-driven change Discovery. In A. Montoyo, R. Munoz, & E. Metais (Eds.), Natural Language Processing and Information Systems (LNCS: Vol. 3513, pp. 227-238), Berlin, Germany: Springer. doi: 10.1007/b136569

Clark, P., Thompson, J., Holmback, H., & Duncan, L. (2000). Exploiting a thesaurus-based semantic net for knowledge-based search. In R.S. Engelmore, & H. Hirsh (eds.) *Proceedings of the Twelfth Conference on Innovative Application of Artificial Intelligence* (pp. 988–995). Menlo Park, CA: Association for the Advancement of Artificial Intelligence Press.

Clifford, J., Croker, A., Grandi, F., & Tuzhilin, A. (1995, September 17-18). On Temporal Grouping. In *Proceedings of the International Workshop on Temporal Databases*, Zürich, Switzerland, (pp. 194–213).

Coenen, F. & Bench-Capon, T. (1993). Maintenance of knowledge-based systems. *The A.P.I.C. Series*, 40.

Colmerauer, A., & Roussel, P. (1992). The birth of Prolog. In The second ACM SIGPLAN conference on History of programming languages, (pp. 37-52).

Corcho, O., & Gomez-Perez, A. (2000). A roadmap to ontology specification languages. In *Knowledge engineering and knowledge Management Methods* (pp. 80–96). Models and Tools. doi:10.1007/3-540-39967-4_7

Corcho, O., Fernández-López, M., & Gómez, P. A. (2006). *Ontological Engineering: Principles, Methods, Tools and Languages, Ontologies for Software Engineering and Software Technology* (pp. 1–48). Berlin: Springer.

Corcho, O., Fernandez-Lopez, M., & Gomez-Perez, A. (2003). Methodologies, tools and languages for building ontologies: Where is their meeting point? *Data & Knowledge Engineering*, *46*, 41–64. doi:10.1016/S0169-023X(02)00195-7

Cowie, J., & Lehnert, W. (1996). Information extraction. *Communications of the ACM*, *39*(1), 80–91. doi:10.1145/234173.234209

Dagan, I., Marcus, S., & Markovitch, S. (1993). Contextual word similarity and estimation from sparse data. In *Proceedings of the 31st conference on Association for Computational Linguistics*, (pp. 164–171). Morristown, NJ: Association for Computational Linguistics

de Bruijn, J., Ehrig, M., Feier, C., Martin-Recuerda, F., Scharffe, F., & Weiten, M. (2006). Ontology Mediation, Merging and Aligning. In Davies, J., Studer, R., & Warren, P. (Eds.), *Semantic Web Technologies* (pp. 95–112). Chichester, UK: John Wiley & Sons. doi:10.1002/047003033X.ch6

De Castro, C., Grandi, F., & Scalas, M. R. (1997). Schema versioning for multitemporal relational databases. *Information Systems, 22*(5), 249–290. doi:10.1016/S0306-4379(97)00017-3

De Leenheer, P., & Mens, T. (2008). Ontology Evolution: State of the Art and Future Directions. In Hepp, M., De Leenheer, P., de Moor, A., & Sure, Y. (Eds.), *Ontology Management for the Semantic Web, Semantic Web Services, and Business Applications, from Semantic Web and Beyond: Computing for Human Experience*. Berlin: Springer.

Declerck, T., & Krieger, H.-U. (2006). Translating XBRL Into Description Logic. An Approach Using Protege, Sesame & OWL. In W. Abramowicz, & H. Mayr, (Eds.), *Proceedings of the 9th International Conference on Business Information System.*

Deerwester, S., Dumais, S. T., Furnas, G. W., Landauer, T. K., & Harshman, R. (1990). Indexing by latent semantic analysis. *Journal of the American Society for Information Science American Society for Information Science, 41*(6), 391–407. doi:10.1002/(SICI)1097-4571(199009)41:6<391::AID-ASI1>3.0.CO;2-9

Dehaspe, L., & Toivonen, H. (1999). Discovery of frequent datalog patterns. In Data Mining and Knowledge Discovery.

Desclés, J.-P., & Pascu, A. (2005). Logic of Determination of Objects: the meaning of variable in quantification. In Flairs 2005, Florida, (pp. 610-616). Menlo Park, CA: AAAI Press.

Desmontils, E., & Jacquin, C. (2001). Indexing a Web Site with a Terminology Oriented Ontology. In *International Semantic Web Working Symposium (SWWS'2001)*, Stanford, CA, (pp. 549-565).

Devambu, P., Brachman, R. J., Selfridge, P. J., & Ballard, B. W. (1991). LASSIE: A knowledge based software information system. *Communications of the ACM, 34*(5), 36–49.

Dey, A. K., & Abowd, G. D. (1999). *Towards a Better Understanding of Context and Context-Awareness. Technical report git-gvu-99-22*. Georgia: Institute of Technology.

Dey, N., Boucher, A., & Thonnat, M. (2002). Image formation model of 3D translucent object observed in light microscopy. In *Proceedings of ICIP'02*, (Vol. 2, pp. 469-472).

Dieng, R. (1993). Méthodes et outils d'acquisition des connaissances. In Spérandio, J.-C. (Ed.), *L'ergonomie dans la conception des projets informatiques* (pp. 335–411). Toulouse: Octares.

Dieng, R., Corby, O., Giboin, A., & Ribière, M. (1998). Methods and tools for corporate knowledge management. In 11th workshop on Knowledge Acquisition, Modeling and Management (KAW' 98), Banff Canada.

Dieng, R., Giboin, A., Amergé, C., Corby, O., Després, S., & Alpay, L. (1998, Winter). Building of a Corporate Memory for Traffic Accident Analysis. *AI Magazine, 19*(4), 80–100.

Ding, Y., Fensel, D., Klein, M., & Omelayenko, B. (2002). The Semantic Web: Yet Another Hip. *Data & Knowledge Engineering, 41*(3), 205–227. doi:10.1016/S0169-023X(02)00041-1

Dividino, R., & Sonntag, D. (2008). Controlled ontology evolution through semiotic-based ontology evaluation. In *Proceedings of the 2nd Workshop on Ontology Dynamics, (IWOD-08) at ISWC 08 Conference* (pp. 1-14). Retrieved from http://www.ics.forth.gr/~fgeo/Publications/IWOD-08_Proceedings.pdf

Djedidi, R. & Aufaure, M-A. (2008). Enrichissement d'ontologies: maintenance de la consistance et évaluation de la qualité. *Journées Francophones sur l'Ingénierie des connaissances (IC'2008).*

Djedidi, R., & Aufaure, M.-A. (2008). Ontological knowledge maintenance methodology. In I. Lovrek, R. J. Howlett, & Lakhmi C. Jain (Eds.), Knowledge-Based Intelligent Information and Engineering Systems (LNCS: Vol. 5177, Part I, pp. 557-564). Berlin, Germany: Springer. doi: 10.1007/978-3-540-85563-7

Djedidi, R., & Aufaure, M.-A. (2009). Ontology change management. In A. Paschke, H. Weigand, W. Behrendt, K. Tochtermann, T. Pellegrini (Eds.), Semantic Systems (I-Semantics 09), Journal of Universal Computer Science (JUCS), ISBN 978-3-85125-060-2, pp 611-621, Verlag der Technischen Universitt Graz.

Djedidi, R., & Aufaure, M.-A. (2010). Onto-Evo^{a}l an Ontology Evolution Approach Guided by Pattern Modelling and Quality Evaluation. *Proceedings of the the 6th International Symposium on Foundations of Information and Knowledge Systems* (FoIKS 2010), (LNCS: Vol. 5956, pp. 286-305). Berlin, Germany: Springer. doi: 10.1007/978-3-642-11829-6

Do, H. H., Melnik, S., & Rahm, E. (2002). Comparison of schema Matching Evaluations. In GI-Workshop Web and Databases, (pp 221-237), Erfurt, Germany.

Doms, A., & Schroeder, M. (2005). GoPubMed: exploring PubMed with the gene ontology. *Nucleic Acids Research*,33.

Donderler, M. E., Saykol, E., Arslan, U., Ulusoy, O., & Gudukbay, U. (2005). BilVideo: Design and implementation of a video database management system. *Multimedia Tools and Applications*, *27*(1), 79–104. doi:10.1007/s11042-005-2715-7

Drucker, P. (2000). *Need to Know: Integrating e-Learning with High Velocity Value Chains*. A Delphi Group White.

Ducournau, R., Euzenat, J., Masini, G., & Napoli, A. (1998). *INRIA Langages et modèles à Objets Etat des recherches et perspectives*. Collection Didactique.

Duineveld, A. J., Stoter, R., Weiden, M. R., Kenepa, B., & Benjamins, V. R. (2000). WonderTools? A comparative study of ontological engineering tools. *International Journal of Human-Computer Studies*, *52*(6), 1111–1133. doi:10.1006/ijhc.1999.0366

Eder, J., & Wiggisser, K. (2007). Change detection in ontologies using DAG comparison. In J.Krogstie, A.L. Opdahl, & G. Sindre, (Eds.), Advanced Information Systems Engineering (LNCS: Vol. 4495, pp. 21-35). Berlin, Germany: Springer. doi: 10.1007/978-3-540-72988-4

Efthimiadis, E. (1996). Query expansion. In Williams, M. E. (Ed.), *Annual Review of Information Systems and Technologies* (*Vol. 31*, pp. 121–187). ARIST.

Ehrig, M., & Staab, S. (2004). QOM-Quick Ontology Mapping. In *Proceedings of the Third International Semantic Web Conference (ISWC)* (pp. 683-697), Hiroshima, Japan.

Ehrig, M., & Sure, Y. (2004). *Ontology Mapping – An Integrated Approach* (*Vol. 3053*). Lecture Notes in Computer Science.

Faatz, A., & Steinmetz, R. (2002). Ontology enrichment with texts from the WWW. In *Semantic Web Mining*. Helsinki, Finland: Second Workshop at ECML/PKDD.

Farquhar, A., Fikes, R., & Rice, J. (1997). Tools for Assembling Modular Ontologies in Ontolingua. In [Menlo Park, CA: AAAI Press.]. *Proceedings of AAAI*, *97*, 436–441.

Farrar, S., & Langendoen, D. T. (2003). A linguistic ontology for the Semantic Web. *GLOT International*, *7*(3), 97–100.

Faure, D., & Nédellec, C. (1998). A corpus-based conceptual clustering method for verb frames and ontology acquisition. In LREC workshop on adapting lexical and corpus resources to sublanguages and applications, (pp. 707–728).

Faure, D., & Nédellec, C. (1998). ASIUM: learning subcategorization frames and restrictions of selection. In Y. Kodrato, (ed.), *10th European Conference on Machine Learning (ECML 98), Workshop on Text Mining.*

Fayyad, U., Piatetsky-Shapiro, G., Smyth, P., & Uthurusamy, R. (Eds.). (1996). *Advances in Knowledge Discovery and Data Mining*. Menlo Park, CA: AAAI Press/The MIT Press.

Fensel, D. (2000). *Ontologies: A Silver Bullet for Knowledge Management and Electronic Commerce*. Berlin: Springer.

Fensel, D., & Bussler, C. (2002). The Web Service Modeling Framework WSMF. [New York: Elsevier Sciences,

B. V.]. *Electronic Commerce Research and Applications, 1*(2), 113–137. doi:10.1016/S1567-4223(02)00015-7

Fensel, D., Angele, J., Decker, S., Erdmann, M., Schnurr, H., Staab, S., et al. (1999). On2broker: Semantic-Based Access to Information Sources at the WWW. In *Proceedings of the World Conference on the WWW and Internet WebNet 99*, Honolulu, HI.

Fensel, D., Decker, S., Erdmann, M., & Studer, R. (1998). Ontobroker: Or How to Enable Intelligent Access to the WWW. In B. Gaines & M. Musen (eds.), *Proc of the 11th Workshop on Knowledge Acquisition, Modeling and Management KAW'98,* Banff, Canada, April 18-23. Retrieved from http://ksi.cpsc.ucalgary.ca/KAW/KAW98/KAW98Proc.html

Fernandez, M., Gomez-Perez, A., & Juristo, N. (1997 March). METHONTOLOGY: From Ontological Arts Towards Ontological Engineering. In *Proceedings of the AAAI97 Spring Symposium Series on Ontological Engineering*, Stanford, CA, (pp. 33-40).

Fernandez-Lopez, M. (1999). Overview of Methodologies for Building Ontologies. In *IJCAI'99 Workshop on Ontologies and Problem-Solving Methods: Lessons Learned and Future Trends*, Stockholm, Sweden, Fernandez-Lopez, M., Gomez-Perez, A., Pazos-Sierra, A. & Pazos-Sierra, J. (1999). Building a chemical ontology using METHONTOLOGY and the ontology design environment. *IEEE Intelligent Systems & their applications, 4* (1), 37–46.

Fernandez-Lopez, M., Gomez-Perez, A., & Juristo, N. (1997). Methontology: From Ontological Art Toward Ontological Engineering. In *Spring Symposium Series on Ontological Engineering, AAAI97*, Stanford, CA.

Ferrandina, F., & Lautemann, S. E. (1996). An integrated approach to schema evolution for object databases. In *Proceedings of the Internation Conference on Object Oriented Information Systems (OOIS 1996)*, (pp. 280-294), London.

Fikes, R., & Farquhar, A. (1999). Distributed Repositories of Highly Expressive Reusable Ontologies. *IEEE Intelligent Systems, 14*(2), 73–79. doi:10.1109/5254.757634

Fisher, D. H. (1998). From thesauri towards ontologies? In W.M. el Hadi, J. Maniez & S.A. Pollitt (Eds.), *Structures and relations in knowledge organization: proceedings of the fifth International ISKO Conference* (pp. 18–30). Wurzberg, Germany: Ergon Verlag.

Flouris, G. & Plexousakis, D. (2005). *Handling ontology change: survey and proposal for a future research direction.* Rapport technique FORTH-ICS/TR-362.

Flouris, G. (2006). *On belief change and ontology evolution.* Ph.D. Thesis, University of Crete, Department of Computer Science, Heraklion, Greece.

Flouris, G., & Plexousakis, D. (2006). Bridging ontology evolution and belief change. Advances in Artificial Intelligence (LNCS: Vol. 3955, pp. 486-489). Berlin, Germany: Springer. doi: 10.1007/11752912_51

Flouris, G., Plexousakis, D., & Antoniou, G. (2005). On applying the AGM theory to DLs and OWL. In Y.Gil, E. Motta, V. Benjamins, & M. Musen, (Eds.), The Semantic Web–ISWC 2005 (LNCS: Vol. 3729, pp. 216-231). Berlin, Germany: Springer. doi: 10.1007/11574620

Flouris, G., Plexousakis, D., & Antoniou, G. (2006). Evolving Ontology Evolution. In *Proceedings of the 32nd International Conference on Current Trends in Theory and Practice of Computer Science (SOFSEM 06)*, (pp. 14-29), Czech Republic.

Fonseca, F. (2007). The Double Role of Ontologies in Information Science Research. *Journal of the American Society for Information Science and Technology, 58*(6), 786–793. doi:10.1002/asi.20565

Foo, N. (1995). Ontology revision. In G. Ellis, R. Levinson, W. Rich, & J. F. Sowa (Eds.), *Proceedings of The 3rd International Conference on Conceptual Structures* (pp. 16-31). London: Springer-Verlag.

Foote, J. (1999). An overview of audio information retrieval. *Multimedia Systems, 7*(1), 2–10. doi:10.1007/s005300050106

Fox, E. A. (1980). Lexical relations: Enhancing effectiveness of information retrieval systems. *SIGIR Forum, 15*(3), 5–36. doi:10.1145/1095403.1095404

Franconi, E., Grandi, F., & Mandreoli, F. (2000). A semantic approach for schema evolution and versioning in object-oriented databases. In Computational Logic — CL 2000 (LNCS: Vol. 1861, pp. 1048-1062). Berlin: Springer. doi: 10.1007/3-540-44957-4

Frege, G. (1893) *Logical Investigations* (Geach, P., Ed.). London: Blackwell.

Freitas, V. (2000). *Autoria Adaptativa de Hipermídia Educacional.* Master's thesis, Instituto de Informática, UFRGS, Porto Alegre, National Library Of Medicine.

Freund, M., Desclés, J.-P., Pascu, A., & Cardot, J. (2004). Typicality, contextual inferences and object determination logic. In Flairs 04, Miami, Florida, (pp. 491-495). Menlo Park, CA: AAAI Press.

Gabel, T., Sure, Y., & Voelker, J. (2004). *KAON – ontology management infrastructure*. SEKT informal deliverable 3.1.1.a, Institute AIFB, University of Karlsruhe, Gaines, B. (1997). Editorial: Using Explicit Ontologies in Knowledge-based System Development. *International Journal of Human-Computer Systems, 46*(2-3), 181.

Gallego-Carrillo, M., Garcia-Alcaide, I., & Montalvo-Herranz, S. (2005). Applying Hierarchical MVC Architecture to High Interactive Web Applications. In *Proceedings of the Third International Conference I. TECH*, (pp. 110--115).

Gandon, F. (2002). *Distributed Artificial Intelligence and Knowledge Management: ontologies and multi-agent systems for a corporate semantic web.* Doctoral dissertation, Institut National de Recherche en Informatique et Automatique (INRIA) and University of Nice Sophia Antipolis, France.

Gandon, F., & Dieng-Kuntz, R. (2001). Ontologie pour un système multi-agents dédié à une mémoire d'entreprise. In *Proceedings of the journées francophones d'Ingénierie des Connaissances IC'2001*, (pp. 1-20). Grenoble, France: Presses Universitaires de Grenoble.

Gangemi, A., Guarino, N., Masolo, C., & Oltramari, A. (2003). Sweetening WordNet with DOLCE. *AI Magazine, 24*(3), 13–24.

Gangemi, A., Pisanelli, D. M., & Steve, G. (1998). Ontology Integration: Experiences with Medical Terminologies. In *Formal Ontology in Information Systems: Proceedings of the First International Conference (FOIS'98),* June 6-8, Trento, Italy. Amsterdam: IOS Press, Inc.

Gao, H. T., Hayes, J. H., & Cai, H. (2005). Integrating biological research through web services. *Computer, 38*(3), 26–31. doi:10.1109/MC.2005.97

García, R., & Celma, O. (2005) Semantic integration and retrieval of multimedia metadata. In *Proceedings of the Fifth International Workshop on Knowledge Markup and Semantic Annotation at the Fourth International Semantic Web Conference,* Galway, Ireland.

Gauch, S., & Smith, J. B. (1993). An expert system for automatic query reformation. *Journal of the American Society for Information Science American Society for Information Science, 44*(3), 124–136. doi:10.1002/(SICI)1097-4571(199304)44:3<124::AID-ASI2>3.0.CO;2-C

Gaudan, S., Kirsch, H., & Rebholz-Schuhmann, D. (2005). Resolving abbreviations to their senses in Medline. *Bioinformatics (Oxford, England), 21*(18), 3658–3664. doi:10.1093/bioinformatics/bti586

Gelfond, M. (2008). Answer sets. In Harmelen, F., Lifschitz, V., & Porter, B. (Eds.), *Handbook of Knowledge Representation* (pp. 285–316). New York: Elsevier. doi:10.1016/S1574-6526(07)03007-6

Genesereth, M. (n.d.). *Knowledge Interchange Format (KIF).* Retrieved from http://www.ksl.stanford.edu/know- ledge-sharing/kif/

Genesereth, M. R., & Nilsson, N. J. (1987). *Logical Foundations of Artificial Inteligence*. San Francisco: Morgan Kaufmann Publishers, Goh, C.H., Bressan, S., Madnick, S. & Siegel, M. (1999). Context Interchange: New Features and Formalisms for the Intelligent Integration of Information. *ACM Transactions on Information Systems, 17*(3), 270–293.

Ghebghoub, O., Abel, M., & Moulin, C. (2008). Semantic Indexing of e-Learning Resources. In *3rd International Conference Information and Communication Technolo-*

gies: From Theory to Applications. ICTTA 2008, (pp. 1--6), Damascus, Syria.

Giboin, A., Gandon, F., Corby, O., & Dieng, R. (2002). Assessment of Ontology-based Tools: Systemizing the Scenario Approach. In J. Angele, & Y. Sure, (Eds.) *Proceedings of EON2002: Evaluation of Ontology-based Tools Workshop, 13th International Conference on Knowledge Engineering and Knowledge Management EKAW 2002*, Siguenza (Spain), September 30th, (pp. 63-73).

Goble, C., Kesselman, C., & Sure, Y. (Eds.). (2005). In Semantic Grid - Convergence of Technologies, number 05271 in Dagstuhl Seminar Proceedings.

Goh, S. C. (2002). Managing Effective Knowledge Transfer: An Integrative Framework and Some Practice Implications. *Journal of Knowledge Management*, *6*(1), 23–30. doi:10.1108/13673270210417664

Golbeck, J., Fragoso, G., Hartel, F., Hendler, J., Parsia, B., & Oberthaler, J. (2003). The National Cancer Institute's thesaurus and ontology. *Journal of Web Semantics*, *1*(1), 1–5. doi:10.1016/j.websem.2003.07.007

Gomez-Perez, A. (1996). A framework to verify knowledge sharing technology. *Expert Systems with Applications*, *11*(4), 519–529. doi:10.1016/S0957-4174(96)00067-X

Gomez-Perez, A. (1998). Knowledge sharing and reuse. In Liebowitz, J. (Ed.), *Handbook of Expert Systems*. New York: CRC.

Gomez-Perez, A. (2001). Evaluation of ontologies. *International Journal of Intelligent Systems*, *16*(3), 1–10. doi:10.1002/1098-111X(200103)16:3<391::AID-INT1014>3.0.CO;2-2

Gómez-Pérez, A., & Manzano-Macho, D. (Eds.). (2003). *Deliverable 1.5: A survey of ontology learning methods and techniques.* Retrieved July 6, 2009, from the OntoWeb Consortium Web site http://www.sti-innsbruck.at/fileadmin/ documents/deliverables/Ontoweb/D1.5.pdf

Gomez-Perez, A., & Rojas, M. D. (1999). Ontological reengineering and reuse. In D. Fensel, R. Studer (Eds.), *11th European Workshop on Knowledge Acquisition, Modeling and Management* (EKAW_99, LNAI, Vol. 1621, pp.139–156). Berlin: Springer.

Gómez-Pérez, A., Fernandez, M., & De Vicente, A. (1996). Towards a Method to Conceptualize Domain Ontologies. In *Workshop on Ontological Engineering at ECAI'96*, (pp. 41-51).

Gómez-Pérez, A., Fernández-López, M., & Corcho, O. (2003). Methodologies and Methods for Building Ontologies. In *Ontological Engineering* (pp. 107–198). London: Springer-Verlag.

Gomez-Perez, A., Fernandez-Lopez, M., & Corcho, O. (2004). *Ontological Engineering.* Berlin: Springer.

Gomez-Perez, A., Fernandez-Lopez, M., & De Vicente, A. (1996). Towards a Method to Conceptualize Domain Ontologies. In *ECAI96 Workshop on Ontological Engineering*, (pp. 41–51), Budapest, Hungary.

González-Pérez, C., & Henderson-Sellers, B. (2006). An Ontology for Software Development Methodologies and Endeavours. In *Ontologies for Software Engineering and Software Technology* (pp. 123–151). Berlin: Springer. doi:10.1007/3-540-34518-3_4

Grau, B. C., Parsia, B., Sirin, E., & Kalyanpur, A. (2005). Automatic Partitioning of OWL Ontologies Using ε-Connections. In *Proceedings of the International Workshop on Description Logics (DL)* Edinburgh, Scotland, UK.

Gruber, T. (1991). The role of a common ontology in achieving sharable, reusable knowledge bases. In *Proceedings of the 2nd International Conference on Principles of Knowledge Representation and Reasoning*, Cambridge, UK, (pp. 601-602).

Gruber, T. (1993). A translation approach to portable ontologies. *Knowledge Acquisition*, *5*(2), 199–220. doi:10.1006/knac.1993.1008

Gruber, T. R. (1993). Toward Principles for the Design of Ontologies Used for Knowledge Sharing. In Guarino, N., & Poli, R. (Eds.), *Formal Ontology in Conceptual Analysis and Knowledge Representation.* Amsterdam: Kluwer Academic Representation.

Gruninger, M., & Fox, M. S. (1995). Methodology for the design and evaluation of ontologies. In *Workshop on Basic Ontological Issues in Knowledge Sharing*, Montreal, Canada.

Gruninger, M., & Lee, J. (2002). Ontology Applications and Design. *Communications of the ACM*, *45*(2), 1–2.

Guarino, N. (1992). Concepts, Attributes, and Arbitrary Relations: Some Linguistic and Ontological Criteria for Structuring Knowledge Bases. *Data & Knowledge Engineering*, *8*, 249–261. doi:10.1016/0169-023X(92)90025-7

Guarino, N. (1997). *Semantic matching: Formal ontological distinctions for information organization, extraction, and integration* (pp. 139–170). SCIE.

Guarino, N. (1997). Understanding, Building and Using Ontologies. A Commentary to Using Explicit Ontologies in KBS Development, by van Heijst, Schreiber, and Wielinga. *International Journal of Human-Computer Studies*, *46*(2/3), 293–310. doi:10.1006/ijhc.1996.0091

Guarino, N. (1998). Formal Ontology and Information Systems. In N. Guarino (Ed.), *Proceedings of the International Conference on Formal Ontology in Information Systems* (pp. 3-17). Amsterdam: IOS Press.

Guarino, N., & Giaretta, P. (1995). Ontologies and Knowledge Bases: Towards a Terminological Clarification. In Mars, N. J. I. (Ed.), *Towards Very Large Knowledge Bases*. Amsterdam: IOS Press.

Guarino, N., & Welty, C. (2000). A formal ontology of properties. In *12th International Conference in Knowledge Engineering and Knowledge Management (EKAW_00)*, (LNAI, Vol. 1937, pp. 97–112). Berlin: Springer.

Guarino, N., & Welty, C. (2000). Ontological analysis of taxonomic relationships. In *19th International Conference on Conceptual Modeling (ER_00)*, (LNCS, Vol. 1920, pp. 210–224). Berlin: Springer.

Guarino, N., & Welty, C. (2000). Towards a methodology for ontology-based model engineering. In *Proceedings of ECOOP-2000 Workshop on Model Engineering*, Cannes, France.

Guarino, N., & Welty, C. (2002). Evaluating ontological decisions with OntoClean. In *Communication of the ACM (CACM)*, *45*(2), 61-65. New York: ACM. Retrieved from http://doi.acm.org/10.1145/503124.503150

Guarino, N., Masolo, C., & Vetere, G. (1999). Ontoseek: Content-based access to the web. *IEEE Intelligent Systems*, *14*(3), 70–80. doi:10.1109/5254.769887

Gupta, K. M., Aha, D. W., Marsh, E., & Maney, T. (2002). An architecture for engineering sublanguage WordNets. In *Proceedings of the First International Conference On Global WordNet*, (pp. 207-215). Mysore, India: Central Institute of Indian Languages.

Haase, P., & Stojanovic, L. (2005). Consistent Evolution of OWL Ontologies. In A. Gomez-Perez & J. Euzenat (Eds.), The Semantic Web: Research and Applications (LNCS, vol.3532, pp. 182-197). Berlin: Springer. doi: 10.1007/b136731

Haase, P., & Sure, Y. (2004). D3.1.1.b State of the art on ontology evolution. *SEKT Deliverable*. Retrieved from http://www.aifb.uni-karlsruhe.de/WBS/ysu/publications/SEKT-D3.1.1.b.pdf

Haase, P., & Sure, Y. (2004). *State of the Art on Ontology Evolution*. Retrieved from http://www.aifb.uni-karlsruhe.de/WBS/ysupublications/SEKT-D3.1.1.b.pdf

Haase, P., & Völker, J. (2008). Ontology learning and reasoning – dealing with uncertainty and inconsistencies. In P. C. G. Costa, C. d'Amato, N. Fanizzi, K. B. Laskey, K. J. Laskey, T. Lukasiewicz et al. (Eds.), Uncertainty Reasoning for the Semantic Web I (LNCS: Vol. 5327, pp. 366-384). Berlin, Germany: Springer. doi: 10.1007/978-3-540-89765-1

Haase, P., Van Harmelen, F., Huang, Z., Stuckenschmidt, H., & Sure, Y. (2005). A Framework for handling inconsistency in changing ontologies. In Y. Gil, E. Motta, V. Benjamins, & M. Musen, (Eds.), The Semantic Web – ISWC 2005 (LNCS: Vol. 3729, pp. 353-367). Berlin, Germany: Springer. doi: 10.1007/11574620

Hahn, U., & Markó, K. (2001). Joint knowledge capture for grammars and ontologies. In *Proceedings of the First*

International Conference on Knowledge Capture K-CAP 2001, Victoria, Canada.

Hahn, U., & Schnattinger, K. (1998). Towards text knowledge engineering. In *Proceedings of the fifteenth national/tenth conference on Artificial intelligence/ Innovative applications of artificial intelligence*, (pp. 524–531). Menlo Park, CA: American Association for Artificial Intelligence.

Haidar, B., Joly, P., & Haidar, S. A Graph-based Approach to Automatically Chain Distributed Multimedia Indexing Services. In *Proceedings of 9th Int. Conf. on Internet and Multimedia Systems and Applications*, Switzerland. Boston: ACTA Press.

Halpin, H., Robu, V., & Shepherd, H. (2007). The Complex Dynamics of Collaborative Tagging. In *WWW 2007*. New York: ACM Press.

Hanks, S., & McDermott, D. (1987). Nonmonotonic logic and temporal projection. *Artificial Intelligence*, *33*(3), 379–412. doi:10.1016/0004-3702(87)90043-9

Harriche, K., Hatchabi, M., & Maiga, A. (2008). *Vers une approche d'integration semantique via les web services. Research report.* Ibn Khaldoun University of Tiaret - Algeria.

Hausenblas, M., Troncy, R., Halaschek-Wiener, C., Bürger, T., Celma, O., Boll, S., & Mannens, E. (2007). Multimedia Semantics on the Web: Vocabularies. *W3C technical reports,* Boston. Retrieved from http://www.w3.org/2005/Incubator/mmsem/wiki/Vocabularies

He, B., & Chen-Chan Chang, K. (2003). Statistical Schema Matching across Web Query Interfaces. In *Proceedings of the ACM SIGMOD International Conference on Management of Data* (pp. 217-228), San Diego, CA.

He, B., & Chen-chuan Chang, K.(2006). Automatic Complex Schema Matching Across Web Query Interfaces: A Correlation Mining Approach. In *ACM Transactions on Database Systems* (*TODS*) (pp. 346-395). New York: ACM Press.

He, B., Chen-Chan Chang, K., & Han, J. (2004). Discovering complex matchings across Web Query Interfaces: A Correlation Mining Approach. In *Proceedings of the Tenth ACM SIGKDD International Conference on Knowledge Discovery and Data Mining (KDD),* (pp. 148-157). New York: ACM Press.

He, H., Meng, W., Yu, C., & Wu, Z. (2005). WISE-Integrator: A System for Extracting and Integrating Complex Web Search Interfaces of the Deep Web. In *Proceedings of the 31st International Conference on Very Large Data Bases (VLDB),* (pp. 1314-1317), Trondheim, Norway.

Hearst, M. A. (1992). *Automatic acquisition of hyponyms from large text corpora*. Technical Report S2K-92-09.

Hearst, M. A. (1998). Automated Discovery of WordNet Relations. In Fellbaum, C. (Ed.), *WordNet: An Electronic Lexical Database* (pp. 132–152). Cambridge, MA: MIT Press.

Hearst, M. A., & Pedersen, J. O. (1996). *Reexamining the cluster hypothesis: Scatter/gather on retrieval results* (pp. 76–84). SIGIR.

Hearst, M. A., Karger, D. R., & Pedersen, J. O. (1995). Scatter/gather as a tool for the navigation of retrieval results. In *Working Notes AAAI Fall Symp.* AI Applications in Knowledge Navigation.

Heflin, J. (n.d.). Towards the Semantic Web: Knowledge and representation in a dynamic, distributed environment. *Faculty of the Graduate School of the University of Maryland.*

Heflin, J., & Hendler, J. (2000). Dynamic Ontology on the Web. In *Proceedings of the 17th National Conference on Artificial Intelligence*, (pp. 443-449). Menlo Park, CA: AAAI/MIT.

Heflin, J., Hendler, J., & Luke, S. (1998). Reading between the lines: Using SHOE to discover implicit knowledge from the Web. In *proc. of the AAAI Workshop on Artificial Intelligence and Information Integration*, WS-98-14, (pp. 51-57). Menlo Park, CA: AAAI Press.

Heflin, J., Hendler, J., & Luke, S. (1999). Coping with changing ontologies in a distributed environment. In *Proceedings of the Workshop on Ontology Management of the 16th National Conference on Artificial Intelligence*

(AAAI-99), WS-99-13, 74-79. Menlo Park, CA: AAAI Press.

Hentschel, C., Nurnberger, A., Schmitt, I., & Stober, S. (2007). SAFIRE: Towards Standardized Semantic Rich Image Annotation. In S. Marchand-Maillet et al. (Eds.), *Proceedings of AMR 2006*, (LNCS 4398, pp. 12–27). Berlin: Springer.

Hepp, M., & de Bruijn, J. (2007). Gentax: A generic methodology for deriving OWL and RDF-S ontologies from hierarchical classifications, thesauri, and inconsistent taxonomies. In *Proceedings of the 4th European Semantic Web Conference*, (LNCS, Vol. 4519, pp. 129–144). Berlin: Springer-Verlag.

Hofstadter, D. R. (1999, January). Gödel, Escher, Bach: An Eternal Golden Braid, (20th anniversary ed.). New York: Basic Books.

Hole, W. T., & Srinivasan, S. (2002). *Discovering missed synonymy in a large concept-oriented metathesaurus*. AMIA.

Hovy, E. H., & Lin, C.-Y. (1999). Automated Text Summarization in SUMMARIST. In Maybury, M., & Mani, I. (Eds.), *Advances in Automatic Text Summarization*. Cambridge, MA: MIT Press.

Hu, W., & Qu, Y. (2006). Block Matching for Ontologies. In *Proceedings of the 5th International Semantic Web Conference (ISWC)*, (pp. 300-313), Athens, GA.

Hu, W., Zhao, Y., & Qu, Y. (2008). Matching large ontologies: A divide-and-conquer approach. *Journal of Data Knowledge Engineering*, *67*, 140–160. doi:10.1016/j.datak.2008.06.003

Huersch, W. (1997). Maintaining consistency and behaviour of object-oriented systems during evolution. In *Proceedings of the ACM Conference on Object-Oriented Programming, Systems, Languages and Applications (OOPSLA '97), ACM SIGPLAN Notices, 32*(10), 1-21.

Huhns, M. N., & Singh, M. (1997). Ontologies for Agents. *IEEE Internet Computing*, *1*, 81–83. doi:10.1109/4236.643942

Hunter, J. (2001). Adding multimedia to the semantic web – building an MPEG-7 Ontology. *International Semantic Web Working Symposium (SWWS)*, Stanford.

Hunter, J., & Little, S. (2005). A framework to enable the semantic inferencing and querying of multimedia content. *Int. J. Web Engineering and Technology*, *2*(2/3), 264–286. doi:10.1504/IJWET.2005.008487

Hutchins, E. (1995). *Cognition in the Wild*. Cambridge, MA: MIT Press.

Hwang, C. H. (1999). Incompletely and imprecisely speaking: Using dynamic ontologies for representing and retrieving information. In *Proceedings of the 6th International Workshop on Knowledge Representation meets Databases (KRDB'99)*, Linköping, Sweden, July 29-30.

IEEE. (1990). *IEEE Standard Computer Dictionary: A Compilation of IEEE Standard Computer Glossaries*. New York: Author.

INTEROP. (2004). *Ontology Interoperability*. State of the Art Report (SOA), WP8ST3 Deliverable, IST-508011.

Izza, S., Vincent, L. & Burlat, P. (2006). Dealing with Semantic Application Integration With in Large and Dynamic Enterprises. *International Journal of Cooperative Information systems, 15*(4), 507-533.

Jacquemin, C. (2001). *Spotting and discovering terms through natural language processing*. Cambridge, MA: The MIT Press.

Jasper, R. (1999). Uschold M., A Framework for Understanding and Classifying Ontology Applications. In *IJCAI-99 Workshop on Ontologies and Problem-Solving Methods (KRR5)*.

Jaziri, W. (2009). A methodology for ontology evolution and versioning. *The Third International Conference on Advances in Semantic Processing (SEMAPRO 2009)*, (pp. 15-21), Sliema, Malta.

Jimeno Yepes, A., Jimenez-Ruiz, E., Berlanga-Llavori, R., & Rebholz-Schuhmann, D. (2008). Use of shared lexical resources for efficient ontological engineering. In Semantic Web Applications and Tools for Life Sciences.

Jimeno-Yepes, A., Berlanga-Llavori, R., & Rebholz-Schuhmann, D. (In press). Ontology refinement for improved information retrieval. In Information Processing & Management: Special Issue on Semantic Annotations in Information Retrieval.

Jones, D., Bench-Capon, T., & Visser, P. (1999). Methodologies for Ontology Development. In *Proceedings of the IJCAI-99 workshop on Ontologies and Problem -Solving Methods.*

Jouis, C. (2002). Logic of Relationships. In Green, R., Bean, C. A., & Myaeng, S. H. (Eds.), *The Semantics of Relationships: An Interdisciplinary Perspective* (pp. 127–140). Dordrecht: Kluwer Academic Publishers.

Kalfoglou, Y., & Robertson, D. (1999). Managing Ontological Constraints. In *Proceedings of the IJCAI99 Workshop on Ontologies and Problem-Solving Methods: Lessons Learned and Future Trends*, Stockholm, Sweden.

Kalfoglou, Y., & Robertson, D. (1999). Use of formal ontologies to support error checking in specifications. In D. Fensel, R. Studer (Eds.), *11th European Workshop on Knowledge Acquisition, Modeling and Management (EKAW_99)*, (LNAI, Vol. 1621, pp. 207–224). Berlin: Springer.

Kashyap, V., & Sheth, A. (1996). Semantic and schematic similarities between database objects: a context-based approach. *The VLDB Journal*, *5*, 276–304. doi:10.1007/s007780050029

Kashyap, V., & Sheth, A. (1998). Semantic heterogeneity in global information systems: the role of metadata, context and ontologies. In Papazoglou, M., & Schlageter, G. (Eds.), *Cooperative Information Systems: Current Trends and Directions* (pp. 139–178). New York: Academic Press.

Kassel, G. (2002). OntoSpec: une méthode de spécification semi-informelle d'ontologies. In Actes des Journées Francophones d'Ingénierie des Connaissances: IC'2002, Rouen, France, (pp. 75-87).

Katz, R. H. (1990). Toward a Unified Framework for Version Modeling in Engineering Databases. *ACM Computing Surveys*, *22*(4), 375–408. doi:10.1145/98163.98172

Katz, R. H., & Lehman, T. J. (1984). Database Support for Versions and Alternative of Large Design Files. *IEEE Transactions on Software Engineering*, *10*(2), 191–200. doi:10.1109/TSE.1984.5010222

Kawtrakul, A., Imsombut, A., Thunkijjanukit, A., Soergel, D., Liang, A., Sini, M., et al. (2005). *Automatic Term Relationship Cleaning and Refinement for AGROVOC, Workshop on The Sixth Agricultural Ontology Service*. Vila Real, Portugal.

Kayser, D. (1997). *La représentation des connaissances.*

Keita, A. K., Catherine, R., & Robert, L. (2006). Un outil d'aide à la construction d'ontologies préconsensuelles: Towntology project. In *Acts of Congress computer science Organizations and Information Systems and Decision XXIVème* (pp. 911–926). Hammamet, Tunisia: INFORSID.

Kellert, P., & Toumani, F. (2003). *Les web services sémantiques*. Research Report *LIMOS/RR03-15*. Computer science Laboratory, modeling and optimization systems, Clermont-Ferrand, France.

Kelley, J. L. (1975). *General Topology*. Berlin: Springer-Verlag.

Khan, L. R., & Luo, F. (2002). Ontology construction for information selection. In *ICTAI '02: Proceedings of the 14th IEEE International Conference on Tools with Artificial Intelligence (ICTAI'02),* (pp. 122). Washington, DC: IEEE Computer Society.

Khan, L., & Luo, F. (2002). Ontology Construction for Information Selection. In *Proceedings of the 14th IEEE International Conference on Tools with Artificial Intelligence*, (pp. 122-127). Washington DC.

Kietz, J. U., Maedche, A., & Volz, R. (2000). A method for semi-automatic ontology acquisition from a corporate intranet. In *EKAW_00 Workshop on Ontologies and Texts, CEUR Workshop Proceedings*, Juan-Les-Pins, (Vol. 51).

Kim, W., Banerjee, J., Chou, H.-T., & Garza, J. F. (1990). Object-Oriented Database Support for CAD. *Com-*

puter Aided Design, *22*(8), 469–479. doi:10.1016/0010-4485(90)90063-I

Kim, W., Bertino, E., & Garza, J. F. (1989). Composite objects revisited. In *Proceedings of the 1989 ACM SIGMOD International Conference on Management of Data*, Portland, Oregon, 31 May – 2 June, pp. 337–347.

King, R., Popitsch, N., & Westermann, U. (2007). METIS: a flexible foundation for the unified management of multimedia assets. *Multimedia Tools and Applications*, *33*, 325–349. doi:10.1007/s11042-007-0101-3

Klein, M. (2001). Combining and Relating Ontologies: an Analysis of Problems and Solutions. In A. Gomez-Perez, M. Gruninger, H. Stuckenschmidt & M. Uschold (Ed.), *IJCAI-2001 Workshop on Ontologies and Information Sharing,* (pp. 53-62).

Klein, M. (2002). *Versioning of distributed ontologies.* EU/IST Project WonderWeb.

Klein, M. (2004). *Change management for distributed ontologies.* Ph.D. Thesis, Dutch Graduate School for Information and Knowledge Systems, Germany.

Klein, M., & Fensel, D. (2001). Ontology versioning on the semantic web. In I. F. Cruz, S. Decker, J. Euzenat, & D. L. McGuinness (Eds.), *Proceedings of the first International Semantic Web Working Symposium (SWWS'01)* (pp. 75-91), Stanford University, CA.

Klein, M., & Noy, N. F. (2003). A component-based framework for ontology evolution. In F. Giunchiglia, A. Gomez-Perez, A. Pease, H. Stuckenschmidt, Y. Sure, & S. Willmott (Eds.). *Proceedings of the IJCAI-2003 Workshop on Ontologies and Distributed Systems* (CEUR Workshop Proceeding series: Vol. 71). Retrieved from http://sunsite.informatik.rwth-aachen.de/Publications/CEUR-WS/Vol-71/Klein.pdf

Klein, M., Fensel, D., Kiryakov, A., & Ognyanov, D. (2002). Ontology versioning and change detection on the web. Knowledge Engineering and Knowledge Management: Ontologies and the Semantic Web (LNCS: Vol. 2473, pp. 197-212). Berlin, Germany: Springer. doi:10.1007/3-540-45810-7

Klein, M., Kiryakov, A., Ognyanov, D., & Fensel, D. (2002). Finding and characterizing changes in ontologies. LNCS: Vol. 2503. Conceptual Modeling — ER 2002 (pp. 79–89). Berlin, Germany: Springer. doi:10.1007/3-540-45816-6

Klusch, M., Pries, B., & Sycara, K. (2006). Automated semantic web service discovery with OWLS-MX. In *Proc. 5th International Joint Conference on Autonomous Agents and Multiagent Systems,* (pp. 915-922), Hakodate, Japan. New York: ACM.

Knublauch, H., Musen, M. A., & Rector, A. L. (2004). Editing description logic ontologies with the Protege OWL plugin. In *Proceedings of the International Workshop on Description Logics (DL).*

Köhler, J., Munn, K., Ruegg, A., Skusa, A., & Smith, B. (2006). Quality control for terms and definitions in ontologies and taxonomies. *BMC Bioinformatics*, *7*, 212. doi:10.1186/1471-2105-7-212

Kopecky, J., Vitvar, T., Bournez, C., & Farrell, J. (2007). SAWSDL: Semantic Annotations for WSDL and XML Schema. *IEEE Internet Computing*, *11*(6), 60–67. doi:10.1109/MIC.2007.134

Korfhage, R. (1997). *Information Storage and Retrieval.* Chichester, UK: John Wiley & Sons.

Kourtesis, D., & Paraskakis, I. (2008). Combining SAWSDL, OWL-DL and UDDI for Semantically Enhanced Web Service Discovery. In M. Hauswirth, M. Koubarakis, and S. Bechhofer, (Eds.), *Proc. 5th European Semantic Web Conference,* (LNCS 5021, pp. 614-628). Berlin: Springer.

KRAFT. (2004). *Knowledge Re-use and Functional Transformation.* Retrieved from http://www.cs.cf.ac.uk/kraft/

Kraus, S., Lehmann, D., & Magidor, M. (1990). Nonmonotonic reasoning, preferential models, and cumulative logics. *Artificial Intelligence*, *44*, 167–207. doi:10.1016/0004-3702(90)90101-5

Kuck, J., & Reichartz, F. (2007). A collaborative and feature-based approach to context-sensitive service discovery. In *16th International World Wide Web Conference, Workshop on Emerging Applications for Wireless and Mobile Access*, Banff, Alberta, Canada.

Kuratowski, C. (1958). *Topologie. Panstwowe Wydawnie two Naukowe*. Varsovie.

La France, M. (1992). Questioning Knowledge Acquisition. In Lauer, T. W., Peacock, E., & Graesser, A. C. (Eds.), *Questions And Information Systems*. Mahwah, NJ: L. Erlbaum associates.

Lacasta, J., Nogueras-Iso, J., López-Pellicer, F. J., Muro-Medrano, P. R., & Zarazaga-Soria, F. J. (2007). ThManager: An Open Source Tool for creating and visualizing SKOS. *Information Technology and Libraries*, *26*(4), 40–53.

Lagoze, C., & Hunter, J. (2001). The ABC ontology and model (v3.0). *Journal of Digital Information*, *2*,Retrieved from http://jodi.ecs.soton.ac.uk/Articles/v02/i02/Lagoze/.

Lakshmanan, L., Ng, R., Han, J., & Pang, A. (1999). Optimization of constrained frequent set queries with 2-variable constraints. In the ACM SIGMOD Intl. Conf. on Management of Data, (pp. 157–168).

Lambrix, P., Habbouche, M., & Perez, M. (2003). Evaluation ofOntology Development Tools for Bioinformatics. *Bioinformatics (Oxford, England)*, *19*(12), 1564–1571. doi:10.1093/bioinformatics/btg194

Lara, R., Cantador, I., & Castells, P. (2006). BRL Taxonomies and OWL Ontologies for Investment Funds R. In *ER* (pp. 271–280). Workshops.

Larson, J., Navathe, S., & Elmrasri, R. (1989). A theory of attribute equivalence in databases with application to schema integration. *IEEE Transactions on Software Engineering*, *15*, 449–463. doi:10.1109/32.16605

Lassila, O., & MacGuinness, D. (2001). *The Role of Frame-Based Representations on the Semantic*. Web Knowledge Systems Laboratory Report. Retrieved July 6, 2009 from http://www.ksl.stanford.edu/KSL_Abstracts/KSL-01-02.html

Lavrenko, V., & Croft, W. B. (2001). Relevance based language models. In *Proceedings of the 24th annual international ACM SIGIR conference on Research and development in information retrieval.*

Leblanc, A., & Abel, M.-H. (2007a). Using organizational memory and forum in an organizational learning. In *IEEE/ACM ICDIM'07, The Second International Conference on Digital Information Management*, (pp. 266--271).

Leblanc, A., & Abel, M.-H. (2007b). E-memorae2.0: An e-learning environment as learners communities support. *IJCSA*, *5*(1), 108–123.

Leblanc, A., Abou Assali, A., Abel, M.-H., & Lenne, D. (2008). Use of topic maps to support learning organisational memory. In *Proceedings of the Fourth International Conference on Topic Maps Research and Applications, TMRA 2008 Subject-centric Computing*, Leipzig, Germany, (15--17).

Leclère, L., Trichet, T., & Frûst, F. (2002). Construction of an ontology related to the projective geometry. In RFIA 13th congrès des Reconnaissance des Frames et Intelligence Artificielle, France.

Lee, J. B., Kim, J. J., & Park, J. C. (2006). Automatic extension of gene ontology with flexible identification of candidate terms. *Bioinformatics (Oxford, England)*, *22*(6), 665–670. doi:10.1093/bioinformatics/btl010

Lee, J. K., Oh, J.-H., & Hwang, S. (2005). Strg-index: Spatio-temporal region graph indexing for large video databases. In *Proceedings of SIGMOD Conference*, (pp. 718–729).

Lee, J., Madnick, S., & Siegel, M. (1996). Conceptualizing semantic interoperability: a perspective from the knowledge level. *International Journal of Cooperative Information Systems*, *5*(4), 367–393. doi:10.1142/S0218843096000142

Lenat, D. B., & Guha, R. V. (1990). *Building Large Knowledge-Based Systems: Representation and Inference in the Cyc Project*. Boston: Addison-Wesley.

Leung, N. K. Y., & Lau, S. K. (2006). Relieving the Overloaded Help Desk: A Knowledge Management Approach. *Communications of International Information Management Association*, *6*(2), 87–98.

Leung, N. K. Y., Lau, S. K., & Fan, J. P. (2007). An Ontology-based Knowledge Network to Reuse Inter-organization Knowledge. In M. Toleman, A. Cater-Steel, & D. Roberts (Ed.), *Proceedings of the 18th Australasian Conference on Information Systems*. The University of Southern Queensland.

Li, H. & Abe, N. (1998). *Word clustering and disambiguation based on cooccurrence data*. CoRR, cmp-lg/9807004.

Lin, C.-Y., & Hovy, E. H. (2000). The Automated Acquisition of Topic Signatures for Text Summarization. In *Proceedings of the COLING Conference*, Strasbourg, France.

Lin, D. (1998). Automatic retrieval and clustering of similar words. In *Proceedings of the 17th international conference on Computational linguistics*, (pp. 768–774). Association for Computational Linguistics.

Liu, Z., & Chu, W. W. (2005). Knowledge-based query expansion to support scenario-specific retrieval of medical free text. In *SAC '05: Proceedings of the 2005 ACM symposium on Applied computing*, (pp. 1076–1083). New York: ACM Press.

Loke, S. W., & Davison, A. (1998). LogicWeb: Enhancing the Web with Logic Programming. *The Journal of Logic Programming*, *36*, 195–240. doi:10.1016/S0743-1066(98)00002-8

Lonsdale, D., Ding, Y., Embley, D. W., & Melby, A. (2002). Peppering Knowledge Sources with SALT, Boosting Conceptual Content for Ontology Generation. In *Proceedings of the AAAI Workshop on Semantic Web Meets Language Resources*, Edmonton, Alberta, Canada.

Lu, J., Wang, S., & Wang, J. (2005). An experiment on the Matching and Reuse of XML Schemas. In *Proceedings of the 5th International Conference on Web engineering (ICWE)*. (pp. 273-284), Sydney, Australia.

Luke, S., & Heflin, J. (2000, February). *SHOE 1.01 Proposed specification*. SHOE Project. Retrieved from http://www.cs.umd.edu/projects/plus/SHOE/

Luong, P. (2007). *Gestion de l'évolution d'un web sémantique d'entreprise*. PhD Thesis, Ecole des Mines de Paris, Paris.

Lussier, Y. A., & Li, J. (2004). Terminological Mapping for High Throughput Comparative Biology of Phenotypes. In R. Altman, A. Dunker, L. Hunter, T. Jung and T. Klein (Ed.), *Proceedings of the Pacific Symposium on Biocomputing* (pp. 202-213). World Scientific.

Madhavan, J., Bernstein, P. A., & Rahm, E. (2001). Generic schema matching with cupid. In *Proceedings of 27th International Conference on Very Large Data Bases*, September 11-14, 2001, Roma, Italy (p. 49-58).

Madhavan, J., Bernstein, P. A., Doan, A., & Halevy, A. Y. (2005). Corpus-based Schema Matching. In *Proceedings of the 21st International Conference on Data Engineering (ICDE)*, (pp. 57-68), Tokyo, Japan.

Madhavji, N. H., Fernandez-Ramil, J., & Perry, D. E. (2006). *Software evolution and feedback: Theory and practice*. New York: Wiley. doi:10.1002/0470871822

Maedche, A., & Staab, S. (2000). Discovering Conceptual Relations from Text. In W. Horn, (Ed.), *Proceedings of the 14th European Conference on Artificial Intelligence*, (pp. 321–325). Amsterdam: IOS Press.

Maedche, A., & Staab, S. (2001). Ontology learning for the semantic web. *IEEE Intelligent Systems*, *16*(2), 72–79. doi:10.1109/5254.920602

Maedche, A., & Staab, S. (2003). Ontology Learning. In S. Staab & R. Studer (eds.), *Handbook on Ontologies in Information Systems*. Berlin: Springer. Retrieved from http://www.aifb.unikarlsruhe.de/WBS/sst/Research/Publications/handbook-ontology-learning.pdf

Maedche, A., & Volz, R. (2001). The ontology extraction and maintenance framework text-toonto. In *Proceedings of the ICDM'01 Workshop on Integrating Data Mining and Knowledge Management*.

Mäedche, A., & Volz, R. (2001). The ontology extraction and maintenance framework text-to-onto. In F. J. Kurfess, & M. Hilario, (Eds.), *Proceedings of the Workshop on Integrating Data Mining and Knowledge Management*, at

The 2001 IEEE International Conference on Data Mining ICDM'01, San Jose, CA. Retrieved from http://users.csc.calpoly.edu/~fkurfess/Events/DM-KM-01/Volz.pdf

Maedche, A., Motik, B., & Stojanovic, L. (2003). Managing multiple and distributed ontologies in the semantic web. *The VLDB Journal, 12*(4), 286–302. doi:10.1007/s00778-003-0102-4

Mäedche, A., Stojanovic, L., Studer, R., & Volz, R. (2002). Managing multiple ontologies and ontology evolution in OntoLogging. In *Proceedings of the Conference on Intelligent Information Processing* (IIP-2002), Part of the IFIP World Computer Congress WCC2002, (pp. 51-63). Montreal, Canada.

Manola, F., & Miller, E. (Eds.). (2004). *RDF Primer, W3C, W3C Recommendation 10 February 2004*. Retrieved July 6, 2009, from W3C Web site http://www.w3.org/TR/2004/REC-rdf-primer-20040210/

Marques, O., & Barman, N. (2003). Semi-automatic semantic annotation of images using machine learning techniques. *International Semantic Web Conference*, Florida, October 2003.

Martin, D., Burstein, M., Mcdermott, D., Mcilraith, S., Paolucci, M., & Sycara, K. (2007). Semantics to Web Services with OWL-S. *World Wide Web (Bussum), 10*(3), 243–277. doi:10.1007/s11280-007-0033-x

Martin, P. (1996). *Exploitation de Graphes Conceptuels et de documents structurés et Hypertextes pour l'acquisition de connaissances et la recherche d'informations*. Ph.D. Thesis, University Nice - Sophia Antipolis, October 14, 1996.

Martin, P., & Eklund, P. (2000). Knowledge retrieval and the world wide web. In Dieng, R. (Ed.), *IEEE Intelligent Systems* (pp. 18–25). Special Issue on Knowledge Management and Knowledge Distribution Over the Internet.

Maximilien, E. M., & Singh, M. P. (2004). A framework and Ontology for Dynamic Web Services Selection. *IEEE Internet Computing*, 5.

Mayfield, J. (2003). Information retrieval on the semantic web: Integrating inference and retrieval. In *Proceedings of the SIGIR Workshop on the Semantic Web.*

McCarthy, J. (1980). Circumscription — A Form of Nonmonotonic Reasoning. *Artificial Intelligence, 13*, 27–39. doi:10.1016/0004-3702(80)90011-9

McCarthy, J. (1986). Application of circumscription to formalizing common-sense knowledge. *Artificial Intelligence, 28*, 86–116. doi:10.1016/0004-3702(86)90032-9

McGuinness, D. (2002). Description Logics for Configuration. In F. Baader, D. McGuinness, D. Nardi, P. Patel-Schneider, (Ed.), *The Description Logic Handbook: Theory, Implementation, and Applications*. Cambridge, UK: Cambridge University Press. Retrieved from http://www.ksl.stanford.edu/people/dlm/papers/dlhb-configuration.pdf

McGuinness, D. L., Fikes, R., Rice, J., & Widler, S. (2000). An Environment for Merging and Testing Large Ontologies. In Cohn, A., Giunchiglia, F., & Selman, B. (Eds.), *KR2000: Principles of Knowledge Representation and Reasoning* (pp. 483–493). San Francisco: Morgan Kaufmann Publishers.

McIlraith, S., Son, T., & Zeng, H. (2001). Semantic Web Services. *IEEE Intelligent Systems, 16*(2), 46–53. doi:10.1109/5254.920599

Meersman, R. (2001). *Ontologies and databases: More than a fleeting resemblance*. Rome: OES/SEO Workshop.

Melnik, S., Garcia-Molina, G., & Rahm, E. (2002). Similarity Flooding: A Versatile Graph Matching Algorithm and Its Application to Schema Matching. In *Proceedings of the 18th International Conference on Data Engineering*, San Jose, CA, (pp. 117-128).

Mena, E., Illarramendi, A., Kashyap, V. & Sheth, A. (2000). OBSERVER: An Approach for Query Processing in Global Information Systems based on Interoperation across Pre-existing Ontologies. *International journal on Distributed and Parallel Databases (DAPD), 8*(2), 223-272.

Mena, E., Kashyap, V., Illarramendi, A., & Sheth, A. (1998). *Domain Specific Ontologies for Semantic Information Brokering on the Global Information Infrastructure, Formal Ontology in Information Systems.* Amsterdam: IOS Press.

Mena, E., Kashyap, V., Sheth, A., & Illarramendi, A. (1996). OBSERVER: An approach for query processing in global information systems based on interoperability across pre-existing ontologies. In *Proceedings 1st International Conference on Cooperative Information Systems (IFCIS)*, Brussels.

Mendes, O., & Abran, A. (2004). Software Engineering Ontology: A Development Methodology. *Metrics News, 9*(1), 68–76.

Mens, T. (2002). A State-of-the-Art Survey on Software Merging. *Transactions on Software Engineering, 28*(5), 449–462. doi:10.1109/TSE.2002.1000449

Mens, T., & Demeyer, S. (2007). *Software Evolution.* Berlin: Springer.

Mentzas, G. (2002). *Knowledge asset management: beyond the process-centered and product-centered approaches.* Berlin: Springer.

Mentzas, G., Aposolou, D., Young, R., & Abecker, A. (2001). Knowledge Networking: a Holistic Solution for Leveraging Corporate Knowledge. *Journal of Knowledge Management, 5*(1), 94–106. doi:10.1108/13673270110384446

Menzis, T. (1999). Knowledge maintenance: The state-of-the-art. *The Knowledge Engineering Review, 14*(1), 1–46. doi:10.1017/S0269888999134052

METEOR-S. (2005). *Deliverables of METEOR-S project.* Retrieved from http://lsdis.cs.uga.edu/Projects/METEOR-S/

Mhiri, M. & Gargouri, F. (2009). Méthodologie de construction des ontologies pour la résolution de conflits des Systèmes d'Information. *Revue Technique et Science Informatiques, 28.*

Mhiri, M., Mtibaa, A., & Gargouri, F. (2005). OntoUML: Towards a language for the specification of information systems' ontologies. In T*he Seventeenth International Conference on Software Engineering and Knowledge Engineering,* Taipei, Taiwan – China, (pp. 743-746).

Mika, P. (2005). Ontologies are Us: a Unified Model of Social Networks and Semantics. In ISWC, (LNCS 3729, pp. 522-536). Berlin: Springer.

Minsky, M. (1975). A framework for representing knowledge. In The Psychology of Computer Vision, (pp. 211-277)

Missikoff, M., Navigli, R., & Velardi, P. (2002). The Usable Ontology: An Environment for Building and Assessing a Domain Ontology. In *International Semantic Web Conference (ISWC'2002)*, Sardinia, Italia.

Missikoff, M., Velardi, P., & Fabriani, P. (2003). Text mining techniques to automatically enrich a domain ontology. []. Amsterdam: Kluwer Academic Publishers.]. *Applied Intelligence, 18*, 323–340. doi:10.1023/A:1023254205945

Mitra, M., Singhal, A., & Buckley, C. (1998). Improving automatic query expansion. In Research and Development in Information Retrieval, (pp. 206–214).

Mizoguchi, R., Ikeda, M., & Sinitsa, K. (1997). Roles of Shared Ontology in AI-ED Research, Intelligence, Conceptualization, Standardization, and Reusability. In *Proc. of AIED-97*, (pp.537-544), also as Technical Report AI-TR-97-4, I.S.I.R., Osaka University

Moguillansky, M. O., Rotstein, N. D., & Falappa, M. A. (2008). A theoretical model to handle ontology debugging and change through argumentation. In *Proceedings of the 2nd Workshop on Ontology Dynamics, (IWOD-08) at ISWC 08 Conference* (pp. 29-42). Retrieved from http://www.ics.forth.gr/~fgeo/Publications/IWOD-08_Proceedings.pdf

Mohan, P., & Brooks, C. (2003). Learning objects on the semantic Web. In *Advanced Learning Technologies, Proceedings. The 3rd IEEE International Conference on*, (pp. 195--199).

Moldovan, D. I., & Girju, R. C. (2001). An interactive tool for the rapid development of knowledge Bases.

[IJAIT]. *International Journal of Artificial Intelligence Tools, 10*(1-2).

Moore, R. (1985). Semantical considerations on nonmonotonic reasoning. *Artificial Intelligence, 25*(1), 75–94. doi:10.1016/0004-3702(85)90042-6

Morgensteirn, L., & Guerreiro, R. (1993). Epistemic logics for multiple agent nonmonotonic reasoning. In *Proceedings of the Second Symposium on logical Formalizations of Commonsense Reasoning (CS-93).*

Morgenstern, L. (1999). Nonmonotonic Logics. In R.A. Wilson & F. Keil (Ed.), *MIT Encyclopedia of the Cognitive Sciences*. Cambridge, MA: The MIT Press. Retrieved from http://www-formal.stanford.edu/leora/nonmon.pdf

Motelet, O. (2007). *Improving Learning-Object Metadata Usage during Lesson Authoring*. PhD thesis, Chile university, October 2007.

Mtibaa, A., & Jaziri, W. (2009). Contexte: vision d'un acteur de son système d'information à travers une ontologie. In *Second International Conference on Web and Information Technologies (ICWIT '09),* (pp. 17-31), Kerkennah, Tunisia.

Mtibaa, A., Jaziri, W., & Gargouri, F. (2008). Proposition d'une extension de l'ontologie de domaine pour supporter la multitude de contexte lors de la spécification des besoins. *International Conference on Web and Information Technologies (ICWIT '08),* Sidi Bel Abbes, Algeria.

Munkata, T. (1998). Rough sets. In *Fundamentals of the New Artificial Intelligence* (pp. 140–182). New York: Springer-Verlag.

Muñoz, L. S. (2004). *Ontology-based Metadata for e-learning Content*. Porto Alegre: Master's Degree in Computer Science.

Nack, F., Van Ossenbruggen, J., & Hardman, L. (2005, January - March). That Obscure Object of Desire: Multimedia Metadata on the Web (Part II). *IEEE MultiMedia, 12*(1), 54–63. doi:10.1109/MMUL.2005.12

Najjar, J. & Duval, E. (2006). Actual use of learning objects and metadata: An empirical analysis. *Technical Committee on Digital Libraries Bulletin (TCDL), 2*(2).

Naphade, M., Smith, J. R., Tesic, J., Chang, S., Hsu, W., Kennedy, L., et al. (2006). A Large-Scale Concept Ontology for Multimedia. *IEEE MultiMedia, 13* (3), 86-91. Retrieved from DOI=http://dx.doi.org/10.1109/MMUL.2006.63

Navathe, S., & Gadgil, S. (1982). A methodology for view integration in logical database design. In *Eighth International Conference on very Large Data Bases*, Mexico City, Mexico.

Navigli, R., & Velardi, P. (2002). Automatic adaptation of wordnet to domains. In *Proceedings of 3rd International Conference on Language Resources and Evaluation.*

Neches, R., Fikes, R. E., Finin, T., Gruber, T. R., Senator, T., & Swartout, W. R. (1991). Enabling technology for knowledge sharing. *AI Magazine, 12*(3), 36–56.

Newman, V. (1997). Redefining Knowledge Management to Deliver Competitive Advantage. *Journal of Knowledge Management, 1*(2), 123–128. doi:10.1108/EUM0000000004587

Ng, R., Lakshmanan, L., Han, J., & Pang, A. (1998). Exploratory mining and pruning optimizations of constrained association rules. In *Proc. of the ACM SIGMOD Intl. Conf. On Management of Data*, (pp. 13–24).

Nie, J. Y., & Jin, F. (2002). Integrating logical operators in query expansion in vector space model. In *Workshop on Mathematical/Formal Methods in Information Retrieval, 25thACM-SIGIR*, Tampere, Finland, (Vol. 8).

Niles, I., & Pease, A. (2001). Towards a Standard Upper Ontology. In C. Welty & B. Smith (Ed.), *Proceedings of the International Conference on Formal Ontology in Information Systems* (pp. 2-9). New York: ACM.

Nilsson, M., Palmer, M., & Brase, J. (2003). The LOM RDF Binding-Principles and Implementation. In *Proceedings of the Third Annual ARIADNE conference, 140.*

Ning, X., Jin, H., & Wu, H. (*2006*). Semrex: Towards large-scale literature information retrieval and browsing with semantic association. In *Proceedings of IEEE International Conference on e-Business Engineering*

(ICEBE 06), (pp. 602–609). Washington, DC: IEEE Computer Society.

Nobécourt, J. (2000). A method to build formal ontologies from text. In *EKAW-2000 Workshop on ontologies and text*, France.

Nonaka, I., & Takeuchi, H. (Eds.). (1995). *The Knowledge-Creating Company: How Japanese Companies Create the Dynamics of Innovation.* Oxford, UK: Oxford University Press.

Nonaka, I., Toyama, R., & Konno, N. (2001). SECI, Ba and Leadership a Unified Model of Dynamic Knowledge Creation. In S. Little, P. Quintas & T. Ray (Ed.), Managing Industrial Knowledge Creation, Transfer and Utilization (pp. 13-43). Milton Keynes, UK: The Open University.

Noy, N. F., & Hafner, C. D. (1997). *The State of the Art in Ontology Design - A Survey and Comparative Review* (pp. 53–74). Menlo Park, CA: AAAI.

Noy, N. F., & Klein, M. (2003). Tracking complex changes during ontology evolution. In *Collected Posters ISWC 2003*, Sanibal Island, FL. Retrieved from http://www.stanford.edu/~natalya/papers/trackingChangesPoster.pdf

Noy, N. F., & Klein, M. (2004). Ontology evolution: Not the same as schema evolution. *Knowledge and Information Systems, 6*(4), 428–440. doi:10.1007/s10115-003-0137-2

Noy, N. F., & McGuinness, D. (2001). *Ontology development. 101: A guide to creating your first ontology.* Stanford Knowledge Systems Laboratory Technical Report KSL-01-05 and Stanford Medical Informatics Technical Report SMI-2001-0880. Retrieved from http://protege.stanford.edu/publications/ontology_development/ontology101.html

Noy, N. F., & Musen, M. A. (1999). An algorithm for merging and aligning ontologies: automation and tool support. In *the Proceedings of the Workshop on Ontology Management at Sixteenth National Conference on Artificial Intelligence (AAAI)*, Orlando, FL.

Noy, N. F., & Musen, M. A. (2000). PROMPT: Algorithm and Tool for Automated Ontology Merging and Alignment. In *17th National Conference on Artificial Intelligence (AAAI_00),* Austin, TX.

Noy, N. F., & Musen, M. A. (2003). The PROMPT suite: Interactive tools for ontology merging and mapping. *International Journal of Human-Computer Studies, 59*(6), 983–1024. doi:10.1016/j.ijhcs.2003.08.002

Noy, N. F., & Musen, M. A. (2004). Ontology versioning in an ontology management framework. *IEEE Intelligent Systems, 19*(4), 6–13. doi:10.1109/MIS.2004.33

Noy, N. F., Chugh, A., Liu, W., & Musen, M. A. (2006). A framework for ontology evolution in collaborative environments. In The Semantic Web - ISWC 2006 (LNCS: Vol. 4273, pp. 544-558). Berlin, Germany: Springer. doi: 10.1007/11926078

Noy, N. F., Fergerson, R., & Musen, M. (2000). The Knowledge Model of Protege-2000: Combining Interoperability and Flexibility. In R. Dieng & O. Corby, (Eds.), Knowledge Acquisition, Modeling and Management, (LNCS,Vol. 1937, pp. 17–32). New York: Springer.

Noy, N. F., Kunnatur, S., Klein, M., & Musen, M. A. (2004). Tracking changes during ontology evolution. The Semantic Web – ISWC 2004 (LNCS: Vol. 3298, 259-273). Berlin, Germany: Springer. doi: 10.1007/b102467

Noy, N., & Musen, M. (2002). PROMPTDIFF: A fixed-point algorithm for comparing ontology versions. In *Proceedings of the 18th National Conference on Artificial Intelligence and Fourteenth Conference on Innovative Applications of Artificial Intelligence,* Edmonton, Alberta, Canada. Menlo Park, CA: AAAI Press.

O'Brien, P., & Abidi, S. S. R. (2006). Modeling intelligent ontology evolution using biological evolutionary processes. In *Proceedings of the IEEE International Conference on Engineering of Intelligent Systems ICEIS 06.* Islamabad, Pakistan. doi: 0.1109/ICEIS.2006.1703172

Oberle, D., Volz, R., Motik, B., & Staab, S. (2004). An extensible ontology software environment. In S. Staab, & R. Studer (Eds.), *International Handbooks on Information Systems, Handbook on Ontologies,* (pp. 311-333). Berlin:

Springer. Retrieved from http://www.aifb.uni-karlsruhe.de/WBS/dob/pubs/handbook2003a.pdf

Oinas-Kukkonen, H. (2005). Towards Evaluating Knowledge Management through the 7C Model. In *Proceedings of the European Conference on Information Technology Evaluation*. University of Oulu.

Oliver, D. E., Shahar, Y., Musen, M., & Shortliffe, E. H. (1999). Representation of change in controlled medical terminologies. *AI in Medicine*, *15*(1), 53–76.

Ouksel, A., & Naiman, C. (1994). Coordinating context building in heterogeneous information systems. *Journal of Intelligent Information Systems*, *3*(2), 151–183. doi:10.1007/BF00962977

OWL-SC. (2004). OWL-S: Semantic Markup for Web Services. *OWL-S Coalition*. Retrieved from http://www.daml.org/services/

Özsoyoğlu, G., & Snodgrass, R. T. (1995). Temporal and real-time databases: A survey. *IEEE Transactions on Knowledge and Data Engineering*, *7*(4), 513–532. doi:10.1109/69.404027

Paolucci, M., Kawamura, T., Payne, T. R., & Sycara, K. P. (2002). Semantic Matching of Web Services Capabilities. In *Proc. 1st International Semantic Web Conference on The Semantic Web*, (LNCS 2342 pp. 333-347). London: Springer.

Papazoglou, M. P., Traverse, P., Dustdar, S., & Leymann, F. (2007). Service-oriented computing: State of the art and research challenges. *Computer*, *40*(2), 38–45. doi:10.1109/MC.2007.400

Parsia, B., Sirin, E., & Kalyanpur, A. (2005). Debugging OWL ontologies. In *Proceedings of the 14th International World Wide Web Conference (WWW2005)*, 633-640. Retrieved from http://www2005.org/cdrom/docs/p633.pdf

Patterson, D., Rooney, N., Dobrynin, V. & Galushka, M. (2005). Sophia: A novel approach for textual case-based reasoning. *IJCAI*, 15–20.

Paulheim, H. (2008). On Applying Matching Tools to Large-Scale Ontologies. In *Proceedings of the 3rd International Workshop on Ontology Matching (OM-2008) Collocated with the 7th International Semantic Web Conference (ISWC-2008)*, Karlsruhe, Germany.

Pearson, J. (1998). *Terms in Context. Studies in Corpus Linguistics, 1*. Philadelphia: John Benjamins.

Pei, J., Hong, J., & Bell, D. A. (2006). A Novel Clustering-based Approach to Schema Matching. In *Proceedings of the 4th International Conference on Advances in Information Systems (ADVIS)*, (pp. 60-69), Izmir, Turkey.

Peirce, C. S. (1867), On a New List of Categories. In *Proceedings of the American Academy of Arts and Sciences, 7*(1868), 287–298.

Petasis, B., Karkaletsis, V., & Paliouras, G. (2007). D4.3: Ontology Population and Enrichment: State of the Art. *BOEMIE Deliverable*. Retrieved from http://www.boemie.org/sites/default/files/D%204.3.pdf

Peters, R. J., & Oezsu, M. (1997). An axiomatic model of dynamic schema evolution in object base management systems. *ACM Transactions on Database Systems*, *22*(1), 75–114. doi:10.1145/244810.244813

Pezik, P., Jimeno-Yepes, A., Lee, V., & Rebholz-Schuhmann, D. (2008). *Static dictionary features for term polysemy identification. Building and evaluating resources for biomedical text mining*. LREC Workshop.

Phaedra, M., & Permanand, M. (2006). *Incorporating multiple ontologies into the ieee learning object metadata standard* (pp. 143–154). CSWWS.

Pierra, G. (2005). Context-explication in conceptual ontologies: PLIB ontologies and their use for industrial data. *Journal of Advanced Manufacturing Systems*.

Pierra, G. (2008). Context representation in domain ontologies and its use for semantic integration of data. *Journal on Data Semantics*, 174-211.

Pinto, H. S., & Martins, J. P. (2000). Reusing Ontologies. In *Proceedings of the AAAI Spring Symposium Series, Workshop on Bringing Knowledge to Business Processes* (pp. 77–84). Menlo Park, CA: AAAI Press.

Pinto, H. S., & Martins, J. P. (2001). A Methodology for Ontology Integration. In *Proceedings of the 1st International Conference on Knowledge Capture* (pp. 131-138). New York: ACM.

Plessers, P. (2006). *An approach to web-based ontology evolution.* Ph.D. Thesis, University of Brussels, Belgium.

Plessers, P., & De Troyer, O. (2005). Ontology change detection using a version log. In Y.Gil, E. Motta, V.R. Benjamins, & M. Musen (Eds.), The Semantic Web – ISWC 2005 (LNCS: Vol. 3729, pp. 578-592). Berlin, Germany: Springer-Verlag. doi: 10.1007/11574620

Plessers, P., & De Troyer, O. (2006) Resolving inconsistencies in evolving ontologies. In Y. Sure, & J. Domingue (Eds.), *The Semantic Web: Research and Applications, Proceedings of the 3rd European Semantic Web Conference ESWC 2006* (LNCS: Vol.4011, pp. 200-214). Berlin, Germany: Springer. doi: 10.1007/11762256

Plessers, P., De Troyer, O., & Casteleyn, S. (2007). Understanding ontology evolution: A change detection approach. *Journal of Web Semantics: Science, Services and Agents on the World Wide Web, 5*, 39-49. Retrieved from http://www.sciencedirect.com

Polsani, P. (2003). Use and Abuse of Reusable Learning Objects. *Journal of Digital Information*, (4): 2003–02.

Porzel, R., & Malaka, R. (2004). A Task-based Approach for Ontology Evaluation. In *ECAI Workshop on Ontology Learning and Population*, Valencia, Spain.

Pratt, W., & Wasserman, H. (2000). QueryCat: Automatic Categorization of MEDLINE Queries. *Journal of the American Medical Informatics Association, 7*, 655–659.

Pratt, W., Hearst, M. A., & Fagan, L. M. (1999). *A knowledge-based approach to organizing retrieved documents* (pp. 80–85). AAAI/IAAI.

Predoiu, L., Feier, C., Scharffe, F., de Bruijn, J., Martin-Recuerda, F., Manov, D., & Ehrig, M. (2006). *State-of-the-art Survey on Ontology Merging and Aligning V2, EU-IST Integrated Project (IP) IST-2003-506826 SEKT: Semantically Enabled Knowledge Technologies.* Innsbruck, Austria: University of Innsbruck.

Psyché, V., Bourdeau, J., & Mizoguchi, R. (2003). Ontology Development at the Conceptual Level for Theory-Aware ITS Authoring Systems. Journal of AI in Education, Shaping the future of Learning through Intelligent Technologies, 491-493.

Qiu, Y., & Frei, H. P. (1993). Concept-based query expansion. In *Proceedings of SIGIR-93, 16th ACM International Conference on Research and Development in Information Retrieval*, (pp. 160–169), Pittsburgh.

Quillian, M. (1967). Word Concepts: A Theory and Simulation of some Basic Semantic Capabilities. *Behavioral Science*, *12*, 410–430. doi:10.1002/bs.3830120511

Quillian, M. R. (1966). *Semantic Memory.* Report AD-641671, Clearinghouse for Federal Scientific and Technical Information.

Rabarijaona, A., Dieng, R., & Corby, O. (1999). Building a XML-based Corporate Memory. In *Proceedings of the IJCAI'99 Workshop on Knowledge Management and Organizational Memories*, Stockholm, Sweden.

Rahm, E., & Bernstein, P. (2001). A Survey of Approaches to Automatic Schema Matching. *The International Journal on Very Large Data Bases*, *10*(1), 334–350. doi:10.1007/s007780100057

Rahm, E., Do, H. H., & Maßmann, S. (2004). Matching Large XML Schemas. In *SIGMOD Record* (pp. 26–31). New York: ACM Press.

Rao, J., & Su, X. (2004). A survey of automated Web service composition methods. In *Proceedings of the First International Workshop on Semantic Web Services and Web Process Composition,* (SWSWPC'2004, LNCS), San Diego, CA. Berlin: Springer-Verlag.

Rebholz-Schuhmann, D., Kirsch, H., & Nenadic, G. (2006). *IeXML: towards an annotation framework for biomedical semantic types enabling interoperability of text processing modules. SIG BioLink.* ISMB.

Rebholz-Schuhmann, D., Kirsch, H., Arregui, M., Gaudan, S., Riethoven, M., & Stoehr, P. (2007). EBIMed –

Text crunching to gather facts for proteins from Medline. *Bioinformatics (Oxford, England)*, *23*(2), e237–e244. doi:10.1093/bioinformatics/btl302

Reigeluth, C. (1996). A New Paradigm of ISD? *Educational Technology*, *36*, 13–20.

Reiter, R. (1978). On closed world data bases. In Gallaire, H., & Minker, J. (Eds.), *Logic and Data Bases* (pp. 119–140). New York: Plenum.

Reiter, R. (1980). A logic for default reasoning. *Artificial Intelligence*, *13*, 81–132. doi:10.1016/0004-3702(80)90014-4

Rifaieh, R. (2004).*Utilisation des ontologies contextuelles pour le partage sémantique entre les systèmes d'information dans le entreprises.* PhD Thesis, INSA-Lyon, France.

Rifaieh, R., Arara, A., & Benharkat, N. (2004). A view of Enterprise Information Systems based on Contextual Ontologies. *IEEE International Conference on Computational Cybernetics (ICCC'2004).*

Riloff, E., & Jones, R. (1999). Learning Dictionaries for Information Extraction by Multi-level strapping, Boot-. In *Proceedings of the Sixteenth National Conference on Artificial Intelligence*, (pp. 1044–1049). Menlo Park, CA: The AAAI Press/MIT Press.

Riloff, E., & Shepherd, J. (1997). A corpus-based approach for building semantic lexicons. In C. Cardie & R. Weischedel, (Eds.), *Proceedings of the Second Conference on Empirical Methods in Natural Language Processing*, (pp. 117–124). Somerset, NJ: Association for Computational Linguistics.

Roark, B., & Charniak, E. (1998). *Noun-phrase cooccurence statistics for semiautomatic semantic lexicon construction* (pp. 1110–1116). COLING-ACL.

Roberts, D. D. (1973). *The Existential Graphs of Charles S. Peirce*. The Hague, The Netherlands: Mouton.

Roddick, J. F. (1996). A survey of schema versioning issues for database systems. *Information and Software Technology*, *37*(7), 383–393. doi:10.1016/0950-5849(95)91494-K

Roelleke, T. (2003). A frequency-based and a poisson-based definition of the probability of being informative. In *Proceedings of the 26th annual international ACM SIGIR conference on research and development in information retrieval*, (pp. 227–234). New York: ACM Press.

Rogozan, D., Paquette, G., & Rosca, I. (2004). Gestion de l'évolution de l'ontologie utilisée comme référentiel sémantique dans un environnement de téléapprentissage. In *TICE International Symposium on Technologies de l'Information et de la Communication dans les Enseignements d'ingénieurs et dans l'industrie*, France.

Roitman, H., & Gal, A. (2006). Fully automatic extraction and consolidation of ontologies from web sources using sequence semantics. In *EDBT workshops* (pp. 573–576). Ontobuilder.

Roman, D., & Keller, U., H. Lausen H., De Bruijn, J., Lara R., Stollberg,M., Polleres, A., Feier, C., Bussler, C. & Fensel, D. (2005). Web Service Modeling Ontology. *Applied Ontology*, *1*(1), 77–106.

Rosenfeld, B., Feldman, R., Fresko, M., Schler, J., & Aumann, Y. (2004). *Teg: a hybrid approach to information extraction* (pp. 589–596). CIKM.

Rosse, C., Kumar, A., Mejino, J. L. V., Jr., Cook, D. L., Detwiler, L. T., & Smith, B. (2005). A strategy for improving and integrating biomedical ontologies. In proceedings of AMIA, Washington, DC.

Roux, C., Proux, D., Rechermann, F., & Julliard, L. (2000). An ontology enrichment method for a pragmatic information extraction system gathering data on genetic interactions. In *Proceedings of the ECAI2000 Workshop on Ontology Learning (OL'2000)*, Berlin, Germany, Sánchez, D.M, Cavero, J.M. & Marcos, E. (2005). On models and ontologies. In *First International Workshop on Philosophical Foundations of Information Systems Engineering*, Porto, Portugal.

Safar, B., Kefi, H., & Reynaud, C. (2004). OntoRefiner, a user query refinement interface usable for Semantic Web Portals. In *Applications of Semantic Web technologies to web communities*. Workshop ECAI.

Salvat, E., & Mugnier, M. L. (1996) Sound and complete forward and backward chainings of graph rules. In *Proc. of the 4th ICCS'96*, Sydney, Australia, (LNCS 1115, pp. 248-262). Berlin: Springer-Verlag.

Sassi, N., & Jaziri, W. (2009). Vérification de la cohérence de l'ontologie par analyse des effets de bords. *Proceedings of the Second International Conference on Web and Information Technology (ICWIT 2009)*, Kerkennah, Tunisia, Sassi, N., Jaziri, W. & Gargouri, F. (2008). Formalisation of evolution Changes to update domain ontologies. In *Proceedings of International Arab Conference on Information Technology (ACIT'2008)*, Hammamet, Tunisia.

Sassi, N., Jaziri, W., & Gargouri, F. (2009). Anticipatory approach to maintain consistency in ontology versions. In *The Third International Conference on Advances in Semantic Processing (SEMAPRO 2009)*, (pp. 7-14), Sliema, Malta.

Sassi, N., Jaziri, W., & Gargouri, F. (2009). How to evolve ontology and maintain its coherence: A corrective operations-based approach. In *The International Conference on Knowledge Engineering and Ontology Development (IC3K-KEOD 2009)*, Madeira, Portugal.

Scharffe, F., de Bruijn, J., & Foxvog, D. (2006). *Ontology Mediation Patterns Library V2. EU-IST Integrated Project (IP) IST-2003-506826 SEKT: Semantically Enabled Knowledge Technologies.* Innsbruck, Austria: University of Innsbruck.

Schlobach, S., & Cornet, R. (2003). Non-standard reasoning services for the debugging of description logic terminologies. In *Proceedings of the 18th International Joint Conference on Artificial Intelligence (IJCAI-03).*

Schmidt, A., Aidoo, K., Takaluoma, A., Tuomela, P., Van Laerhoven, K., & Van de Velde, W. (1999). Advanced interaction in context. In *Proceedings of First International Symposium on Handheld and Ubiquitous Computing, HUC'99*, (pp. 89-101). Karlsruhe, Germany: Springer Verlag.

Schreiber, A., Wielinga, B., & Jansweijer, W. (1995). *The KACTUS view on the 'O' word. Technical Report, ESPRIT Project 8145 KACTUS.* The Netherlands: University of Amsterdam.

Schwartz, A. S., & Hearst, M. A. (2003). A simple algorithm for identifying abbreviation definitions in biomedical text. *Pacific Symposium on Biocomputing, 451*, 62.

Sciore, E., Siegel, M., & Rosenthal, A. (1994). Using Semantic values to Facilitate Interoperability Among Heterogeneous Information Systems. *ACM Transactions on Database Systems, 19*(2), 254–290. doi:10.1145/176567.176570

Sebillotte, S. (1991). *Décrire des tâches selon les objectifs des opérateurs de l'Interview à la Formalisation.* Technical report from INRIA n°125.

Sellami, S., Benharkat, A.-N., & Amghar, Y. (2008) From simple to large scale Matching: A hybrid approach. In *23rd International Symposium on Computer and Information Sciences (ISCIS)*, Istanbul, Turkey, (pp. 1-4).

Selman, B., & Levesque, H. (1993). The complexity of path-based defeasible inheritance. *Artificial Intelligence, 62*(2), 303–340. doi:10.1016/0004-3702(93)90081-L

Sheth, A. (1999). Changing Focus on Interoperability in Information Systems: from System, Syntax, Structure to Semantics. In Goodchild, M., Egenhofer, M., Fegeas, R., & Kottman, C. (Eds.), *Interoperating Geographic Information Systems* (pp. 5–30). Amsterdam: Kluwer Academic Publishers.

Shvaiko, P. (2004) A Classification of Schema-based Matching Approaches. In S. McIlraith, D. Plexousakis & F. van Harmelen, (Ed.), *Proceedings of the Meaning Coordination and Negotiation Workshop at the 1st International Semantic Web Conference*. Heidelberg, Germany: Springer.

Shvaiko, P., & Euzenat, J. (2005). A Survey of Schema-based Matching approaches. *Journal on Data Semantics IV, 3730*, 146–171. doi:10.1007/11603412_5

Sicilia, M. & Lytras, M. (2005). The semantic learning organization. *The Learning Organization: An International Journal*, 12.

Siorpaes, K., & Hepp, M. (2008). Games with a Purpose for the Semantic Web. *IEEE Intelligent Systems, 23*(3), 50–60. doi:10.1109/MIS.2008.45

Sirin, E. (2003). Semi-automatic composition of Web services using semantic description. In *Proceedings of Web Services Modelling, Architectures and Infrastructures workshop in conjunction with ICEIS'2003, 12(8), 72-7.* Hendler J, Parsia B.

Sirin, E., & Parsia, B. (2004). Pellet: An OWL DL reasoner. In V. Haaslev & R. Moller (Eds.), *Proceedings of the International Workshop on Description Logics (DL2004)*, (CEUR Workshop Proceedings: Vol. 104), Whistler, BC, Canada. Retrieved from http://sunsite.informatik.rwth-aachen.de/Publications/CEUR-WS/Vol-104/30Sirin-Parsia.pdf

Slimani, T., Ben-Yaghlane, B., & Mellouli, K. (2004). Approche binaire pour la génération de fortes règles d'associations. In EGC 2004, Revue des Nouvelles Technologies de l'Information, (pp. 329-340). Clermont Ferrand, France: Cépadues edition.

Slimani, T., Ben-Yaghlane, B., & Mellouli, K. (2008). PmSPARQL: Extended SPARQL for multi-paradigm Path Extraction. *International Journal of Computer, Information, and Systems Science, and Engineering, 2*(3), 179–190.

Slimani, T., Ben-Yaghlane, B., & Mellouli, K. (2008). SWAR: Modèle de génération des règles d'associations sémantiques à partir d'une base d'association. In *Proceedings of EGC 08, Atelier de fouille de données complexes*, Sophia Antipolis, (pp. 83-94).

Smiljanic, M., Keulen, M., & Jonker, W. (2006). Using Element Clustering to Increase the Efficiency of XML Schema Matching. In *Proceedings of the 22nd International Conference on Data Engineering Workshops (ICDE Workshops).*

Smith, B. (2003). Ontology. In Floridi, L. (Ed.), *Blackwell Guide to the Philosophy of Computing and Information* (pp. 155–166). London: Wiley Blackwell.

Smith, B., & Rosse, C. (2004). The role of foundational relations in the alignment of biomedical ontologies. *Medinfo, 11*(Pt 1), 444–448.

Smith, E. A. (2001). The Role of Tacit and Explicit Knowledge in the Workplace. *Journal of Knowledge Management, 5*(4), 311–321. doi:10.1108/13673270110411733

Soergel, D., Lauser, B., Liang, A., Fisseha, F., Keizer, J., & Katz, S. (2004). Reengineering Thesauri for New Applications: the AGROVOC Example. *Journal of Digital Information, 4*(4), 1–19.

Sombé, L. (1989). *Raisonnements sur des informations incomplètes en intelligence artificielle: comparaison de formalismes à partir d'un exemple*. Éditions Teknea.

Song, D., Cho, M., Choi, C., Shin, J., Park, J., & Kim, P. (2006). Knowledge Representation for Video Assisted by Domain-Specific Ontology. In A. Hoffmann et al. (Eds.), *Proceedings of PKAW 2006*, (LNAI 4303, pp. 144 – 155). Berlin: Springer-Verlag.

Sowa, J. (2000). *Knowledge Representation: Logical, Philosophical, and Computational Foundations.* Pacific Grove, CA: Brooks Cole Publishing Co. Retrieved from http://www.jfsowa.com/ontology/index.htm

Sowa, J. F. (1984). *Conceptual structures: information processing in mind and machine*. Boston: Addison-Wesley Longman Publishing Co., Inc.

Sowa, J. F. (1996) Ontologies for Knowledge Sharing. *Terminology and Knowledge Engineering Congress (TKE '96).*

Sowa, J.-F. (2000). Ontology, Metadata, and Semiotics Proceedings of ICCS'2000 in Darmstadt, Germany, on August 14, 2000. In B. Ganter & G. W. Mineau, (eds.), *Conceptual Structures: Logical, Linguistic, and Computational Issues*, (LNAI 1867, pp. 55-81). Berlin: Springer-Verlag.

Sowa, J.-F. (2000). *Guided Tour of Ontology.* Retrieved from http://www.jfsowa.com/ontology/guided.htm

Sowa, J.-F. (2002). Conceptual Graphs Standard. *ISO/JTC1/SC 32/WG2N000.* Retrieved from http://users.bestweb.net/~sowa/cg/cgstand.htm

Spaccapietra S., Parent C. & Dupont Y. (1992). Model Independent Assertions for Integration of Heterogeneous Schemas. *VLDB Journat*, 81-126.

Spaccapietra, S., Parent, C., Vangenot, C., & Cullot, N. (2004). On Using Conceptual Modeling for Ontologies. In *WISE 2004 Workshops, LNCS* (pp. 22–33). Berlin: Springer.

Spasic, I., Ananiadou, S., McNaught, J., & Kumar, A. (2005). Text mining and ontologies in biomedicine: Making sense of raw text. *Briefings in Bioinformatics, 6*(3), 239–251. doi:10.1093/bib/6.3.239

Spyns, P., & Meersman, R. (2002). Jarrar M., Data modelling versus ontology engineering. *SIGMOD Record, 31*(4), 12–17. doi:10.1145/637411.637413

Srikant, R., & Agrawal, R. (1997). Mining generalized association rules. *Future Generation Computer Systems, 13*(2–3), 161–180. doi:10.1016/S0167-739X(97)00019-8

Srikant, R., Vu, Q., & Agrawal, R. (1997). Mining association rules with item constraints. In *Proc. of the 3rd Intl. Conf. on Knowledge Discovery and Data Mining,* (pp. 67–73).

Staab, S., Schnurr, H. P., Studer, R., & Sure, Y. (2001). Knowledge processes and ontologies. *IEEE Intelligent Systems, 16*(1), 26–34. doi:10.1109/5254.912382

Stamou, G., van Ossenbruggen, J., Pan, J. Z., & Schreiber, G. (2006, January-March). Multimedia Annotations on the Semantic Web. In IEEE Multimedia.

Stoilos, G., Stamou, G., & Kollias, S. (2005). A string metric for ontology alignment. In *4th International Semantic Web Conference (ISWC),* Galway.

Stojanovic, L. (2004). *Methods and Tools for Ontology Evolution.* PhD Thesis, University of Karlsruhe, Germany.

Stojanovic, L., & Motik, B. (2002). Ontology evolution within ontology editors. In *Proceedings of the OntoWeb-SIG3 Workshop Evaluation of Ontology-based Tools (EON2002) at the 13th International Conference on Knowledge Engineering and Knowledge Management* (EKAW 2002), (CEUR-WS: Vol. 62, pp. 53-62). Retrieved from http://ftp.informatik.rwth-aachen.de/Publications/CEUR-WS/Vol-62/EON2002_Stojanovic.pdf

Stojanovic, L., Maedche, A., Motik, B., & Stojanovic, N. (2002). User-driven ontology evolution management. *Knowledge Engineering and Knowledge Management: Ontologies and the Semantic Web, Proceedings of the 13th International Conference* EKAW2002 (LNCS: Vol. 2473, pp. 285-300). Berlin, Germany: Springer-Verlag. doi: 10.1007/3-540-45810-7

Stojanovic, L., Maedche, M., Stojanovic, N., & Studer, R. (2003). Ontology evolution as reconfiguration-design problem Solving. In *Proceedings of the 2nd international conference on Knowledge capture K-CAP'03* (pp.162-171). New York: ACM.

Stojanovic, L., Staab, S., & Studer, R. (2001). eLearning based on the Semantic Web. In *WebNet2001-World Conference on the WWW and Internet.*

Stojanovic, L., Stojanovic, N., & Handschuh, S. (2002). Evolution of the Metadata in the Ontology based Knowledge Management Systems. In *Proceedings of the 1st German Workshop on Experience Management,* (pp. 65 – 77), Germany.

Stojanovic, L., Stojanovic, N., & Handschuh, S. (2002). Evolution of the metadata in the ontology-based knowledge management systems. In *GI 2002, Proceedings of the 1st German Workshop on Experience Management GWEM 2002* (LNAI: Vol.10 pp. 65-77).

Stojanovic, L., Stojanovic, N., Gonzalez, J., & Studer, R. (2003a). OntoManager— A System for the usage-based ontology management. In On The Move to Meaningful Internet Systems 2003: CoopIS, DOA, and ODBASE (LNCS: Vol. 2888, pp 858–875). Berlin, Germany: Springer. 10.1007/b94348

Stojanovic, N., Stojanovic, L., & Volz, R. (2002c). A Reverse engineering approach for migrating data-intensive web sites to the semantic web. In *Proceedings of the Conference on Intelligent Information Processing* (IIP-2002), Part of the IFIP World Computer Congress WCC2002, (pp.141-154), Montreal, Canada.

Stuckenschmidt, H., & Klein, M. (2004). Structure-based Partitioning of large concept hierarchies. In *Proceedings of the 3rd International Semantic Web Conference (ISWC),* (pp. 289-303), Hiroshima, Japan.

Studer, R., Benjamins, V. R., & Fensel, D. (1998). Knowledge engineering: principles and methods. *Data & Knowledge Engineering, 25*(1-2), 161–197. doi:10.1016/S0169-023X(97)00056-6

Stumme, G., Hotho, A., & Berendt, B. (2006). Semantic Web Mining: State of the art and future directions. *Journal of Web Semantics, 4*(2), 124–143. doi:10.1016/j.websem.2006.02.001

Su, W., Wang, J., & Lochovsky, F. (2006a). Holistic Schema Matching for Web Query Interface. In *Proceedings of the 10th International Conference on Extending Database Technology (EDBT),* (pp. 77-94), Munich, Germany.

Su, W., Wang, J., & Lochovsky, F. (2006b). Holistic Query Interface Matching using Parallel Schema Matching. In *Proceedings of the 22nd International Conference on Data Engineering (ICDE),* Atlanta, GA.

Sure, Y. (2002). On-to-knowledge – ontology based knowledge management tools and their application. *German Journal Kuenstliche Intelligenz,* 35-37.

Sure, Y., Staab, S., & Studer, R. (2004). On-To-Knowledge Methodology (OTKM). In Staab, S., & Studer, R. (Eds.), *Handbook on Ontologies* (pp. 117–132). Berlin: Springer Verlag.

Swartout, B., Ramesh, P., Knight, K., & Russ, T. (1997). Toward Distributed Use of Large-Scale Ontologies. In *AAAI Symposium on Ontological Engineering,* Stanford, CA.

Sycara, K., Paolucci, M., Ankolekar, A., & Srinivasan, N. (2003, December). Automated discovery, interaction and composition of Semantic Web services. *Web Semantics: Science. Services and Agents on the World Wide Web, 1*(1), 27–46. doi:10.1016/j.websem.2003.07.002

Talens, G., Oussalah, C., & Colinas, M. F. (1993, August 24-27). Versions of simple and composite object. In *Proceedings of the 19th International Conference on Very Large Data Bases (VLDB 1993)*, Dublin, Ireland, (pp. 62–72).

Tansel, A. U., Clifford, J., Gadia, S. K., Jajodia, S., Segev, A., & Snodgrass, R. T. (Eds.). (1993). Temporal Databases: Theory, Design and Implementation. Redwood City, CA: Benjamin/Cummings Publishing Company.

Roques, P., & Vallée, F. (2003). UML en action (2nd Ed.). Paris: Eyrolles edition.

Thompson, J. D., Holbrook, S. R., Katoh, K., Koehl, P., Moras, D., Westhof, E., & Poch, O. (2005). Mao: a multiple alignment ontology for nucleic acid and protein sequences. *Nucleic Acids Research, 33*, 4164. doi:10.1093/nar/gki735

Tsinaraki, C. Polydoros, P. & Christodoulakis, S. (2004, July 21-23). Interoperability support for Ontology-based Video Retrieval Applications. In *Proc. of 3rd International Conference on Image and Video Retrieval (CIVR 2004)*, Dublin, Ireland.

Tsujii, J. I., & Ananiadou, S. (2005). Thesaurus or logical ontology, which do we need for mining text? *Language Resources and Evaluation, 39*(1), 77–90. doi:10.1007/s10579-005-2697-0

Tzouvaras, V., Troncy, R., & Pan, J. (2007). *Multimedia Annotation Interoperability Framework.* W3C Technical reports, Boston. Retrieved from http://www.w3.org/2005/Incubator /mmsem/XGR-interoperability/

Uschold, M. (1996). Towards A Unified Methodology. In *Expert Systems.* Cambridge: Building Ontologies.

Uschold, M. E. (1998). Knowledge level modelling: Concepts and terminology. *The Knowledge Engineering Review, 13*(1), 5–29. doi:10.1017/S0269888998001040

Uschold, M., & Gruninger, M. (1996). Ontologies: Principles methods and applications. *The Knowledge Engineering Review, 11*(2), 93–155. doi:10.1017/S0269888900007797

Uschold, M., & Jasper, R. (1999). A framework for understanding and classifying ontology applications. In *Proceedings of the IJCAI99 Workshop on Ontologies and Problem-Solving Method*, Stockholm, Sweden.

Uschold, M., & King, M. (1995). Towards a Methodology for Building Ontologies. *IJCAI95 Workshop on Basic Ontological Issues in Knowledge Sharing*, Montreal.

Valente, A., Russ, T., Macgregor, R., & Swartout, W. (1999). Building and (Re)Using an Ontology of Air Campaign Planning. *IEEE Intelligent Systems*, *14*(1), 27–36. doi:10.1109/5254.747903

van Assem, M., Malaisé, V., Miles, A., & Schreiber, G. (2006). A Method to Convert Thesauri to SKOS. In *Proceedings of the 3rd European Semantic Web Conference*, (LNCS, Vol. 4011, pp. 95-109).

van Assem, M., Menken, M. R., Schreiber, G., Wielemaker, J., & Wielinga, B. (2004). A method for converting thesauri to RDF/OWL. In S.A. McIlraith, D. Plexousakis, & F. van Harmelen, (eds), *Proceedings of the Third International Semantic Web Conference*, (LNCS, Vol. 3298, pp. 17-31).

Van Damme, C., Hepp, M., & Siorpaes, K. (2007). An integrated approach for turning folksonomies into ontologies. In *Bridging the Gap between Semantic Web and Web 2.0 (SemNet 2007)* (pp. 57–70). Folksontology.

Van de Riet, R., Burg, H., & Dehne, F. (1998). Linguistic Issues in Information System Design. In *Formal Ontology in Information System*. Amsterdam: IOS Press.

Van der Vet, P. E., & Mars, N. J. I. (1998). Bottom-Up Construction of Ontologies. *IEEE Transactions on Knowledge and Data Engineering*, *10*(4), 513–526. doi:10.1109/69.706054

Van Heijst, G., Schreiber, A., & Wielinga, B. (1997). Using Explicit Ontologies in KBS Development. *International Journal of Human-Computer Studies*, *46*(2/3), 183–292. doi:10.1006/ijhc.1996.0090

van Rijsbergen, C. J. (1979). *Information retrieval* (2nd ed.). London: Butterworths.

Viola, P. & Jones, M. (2001). Robust real-time object detection. *International Journal of Computer Vision*.

Visser, P. R. S., & Cui, Z. (1998). On Accepting Heterogeneous Ontologies in Distributed Architectures. In *Proceedings of the ECAI'98 workshop on Applications of Ontologies and Problem-solving methods* (pp. 112-119).

Visser, P. R. S., Jones, D. M., Bench-Capon, T. J. M., & Shave, M. J. R. (1998). Assessing Heterogeneity by Classifying Ontology Mismatches. In N. Guarino (Ed.), *Proceedings of the International Conference on Formal Ontology in Information Systems* (pp. 148-162). Amsterdam: IOS Press.

Völker, J., Hitzler, P., & Cimiano, P. (2007b). Acquisition of OWL DL axioms from lexical resources. In E. Franconi, M. Kifer, & W. May (Eds.), *The Semantic Web: Research and Applications, Proceedings of the 6th International Semantic Web Conference, ISWC 2007* (LNCS: Vol. 4519, pp. 670-685). Berlin, Germany: Springer. doi: 10.1007/978-3-540-72667-8

Völker, J., Vrandecic, D., & Sure, Y. (2005). Automatic evaluation of ontologies (AEON). In Y. Gil, E. Motta, V. R. Benjamins, & M. A. Musen (Eds.), The Semantic Web–ISWC 2005 (LNCS: Vol. 3729, pp.716-731). Berlin, Germany: Springer. doi: 10.1007/11574620

Völker, J., Vrandecic, D., Sure, Y., & Hotho, A. (2007a). Learning disjointness. In E. Franconi, M. Kifer, & W. May (Eds.), *The Semantic Web: Research and Applications, Proceedings of the 6th International Semantic Web Conference, ISWC 2007* (LNCS: Vol. 4519, pp. 175-189). Berlin, Germany: Springer. doi: 10.1007/978-3-540-72667-8 Retrieved from http://www.eswc2007.org/pdf/eswc07-voelker1.pdf

Voorhees, E. M. (1994). Query expansion using lexical-semantic relations. In *Proceedings of the 17th annual international ACM SIGIR conference on Research and development in information retrieval*, (pp. 61–69). New York: Springer-Verlag, Inc.

Wache, H., Scholz, T., Stieghahn, H., & Konig-Ries, B. (1999). An integration method for the specification of rule–oriented mediators. In [Kyoto, Japan.]. *Proceedings of DANTE*, *99*, 109–112.

Wache, H., Vogele, T., Visser, U., Stuckenschmidt, H., Schuster, G., Neumann, H., & Hubner, S. (2001) Ontology-based Integration of Information – A Survey of Existing Approaches. In H. Stuckenschmidt (Ed.), *Proceedings of the IJCAI workshop on Ontologies and Information Sharing* (pp. 108–117).

Wagner, A. (2000). Enriching a lexical semantic net with selectional preferences by means of statistical corpus analysis. In *Proceedings of the ECAI-2000 Workshop on Ontology Learning*, (pp. 37-42), Berlin.

Wagner, S. M., & Buko, C. (2005). An Empirical Investigation of Knowledge-sharing in Networks. *The Journal of Supply Chain Management*, *41*(4), 17–31. doi:10.1111/j.1745-493X.2005.04104003.x

Wang, H., Horridge, M., Rector, A., Drummond, N., & Seidenberg, J. (2005). Debugging OWL-DL ontologies: A heuristic approach. In Y. Gil, E. Motta, V. R. Benjamins, & M. A. Musen (Eds.), The Semantic Web – ISWC 2005 (LNCS: Vol. 3729, pp. 745-757). Berlin, Germany: Springer. doi: 10.1007/11574620

Wang, Z., Wang, Y., Zhang, S., Shen, G., & Du, T. (2006). Matching Large Scale Ontology Effectively. In *Proceedings of the First Asian Semantic Web Conference (ASWC),* (pp. 99-106), Beijing, China.

Weber, R. (1997). *Ontological Foundations of Information Systems*. Coopers and Lybrand.

Widdows, D. (2003). *Unsupervised methods for developing taxonomies by combining syntactic and statistical information*. HLT-NAACL.

Wielinga, B. J., Schreiber, A. T., Wielemaker, J., & Sandberg, J. A. C. (2001). From Thesaurus to Ontology. In *Proceedings of the 1st international conference on Knowledge capture* (pp. 194–201). New York: Association for Computing Machinery Press.

Wilbur, W. J., & Kim, W. (2001). Flexible Phrase Based Query Handling Algorithms. In. *Proceedings of the ASIST Annual Meeting*, *38*, 438–449.

Wiley, D. (2002). *Connecting learning objects to instructional design theory: A definition, a metaphor, and a taxonomy* (pp. 571–577). The Instructional Use of Learning Objects.

Wong, S., Michael, K., Wu, D., & Butz, C. J. (2003). Probabilistic Reasoning in Bayesian Networks: A Relational Database Approach. In Advances in Artificial Intelligence, (LNCS Vol. 2671, pp. 583–590).

WSDL-S. (2005). *Web Service Semantics - WSDL-S, Version 1.0*. Recovered W3C Member. Retrieved from http://www.w3.org/Submission/WSDL-S/

WSMF. (2005). *Semantic Web Services Framework (SWSF) Overview*. Recovered W3C Member Submission. Retrieved from http://www.w3.org/Submission/SWSF/

WSMO. (2005). *Web Service Modeling Ontology Project Deliverables D1v1.0*. WSMO Project, CMS-WG Final Draft. Retrieved from http://cms-wg.sti2.org/TR/d1/v1.0/

Xiong, H., Tan, P. N., & Kumar, V. (2003). Mining strong affinity association patterns in data sets with skewed support distribution. In *IEEE International Conference on Data Mining*, (pp. 387–394). Washington, DC: IEEE Computer Society.

Xu, F., Kurz, D., Piskorski, J., & Schmeier, S. (2002). A Domain Adaptive Approach to Automatic Acquisition of Domain Relevant Terms and their Relations with Bootstrapping. In *Proceedings of LREC 2002, the third international conference on language resources and evaluation*, Las Palmas, Canary island, Spain.

Xu, J., & Croft, W. B. (1996). Query expansion using local and global document analysis. In *Proceedings of the Nineteenth Annual International ACM SIGIR Conference on Research and Development in Information Retrieval*, (pp. 4–11).

Xu, R., & Wunsch, D. (2005). Survey of Clustering Algorithms. *IEEE Transactions on Neural Networks*, 645–678. doi:10.1109/TNN.2005.845141

Xuan, D. N., Bellatreche, L., & Pierra, G. (2006). A versioning management model for ontology-based data warehouses. In *Proceedings of 8th International Conference on Data Warehousing and Knowledge Discovery DaWak'06*. Krakow, Poland. Retrieved from http://www.lisi.ensma.fr/ftp/pub/documents/papers/2006/2006-TICDWKD-XUAN.pdf

Yildirim, Y., & Yazici, A. (2007). Ontology-Supported Video Modeling and Retrieval. In S. Marchand-Maillet et al, (Eds.), *Proceedings of AMR 2006*, (LNCS 4398, pp. 28–41). Berlin: Springer.

Yu, A. C. (2006). Methods in biomedical ontology. *Journal of Biomedical Informatics, 39*(3), 252–266. doi:10.1016/j.jbi.2005.11.006

Zadeh, L. A. (1996). *Fuzzy Sets, Fuzzy Logic, Fuzzy Systems*. Amsterdam: World Scientific Press.

Zhang, S., & Bodenreider, O. (2004). Investigating implicit knowledge in ontologies with application to the anatomical domain. In *Pacific Symposium on Biocomputing*, (pp. 250–261).

Zhuge, H. (2004). *The knowledge grid*. Amsterdam: World Scientific.

Zhuge, H. (2007). *Autonomous semantic link network model for the Knowledge Grid, Concurrency and Computation: Practice and Experience* (pp. 1065-1085), 7(19).

Zicari, R. (1991). A framework for schema updates in an object-oriented database system. In *Proceedings of the Seventh International Conference on Data Engineering (ICDE'91)*, (pp. 2-13), Kobe, Japan.

About the Contributors

Faïez Gargouri is Professor at the Institut Supérieur d'Informatique et de Multimédia de Sfax (Tunisia), where he is the head. He obtained his Master's degree in Computer Science from the Paris 6 University (1990) and a PhD from the Paris 5 University (1995). In 2002, he obtained an Habilitation Universitaire en Informatique from the Faculté des Sciences de Tunis (Tunisia). His research interest focus of information systems field: Design, Quality measurement, Verification, Data Warehousing, Multimedia, Knowledge management, Ontology. He published more than 100 papers in journals, international conferences and he is a PC member of multiple international conferences.

Wassim Jaziri received his Ph.D. degree in Computer Science in 2004 from INSA-Rouen, France. In October 2004, he joined the CIRAD Research Centre (Montpellier, France), as a post-doctorate researcher. Actually, He is Assistant Professor in Computer Science at ISIM-Sfax, Tunisia. He is a member of the Miracl Laboratory (Multimedia, InfoRmation systems and Advanced Computing Laboratory). His main interests are Geographic Information Systems, Spatio-temporal Database, Ontology and Optimization. Wassim Jaziri was involved in several projects and has published papers in a number of international journals and conferences. He is also a PC member of multiple international conferences.

* * *

Youssef Amghar is Professor in the field of information systems at the University of Lyon. His field of teaching concern project management, databases and development processes. His is an active member of SOC team. His current research interests include business process, services, interoperability of applications and legal documents modeling. He is author of several papers related to these research activities and managed some projects about decisions support and e-learning.

Marie-Aude Aufaure obtained her PhD in computer science from the University of Paris 6 in 1992, and her HDR from the University Claude Bernard Lyon 1 in 2002. From 1993 to 2001, she was associated-professor at the University of Lyon; then, she has integrated a French research centre in computer science (INRIA) during two years. She was professor at Supélec from 2003 to 2008 and scientific partner of the Inria Axis project. Now, she is full professor at Centrale Paris (MAS Laboratory) and head of the SAP Business Objects Chair on Business Intelligence. Her research interests deals with the analysis, retrieval and querying of unstructured data, and the combination of structured and unstructured data from a Business Intelligence perspective. She also works on ontology life-cycle (construction and evolution). Another research interest concerns semantic information retrieval and personalization, especially for geographical data. Her work was published in international journals, books and conferences. URL: http://perso.ecp.fr/~aufaurema/.

Mostefa Belarbi graduated with Engineer Degree in Computer Science from Institute of Computer Science, University of Senia - Oran, in 1988. He received the Master degree in Software Engeeniring from the University of Sciences and Technology –USTO- Oran, Algeria in 1997. He received the PhD degree in Computer Science from the National Institute of Applied Sciences – INSA in Lyon, France in 2003. He is currently an Associate Professor at Ibn Khaldoun University of Tiaret, Algeria. His research interests include Real-Time and Embedded Systems, Formal Validation, Information Systems, Web Engineering, Pervasive information systems.

Aïcha-Nabila Benharkat is Associate Professor at the Computer Science Department (INSA de LYON) since 1992.In this role, she worked on the Integration of heterogeneous databases using Description Logics. Currently, her research interests include the interoperability domain, schema matching techniques in small and large scale, business process as well as the interoperability in service oriented information system and quality and Web services discovery. Aïcha-Nabila BENHARKAT participated to the organisation of scientific events (the international conference. She is member of several conference program committees.

Rafael Berlanga-LLavori is an associate professor in the Computer Science career at University Jaume I, Spain for 12 years. He received the B.S. degree from Universidad de Valencia in Physics, and the Ph.D. degree in Computer Science in 1996 from the same university. He is author of several articles in international journals, such as BMC Bioinformatics, Information Processing and Management and Information Science among others, as well as numerous communications in international conferences such as DEXA, ECIR, CIARP, etc. His current research interests are knowledge bases, information retrieval, and temporal reasoning.

Rafik Bouaziz was a consulting Engineer in the Organization and Computer Science and a head of the department of Computer Science at CEGOS-TUNISIA between 1979 and 1986. Then, he integrated higher education area as an associate Professor in the Faculty of Economics and Management of Sfax (Tunisia). During the preparation of his PhD, he worked on temporal data management and historical record of data in Information Systems. Currently, his main research topics of interest are temporal databases, real-time databases, information systems, data warehousing and workflows.

Aicha Boubekeur graduated with Engineer Degree in Computer Science from Univrsity of Senia, Oran, in 1991. She received the Master of Science degree from the University of Mascara, Algeria, in 2007. She prepares Phd Thesis in computer science at the USTO University of Oran, Algeria. Currently, she is an Associate Professor at Ibn Khaldoun University, Tiaret, Algeria. His research interests include system interoperability, databases, ontology engineering, web-based information systems, semantic web and semantic web services.

Julien Bourdaillet received his PhD from the University Pierre and Marie Curie in December 2007. Its research interests focus on Natural Language Processing, Machine Translation, Machine Learning, Stringology, Bioinformatics and Linguistics. During his thesis he proposed a novel algorithm for aligning different versions of large texts, such as whole books, which permits to discover moves of small text phrases. This works borrows from bioinformatics the use of character sequence alignment and adapt it to process natural language sequences without linguistics resources. This algorithm works for any alphabetic language, achieves scalability and noise robustness. Currently, he holds a postdoctoral position at the University of Montreal where he works on Machine Translation. He is interested in word alignment algorithms for translation spotting and phrase table construction.

Zouhaier Brahmia is a PhD student in computer science at the Faculty of Economic Sciences and Management of Sfax (University of Sfax, Tunisia), working on schema versioning in multi-temporal databases. He received his master degree in computer science from the same faculty in 2005. His scientific interests are in the fields of temporal databases, native XML databases, and schema versioning.

Abdallah Chouarfia graduated with Engineer Degree in Computer Science from National Institute of Computer Science, Algiers, in 1981. He received the PhD degree in Computer Science from the Paul Sabatier University in Toulouse, France in 1983. He was a Senior Lecturer in the Computer Science Department at Science and Technology University, Oran Algeria from 1987-2010. He is currently an Associate Professor at Science and Technology University, Oran Algeria. His research interests include software engineering, information systems interoperability, networking, web engineering, pervasive systems.

Rim Djedidi is a PhD student in computer science at Paris-Sud XI University (France); working at the computer science department of Supélec engineering school. She will be defending her thesis by the end of 2009. Her subject is about developing domain ontology through an iterative and incremental process that helps correcting, refining and enriching ontology. Since 2007, she is a member of the DAFOE project (Differential and Formal Ontology Editor) of the French National Research Agency ANR. In 2005, she obtained a research master diploma in document and software engineering at La Manouba University (Tunisia). She worked on developing a handbook of e-Business models with the collaboration of HEC Lausanne School (Switzerland). In 2002, she obtained a master's degree in computer science applied to management, at Tunis University (Tunisia). She worked as a non-permanent assistant professor in Tunisian and French schools. Her work was published in francophone and international conferences.

Gilles Falquet holds MS and PhD degrees in Computer Sciences from the University of Geneva (Switzerland). He is currently an associate professor at the Department of Information Systems of the University of Geneva. His research interests include: semantic digital libraries and virtual documents, ontology management systems, point of views in ontologies, knowledge-based indexing and information retrieval, and knowledge visualization. He has co-authored numerous publications on these topics in books, journals or conference proceedings.

Joshua Fan has a Bachelor of Mathematics (Computer Science) degree, a Bachelor of Engineering (Electrical, Honours) degree and a PhD degree in Information Systems and Operations Research from the University of Wollongong. Since joining the University of Wollongong, Joshua has taught logistics and supply chain management, project management, electronic business, information systems analysis and design, management information systems, operations and production management. Prior to joining the Graduate School of Business, Joshua was the postgraduate studies advisor (Information Systems) for the School of Economics and Information Systems of the University of Wollongong. Joshua has supervised a number of research students and he undertakes research collaborations with colleagues at Anderson Graduate School of Management, University of California (Riverside) in the United States and Assumption University in Bangkok, Thailand. His research interests centre on operations research, supply chain management, electronic business and workflow optimizations.

Jean-Gabriel Ganascia first studied mathematics and physics to be an engineer. In parallel he studied Philosophy - *licence de Philosophie* université Paris I (Sorbonne) - and computer science - DEA Paris VI university. He got his *Doctorat d'ingénieur* in 1983 and his *Thèse d'état* in 1987. He was successively named assistant professor at Orsay University (Paris XI) (1982), *Maître de conference* at Orsay University (1987) and Professor at Paris VI University (1988). He acted as a program leader in the CNRS executive from November 1988 to April 1992 before moving to direct the Cognitive Science Coordinated Research Program and head the Cognition Sciences Scientific Interest Group since January 1993 until 2000. Now, he leads the ACASA team in the LIP6 that is the computer science laboratory of Paris VI University.

Fabien Gandon is a research scientist at INRIA Sophia Antipolis. He has PhD and an HDR (Habilitation to Direct Research) in Computer Science from the University of Nice-Sophia-Antipolis. He worked on Knowledge engineering, Ontologies, Semantic Web, Multi- Agents Systems, Mobile devices, Context-Awareness and Privacy. He previously worked for the Mobile Commerce Laboratory of Carnegie Mellon in Pittsburgh, PA. He published more than 34 articles in journals, books or conferences and has been reviewer of more than 23 journals, conferences or workshops. His professional interests include: Knowledge engineering, Ontologies, Semantic Web, Multi-Agents Systems, Context-Awareness and Privacy. He is involved in the W3C, in working groups like GRDDL and RDFa.

Jacques Guyot holds MS and PhD degrees in Computer Sciences from the University of Geneva (Switzerland). He is currently a scientific collaborator at the Department of Information Systems of the University of Geneva and an independent consultant in information retrieval and text mining. His research interests include: ontology management systems, knowledge-based indexing and information retrieval. He has co-authored numerous publications on these topics in books, journals or conference proceedings.

Bassel Habib is a PhD student preparing his 3rd year at the University Pierre and Marie Curie (Paris 6). His thesis topic is "CYBERNARD: The Rational Reconstruction of the scientific approach of Claude Bernard". This thesis focuses on the scientific discovery and interests, in particular, in the scientific work of Claude Bernard. In this context, a virtual laboratory has been built in which several virtual experiments can be reproduced. This work is registered under a collaborative project between three teams: the team of the LIP6 (Laboratoire de Paris 6) scientists ACASA, epistemologists team at the Ecole Normale Superieure (ENS) and the team of language Institut des Textes et Manuscrits modernes (ITEM).

Antonio Jimeno-Yepes has a Master's Degree in Computer Science (Univeritat Jaume I, Castellon, Spain), a Master's Degree in Intelligent Systems (Universitat Jaume I) and a Ph.D. in Computer Science (Universitat Jaume I). He worked in CERN (European Organization for Nuclear Research) in the LHC layout database and in several projects within the document management systems EDMS. In 2006 he joined the Rebholz Group at the European Bioinformatics Institute.

Christophe Jouis is *Maître de conference* (Associate Professor) at the Universite Paris Sorbonne Nouvelle since 1999 and is attached for research at LIP6 (Laboratoire d'Informatique de Paris 6), since 2005. From 2000 to 2004, Christophe Jouis was Associate Professor at the Universite du Québec à Trois-Rivières (UQTR), Quebec, Canada, at the Department of Mathematics and Informatics (DMI). During this period, he approached the methods of analysis and representation of knowledge from text using the models developed at DMI: Categorical Grammars, Combinatory logic of Curry. He worked for several years on the semantic representations of scientific texts and / or techniques. Among the issues we have the following: (1) the semantics of the semantic relationships between concepts (i.e., for each relationship, the number and type of its arguments, its algebraic properties, etc.). One possible approach to this problem is to organize the semantic relationships in a typology based on logical properties and (2) a contribution to the study of formal ontologies: the problem of atypical entities.

Javier Lacasta holds MS and PhD degree in Computer Science from the University of Zaragoza (Spain). In May 2002 he started collaborating with the Advanced Information Systems Laboratory, and since September 2006 he is an assistant professor at the Computer Science and Systems Engineering Department of the University of Zaragoza. His research interests are focused on the multidisciplinary area of Spatial Data Infrastructures, working on metadata management, information retrieval and ontologies. Within this context, he has co-authored numerous publications in books, journals or conference proceedings; and has collaborated in several R+D projects. During 2007 he was a visiting researcher in the "Hypermedia Research Unit" at the University of Glamorgan.

Sim Kim Lau is a Senior Lecturer in the Information Systems Discipline, School of Information Systems and Technology at the University of Wollongong. She has published widely in refereed Information Systems journals and international conferences. Her research interest is in the area of application of artificial intelligence and optimization techniques to business problems, and more recently in the area of semantic web and ontology. Her teaching covers knowledge-based information systems and decision support systems.

Nelson Leung has a Bachelor of Information Technology from the Queensland University of Technology, a Master of Information Systems from the Griffith University and a Master of Information Systems (Research) from the University of Wollongong. Nelson is now working as a full-time BIS Lecturer at the RMIT International University Vietnam. He is also a PhD candidate at the University of Wollongong, an Adjunct Researcher at the Payap University and an Associate Editor for Interdisciplinary Journal of Information, Knowledge, and Management (IJIKM). Nelson has published widely in refereed book, journals and international conferences. Prior to that, he was an Instructor at the University of Wollongong.

Mimoun Malki graduated with Engineer Degree in Computer Science from National Institute of Computer Science, Algiers, in 1983. He received the Master of Science degree and the PhD degree in Computer Science from the University of Sidi Bel-Abbes, Algeria, in 1992 and 2002, respectively. He was a Senior Lecturer in the Computer Science Department at the University of Sidi Bel-Abbes from

1986 to 2002. Currently, he is an Associate Professor at Djillali Liabes University, Sidi Bel-Abbes Algeria. His research interests include databases, ontology engineering, web-based information systems, semantic web, web services and web reengineering.

Javier Nogueras-Iso holds MS and PhD degrees in Computer Science from the University of Zaragoza (Spain). After working for the Economic and Social Committee of the European Communities (Brussels) in 1998, he started his research at the Computer Science and Systems Engineering Department of the University of Zaragoza., where he is currently an Associate Professor. Additionally, he is a member of the Aragón Institute for Engineering Research and has completed a postdoctoral stay at the Institute of Environment and Sustainability of the Joint Research Centre (Ispra, Italy). His research interests are focused on Spatial Data Infrastructures, and Geographic Information Retrieval. Within this context, he has co-authored numerous publications in books, journals or conference proceedings; and has collaborated in several R+D projects.

Yann Pollet is presently professor, chair of System Integration at Cnam (National Conservatory of Arts and Crafts, a French higher education and research Institution), Paris. He is an engineer, having got in 1979 a degree from the Ecole Nationale Supérieure de Techniques Avancées (ENSTA, French Grande Ecole) He also got a master degree in Management at IAE (Upper level Company Administration School) in 1980 and wrote his PhD thesis on Advanced Data Management Systems for multimedia and geographical data. Working from 1981 in the domain of Command Control, Communication and Intelligence Systems (C3I), he was manager of the R&D Division of the Matra Sytèmes & Information company from 1991 to 2002, and was involved as technical expert in Architectural issues in many system development projects. He jointed Cnam in 2002 as chair professor. His main areas of interest are Systems Analysis & Design, System Integration, Semantic Web and Knowledge Management.

Dietrich Rebholz-Schuhmann received a master in medicine (Univ. Duesseldorf, 1988), a Ph.D. in immunology (Univ. Duesseldorf, 1989) and a master in computer science (Univ. Passau, 1993). Afterwards, he worked as a senior scientist at the gsf (Munich) in the field of image analysis and 3D visualization. From 1998 to 2003 he headed a R&D team at LION bioscience AG, Heidelberg working on novel text mining solutions. Since 2003 he is research team leader at the European Bioinformatics Institute, Hinxton (Uk) doing research in biomedical literature analysis. During his research career, Dr. Rebholz-Schuhmann established several world-class bioinformatics service, managed several large-scale research projects (Bio-Path, BootStrep, CALBC), successfully published well-cited papers in different scientific domains and is teaching at regular intervals his research work on workshops at the EBI and on international conferences.

Najla Sassi is a PhD student in Computer Science, working on ontology evolution and versioning. She received her Master's degree in computer science from the Institut National des Sciences Appliquées (Rouen, France) in 2004. Actually, She is Assistant in Computer Science at the Institut Supérieur des Sciences Appliquées et de Technologie (Mahdia, Tunisia) and a member of the Information Systems Team of the Miracl Laboratory (Multimedia, InfoRmation systems and Advanced Computing Laboratory). His main interests are Artificial Intelligence, Ontology and Information Systems.

Sana Sellami is a PhD candidate at the INSA of Lyon (National Institute of Applied Sciences Lyon France). She received her Master degree from Computer Science from INSA of Lyon in 2006 and Engineering degree form INSAT, Tunisia in 2006. Her main research interests are related to Large scale XML schemas matching domain.

Jacques Teller is director of LEMA (Local Environment Management and Analysis) and professor of urban design and planning at the University of Liege. He is graduated as civil engineer in architecture and his thesis was dedicated to the regulation of the urban form, and more specifically to the development of performance- based morphological indicators. He coordinated several European research projects: SUIT (FP5), APPEAR (FP5) & PICTURE (FP6). He was the Chairman of the COST Action C21 dedicated to Urban ontologies for an improved communication in urban development projects (http://www.towntology.net/). He is now involved in the development of an inter-disciplinary research centre dedicated to the urban, territorial and rural environment (LEPUR).

Index

A

abstract concepts 4
abstraction level 291
additive changes 185
administrative metadata 286
advanced authoring format (AAF) 287
algebraic approach 98, 101, 116
algorithm pattern 294
annotation category 165
AnsProlog 78, 83, 84, 92
application domain 257
application ontologies 40, 41, 42
a priori 82
artifacts 35, 50, 60, 65
artificial intelligence (AI) 1, 2, 3, 34, 35, 80, 97, 143, 303, 316
artificial intelligence systems 143
ascendant algorithm 114
assertional box (A-Box) 18
assisted navigation 12
attribute properties 102, 104, 106
autoepistemic logic 82
automatic invocation 303, 304
automatic processing 247, 248

B

belief revision 83
Bellflower 213, 218
biomedical domain 261, 266, 267, 270, 271, 272, 274, 277
bi-temporal databases 227
black box 153, 156
bootstrap 217
bootstrapping ontology evolution with multimedia information extraction (BOEMIE) 187, 188, 198, 202, 206
bottom-up 44, 50, 186
brainstorming 14

C

candidate solution 112, 114
canonical decomposition 105
centroid 217
change capturing 189
change classification 184, 189
change detection 182, 185, 194, 195, 196, 197, 204, 206
change effects 180, 182, 184, 185, 189
change implementation 185, 194
change propagation 233
change representation 179, 189, 196
change semantics 189
Chimaera 146
circumscription 83, 97
classification category 165
clustering based approach 215
coherence analysis 233
common ontology for multimedia (COMM) 290, 292, 293, 294, 295, 297, 300
complex mappings 109
composition operator 104, 105
computable models 1, 2
computational semantics 16
concept analysis 11
conceptual base 31
conceptual graphs (CG) 17, 52

conceptualization 3, 10, 11, 12, 13, 15, 16, 32, 33, 35, 36, 39, 42, 46, 51, 53, 55, 58, 143, 146, 147, 160, 180, 181, 183, 186, 188, 193, 247, 249, 255
consistency maintenance 182, 185, 188, 194, 200
constraint-based technique 146
content metadata 286
contextualization 28, 56, 57, 64
contextualized services 12
Creative Commons 9, 14
cross-language information retrieval (CLIR) 43

D

data analysis 98
database management system (DBMS) 237, 238
database schemas 33, 34, 59, 61, 66, 180, 181, 184, 202, 225, 226, 227, 228, 230, 231, 236, 237
data-driven 186, 196, 197, 203
data dynamics 181
data integration 210, 211, 212
data mining 11, 14
data retrieval 98
datatype properties 101, 102, 111, 116
data versioning 227, 229
deep Web 208, 211, 215
default logic 81, 82, 84
demarcated 85
derived classes 103, 104, 107
derived property 102, 103, 113
description logics 3, 11, 12, 18, 22, 53
descriptive conflicts 30
design rationale 1, 2
digital libraries 208
directed relationships 102
disseminate 11, 12, 147, 148, 160
distributed change management 188
distributed ontology 184, 188, 196, 200
distributed resources 99
domain experts 254, 258
domain ontologies 40, 41, 42, 73, 101, 102, 105, 172, 175
dual correlation mining (DCM) 215, 219
Dublin Core 9, 14
dynamic learning 162
dynamic learning content 162

E

educational category 163, 164, 165
e-learning 162, 164, 165, 171, 176
e-learning application 162, 165
entailment relations 83
enterprise modeling 35, 43
evolution changes 226, 233, 243
exchangeable image file format (EXIF) 286, 288, 290, 302
execution graph 108, 111, 112
extended property 102
extended relationship 106, 107, 109
externalization 148, 152
external knowledge sources 142
external resources 220, 221

F

Falcon-Ao 218
first order logic (FOL) 16, 52
folksonomies 248
formal change specification 185
formal concept analysis (FCA) 189
formalism 1, 3, 8, 12, 16, 17, 18, 19, 20
formalization 3, 8, 11, 15, 16, 18, 247, 254, 256
formal knowledge 181
formal ontologies 147
friend of a friend (FOAF) 9, 19

G

general category 164, 167, 175
genericity 41
generic ontologies 42
genetic algorithm (GA) 217, 222, 223
global schema 211
granularity 13, 18
graphical user interface (GUI) 218, 219
group knowledge 147
group memory 171

H

heavy ontologies 8
heavyweight ontologies 32, 37

heterogeneity 142, 143, 144
heterogeneity conflicts 30
heterogeneous information systems 143
heuristic algorithms 216, 217
hierarchical relation 250
hierarchy 38, 39, 46, 53
highly informal ontologies 40
holistic matching 215, 219
holistic schema matching (HSM) 219
homomorphism 108
homonymy 5, 11, 29, 49
hybrid approach 232
hypernyms 251

I

individual memory 172
inference rules 78, 81, 83, 88, 92, 95
information extraction
262, 263, 269, 270, 278, 282
information retrieval 4, 11, 13, 32, 35, 43, 62, 261, 262, 265, 266, 272, 277, 278, 279, 280, 281, 282, 283
information systems 1, 9, 13, 19, 20, 27, 28, 29, 31, 33, 53, 54, 55, 56, 57, 59, 64, 65, 70, 72, 73
integration complexity 303
interactivity 162, 164, 165
inter-element 38
internalization 148, 152
International Standard Book Number (ISBN) 153
International Standard Serial Number (ISSN) 153
interoperability 143, 144, 145, 147, 148, 149, 179, 230, 285, 288, 290, 291, 292, 295, 297, 302
intra-element 38
inverse relationships 8

K

K-means 213, 216, 217
knowledge acquisition 14, 15, 23, 247, 248
knowledge base 34, 35, 36, 38, 39, 50, 64, 181, 192
knowledge-based systems 27, 45, 50, 64
knowledge dissemination 153
knowledge engineering (KE) 32, 34, 42, 43, 49, 53, 54, 62, 64, 65, 66, 67, 69, 70, 71, 72, 74, 75, 303
knowledge management 142, 144, 148, 152, 158, 159, 160
knowledge management system (KMS) 148, 149, 153, 156, 157
knowledge modeling 1, 2, 15, 18
knowledge resources 247, 248
knowledge systems 32, 41

L

language-based technique 146
large scale matching 208, 209, 210, 212, 213, 217, 218, 219, 220, 221, 224
learned ontology model 187, 198
learning disjointness axioms (LeDA) 187
learning expressive ontologies (LExO) 187
learning object indexing 166, 169
learning object (LO)
163, 165, 168, 169, 171, 175
learning object metadata (LOM) 162, 163, 164, 165, 166, 167, 168, 169, 170, 171, 174, 175, 176, 177
learning object repositories (LOR)
165, 168, 169, 171
learning objects indexing tool (LOIT)
168, 169
lifecycle category 164
lightweight ontologies 8, 37
linguistic level 48
linguistic representation 4, 5
linguistic resources technique 146
logic programming 78, 83, 92, 94
logic programming language 78, 83, 92, 94
LOM ontology 162, 163, 166, 167, 168, 170, 175, 176
LOM standard 162, 163, 166, 175

M

machine learning 209, 216
Malasco 218
managing end to end operations (METEOR-S)
306, 307, 308, 320
mapping approach
145, 150, 152, 153, 154, 155

mapping assertion 107, 108, 110, 113
mapping model 101, 103, 107, 108
matching algorithms 210, 217, 219, 221
matching systems 209, 210, 218, 221
material exchange format (MXF) 287
medical events 101, 102, 103
medical subject headings (MeSH) 43, 44, 250
MEMORAe 171, 172, 174, 175, 176
merged ontologies 145, 161
metadata richness 287
metadata schemas 286, 288, 289
meta-metadata category 164
metaontologies 41
metaphysics 2, 3, 10
methodology 27, 34, 44, 45, 46, 47, 48, 49, 50, 51, 61, 70, 73
methontology 45, 47, 68
middleware 149, 309, 310
model-based technique 147
modeling context 32
model of possible ontologies (POM) 186, 197
model-view-controller (MVC) 168, 176
modularization-based ontology matching (MOM) 214
monolithic 32
MPEG-7 285, 286, 287, 288, 290, 291, 292, 293, 294, 295, 301
multimedia indexing 284, 285, 299

N

natural language 5, 8, 11, 12, 14, 17, 19, 23
natural language processing (NLP) 11, 12, 17, 32, 41, 43, 46, 48, 250, 251, 258
negative predicates 78, 81, 92, 95
neural networks 253
non-monotonic reasoning 81, 82, 83, 85
normalization level 48

O

object properties 102, 111
OntoBuilder 219
ontological continuity 184
ontological database 233
ontological knowledge 3, 6, 7, 8, 17, 18
ontological network 78, 79
ontological theory 35
ontology 1, 2, 3, 4, 5, 6, 7, 8, 9, 10, 11, 12, 13, 14, 15, 16, 18, 19, 20, 21, 22, 23, 25, 26, 27, 78, 142, 143, 144, 145, 146, 147, 149, 150, 151, 152, 153, 154, 155, 156, 157, 159, 161
ontology adaptation 181
ontology-based indexing 284
ontology building 28, 39, 46, 50, 51
ontology coherence 226, 233, 235, 243
ontology construction 247, 257
ontology development 261, 262, 276, 278
ontology engineering 1, 2, 13, 14, 28, 58, 60, 63, 64, 65, 75
ontology engineering tools 179, 180
ontology enrichment 46, 49, 74
ontology evolution 179, 180, 181, 182, 183, 185, 186, 187, 189, 191, 192, 193, 194, 195, 196, 198, 200, 201, 202, 203, 204, 205, 206, 226, 230, 231, 232, 233, 235, 236, 243
ontology formalism 78, 303
ontology learning 247, 248, 249, 257, 259, 274, 277
ontology maintenance 46, 48, 49, 63
ontology matching 208, 209, 210, 213, 214, 218, 220, 221
ontology mediation 142, 144, 145, 147, 150, 151, 152, 153, 157
ontology model 179, 182, 183, 189, 190
ontology object 1, 2
ontology refinement 274, 276, 277
ontology representation 46, 49
ontology representation language 179, 180
ontology revision 181, 193
ontology usage patterns 189
ontology versioning 181, 185, 196, 225, 226, 230, 231, 232, 236
ontology versioning system 225, 226
ontology Web language for services (OWL-S) 306, 307, 308, 309, 320
optimal solution 111, 112, 114, 115
optimization techniques 208, 209, 216, 217, 221
orchestration 98, 111, 112, 115, 116, 117
organizational knowledge 142, 147, 148, 149, 150, 152, 153, 155, 156, 157, 160

organizational ontology 152
organization memory 172, 173

P

pair-wise matching 212
parallel schema matching (PSM) 219
partial realization 103, 108, 115
partitioning strategy 214
posteriori 190
problem solving 41, 66
problem solving methods (PSMs) 43, 44
projection 104, 106
project memory 172
Prolog 83, 84, 96
propositional logic 52, 83
PROTOPLASM 218
prototype 98, 117
purpose dimension 41

Q

quality of service (QoS) 99, 107, 108, 110, 111, 112, 113, 114, 115
query formulation 43
quick ontology matching (QOM) 218, 222

R

RDF binding 165, 166
relational data model 227
relation category 165
relationship composition with structured expressions (RCS) 101, 103
relationship restriction 104, 105
repository of structures technique 147
representational vocabulary 143
research innovation 9
resource description framework (RDF) 165, 166, 167, 169, 177, 249, 250, 259, 260, 286, 289, 290
restricted mappings 107
restriction filter 105
retrieval systems 248
reusability 41
right category 164
rigorously formal ontologies 40

S

scalability 209, 210, 211, 212, 215, 217, 221
scalable matching 209, 210, 212, 218, 221
schema evolution 228, 231, 236, 244
schema matching 208, 209, 210, 211, 212, 213, 215, 218, 221, 222, 223
schema versioning 226, 227, 228, 229, 230
semantic analysis 43
semantic annotations 292, 295, 298
semantic conflicts 29, 30, 31, 32
semantic gap 284, 294
semantic integration 304, 305, 307, 310, 314, 318
semantic metadata 293, 295
semantic networks 35, 52, 53
semantics 1, 9, 10, 12, 13, 15, 16, 17, 19, 98, 99, 101, 103, 107, 113, 117
semantics-based computing 99
semantic service-oriented architecture (SSOA) 99
Semantic Web 144, 158, 160, 165, 171, 176, 177, 208, 210, 211, 212, 213, 221, 222, 223, 284, 285, 289, 298, 300, 301, 302
Semantic Web services (SWS) 100, 304, 306, 307, 308, 310, 313, 319, 321
semi-automatically 150
semi-formal ontologies 40
service descriptions 99
service discovery 100, 101, 117
service oriented architecture (SOA) 303, 304, 308, 309, 318, 321
service-oriented computing (SoC) 99
shared ontology 144, 150, 152, 153, 154, 155
simple knowledge organization system (SKOS) 249, 250, 255, 259, 260
socialization 148, 152
software components 1, 3, 20
software framework 101, 115
software tools 27, 32, 40
source ontologies 145, 146, 147, 152, 161
source ontology 152
specialized multimedia ontologies 285, 287
specification of change 232, 236

stakeholders 13, 14, 15
statistical algorithms 209, 216, 217
statistical strategy 215
string-based technique 146
structural conflicts 29, 30
structure informal ontologies 40
subsumption 5, 13, 16, 17, 18, 21
subsymbolic 291
subtractive changes 185
suggested upper merged ontology (SUMO) 145
symbolic representation 4
synergies 208
synonymy 5, 11, 29, 49
syntactic conflicts 29
system engineering 40, 64

T

tableau algorithms 11
tacit knowledge 11
task ontologies 40
taxonomical knowledge 7
taxonomy-based technique 147
team memory 171
technical category 164
technical metadata 286
temporal data 225, 226, 227, 228
temporal database 227
temporal reasoning 87
terminological box (T-Box) 18
text analysis 11
thesaurus 248, 249, 250, 251, 252, 253, 254, 255, 256, 257, 258, 259
top-down 44, 50, 186
topic maps 172, 176
top-level ontologies 40
topology 78, 80, 81, 84, 85, 86, 87, 94, 95, 96
TOVE project 9
transaction time 225, 226, 227, 229
transaction time databases 227
transformation methodology 254
transitivity 235
typical object 80
typology 4, 30, 39, 40

U

unambiguous 3, 5, 6
uniform optimal solution 111, 112, 114
union operator 104, 105
universal description, discovery, and integration (UDDI) 304, 308
URBAMET 249, 251, 252, 253, 254, 257, 258, 259
urban domain 247, 248, 249, 255, 257, 258
urban planning 255, 257
URBISOC 249, 254, 255, 257, 258
usage integration 12
user interaction 248
utility function 111

V

valid time 225, 226, 227, 228, 229
valid time databases 227
value context 32, 54
versioning management 226, 243
versioning support 230
version logging 185
VERSIOS 237, 243

W

Web ontology language (OWL) 249, 250, 256, 259, 260
Web service modeling ontology (WSMO) 100, 306, 307, 308, 309, 310, 313, 314, 315, 316, 317, 318, 321
Web service modelling language (WSMF) 307, 308, 309, 320
workflows 98, 100
working groups 164

X

XML binding 165